P9-DDH-768

Glencoe
Mathematics
Applications and Connections

Course 2

$$\frac{1}{3} = \frac{t}{60}$$

 **Glencoe**
McGraw-Hill

New York, New York · Columbus, Ohio · Woodland Hills, California · Peoria, Illinois

Visit the Glencoe Mathematics Internet Site for
Mathematics: Applications and Connections at

www.glencoe.com/sec/math/mac/mathnet

You'll find:

Chapter Review

Test Practice

Data Collection

Games

*inter*NET CONNECTION links to websites relevant to
Chapter Projects, Interdisciplinary Investigations, exercises

and much more!

About the Timepieces Hologram
The optimum viewing angle for the timepieces hologram on the cover of this textbook is a 45° angle. For best results, view the hologram at this angle under a direct light source, such as sunlight or incandescent lighting.

Glencoe/McGraw-Hill

A Division of The **McGraw-Hill** *Companies*

Send all inquiries to:
Glencoe/McGraw-Hill
8787 Orion Place
Columbus, OH 43240-4027

ISBN: 0-07-822859-X

8 9 10 027/043 08 07 06 05 04 03

Dear Students, Teachers, and Parents,

Mathematics students are very special to us! That's why we wrote **Mathematics: Applications and Connections,** a math program designed specifically for you. The exciting, relevant content and up-to-date design will hold your interest and answer the question "When am I ever going to use this?"

As you page through your text, you'll notice the variety of ways math is presented for you. You'll see real-world applications as well as connections to other subjects like science, history, language arts, and music. You'll have opportunities to use technology tools such as the Internet, CD-ROM, graphing calculators, and computer applications like spreadsheets.

You'll appreciate the easy-to-follow lesson format. Each new concept is introduced with an interesting application or connection followed by clear explanations and examples. As you complete the exercises and solve interesting problems, you'll learn a great deal of useful math. You'll also have the opportunity to complete relevant Chapter Projects, Hands-On Labs, and Interdisciplinary Investigations. Test Practice, Test-Taking Tips, and Reading Math Study Hints will help you improve your test-taking skills.

Each day, as you use **Mathematics: Applications and Connections,** you'll see the practical value of math. You'll quickly grow to appreciate how often math is used in ways that relate directly to your life. If you don't already realize the importance of math in your life, you soon will!

Sincerely, The Authors

Kay McClain

Patricia S. Wilson

Patricia Frey

Linda Dritsas

Barbara Smith

Jack M. Ott

Ron Pelfrey

David Molina

Beatrice Moore-Harris

Authors

William Collins
Director of The Sisyphus
Mathematics Learning Center
W. C. Overfelt High School
San Jose, CA

Linda Dritsas
District Coordinator
Fresno Unified School District
Fresno, CA

Patricia Frey
Mathematics Department
   Chairperson
Buffalo Academy for Visual
   And Performing Arts
Buffalo, NY

*"**Mathematics: Applications and Connections** helps students make the connection between mathematics and the real world. Applications lead students through the classroom door into the world of art, geography, science, and beyond."*—**Linda Dritsas**

Arthur C. Howard
Program Director for Secondary
   Mathematics
Aldine Independent School District
Houston, TX

Kay McClain
Lecturer
George Peabody College
Vanderbilt University
Nashville, TN

David Molina
Adjunct Professor of Mathematics
   Education
The University of Texas at Austin
Austin, TX

Beatrice Moore-Harris
Staff Development Specialist
Bureau of Education and
    Research
Houston, TX

Jack M. Ott
Distinguished Professor of
    Mathematics Education
University of South Carolina
Columbia, SC

Ronald Pelfrey
Mathematics Consultant
Lexington, KY

*"**Mathematics: Applications and Connections** is designed to help middle school students develop mathematical power—the ability to use what they know, and to give them a good start into higher level mathematics. This text also helps students learn reasoning skills, make connections to the real world, become expert problem solvers, and explain their work to others."—***Jack Price***

*"The strongest focus of any middle school mathematics program should be on problem solving. **Mathematics: Applications and Connections** not only provides such a focus, but it makes the problem solving alive for students through its applications and connections."—***Ronald Pelfrey***

Jack Price
Professor, Mathematics Education
California State Polytechnic
    University
Pomona, CA

Barbara Smith
Mathematics Supervisor
Unionville-Chadds Ford
    School District
Kennett Square, PA

Patricia S. Wilson
Associate Professor of
    Mathematics Education
University of Georgia
Athens, GA

Academic Consultants and Teacher Reviewers

Each of the Academic Consultants read all 39 chapters in Courses 1, 2, and 3, while each Teacher Reviewer read two chapters. The Consultants and Reviewers gave suggestions for improving the Student Editions and the Teacher's Wraparound Editions.

ACADEMIC CONSULTANTS

Richie Berman, Ph.D.
Mathematics Lecturer and Supervisor
University of California, Santa Barbara
Santa Barbara, California

Robbie Bonneville
Mathematics Coordinator
La Joya Unified School District
Alamo, Texas

Cindy J. Boyd
Mathematics Teacher
Abilene High School
Abilene, Texas

Gail Burrill
Mathematics Teacher
Whitnall High School
Hales Corners, Wisconsin

Georgia Cobbs
Assistant Professor
The University of Montana
Missoula, Montana

Gilbert Cuevas
Professor of Mathematics Education
University of Miami
Coral Gables, Florida

David Foster
Mathematics Director
Robert Noyce Foundation
Palo Alto, California

Eva Gates
Independent Mathematics
   Consultant
Pearland, Texas

Berchie Gordon-Holliday
Mathematics/Science Coordinator
Northwest Local School District
Cincinnati, Ohio

Deborah Grabosky
Mathematics Teacher
Hillview Middle School
Whittier, California

Deborah Ann Haver
Principal
Great Bridge Middle School
Virginia Beach, Virginia

Carol E. Malloy
Assistant Professor, Math Education
The University of North Carolina,
   Chapel Hill
Chapel Hill, North Carolina

Daniel Marks, Ed.D.
Associate Professor of Mathematics
Auburn University at Montgomery
Montgomery, Alabama

Melissa McClure
Mathematics Consultant
Teaching for Tomorrow
Fort Worth, Texas

TEACHER REVIEWERS

Course 1

Carleen Alford
Math Department Head
Onslow W. Minnis, Sr. Middle School
Richmond, Virginia

Margaret L. Bangerter
Mathematics Coordinator K-6
St. Joseph School District
St. Joseph, Missouri

Diana F. Brock
Sixth and Seventh Grade Math Teacher
Memorial Parkway Junior High
Katy, Texas

Mary Burkholder
Mathematics Department Chair
Chambersburg Area Senior High
Chambersburg, Pennsylvania

Eileen M. Egan
Sixth Grade Teacher
Howard M. Phifer Middle School
Pennsauken, New Jersey

Melisa R. Grove
Sixth Grade Math Teacher
King Philip Middle School
West Hartford, Connecticut

David J. Hall
Teacher
Ben Franklin Middle School
Baltimore, Maryland

Ms. Karen T. Jamieson, B.A., M.Ed.
Teacher
Thurman White Middle School
Henderson, Nevada

David Lancaster
Teacher/Mathematics Coordinator
North Cumberland Middle School
Cumberland, Rhode Island

Jane A. Mahan
Sixth Grade Math Teacher
Helfrich Park Middle School
Evansville, Indiana

Margaret E. Martin
Mathematics Teacher
Powell Middle School
Powell, Tennessee

Diane Duggento Sawyer
Mathematics Department Chair
Exeter Area Junior High
Exeter, New Hampshire

Susan Uhrig
Teacher
Monroe Middle School
Columbus, Ohio

Cindy Webb
Title 1 Math Demonstration Teacher
Federal Programs LISD
Lubbock, Texas

Katherine A. Yule
Teacher
Los Alisos Intermediate School
Mission Viejo, California

Course 2

Sybil Y. Brown
Math Teacher Support Team-USI
Columbus Public Schools
Columbus, Ohio

Ruth Ann Bruny
Mathematics Teacher
Preston Junior High School
Fort Collins, Colorado

BonnieLee Gin
Junior High Teacher
St. Mary of the Woods
Chicago, Illinois

Larry J. Gonzales
Math Department Chair
Desert Ridge Middle School
Albuquerque, New Mexico

Susan Hertz
Mathematics Teacher
Revere Middle School
Houston, Texas

Rosalin McMullan
Mathematics Teacher
Honea Path Middle School
Honea Path, South Carolina

Mrs. Susan W. Palmer
Teacher
Fort Mill Middle School
Fort Mill, South Carolina

Donna J. Parish
Teacher
Zia Middle School
Las Cruces, New Mexico

Ronald J. Pischke
Mathematics Coordinator
St. Mary of the Woods
Chicago, Illinois

Sister Edward William Quinn I.H.M.
Chairperson Elementary Mathematics
Curriculum
Archdiocese of Philadelphia
Philadelphia, Pennsylvania

Marlyn G. Slater
Title 1 Math Specialist
Paradise Valley USD
Paradise Valley, Arizona

Sister Margaret Smith O.S.F.
Seventh and Eighth Grade Math Teacher
St. Mary's Elementary School
Lancaster, New York

Pamela Ann Summers
Coordinator, Secondary Math/Science
Lubbock ISD
Lubbock, Texas

Dora Swart
Teacher/Math Department Chair
W. F. West High School
Chehalis, Washington

Rosemary O'Brien Wisniewski
Middle School Math Chairperson
Arthur Slade Regional School
Glen Burnie, Maryland

Laura J. Young, Ed.D.
Eighth Grade Mathematics Teacher
Edwards Middle School
Conyers, Georgia

Susan Luckie Youngblood
Teacher/Math Department Chair
Weaver Middle School
Macon, Georgia

Course 3

Beth Murphy Anderson
Mathematics Department Chair
Brownell Talbot School
Omaha, Nebraska

David S. Bradley
Mathematics Teacher
Thomas Jefferson Junior High School
Salt Lake City, Utah

Sandy Brownell
Math Teacher/Team Leader
Los Alamos Middle School
Los Alamos, New Mexico

Eduardo Cancino
Mathematics Specialist
Education Service Center, Region One
Edinburg, Texas

Sharon Cichocki
Secondary Math Coordinator
Hamburg High School
Hamburg, New York

Nancy W. Crowther
Teacher, retired
Sandy Springs Middle School
Atlanta, Georgia

Charlene Mitchell DeRidder, Ph.D.
Mathematics Supervisor K-12
Knox County Schools
Knoxville, Tennessee

Ruth S. Garrard
Mathematics Teacher
Davidson Fine Arts School
Augusta, Georgia

Lolita Gerardo
Secondary Math Teacher
Pharr San Juan Alamo High School
San Juan, Texas

Donna Jorgensen
Teacher of Mathematics/Science
Toms River Intermediate East
Toms River, New Jersey

Statha Kline-Cherry, Ed.D.
Director of Elementary Education
University of Houston – Downtown
Houston, Texas

Charlotte Laverne Sykes Marvel
Mathematics Instructor
Bryant Junior High School
Bryant, Arkansas

Albert H. Mauthe, Ed.D.
Supervisor of Mathematics
Norristown Area School District
Norristown, Pennsylvania

Barbara Gluskin McCune
Teacher
East Middle School
Farmington, Michigan

Laurie D. Newton
Teacher
Crossler Middle School
Salem, Oregon

Indercio Abel Reyes
Mathematics Teacher
PSJA Memorial High School
Alamo, Texas

Fernando Rosa
Mathematics Department Chair
Edinburg High School
Edinburg, Texas

Mary Ambriz Soto
Mathematics Coordinator
PSJA I.S.D.
Pharr, Texas

Judy L. Thompson
Eighth Grade Mathematics Teacher
Adams Middle School
North Platte, Nebraska

Karen A. Universal
Eighth Grade Mathematics Teacher
Cassadaga Valley Central School
Sinclairville, New York

Tommie L. Walsh
Teacher
S. Wilson Junior High School
Lubbock, Texas

Marcia K. Ziegler
Mathematics Teacher
Pharr-San Juan- Alamo North High School
Pharr, Texas

Student Advisory Board

The Student Advisory Board gave the editorial staff and design team feedback on the design, content, and covers of the Student Editions. We thank these students from Crestview Middle School in Columbus, Ohio, and McCord Middle School in Worthington, Ohio, for their hard work and creative suggestions in making *Mathematics: Applications and Connections* more student friendly.

Mara Flood

Anna Hoza

Neil Gardner

Lareva Summerall

Mike Leonelli

Chris Franklin

Kelly McLoughlin

Jennifer Shoemaker

Maghee Disch

Table of Contents

CHAPTER 1

Problem Solving, Algebra, and Geometry

Applications, Connections, and
Integration Index, pages xxii–1.

Let the Games Begin

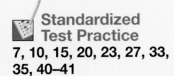

interNET CONNECTION

Standardized Test Practice

MATH IN THE MEDIA

CHAPTER 2

Applying Decimals

Let the Games Begin

- Match-Up, **73**

- Aerospace, **60**

interNET CONNECTION

Standardized Test Practice

CHAPTER 3 Statistics: Analyzing Data

Interdisciplinary Investigation

If the Shoe Fits…, **128**

Let the Games Begin

- Can You Guess?, **107**

interNET CONNECTION

Standardized Test Practice

Applications, Connections, and
Integration Index, pages xxii–1.

Using Number Patterns, Fractions, and Percents

Let the Games Begin

inter NET CONNECTION

Standardized Test Practice

CHAPTER 5

Algebra: Using Integers

Applications, Connections, and
Integration Index, pages xxii–1.

Interdisciplinary Investigation

"A" is for Apple, **264-265**

Let the Games Begin

• Math-O, **237**

interNET CONNECTION

Standardized Test Practice

CHAPTER 7

Applying Fractions

Applications, Connections, and
Integration Index, pages xxii–1.

Using Proportional Reasoning

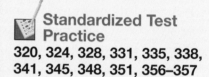

Let the Games Begin

interNET CONNECTION

Standardized Test Practice

MATH IN THE MEDIA

Applications, Connections, and
Integration Index, pages xxii–1.

 **Interdisciplinary Investigation**

Pi for Polygons, **404**

 **Let the Games Begin**

- Tic-Tac Squares, **391**

 **School to Career**

- Fashion, **368**

 interNET CONNECTION

- Chapter Project, **359**
- School to Career, **368**
- Let the Games Begin, **391**
- Data Update, **392**
- Interdisciplinary Investigation, **405**
- Chapter Review, **398**
- Test Practice, **403**

Standardized Test Practice
365, 373, 379, 385, 387, 391, 394, 397, 402–403

CHAPTER 10

Geometry: Exploring Area

CHAPTER 11

Applying Percents

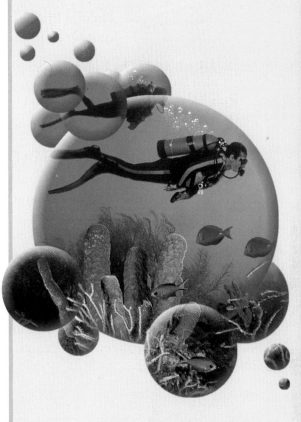

Applications, Connections, and
Integration Index, pages xxii–1.

Let the Games Begin
- Time to Shop, **477**

interNET CONNECTION
- Chapter Project, **449**
- Data Update, **465**
- School to Career, **473**
- Let the Games Begin, **477**
- Chapter Review, **482**
- Test Practice, **487**

MATH IN THE MEDIA
- Smart Shopping, **472**

- Media, **473**

Interdisciplinary Investigation

The Perfect Package, **524**

Let the Games Begin

• Shape-Tac-Toe, **513**

School to Career

• Biochemistry, **507**

interNET CONNECTION

• Chapter Project, **489**
• School to Career, **507**
• Let the Games Begin, **513**
• Data Update, **517**
• Interdisciplinary Investigation, **525**
• Chapter Review, **518**
• Test Practice, **523**

Standardized Test Practice
495, 497, 501, 506, 513, 517, 522–523

MATH IN THE MEDIA

• SHOE, **495**

CHAPTER 13

Exploring Discrete Math and Probability

Let the Games Begin

- Take a Chance, **530**
- Cherokee Butterbean Game, **541**

SCHOOL to CAREER

- Design, **537**

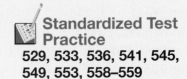

interNET CONNECTION

Standardized Test Practice

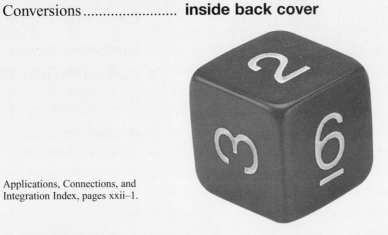

Applications, Connections, and
Integration Index, pages xxii–1.

Applications, Connections, and Integration Index

Arthropods

Insects 88%

Spiders 7%

Crustaceans 4%

Millipedes and Centipedes 1%

CHAPTER 1

Problem Solving, Algebra, and Geometry

What you'll learn in Chapter 1

- to solve problems using the four-step plan,
- to evaluate algebraic expressions using the order of operations,
- to evaluate expressions with exponents,
- to find the areas of rectangles and parallelograms, and
- to choose the best method of computation to solve real-world problems.

CHAPTER Project

FUN WAYS TO BE FIT

In this project, you will use the four-step plan to design a week-long fitness program. You will assume that you eat about 2,400 to 2,800 Calories per day, which is the amount recommended for young adults ages 11 to 14.

Getting Started

- Which activity from the table burns the least Calories per hour? Which activity burns the most Calories per hour? About how many Calories would you burn in one hour of in-line skating?

- To maintain the same weight over a period of time, you must eat and burn about an equal number of Calories. Suppose you weigh 150 pounds and you eat two pieces of pizza having 287 Calories in all. Name an activity that you can do to burn off those Calories.

Technology Tips

- Use a **spreadsheet** to find the number of Calories burned for your activity plan.
- Use a **word processor**.

 **inter NET** **CONNECTION** **Data Update** **For up-to-date information on fitness, visit:**

www.glencoe.com/sec/math/mac/mathnet

Working on the Project

You can use what you'll learn in Chapter 1 to help you design your fitness program.

Page	Exercise
7	11
10	37
23	40
39	Alternative Assessment

Calories Burned During Selected Activities		
Activity	**Calories/Hour (100-lb person)**	**Calories/Hour (150-lb person)**
sleeping, watching TV	60	90
aerobic dance	456	684
basketball	376	564
bicycling (15 mph)	472	708
cross-country skiing	389	583
downhill skiing	280	420
in-line skating	400	600
running (8-minute mile)	595	893
walking (15-minute mile)	195	292
swimming (35 yd/min)	331	497

A Plan for Problem Solving

What you'll learn

You'll learn to solve problems using the four-step plan.

When am I ever going to use this?

Knowing how to solve problems can help you play games such as Clue.

Word Wise

population
sample

When Martin Handford was a boy, he was fascinated by picture books with colorful crowd scenes. As an adult, he is famous as the author of the *Where's Waldo* books in which Waldo is hidden somewhere in detailed crowd scenes.

About how many characters are there in this *Where's Waldo* scene?

Where would you begin to solve this problem? In mathematics, we have a four-step plan to solve problems.

1. **Explore** Determine what information is given in the problem and what you need to find. Do you have all the information you need to solve the problem? Is there too much information?

2. **Plan** After you understand the problem, select a strategy for solving it. There may be several ideas or strategies that you can use. It is usually helpful to make an estimate of what you think the answer should be.

3. **Solve** Solve the problem by carrying out your plan. If your plan doesn't work, try another, and maybe even another.

4. **Examine** Finally, examine your answer carefully. See if it fits the facts given in the problem. Compare it to your estimate. You may also want to solve the problem again in a different way. If the answer is not correct, make a new plan and start again.

Example ——1 Let's try our plan on the opening problem.

Explore *What do you know?*

There are many characters in the picture. They seem to be evenly scattered throughout the picture.

What are you trying to find?

You need to find an *estimate* of the total number of characters in the picture. You do *not* need an exact answer.

Plan The total number of characters is called the **population**. Counting all the characters would take a lot of time. Besides, you only need an estimate, not an exact count.

You could count the number of characters in a small section of the picture. This small section is called a **sample**. Then multiply the number in the sample by the number of sections to get an estimate of the population.

One way to divide the picture into sections is by placing a rectangular grid over the picture.

Sample

Population

Solve There are 18 characters in the sample shown above and 16 sections in the picture. Multiply 16 by 18 to find the population.

$$16 \times 18 = 288$$

There are about 288 characters in the picture.

Examine Is your answer reasonable? You can check by using another small section as your sample. If your two samples are very different, try using a third sample as a check and revise your estimate.

Throughout this textbook, you will be solving many kinds of problems. Some can be solved easily by adding, subtracting, multiplying, or dividing. Others can be solved by using a strategy like finding a pattern, solving a simpler problem, making a model, drawing a graph, and so on. No matter which strategy you use, you can always use the four-step plan to solve a problem.

Example

APPLICATION

2 Entertainment In 1996, there were 1.3 billion movie tickets sold for a record $5.9 billion. The table shows movie admissions since 1946. If the recent trend continues, estimate how many movie tickets will be sold in 2001.

Movie Tickets (billions)	
Year	Number
1946	4.1
1951	2.8
1956	1.9
1961	1.2
1966	1.0
1971	0.8
1976	1.0
1981	1.1
1986	1.0
1991	1.2
1996	1.3

Source: Motion Picture Association

Explore   You know the number of admissions for each year. You need to find any trend in the data and predict how many tickets will be sold in 2001.

Plan   One good way to find a trend in the data is to show the data on a graph. Usually a line graph is used to show trends in data. Then look for a pattern in the graph.

Solve

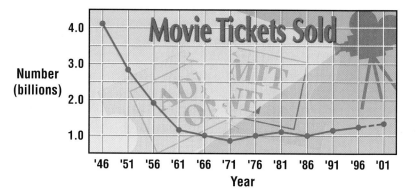

The data shows a steep drop for the years between 1946 and 1961. Then the graph levels off and shows a slight increase until 1996. To predict the number of tickets in 2001, extend the line to 2001 with a dashed line. About 1.4 billion tickets will be sold in 2001.

Year	Number (billions)	
1986	1.0	
1991	1.2	+0.2
1996	1.3	+0.1
2001	1.4	+0.1

Examine   Look for another pattern in the data for the last ten years. The number 1.4 billion seems reasonable.

CHECK FOR UNDERSTANDING

Communicating Mathematics

Math Journal

Read and study the lesson to answer each question.

1. *Tell* why it is important to plan before solving a problem.

2. *Explain* what to do if your plan to solve a problem doesn't work.

3. *Write* two or three sentences in your journal that describe what you expect to learn in this course.

Use the four-step plan to solve each problem.

4. *Travel* The Masons traveled 753 miles to the Great Smoky Mountains for their vacation. They took a different route home and traveled 856 miles. How many miles did they travel in all?

5. *Sports* There will be 460 people at the sports award banquet. If each table seats 8 people, how many tables are needed?

EXERCISES

Use the four-step plan to solve each problem.

6. *Communication* According to a business magazine, the average American makes 184,702 phone calls in a life-time. Of these, 176 are international, 17,827 are long distance, and the others are local calls. How many local calls are made?
   Source: *U.S. News and World Report*

7. *Geometry* Write one sentence describing how the shapes at the right are the same and one sentence describing how they are different.

8. **Standardized Test Practice** In a wheat field near Duns, Scotland, a reproduction of the painting *Sunflowers* was created with 250,000 plants and flowers. The "painting" covered a 46,000-square foot area. What is a reasonable number of flowers and plants per square foot?

   A 5   **B** 50   **C** 100   **D** 500   **E** 1,000

9. *Patterns* Predict how long it would take you to count out loud from one to one million. Explain your reasoning.

10. *Fast Food* The graph shows the amount of fast food that was served daily in the United States in 1990 and 1995.

    a. How many more hamburgers were served in 1995 than in 1990?

    b. If the amount served grows at the same rate, predict how many hamburgers, French fries, and soft drinks will be served in the year 2000.

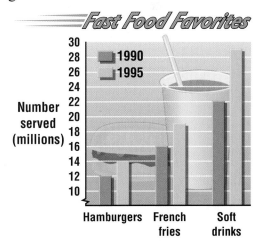

11. ***Working on the*** CHAPTER Project Refer to the table on page 3. Suppose you weigh 150 pounds and want to burn about 1,300 Calories in two hours. Find two different activities you could do for one hour each to burn about 1,300 Calories altogether.

12. *Critical Thinking* Use the digits 1, 2, 3, 4, and 5 to form a two-digit and a three-digit number so that their product is the least product possible. Use each digit only once.

For **Extra Practice**, see page 568.

Order of Operations

What you'll learn

You'll learn to evaluate expressions using the order of operations.

When am I ever going to use this?

Knowing the order of operations will help you find the total cost of adults' and children's admissions to the zoo.

Word Wise

order of operations

Pickup sticks is a game that originated in China many thousands of years ago. If you've played the game before, you know that when the sticks land parallel to each other, they are easy to pick up. If they intersect, it is difficult to pick one up without moving the others.

After all of the sticks have been picked up, you tally your score. Different-colored sticks receive different point values. These values are shown at the right.

Pickup Sticks Scoring	
Color	**Points**
Black	20
Red	10
Blue	6
Green	4
Yellow	2

Suppose you picked up 2 red sticks, 1 blue stick, and 4 yellow sticks. Here's how you can find your score.

$$2 \times 10 + \underbrace{1 \times 6}_{red} + \underbrace{4 \times 2}_{blue} = \underbrace{20}_{yellow} + 6 + 8 \text{ or } 34$$

In the expression above, you multiplied before adding. You were using the **order of operations** that mathematicians have agreed on. The order of operations ensures that expressions have only one value. Grouping symbols, like parentheses, are used to change the order of operations.

Order of Operations	1. Do all operations within grouping symbols first. 2. Multiply and divide in order from left to right. 3. Add and subtract in order from left to right.

Examples

1 Evaluate $(8 - 2) \div 3$.

$(8 - 2) \div 3 = 6 \div 3$ *Subtract first since $8 - 2$ is in parentheses.*

$= 2$ *Divide by 3.*

2 Evaluate $13 + 6 - 4 + 12$.

$13 + 6 - 4 + 12 = 19 - 4 + 12$ *$13 + 6 = 19$*

$= 15 + 12$ *$19 - 4 = 15$*

$= 27$ *$15 + 12 = 27$*

Study Hint

Technology To change the order of operations, use the parentheses keys.

In addition to the symbol $\times$, there are other ways to indicate multiplication. One way is to use a raised dot. Another way is to use parentheses.

$3 \cdot 5$ means 3×5

$2(4 + 5)$ means $2 \times (4 + 5)$

3 Evaluate $3(24 - 7) - 2 \cdot 13$.

$$3(24 - 7) - 2 \cdot 13 = 3(17) - 2 \cdot 13 \quad \textit{Subtract 7 from 24.}$$
$$= 51 - 2 \cdot 13 \quad \textit{Multiply 3 and 17.}$$
$$= 51 - 26 \quad \textit{Multiply 2 and 13.}$$
$$= 25 \quad \textit{Subtract 26 from 51.}$$

INTEGRATION **4** **Geometry** The distance around a geometric figure is called its *perimeter.* In a rectangle, the opposite sides have the same measure.

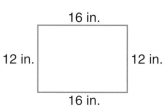

a. Write an expression to find the perimeter of the rectangle.

$2 \cdot 12 + 2 \cdot 16$ *The expression represents two widths plus two lengths.*

b. Evaluate the expression.

$$2 \cdot 12 + 2 \cdot 16 = 24 + 32 \quad \textit{Multiply 2 by 12 and 2 by 16.}$$
$$= 56 \quad \textit{Add 24 and 32.}$$

The perimeter is 56 inches.

CHECK FOR UNDERSTANDING

Communicating Mathematics

Read and study the lesson to answer each question.

1. *Tell* which operation you should do first in the expression $(3 + 5) \times 6$.

2. *Write* an expression to find the perimeter of the rectangle at the right.

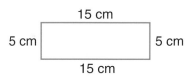

3. *You Decide* Tia thinks that $16 - 8 \div 8$ is 15. Antoine thinks $16 - 8 \div 8$ is 1. Who is correct? Explain your reasoning.

Guided Practice

Name the operation that should be done first in each expression.

4. $12 - 3 \cdot 4$ **5.** $7 + 3(5 - 2)$

Evaluate each expression.

6. $12 - 3(4)$ **7.** $12 \div 3(4)$

8. $3 \cdot 4(5 - 3)$ **9.** $3(4 + 7) - 5 \cdot 4$

10. *Photography* Juanita has 2 rolls of film with 36 exposures and 3 rolls of film with 24 exposures. How many photos can she take with this film?

EXERCISES

Practice

Name the operation that should be done first in each expression.

11. $3 + 5 \cdot 6$ **12.** $10 - (3 + 4)$ **13.** $4 + 2(8 - 6)$

14. $5 + 7(8)$ **15.** $(8 - 4) \div 2$ **16.** $7 \times 9 - (4 + 3)$

Evaluate each expression.

17. $3 + 5 \cdot 4$

18. $(12 - 4) \div 2$

19. $5 \cdot 8 - 3 \cdot 4$

20. $5 - 3 + 1$

21. $16 \div 4 \cdot 2$

22. $12 - 8 \div 4 + 6$

23. $(8 + 3) - 5$

24. $(17 + 3) \div (4 + 1)$

25. $4(6 + 4) \div 2$

26. $24 \div (7 - 3)$

27. $14 - (19 - 19)$

28. $25 \div (9 - 4)$

29. What is the value of $84 - 28 \div (4 \cdot 7)$?

30. Evaluate $82 - 43 - 6 \div 6$.

Copy each sentence below. Insert parentheses to make each sentence true.

31. $2 \cdot 14 - 9 - 17 - 14 = 7$

32. $16 + 5 \times 4 \div 2 = 42$

33. $64 \div 8 + 24 - 1 = 1$

34. $36 \div 3 - 9 \div 3 = 1$

Applications and Problem Solving

35. *Entertainment* An adult ticket to the zoo costs $5, a student ticket costs $3, and a senior ticket costs $4.

    a. Write an expression to find the total cost of 2 adult, 1 senior, and 3 student tickets.

    b. Evaluate the expression.

36. *Geometry* Find the perimeter of the parallelogram at the right.

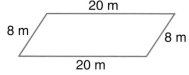

37. *Working on the* **CHAPTER Project** Refer to the table on page 3. Choose a weight of 100 or 150 pounds. Use a combination of at least 3 activities, each for a certain number of hours, that would burn approximately 2,400 Calories. Write an expression that shows the total number of Calories burned.

38. *Critical Thinking* Some calculators are programmed to follow the order of operations. Explain how you could tell whether your calculator follows the order of operations.

Mixed Review

39. **Standardized Test Practice** There are 72 tables set up at the music luncheon. If each table seats 6 people and all of the tables will be filled, how many people are expected to attend the music luncheon? *(Lesson 1-1)*

    A 432   **B** 78   **C** 66   **D** 12   **E** Not Here

40. *Sports* The graph shows the yearly average spent for sports apparel in the United States by each age group. On average, how much more was spent by the 13-to-17 age group than the 25-to-34 age group? *(Lesson 1-1)*

Who Buys Sports Apparel?

Yearly spending by age group

Age group	Amount
13-17	$311
18-24	$221
25-34	$187
35-54	$180
55-up	$155

Source: Sporting Goods Manufacturers Assoc.

For **Extra Practice,** see page 568.

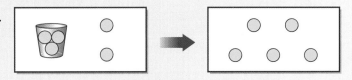

HANDS-ON LAB

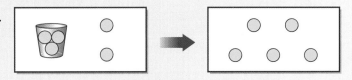

COOPERATIVE LEARNING

1-3A Variables and Expressions

A Preview of Lesson 1-3

☐: cups and counters

☐ integer mat

The backpack on the table has some books inside of it. There are 2 more books on the table. The total number of books is *the sum of 2 and the number of books in the backpack*. This phrase contains a constant that you know, 2, and an unknown value.

In mathematics, you will work with unknown numbers and constants. You can represent unknown numbers with cups and constants with counters.

TRY THIS

Work with a partner.

1 Model *the sum of some number and 2*.
- Use a cup to represent the unknown value and 2 counters to represent the constant.
- Any number of counters may be in the cup. Suppose you put 3 counters in the cup. Instead of an unknown value, you now know the cup has a value 3. When you empty the cup and count all the counters, the expression has a value of 5.

2 Model the phrase *twice some number*.
You don't know the value of the number, so let a cup represent this value. You will need to use 2 cups to represent *twice* some number.

The same number of counters should be in each cup.

ON YOUR OWN

Model each phrase with cups and counters. Then put four counters in each cup. How many counters are there in all? Record your answers by drawing pictures of your models.

1. the sum of 3 and a number

2. 3 times a number

3. 5 more than a number

4. twice a number plus 1

5. Write a sentence to describe what the cup represents.

Integration: Algebra
Variables and Expressions

What you'll learn

You'll learn to evaluate simple algebraic expressions.

When am I ever going to use this?

Knowing how to evaluate algebraic expressions can help you determine the amount of your paycheck.

Word Wise

variable
algebra
algebraic expression
evaluate

Study Hint

Reading Math The letter x is often used as a variable. It is also common to use the first letter of the value you are representing.

Can you twirl a hula hoop? Hula hoops have been around since the 1950s, when they were created by a company in California. At that time, they cost about $2 each and were so popular that 25 *million* were sold in just two months.

You can make a table to show the pattern between the cost of each hula hoop and the total sales.

Number of Hula Hoops	Sales (Dollars)
0	$2 × 0 = $0
1	$2 × 1 = $2
2	$2 × 2 = $4
3	$2 × 3 = $6
4	$2 × 4 = $8

Notice that the cost of each hula hoop is a constant, $2, but the number of hula hoops varies. You can use a placeholder, or **variable**, to represent the number of hula hoops. The expression for the amount earned is $2 × ❑ or $2 × *n*.

cost per hula hoop ⟶ ↓   ↓ ⟵ *number of hula hoops*

$2 \times n$

⟶ *total sales*

The branch of mathematics that involves expressions with variables is called **algebra**. The expression $2 \times n$ is called an **algebraic expression** because it contains variables, numbers, and at least one operation.

You can **evaluate** an algebraic expression by replacing the variable with a number and then finding the value of the numerical expression.

Example ① Evaluate $n + 2$ if $n = 7$.

$$n + 2 = 7 + 2 \quad \textit{Replace n}$$
$$= 9 \quad \textit{with 7.}$$

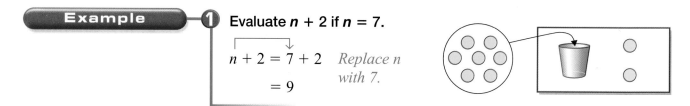

In mathematics, the following symbols are used for multiplication and division with variables.

$3a$ **means** $3 \times a$ or $3 \cdot a$ | $4cd$ **means** $4 \times c \times d$

rs **means** $r \times s$ | $\dfrac{m}{2}$ **means** $m \div 2$

Examples

Study Hint

Reading Math In Example 2, $4a$ means 4 times a, but 48 does not mean 4 times 8. So, there are parentheses around the 8 in 4(8).

CONNECTION

2 Evaluate $4a + 5b$ if $a = 8$ and $b = 3$.

$4a + 5b = 4(8) + 5(3)$ *Replace a with 8 and b with 3.*

$= 32 + 15$ *Multiply first.*

$= 47$ *Add 32 and 15.*

3 Evaluate $\dfrac{xy}{3}$ if $x = 7$ and $y = 9$.

$\dfrac{xy}{3} = \dfrac{(7)(9)}{3}$ *Replace x with 7 and y with 9.*

$= \dfrac{63}{3}$ or 21

4 **Health** The expression $110 + \dfrac{A}{2}$, where A stands for a person's age, is used to estimate a person's normal systolic blood pressure. Estimate the normal blood pressure for an 18-year-old person.

$110 + \dfrac{A}{2} = 110 + \dfrac{18}{2}$ *Replace A with 18.*

$= 110 + 9$ or 119   *Divide 18 by 2. Then add.*

The normal blood pressure for an 18-year-old person is about 119.

HANDS-ON

MINI-LAB

Work with a partner. isometric dot paper

The pattern below is made up of equilateral triangles.

Try This

1. Draw the next three figures in the pattern.

2. Find the perimeter of each figure and record your data in a table. The first three are completed for you.

Number of Triangles	1	2	3	4	5	6
Perimeter	3	4	5	?	?	?

Talk About It

3. Without drawing the figure, determine the perimeter of a figure made up of 10 triangles. Check by making a drawing.

CHECK FOR UNDERSTANDING

Communicating Mathematics

Read and study the lesson to answer each question.

1. *Tell,* in your own words, the difference between numbers and variables.

2. *Explain* how evaluating an expression is similar to using the cups and counters model in Lesson 1-3A.

HANDS-ON MATH

3. Refer to the Mini-Lab on page 13.

   a. If you know the number of triangles in the pattern, explain how you can find the perimeter of the figure.

   b. If n represents the number of triangles, write an expression that represents the perimeter of the figure.

Guided Practice

Evaluate each expression if $x = 6$, $y = 4$, $z = 0$, $a = 3$, $b = 2$, and $c = 7$.

4. $9 - c$

5. $x + y$

6. $4ab$

7. $xy - 4$

8. $10x - 2z$

9. $\frac{2x}{a}$

10. *Fitness* You can estimate how fast you walk in miles per hour by evaluating the expression $\frac{s}{30}$, where s is the number of steps you take in one minute. Estimate your speed if you take 93 steps in one minute.

EXERCISES

Practice

Evaluate each expression if $a = 6$, $b = 3$, and $c = 2$.

11. $a + b$

12. ac

13. $\frac{8b}{2}$

14. $a - b + 5$

15. $ab - 1$

16. $a + c - 2b$

17. $a + b + c$

18. $3a + b$

19. $ab - c$

20. $\frac{a}{c} + 4$

21. $2ab$

22. $\frac{a}{b} + c$

23. $\frac{2a}{3} - c$

24. $2(a + b) - c$

25. $2a - 3b$

26. $5 - \frac{2b}{c}$

27. $\frac{6(a + c)}{b}$

28. $c(b + a) - a$

29. Evaluate $125x + 75x$ if $x = 5$.

Applications and Problem Solving

30. *Earth Science* One way to estimate how many miles a thunderstorm is from you is to count the number of seconds between the lightning and the thunder. Then divide the number of seconds by five.

   a. Write an expression to estimate the distance.

   b. Estimate the distance if you count 6 seconds.

31. *Money Matters* Tammy charges $3 per hour for baby-sitting.
    a. Make a chart that shows how much money Tammy earns for baby-sitting 1, 2, 3, 4, and 5 hours.
    b. Write an expression to find how much money Tammy will earn for any number of hours. Let n be the number of hours she baby-sits.
    c. Suppose Tammy baby-sits for 8 hours. How much money will she earn?

32. *Critical Thinking* Find values of x and y so that the value of $7x + 2$ is greater than the value of $3y + 23$.

Mixed Review

33. *Decorating* Wallpaper for a bedroom costs $16 per roll for the walls and $9 per roll for the border. If the room requires 12 rolls of paper for the walls and 6 rolls for the border, compute the total cost for the wallpaper and border. *(Lesson 1-2)*

34. Evaluate the expression $36 + 9 \div 3$. *(Lesson 1-2)*

35. **Standardized Test Practice** Suppose there are 15 tables set up for the football banquet at Walnut Springs Middle School. Each table seats 8 people. You notice there are 5 empty seats. Which expression could you use to find the number of people seated at the banquet? *(Lesson 1-2)*

    A $15 + 8 - 5$ **B** $15 \times 8 + 5$ **C** $15 \times 8 - 5$

    D $15 + 8 + 5$ **E** $15 \times (8 - 5)$

36. *Sports* A baseball stadium holds 20,000 people. If 3,650 people can be seated in the bleachers, how many seats are available in the rest of the stadium? *(Lesson 1-1)*

For **Extra Practice**, see page 568.

CHAPTER 1   Mid-Chapter Self Test

1. *Transportation* Nine school buses serve Maplewood Middle School. The buses travel a total of 4,485 miles in one week. On average, about how many miles does each bus travel weekly? *(Lesson 1-1)*

2. Evaluate the expression $12 - 6 \div 3 + 8$. *(Lesson 1-2)*

Evaluate each expression if $x = 5$, $y = 4$, and $z = 10$. *(Lesson 1-3)*

3. $5(x + y) - z$ 4. $xy + 2z$

5. *Health* Do you know how much blood is in your body? You can find the approximate number of quarts by evaluating the expression $\frac{w}{30}$, where w is your weight in pounds. Estimate how much blood a 120-pound person has. *(Lesson 1-3)*

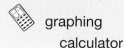

GRAPHING CALCULATORS

1-3B Evaluating Expressions

A Follow-Up of Lesson 1-3

graphing
calculator

You can use a graphing calculator to evaluate algebraic expressions. You can store a value in memory using a variable and then recall that value when evaluating an expression.

TRY THIS

Work with a partner.

Use a graphing calculator to evaluate $4x - 6$ and $5y + 8$ if $x = 3$ and $y = 2$.

Step 1 Store each value in its respective variable. To store the value 3 in the variable x, enter

3 [STO▶] [ALPHA] [X] [ENTER].

To store the value 2 in the variable y, enter

2 [STO▶] [ALPHA] [Y] [ENTER].

Step 2 Evaluate each expression. To evaluate $4x - 6$, enter

4 [ALPHA] [X] [−] 6 [ENTER]. The result is 6.

To evaluate $5y + 8$, enter

5 [ALPHA] [Y] [+] 8 [ENTER]. The result is 18.

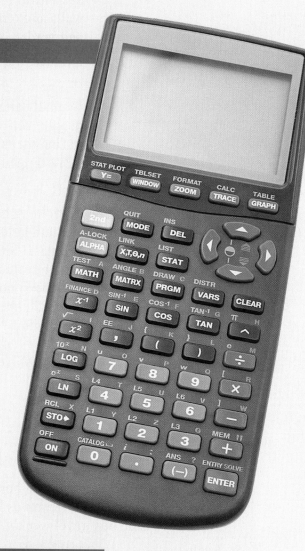

ON YOUR OWN

Use a graphing calculator to evaluate each expression if $m = 6$ and $n = 2$.

1. $8m$

2. $2m + 12$

3. $24 \div 2m$

4. mn

5. $2m - 3n$

6. $3m + 5n - 8$

7. How could you use a graphing calculator to evaluate the expression ab if $a = 4 + 8 \times 2$ and $b = 15 - 4 + 2$?

Integration: Algebra
Powers and Exponents

The Rajah's Rice, A Mathematical Folktale from India is the story of a young girl named Chandra. She loved elephants and helped take care of the Rajah's elephants. In fact, she helped cure the elephants when they were sick. The Rajah was so pleased that he decided to give her a reward.

Chandra also loved mathematics. So she asked for this reward.

"If Your Majesty pleases, place two grains of rice on the first square of this chessboard. Place four grains on the second square, eight on the next, and so on, doubling each pile of rice till the last square."

Do you think this was a good reward?

The chart lists the grains of rice placed on the first eight squares of the chessboard.

Square	1	2	3	4	5	6	7	8
Grains of Rice	2	4	8	16	32	64	128	256

Notice the pattern 2, 4, 8, 16, ... When two or more numbers are multiplied, these numbers are called **factors** of the product. The number of grains of rice can be written using only factors of 2. For example, 16 can be written as $2 \cdot 2 \cdot 2 \cdot 2$. When the same factor is used, you may use an **exponent** to simplify the notation.

$$16 = \underbrace{2 \cdot 2 \cdot 2 \cdot 2}_{four\ factors} = 2^4 \leftarrow exponent$$

four factors   *base* 2^4 *is a power of 2.*

The common factor is called the **base**. Numbers expressed using exponents are called **powers**.

The powers 4^2, 2^3, and 5^4 are read as follows.

Symbols	Words
4^2	four to the second power or four **squared**
2^3	two to the third power or two **cubed**
5^4	five to the fourth power

1 Write 3^5 as a product.

The base is 3. The exponent 5 means 3 is used as a factor 5 times.

$$3^5 = 3 \cdot 3 \cdot 3 \cdot 3 \cdot 3$$

2 Write $10 \cdot 10 \cdot 10$ using exponents.

10 is the base. It is used as a factor 3 times. So, the exponent is 3.

$$10 \cdot 10 \cdot 10 = 10^3$$

When you evaluate expressions with powers, you should evaluate powers before other operations.

Order of Operations	1. Do all operations within grouping symbols first. 2. Do all powers before other operations. 3. Multiply and divide in order from left to right. 4. Add and subtract in order from left to right.

Examples

3 Evaluate 5^4.

$5^4 = 5 \cdot 5 \cdot 5 \cdot 5$ *Definition of power*

$\quad = 625$

Study Hint

Technology Many calculators have a $\boxed{y^x}$ key that allows you to compute exponents. To find 5^4, press 5 $\boxed{y^x}$ 4 $\boxed{=}$. The display shows 625.

4 Evaluate $2 \cdot 6 + 4^2$.

$2 \cdot 6 + 4^2 = 2 \cdot 6 + 16$ *Evaluate the power first, $4^2 = 4 \cdot 4$ or 16.*

$\qquad\qquad = 12 + 16$ *Multiply.*

$\qquad\qquad = 28$ *Add.*

5 Write m^3 as a product.

$m^3 = m \cdot m \cdot m$

6 Write $x \cdot x \cdot x \cdot x$ using exponents.

$x \cdot x \cdot x \cdot x = x^4$

INTEGRATION **7** **Algebra** Evaluate $y^4 + 10$ if $y = 3$.

$y^4 + 10 = 3^4 + 10$ *Replace y with 3.*

$\qquad\qquad = 81 + 10$ *$3^4 = 81$*

$\qquad\qquad = 91$

CHECK FOR UNDERSTANDING

Communicating Mathematics

Read and study the lesson to answer each question.

1. *Explain* how to find the value of the expression *6 squared*.

2. *Define* *power*. Use the terms *factor*, *base*, and *exponent* in your definition.

Math Journal

3. *Write* a short paragraph explaining why expressions like 10^6 are written with exponents.

Guided Practice

Write each power as a product of the same factor.

4. 7^5

5. z^3

Write each product using exponents.

6. $5 \cdot 5 \cdot 5$

7. $x \cdot x \cdot x \cdot x \cdot x \cdot x$

Evaluate each expression.

8. 8^3

9. $6 \div 3 \cdot 2^3$

Evaluate each expression if $n = 4$, $x = 5$, and $t = 2$.

10. $n^3 - t$

11. $2x + x^2$

12. *Number System* The base-ten number system uses powers of ten to express numbers. Write 1,000,000,000 as a power of ten.

EXERCISES

Practice

Write each power as a product of the same factor.

13. 2^4

14. 9^7

15. 4^5

16. a^3

17. m^4

18. y^2

Write each product using exponents.

19. $12 \cdot 12$

20. $8 \cdot 8 \cdot 8 \cdot 8 \cdot 8$

21. $15 \cdot 15 \cdot 15 \cdot 15$

22. $b \cdot b \cdot b$

23. $n \cdot n$

24. $r \cdot r \cdot r \cdot r \cdot r$

Evaluate each expression.

25. 7^2

26. 1^{10}

27. $5 + 4^2$

28. $(5 + 4)^2$

29. $5 \cdot 3 + 2^3$

30. $(16 \div 4)^3 - 6$

Evaluate each expression if $a = 3$, $b = 9$, and $c = 2$.

31. $a^4 + b$

32. $c^2 + ab$

33. $2a^3$

34. $bc - a^2$

35. $a^2 + b^2$

36. $a^2(a + b)$

37. *Algebra* Find the value of $x^2 + 2x + 1$ if $x = 3$.

Use a calculator to determine whether each sentence is *true* or *false*.

38. $3^7 > 7^3$

39. $14^4 = 182$

40. $6^3 < 4^4$

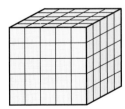

Applications and Problem Solving

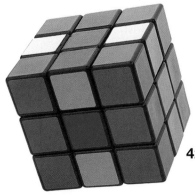

41. *Geometry* Refer to the figures.

    a. Find the number of unit cubes that make up each large cube. Write your answers using exponents.

    b. Why do you suppose the expression 2^3 is sometimes read as 2 *cubed*?

42. *Language Arts* Refer to the beginning of the lesson. Write an expression using exponents to determine how many grains of rice would be on the 64th square of Rajah's chessboard.

43. *Critical Thinking* Based on the pattern shown at the right, write a convincing argument that any number, except 0, raised to the 0 power equals 1.

$$2^4 = 16$$
$$2^3 = 8$$
$$2^2 = 4$$
$$2^1 = 2$$
$$2^0 = ?$$

Mixed Review

44. Standardized Test Practice Cameron is planning how much money he will need to wash his clothes at a laundromat. Each load costs $1.50 to wash and $0.75 to dry. The total cost is represented by the expression $1.50n + 0.75n$, where n is the number of loads. How much would it cost to wash and dry 7 loads? *(Lesson 1-3)*

For **Extra Practice**, see page 569.

A $5.25   **B** $10.50   **C** $12.75   **D** $15.75

45. *Algebra* Evaluate $3a - 2b + 6(a - b)$ if $a = 14$ and $b = 2$. *(Lesson 1-3)*

46. Evaluate $3(6 - 2) + 2 + 4$. *(Lesson 1-2)*

47. *True* or *false*? In the four-step plan, the *Solve* step comes last. *(Lesson 1-1)*

Let the Games Begin

Alge-bridge

Math Skill
Evaluating Expressions

Get Ready This game is for two, three, or four players.

    18 index cards   scissors   one number cube

Get Set Cut the index cards in half so you have 36 smaller cards. On each card, write a different expression containing only one variable.

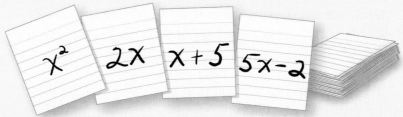

Go

- Deal all the cards to the players. The dealer then rolls the number cube. The number on the number cube is the value of the variable for the first round.

- The player to the left of the dealer puts a card faceup on the table and announces its value. Play continues until all players have placed one card on the table. This is the end of the first round. The person whose card has the greatest value wins all of the cards for that round.

- The player to the left of the dealer rolls the number cube. This is the value of the variable for the next round. Play continues until all cards are played.

- The person who has the most cards at the end of the game is the winner.

 Visit www.glencoe.com/sec/math/mac/mathnet for more games.

1-5

Integration: Algebra
Solving Equations

What you'll learn

You'll learn to solve equations using mental math.

When am I ever going to use this?

Knowing how to solve equations mentally can help you solve problems in science.

Word Wise

equation
solve
solution
modeling

Suppose you had this question on a geography test.

The ___?___ is the body of water on the east coast of Texas.

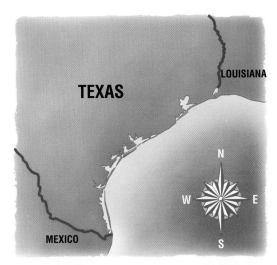

If you fill in the blank with *Gulf of Mexico,* the sentence is true. But if you choose *Pacific Ocean,* the sentence is false. Solving equations in mathematics is like answering fill-in-the-blank questions.

An **equation** is a sentence in mathematics that contains an equals sign.

$$48 + 12 = 60 \qquad 5 \times 4 = 20 \qquad 9 = 81 \div 9$$

Some equations also contain variables. The equation $x + 9 = 17$ is neither true nor false until x is replaced with a number. You **solve** the equation when you replace the variable with a number that makes the equation true. Any number that makes the equation true is called a **solution**. The solution of $x + 9 = 17$ is 8 because $8 + 9 = 17$.

Example ➊ Which of the numbers 10, 11, or 12 is a solution of $9x = 99$?

Replace x with 10.   Replace x with 11.   Replace x with 12.

$$9x = 99 \qquad\qquad 9x = 99 \qquad\qquad 9x = 99$$
$$9(10) \stackrel{?}{=} 99 \qquad 9(11) \stackrel{?}{=} 99 \qquad 9(12) \stackrel{?}{=} 99$$
$$90 = 99 \; \textit{false} \qquad 99 = 99 \; \checkmark\textit{true} \qquad 108 = 99 \; \textit{false}$$

The solution is 11.

Some equations can be solved mentally by using basic facts or arithmetic skills you know.

Example ➋ Solve $8 + t = 15$ mentally.

$$8 + t = 15$$
$$8 + 7 \stackrel{?}{=} 15 \quad \textit{You know that } 8 + 7 = 15.$$
$$15 = 15 \quad \checkmark$$

The solution is 7.

Example

You will learn other methods of solving equations in Chapter 6.

3 Solve $y = \frac{30}{6}$ mentally.

$y = \frac{30}{6}$

$5 \stackrel{?}{=} \frac{30}{6}$ *You know that $\frac{30}{6}$ is 5.*

$5 = 5$ ✓ The solution is 5.

When you write an equation that represents a real-world problem, you are **modeling** the problem.

Example

CONNECTION

4 **Physical Science** Atoms are the "building blocks" of matter. The nucleus of an atom is composed of protons and neutrons. The mass number of an atom is equal to the sum of the number of protons and neutrons.

a. Write an equation to model this situation.

b. A scientist can tell the mass number of an atom by how much it weighs and the number of protons by the atom's electrical properties. A certain isotope of carbon has 6 protons and a mass number of 14. How many neutrons does it have?

a. An equation is $p + n = m$, where p is the number of protons, n is the number of neutrons, and m is the mass number.

b. $p + n = m$

$6 + n = 14$ *Replace p with 6 and m with 14.*

$6 + 8 \stackrel{?}{=} 14$

$14 = 14$ ✓

The solution is 8. Therefore, this isotope of carbon has 8 neutrons.

CHECK FOR UNDERSTANDING

Communicating Mathematics

Read and study the lesson to answer each question.

1. *Explain* what it means to solve an equation.

2. *Tell* the solution of $5x = 45$.

3. *You Decide* Jermaine thinks the solution of $t \div 7 = 49$ is 7; Latisha thinks the solution is 343. Who is correct? Explain your reasoning.

Guided Practice

Name the number that is the solution of the given equation.

4. $s + 8 = 21$; 12, 13, 14

5. $d - 14 = 27$; 39, 40, 41

Solve each equation.

6. $25 + 19 = r$

7. $w \div 4 = 20$

8. $3y = 33$

9. $g + 12 = 30$

10. $12m = 120$

11. $56 \div 7 = a$

12. *Travel* If it takes you 5 hours to travel 250 miles in a car, what is your average speed? Use the equation $250 = 5r$, where r is the average speed.

Practice

Name the number that is the solution of the given equation.

13. $a + 15 = 19$; 4, 5, 6

14. $b - 13 = 29$; 40, 41, 42

15. $11a = 77$; 6, 7, 8

16. $v \div 10 = 4$; 20, 30, 40

17. $33 + t = 51$; 18, 19, 20

18. $13 \cdot 9 = g$; 107, 117, 127

Solve each equation.

19. $x + 35 = 91$

20. $m + 18 = 24$

21. $x - 15 = 71$

22. $15s = 105$

23. $\frac{n}{8} = 9$

24. $\frac{f}{3} = 61$

25. $j + 4 = 14$

26. $10k = 200$

27. $42 \div 7 = t$

28. $p - 18 = 20$

29. $24 + 39 = x$

30. $9a = 108$

31. $d + 25 = 80$

32. $z \div 14 = 8$

33. $13(11) = k$

34. $r - 29 = 117$

35. $a - 75 = 98$

36. $43 + z = 65$

Applications and Problem Solving

37. Sheila was paid $9 per hour and earned $67.50. How many hours did Sheila work? Use the equation $67.50 = 9h$, where h is hours worked.

38. *Physical Science* Refer to Example 4. An isotope of uranium has 92 protons and a mass number of 235. How many neutrons does it have?

39. *Geometry* The perimeter of a square is four times the length of one of its sides. Use the equation $p = 4s$ to find the perimeter of a square whose side has a length of 21 centimeters.

40. *Working on the* CHAPTER Project Refer to the table on page 3. Write and solve an equation to find how many hours you would have to in-line skate to burn 2,400 Calories.

41. *Critical Thinking* Consider the equation $0 + b = c$. What can you say about b and c?

Mixed Review

42. **Standardized Test Practice** Which is equivalent to 4^3? *(Lesson 1-4)*

A $4(3)$ **B** $4 \times 4 \times 4$ **C** $4 + 4 + 4$ **D** $3 \times 3 \times 3 \times 3$

43. *Algebra* Evaluate b^5 if $b = 2$. *(Lesson 1-4)*

44. *Algebra* Evaluate the expression $2x + 3(x + y) - xy$ if $x = 10$ and $y = 2$. *(Lesson 1-3)*

45. Evaluate $5(6 + 3) \div (3 + 2)$. *(Lesson 1-2)*

inter NET
CONNECTION
For the latest statistics about languages, visit:
www.glencoe.com/sec/
math/mac/mathnet

For **Extra Practice**, see page 569.

46. *Population* According to the 1990 census, nearly 32 million people in the United States speak a language other than English in their homes. The chart shows the five most common languages. How many more people speak French than Chinese? *(Lesson 1-1)*

Most Common Foreign Languages Spoken in U.S.		
Rank	Language	Number
1	Spanish	17,339,172
2	French	1,702,176
3	German	1,547,099
4	Italian	1,308,648
5	Chinese	1,249,213

Source: U.S. Census

Integration: Geometry
Fractals and Other Patterns

What you'll learn

You'll learn to find and extend patterns.

When am I ever going to use this?

Knowing how to find patterns can help you solve problems in art and architecture.

Word Wise

fractal

Look around and you can see many patterns that are made of geometric shapes. But there are many things in nature like nautilus shells, clouds, trees, and coastlines that cannot be described by points, lines, and polygons. Look closer, and you can also see patterns in them.

In this century, a new branch of mathematics, called **fractal** geometry, is providing models of nature's designs. One of the most obvious fractal patterns occurs in a tree. In the following Mini-Lab, you will make a fractal tree.

Cultural Kaleidoscope

Fractal geometry began in 1980, when Benoit Mandelbrot discovered a complex geometric structure. Today, it is called the *Mandelbrot Set* in his honor.

HANDS-ON MINI-LAB

Work with a partner.   isometric dot paper

Try This

- The tree shown below starts with a vertical line segment 4 units long to represent the trunk of the tree. From the endpoint, two branches, each half as long as the trunk, are drawn.

- Continue drawing branches until they become too small to draw. Each branch is half as long as the previous branch.

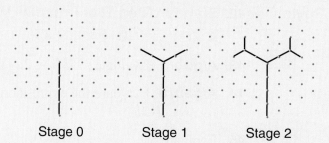

Stage 0   Stage 1   Stage 2

- On isometric dot paper, draw a fractal tree. The trunk of your tree should be 16 units long.

Talk About It

1. Are there parts of your completed tree that look like the entire tree? If so, draw a circle around one such part.

2. Are there other parts of a different size that look like the entire tree? If so, draw a circle around one part.

One of the characteristics of fractals is that they are made up of smaller replicas of the entire shape repeated over and over again in different sizes. Another characteristic is that they are made by a "rule" that describes a pattern. At each stage, you apply the rule to smaller and smaller parts of the figure.

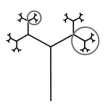

In the following example, you will construct a famous fractal.

Example

INTEGRATION

1 **Geometry** Construct a variation of Sierpinski's triangle using the following rules.

1. Draw a triangle with sides equal in length.
2. Connect the center of the sides with line segments.
3. Shade the middle triangle.
4. Apply Steps 1–3 to each of the remaining unshaded triangles. Continue this process until the triangles become too small to draw.

The first four stages of Sierpinski's triangle are shown below.

Stage 0   Stage 1   Stage 2   Stage 3

Many patterns in real-life are not fractals. One of them involves quilting techniques. A quilt often features a geometric design made by sewing together pieces of fabric.

Example

Real World **APPLICATION**

2 **Quilts** The first five squares in one row of a *Steeple Chase* quilt are shown below. Draw the next two squares in the row.

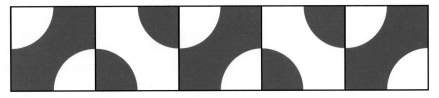

The pattern is made of alternating squares. The inside of each square is dark (D) or light (L). So the pattern is DLDLD. The next two squares should be LD.

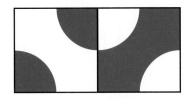

Communicating Mathematics

Read and study the lesson to answer each question.

1. *Explain* why a fern is an example of a fractal.

2. *Choose* the figure that continues the pattern.

a.

b.

c.

d.

HANDS-ON MATH

3. Another famous fractal is a Koch curve. It starts with a line segment. Divide the segment into thirds. Then, construct a triangle with sides equal in length on the middle third of each segment. Then remove the base of the triangle. The first two stages are shown at the right. Draw the next stage of the fractal.

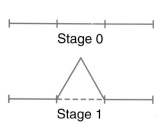

Guided Practice

4. Draw the next two figures that continue the pattern.

5. *Native-American Designs* The Yup'ik live in southwestern Alaska. The borders of their parkas are often decorated with repeated patterns. A traditional Yup'ik pattern is shown. Create your own repeated design.

Practice **Draw the next two figures that continue each pattern.**

6.

7.

8.

9.

10. **Square Carpet** A fractal can be formed by dividing a square into 9 smaller squares and shading the middle square.

   a. Draw the next stage.

   b. Suppose the rule was changed to shade only the four corner squares. Draw the first three stages of the fractal.

Stage 0   Stage 1

11. **Art** The photograph at the right shows how Sierpinski's triangle was used in the design on a twelfth-century pulpit of the Ravello cathedral. Create your own design using Sierpinski's triangle.

12. **Critical Thinking** Explain why the quilt design in Example 2 is *not* a fractal.

Mixed Review

13. **Algebra** Solve $\frac{t}{5} = 15$ mentally. *(Lesson 1-5)*

14. **Standardized Test Practice** Paint for the bedroom costs $12 per gallon, and paint for the bathroom costs $8 per gallon. If it requires 5 gallons of paint for the bedroom and 2 for the bathroom, what is the total cost of paint for the two rooms? *(Lesson 1-2)*

   A $20   **B** $64   **C** $76   **D** $496   **E** Not Here

For **Extra Practice,** see page 569.

MATH ⟩ IN THE MEDIA

1. What pattern in nature is shown in the comic?

2. Explain why the "map" is of no help to Fred.

3. Is this pattern a fractal? Explain your reasoning.

THE FAR SIDE   By GARY LARSON

"Face it Fred—you're lost!"

COOPERATIVE LEARNING

1-7A Area

A Preview of Lesson 1-7

 grid paper

scissors

In this lab, you will investigate the areas of rectangles and parallelograms by using models. The *area* of a geometric figure is the number of square units needed to cover the surface of the figure.

TRY THIS

Work with a partner.

1 On grid paper, draw a rectangle with a length of 6 units and a width of 4 units.

Count the number of squares within the rectangle. Each square represents an area of 1 square unit. So, the area is 24 square units.

ON YOUR OWN

Find the area of each rectangle.

1.

2.

3.

Find the area of each rectangle. Draw rectangles if necessary.

   4. length, 8; width, 3   **5.** length, 5; width, 2
   6. length, 10; width, 8   **7.** length, 9; width, 4
   8. length, 20; width, 10   **9.** length, 15; width, 9

10. Look Ahead Let ℓ represent the length, w represent the width, and A represent the area of a rectangle. Write an equation that shows how to find the area if you know the length and width.

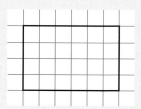

Now, let's find the area of a parallelogram. A *parallelogram* is a four-sided figure whose opposite sides are parallel. One of its sides is the *base*. The distance from the base to the opposite side is called the *height*.

2 On grid paper, draw a parallelogram with a base of 6 units and a height of 4 units.

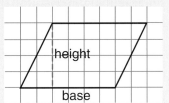

- Draw a line for the height.
- Cut out the parallelogram.

- Then cut the parallelogram along the line for the height as shown.
- Move the triangle to the opposite end of the parallelogram to form a rectangle.

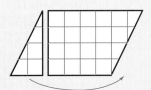

- Count the number of squares within the newly formed rectangle. The area is 24 square units. So, the area of the parallelogram is 24 square units.

11. The rectangle in Example 1 has a length of 6 units and a height of 4 units. The parallelogram in Example 2 has a base of 6 units and a height of 4 units. Compare the area of these two figures.

Find the area of each parallelogram by drawing the figure on grid paper, cutting it out, and counting the squares in the newly formed rectangle.

12. 　　　　**13.** 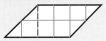　　　　**14.**

15. In Exercises 12–14, how are the base and height of the original parallelogram related to the length and width of the newly formed rectangle?

Find the area of each parallelogram. Use models if necessary.

16. base, 6; height, 3　　　　　　　　**17.** base, 8; height, 4

18. base, 5; height, 5　　　　　　　　**19.** base, 10; height, 6

20. *Look Ahead* Let b represent the base, h represent the height, and A represent the area of a parallelogram. Write an equation that shows how to find the area if you know the base and height.

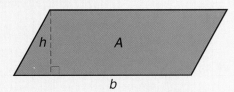

Integration: Geometry
Area

What you'll learn

You'll learn to find the areas of rectangles and parallelograms.

When am I ever going to use this?

Knowing how to find area can help you find the amount of wallpaper needed to decorate a room.

Word Wise

area
rectangle
parallelogram
base
height

If you like to shop at malls, you might want to schedule a visit to Minnesota's Mall of America. Not only does the mega-mall cover more than 4 *million* square feet, it also has a roller coaster and a 74-foot Ferris wheel!

The **area** of a figure is the number of square units needed to cover its surface. For the Mall of America, its area is measured in square feet. The area of a **rectangle** can be found as follows.

Area of Rectangles	**Words:** The area (A) of a rectangle equals the product of its length (ℓ) and width (w).
	Symbols: $A = \ell w$ **Model:**

Model:

w | A |
ℓ

Examples

1 **Find the area of the rectangle at the right.**

$A = \ell w$ *Write the formula for area.*

$A = 23 \cdot 18$ *Replace ℓ with 23 and w with 18.*
           Estimate by rounding:
           $20 \times 20 = 400$

18 ft

23 ft

$23 \times 18 = 414$

$A = 414$ The area is 414 square feet.

2 **Find the width of a rectangle with an area of 30 square inches and a length of 6 inches.**

$A = \ell w$ *Write the formula for area.*

$30 = 6w$ *Replace A with 30 and ℓ with 6.*

$30 \stackrel{?}{=} 6 \cdot 5$ *Solve the equation mentally.*

$30 = 30$ ✓

$w = 5$ The width is 5 inches.

A **parallelogram** is a four-sided figure whose opposite sides are parallel. One of its sides is called its **base**. The distance from the base to the opposite side is called the **height**. The area of a parallelogram is closely related to the area of a rectangle.

Area of Parallelograms	**Words:** The area of a parallelogram (A) equals the product of its base (b) and height (h). **Symbols:** $A = bh$ **Model:**

Examples

3 **Find the area of the parallelogram.**

$A = bh$ *Write the formula for area.*

$A = 30 \cdot 22$ *Replace b with 30 and h with 22. Estimate by rounding:*
$30 \times 20 = 600$

22 cm   25 cm   30 cm

$30 \times 22 = 660$

$A = 660$

The area is 660 square centimeters.

APPLICATION

Real World

4 **Painting** Cecilia wants to paint two walls of her bedroom bright blue. She knows that 1 gallon of paint will cover about 350 square feet of surface. One bedroom wall is 14 feet long and 8 feet high. The other is 12 feet long and 8 feet high. If she wants to put two coats of paint on the walls, will 1 gallon of paint be enough?

area of first wall

$14 \times 8 = 112$

area of second wall

$12 \times 8 = 96$

total area

$112 + 96 = \underbrace{}_{\text{area of both walls}} \times 2 = 416 \underbrace{}_{\text{2 coats of paint}}$

Cecilia needs to have enough paint to cover 416 square feet of surface. One gallon of paint will cover 350 square feet of surface. Therefore, one gallon of paint will not be enough.

Study Hint

Reading Math

Area is expressed in square units. So, the abbreviations for area often use the exponent 2.

square inch → in²

 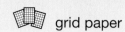

MINI-LAB

Work with a partner.  grid paper

The pattern at the right is made up of unit squares.

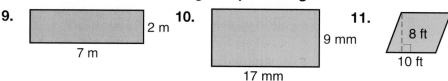

Study Hint

Problem Solving List your data in a table.

Try This

1. Draw the next three figures in the pattern.
2. Find the perimeter and area of each figure.

Talk About It

3. Without drawing the figure, determine the perimeter and area of a square with a length of 10 units.
4. Suppose you have a square with a length of *n* units. Write expressions for the perimeter and area of the square.

CHECK FOR UNDERSTANDING

Communicating Mathematics

Read and study the lesson to answer each question.

1. *Model* a parallelogram with a base of 8 units and a height of 5 units using grid paper.

2. *Tell* how to find the length of the rectangle.

5 m | Area = 50 m²

HANDS-ON MATH

3. If you double the length and width of a rectangle, how does its area change? Explain your reasoning in words or by drawing diagrams.

Guided Practice

Find the area of each rectangle or parallelogram.

4.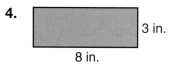
3 in.
8 in.

5.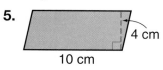
4 cm
10 cm

6. rectangle: ℓ, 1 in.; *w*, 6 in.

7. parallelogram: *b*, 6 ft; *h*, 5 ft

8. *Basketball* The length of a regulation court for professional and college basketball is 94 feet, and the width is 50 feet. What is the area of the court?

EXERCISES

Practice

Find the area of each rectangle or parallelogram.

9.
2 m
7 m

10.
9 mm
17 mm

11.
8 ft
10 ft

12.
5 yd
3 yd

13.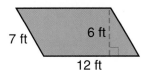
7 ft   6 ft
12 ft

14.
9 in.
8 in.
3 in.

15. rectangle: ℓ, 18 cm; w, 12 cm

16. rectangle: ℓ, 48 in.; w, 20 in.

17. parallelogram: b, 25 ft; h, 15 ft

18. parallelogram: b, 5 cm; h, 15 cm

19. parallelogram: b, 12 yd; h, 11 yd

20. rectangle: ℓ, 16 mi; w, 16 mi

21. *Algebra* What is the length of a rectangle with an area of 35 square yards and a width of 5 yards?

22. *Algebra* What is the base of a parallelogram with an area of 100 square feet and a height of 10 feet?

Applications and Problem Solving

23. *Housing* The graph shows the breakdown of new homes in America based on the number of square feet. One such house is a two-story rectangular house that is 35 feet long and 28 feet wide. Into which category would it be placed?

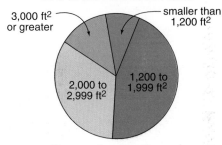

On the Homefront
Size of the American Home

3,000 ft² or greater

smaller than 1,200 ft²

2,000 to 2,999 ft²

1,200 to 1,999 ft²

Source: *American Homestyle*

24. *Geography* The shape of the state of Tennessee resembles a parallelogram. Estimate its area in square miles.

442 miles

115 miles

★ Nashville

Tennessee

25. *Critical Thinking* Draw three different parallelograms, each with an area of 16 square units.

Mixed Review

26. *Geometry* Draw the next two figures that continue the pattern. *(Lesson 1-6)*

27. *Standardized Test Practice* Ali runs a pizza parlor. Her daily cost of operating the parlor consists of a constant cost of $75 for rent, employee wages, and utilities plus $2 for every pizza she makes. If n represents the number of pizzas Ali makes during the day, which expression represents Ali's total daily cost? *(Lesson 1-3)*

  A $75 + 2n$ **B** $77   **C** $2 + 75n$ **D** $75n + 2n$

For **Extra Practice**, see page 570.

28. Evaluate $12(5) - 16 \div 8 + 5$. *(Lesson 1-2)*

1-7B Choose the Method of Computation

A Follow-Up of Lesson 1-7

The student council at Fort Couch Middle School has decided to plant a garden and donate the produce to the local food bank. Let's listen in as two students discuss how to plan the garden.

Winona

We have enough space to plant a garden that is 24 feet long and 16 feet wide. How will we figure out how much fertilizer we need to buy? How about fencing?

I think we should find the exact area and perimeter first. Then we can estimate to figure out how much fertilizer and fencing we'll need.

Antonio

I can use my calculator to multiply 24 and 16. The area is 384 square feet.

I added 24 and 16 in my head and doubled my answer. The perimeter is 80 feet.

THINK ABOUT IT

Work with a partner.

1. **Explain** why it is a good idea to find the exact area and perimeter instead of estimating.

2. **Tell** another tool you might use to find an exact answer.

3. **Write** a sentence explaining how you know when estimation is an acceptable method for solving a problem.

4. **Apply** what you have learned from Antonio and Winona's situation to solve this problem.

   One box of fertilizer will feed 250 square feet of garden. How many boxes should the student council buy? Explain your reasoning. Did you use mental math, estimation, a calculator, or paper and pencil to get your answer?

For **Extra Practice**, see page 570.

ON YOUR OWN

5. When you use the four-step plan for problem solving, you make an estimate of the solution even though you may be asked for an exact solution. *Explain* how estimating is useful when solving problems.

6. Use information from a newspaper or magazine advertisement and write two problems — one that needs an exact answer and one that needs an estimate.

7. *Explain* why an estimate might not be a good way to solve Exercise 23 on page 33.

MIXED PROBLEM SOLVING

STRATEGIES
Look for a pattern.
Solve a simpler problem.
Act it out.
Guess and check.
Draw a diagram.
Make a chart.
Work backward.

Solve. Use any strategy.

8. **Money Matters** Max needs to buy four markers to make posters for a social studies project. He has $4. Does he have enough money if the cost of each marker, including tax, is 89¢?

9. **Construction** Reynaldo wants to build a deck with an area of at least 120 square feet. He has space for a length of up to 14 feet, but no more than 9 feet for the width.

   a. Will he be able to build a deck as large as he wants?

   b. If so, what will be the area of the largest deck possible?

10. **Geography** The Mediterranean Sea is an almost completely closed sea of about 900,000 cubic miles of water. It takes water entering at the Strait of Gibraltar about 150 years to circulate through the sea and leave through the Strait of Gibraltar. About how many cubic miles of water enter the sea in one day?

11. **Money Matters** Hakeem saw an advertisement for piano lessons at $15.95 per lesson. Estimate how much 12 lessons will cost.

12. **Food** Kelly's Family Restaurant offers a Healthy Start Breakfast Special. The menu items and the number of Calories are shown in the chart below. About how many Calories are there in the meal?

Menu Item	Calories
Fresh Fruit Salad	50
Nonfat-Strawberry Yogurt	190
Whole-wheat English Muffin	218
Skim Milk	90

13. **Community Service** There were four drop-off centers for the community food drive. One center collected 2,629 cans of food, the second collected 2,892 cans, the third collected 4,429 cans, and the fourth collected 3,298 cans. The newsletter editor reported that over 13,000 cans of food were collected. Is this answer reasonable? Explain.

14. **Standardized Test Practice** An elephant in a zoo eats 57 cabbages in a week. About how many cabbages does an elephant eat in one year?

  A 7   **B** 700   **C** 1,500

  D 3,000   **E** 21,000

inter NET
CONNECTION Chapter Review For additional review, visit: www.glencoe.com/sec/math/mac/mathnet

Vocabulary

After completing this chapter, you should be able to define each term, concept, or phrase and give an example or two of each.

Number and Operations
base (p. 17)
cubed (p. 17)
exponent (p. 17)
factors (p. 17)
order of operations (p. 8)
powers (p. 17)
squared (p. 17)

Problem Solving
choose the method of computation (p. 34)

Geometry
area (p. 30)
base (p. 31)
fractal (p. 24)
height (p. 31)
parallelogram (p. 31)
rectangle (p. 30)

Probability and Statistics
population (p. 5)
sample (p. 5)

Algebra
algebra (p. 12)
algebraic expression (p. 12)
equation (p. 21)
evaluate (p. 12)
modeling (p. 22)
solution (p. 21)
solve (p. 21)
variable (p. 12)

Understanding and Using the Vocabulary

State whether each sentence is *true* or *false*. If *false*, replace the underlined word or number to make a true statement.

1. The base of 4^7 is <u>4</u>.
2. When the same factor is used, you may use an <u>exponent</u> to simplify the notation.
3. Seven <u>cubed</u> is written as 7^2.
4. Numbers expressed using exponents are called <u>variables</u>.
5. The branch of mathematics that involves expressions with variables is called <u>algebra</u>.
6. An <u>algebraic expression</u> is a sentence in mathematics that contains an equals sign.
7. The solution of $x + 5 = 18$ is <u>13</u>.
8. When you write an equation that represents a real-world problem, you are <u>modeling</u> the problem.
9. The area of a rectangle with a base of 5 inches and a height of 3 inches is <u>8 inches</u>.
10. A parallelogram is a four-sided figure whose opposite sides are <u>equal</u>.
11. The area of a parallelogram equals the product of its base and <u>height</u>.

In Your Own Words

12. ***Explain*** the relationship among factors, exponents, and powers.

Objectives & Examples

Upon completing this chapter, you should be able to:

● solve problems using the four-step plan *(Lesson 1-1)*

Bill practiced the trumpet 3 hours each day for 7 days. How many hours did he practice?

Explore   He practiced 3 hours each day for 7 days.

Plan   Multiply 3 by 7.

Solve $3 \cdot 7 = 21$
He practiced 21 hours.

Examine   Add 3 seven times to find that the answer is correct.

● evaluate expressions using the order of operations *(Lesson 1-2)*

Evaluate $3(9 + 7) - 4 \div 2 + 3$.

$3(9 + 7) - 4 \div 2 + 3$
$\quad = 3(16) - 4 \div 2 + 3$
$\quad = 48 - 2 + 3$ or 49

● evaluate simple algebraic expressions *(Lesson 1-3)*

Evaluate $6x - xy + y$ if $x = 10$ and $y = 3$.

$6x - xy + y = 6(10) - (10)(3) + 3$
$\qquad = 60 - 30 + 3$ or 33

● use powers and exponents in expressions *(Lesson 1-4)*

Evaluate 2^5.

$2^5 = 2 \cdot 2 \cdot 2 \cdot 2 \cdot 2$
$\quad = 32$

Review Exercises

Use these exercises to review and prepare for the chapter test.

Use the four-step plan to solve.

13. *Travel* A car traveling at 60 mph will travel how far in 7 hours?

14. *Money Matters* Ann starts the day with $100. If she spends $35 at the mall, how much is left for groceries?

15. *Books* A school library has 13,274 books; 327 are reference books, 7,015 are nonfiction, and the rest are fiction. How many fiction books are there?

Evaluate each expression.

16. $3 + 7 \cdot 4 - 6$

17. $8(16 - 5) - 6$

18. $12 - 18 \div 9$

19. $83 + 3(4 - 2)$

20. $75 \div 3 + 6(5 - 1)$

21. $10(12 - 2) \div 10$

Evaluate each expression if $p = 12$, $q = 3$, and $r = 5$.

22. $p + q - r$

23. $\dfrac{p + q}{r}$

24. $4q - p$

25. $3(p + q) - r$

26. $25 - 2(p - 2q)$

27. $6qr - p$

Evaluate each expression.

28. 10^3

29. 3^6

30. $3 + 2^4$

31. 15 squared

32. $6 \cdot 2 + 4^2$

33. y^3, if $y = 4$

Objectives & Examples

Review Exercises

● solve equations using mental math
(Lesson 1-5)

Solve $3s = 36$.

$3s = 36$

$3 \cdot 12 = 36$ *You know that 3 · 12 = 36.*

$36 = 36$ ✓

The solution is 12.

Name the number that is a solution of the given equation.

34. $t + 11 = 23$; 11, 12, 13

35. $63 \div 7 = m$; 7, 9, 11

36. $16 - a = 9$; 5, 6, 7

Solve each equation.

37. $t - 12 = 35$

38. $8x = 88$

39. $\frac{m}{4} = 16$

40. $28 + r = 128$

● find and extend patterns *(Lesson 1-6)*

Draw the next two figures that continue the pattern.

The next two figures are

.

Draw the next two figures that continue each pattern.

41.
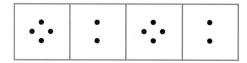

42.
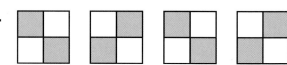

● find the areas of rectangles and parallelograms
(Lesson 1-7)

Find the area of a parallelogram with a base of 6 inches and a height of 3 inches.

$A = bh$

$A = 6 \cdot 3$ *Replace b with 6 and h with 3.*

$A = 18$

The area is 18 square inches.

Find the area of each rectangle or parallelogram.

43.
3 yd
15 yd

44.
4 in.
12 in.

45. rectangle: ℓ, 14 m; w, 12 m

46. parallelogram: b, 2 cm; h, 1 cm

47. rectangle: ℓ, 6 ft; w, 2 ft

48. parallelogram: b, 7 mm; h, 3 mm

Applications & Problem Solving

49. *Choose the Method of Computation* Carlos wants to buy two speakers for his sound system. Each speaker costs $59.99. If Carlos has $112 in his savings account, can he afford to buy the speakers now? *(Lesson 1-7B)*

50. *Money Matters* At a local coffee shop, Mexican coffee is $5 per pound, and Spanish coffee is $9 per pound. How much would you pay for 3 pounds of Mexican coffee and 2 pounds of Spanish coffee? *(Lesson 1-2)*

51. *Life Science* You can estimate the temperature in degrees Fahrenheit by using the expression $\frac{c}{4} + 37$, where c is the number of times a cricket chirps per minute. Find the temperature if a cricket chirps 104 times in one minute. *(Lesson 1-3)*

52. *Sports* The graph shows the number of World Series Championships for selected teams from 1903-1996. How many total championships have been won by these teams? *(Lesson 1-1)*

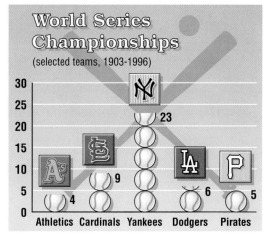

Source: *World Almanac, 1996*

Alternative Assessment

● *Open Ended*

Suppose you are in charge of purchasing the decorations for a school dance. You have decided to buy 4 packages of balloons at $1.39 each, 10 rolls of streamers at 89¢ each, and one dozen table decorations at $2.49 each. You also want to purchase a few bags of confetti that cost $2.25 each. Write an expression that represents the cost of the party items.

If you have $50 to spend, how many packages of confetti can you buy?

● *Completing the* CHAPTER Project

Use the following checklist to make sure your plan is complete.

☑ You have included at least six different activities for the week, the amount of time you do each activity, and how many Calories are burned.

☑ The amount of Calories eaten each day is between 2,400 and 2,800.

☑ Your plan is reasonable.

 PORTFOLIO Select one of the assignments from this chapter and place it in your portfolio. Attach a note to it explaining why you selected it.

A practice test for Chapter 1 is provided on page 607.

Section One: Multiple Choice

There are nine multiple choice questions in this section. Choose the best answer. If a correct answer is *not here* choose the letter for Not Here.

1. What is the value of $x + y + 5$ if $x = 6$ and $y = 15$?

   A 21
   B 25
   C 26
   D 28

2. How many square inches of tile are needed to cover a rectangular art project that measures 18 inches by 10 inches?

   F 28 in²
   G 8 in²
   H 180 in²
   J 90 in²

3. Using this chart, Wesley will write a report on the major rivers of the world.

River	Length (miles)
Amazon	4,000
Chang	3,964
Huang	3,395
Nile	4,160
Ob-Irtysh	3,362

   If he wants to list the three rivers in order from greatest to least, which list should he choose?

   A Ob-Irtysh, Huang, Chang
   B Nile, Chang, Huang
   C Nile, Amazon, Chang
   D Amazon, Chang, Huang

Please note that Questions 4-9 have five answer choices.

4. One page of a textbook is 14 inches long and 11 inches wide. If the only graphic on the page measures 3 inches by 4 inches, which sentence could be used to find x, the amount of space left for text and borders?

   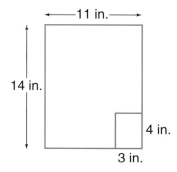

   F $x = (14 + 11) - (4 + 3)$
   G $x = 2(14 + 11) - 2(4 + 3)$
   H $x = 14 \times 11 \times 4 \times 3$
   J $x = (14 \times 11) - (3 \times 4)$
   K $x = \frac{14 \times 11}{4 \times 3}$

5. The student council sells pizzas for a fundraising activity. They charge $6 for each pizza. Which is the best estimate of their earnings if they sell 175 pizzas?

   A $6,000
   B $1,200
   C $1,050
   D $900
   E $750

6.

Scott's Basketball Shooting Record

Field Goals

Games

How many field goals did Scott make in the first five games?

F 4

G 5

H 10

J 19

K 26

7. One can of soup contains 12 ounces. How many ounces are there in 48 cans?

A 4

B 36

C 60

D 576

E Not Here

8. Each tent is put up with 12 poles. How many tents can be put up with 200 poles?

F 16

G 16 R8

H 17

J 20

K Not Here

9. Sonia spent $68 for a tent, $27 for a sleeping bag, and $25 for a backpack. How much more did she spend for the tent than the backpack?

A $93

B $43

C $52

D $120

E Not Here

Test-Taking Tip   THE PRINCETON REVIEW

When taking standardized tests, be careful not to make stray marks on your answer sheet. These marks may be misread by the scoring machine. Any question that appears to have more than one answer will be marked as incorrect.

Section Two: Free Response

This section contains six questions for which you will provide short answers. Write your answers on your paper.

10. What is the value of
$3[2(23 - 11) - (3 + 9)]$?

11. Tai is reading a 258-page novel. If he reads 8 pages an hour, about how long will it take him to read the entire book?

12. Lucy is knitting scarves for her 3 sisters. She needs to buy 9 skeins of yarn. Each skein costs $2.59, including tax. Which information in the problem is *not* needed to find the total cost of the yarn?

13. Find the number of seconds in one week.

14. At the end of a 478-mile trip, the odometer reading in the car was 15,015 miles. What was the reading at the start of the trip?

15. The members of a cycling club are planning a 1,800-mile trip. They believe they can average 15 miles per hour for 6 hours each day. How many days will it take for them to complete the trip?

 **interNET CONNECTION**   **Test Practice** For additional test practice questions, visit:

www.glencoe.com/sec/math/mac/mathnet

Applying Decimals

What you'll learn in Chapter 2

- to compute, estimate, and solve problems using decimals,
- to express decimals in scientific notation and fractions as decimals,
- to find patterns in repeating decimals,
- to use metric units of length, mass, and capacity, and
- to determine reasonable answers in real-world problems.

CHAPTER Project

HOW BIG IS OUR SOLAR SYSTEM?

In this project, you will use decimals to make a model of our solar system. You can use any type of material for your model such as Styrofoam, modeling clay, balls, marbles, or balloons.

Getting Started

- Look at the solar system table. About how many times bigger is Jupiter's diameter than Earth's diameter?
- Use the table to determine which planet's average distance from the Sun is about 30 times Earth's distance from the sun.

Planet	Diameter (Earth = 1)	Average Distance to Sun (AU) (Earth = 1)
Mercury	0.38	0.387
Venus	0.95	0.723
Earth	1.00	1.000
Mars	0.53	1.524
Jupiter	11.19	5.203
Saturn	9.46	9.529
Uranus	4.01	19.191
Neptune	3.88	30.061
Pluto	0.18	39.529

Technology Tips

- Use a **spreadsheet** to convert from miles or kilometers to astronomical units (AU).
- Use an **electronic encyclopedia** to do your research.
- Use a **word processor.**

 **inter NET** **CONNECTION** Data Update **For up-to-date information on the solar system, visit:**

www.glencoe.com/sec/math/mac/mathnet

Working on the Project

You can use what you'll learn in Chapter 2 to help you make your solar system model.

Page	Exercise
46	28
59	32
76	27
83	Alternative Assessment

Comparing and Ordering Decimals

What you'll learn

You'll learn to compare and order decimals.

When am I ever going to use this?

Knowing how to order decimals can help you find books in the library.

Have you ever noticed that contestants on a television word game show often choose the letter R and almost *never* choose Q? They know that the letter R is used more often than the letter Q. The chart shows that an R is used, on average, 6.8 times out of 100 letters, but Q is used only 0.1 time.

Number of Times, per 100, That Each Letter is Used					
A	8.2	J	0.1	S	6.0
B	1.4	K	0.4	T	10.5
C	2.8	L	3.4	U	2.5
D	3.8	M	2.5	V	0.9
E	13.0	N	7.0	W	1.5
F	3.0	O	8.0	X	0.2
G	2.0	P	2.0	Y	2.0
H	5.3	Q	0.1	Z	0.07
I	6.5	R	6.8		

When you use words like *more, less,* or *equal to,* you are comparing numbers. You can compare decimals like 6.8 and 0.1 using a number line.

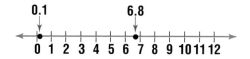

On a number line, numbers to the right are greater than numbers to the left.

6.8 is greater than 0.1.
$6.8 > 0.1$

0.1 is less than 6.8.
$0.1 < 6.8$

Study Hint

Reading Math The symbols $<$ and $>$ always point to the lesser of the two numbers.

You can also compare decimals by comparing the digits in each place-value position. For example, to compare 3.47 and 3.82, align the numbers by their decimal points.

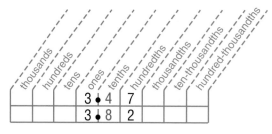

Start at the left and compare the digits in each place-value position. In the ones place, the digits are the same. In the tenths place, $4 < 8$. So, $3.47 < 3.82$.

Example 1

Compare 0.3 and 0.30.

0.3   *In the ones and tenths places,*
0.30   *the digits are the same.*

Annexing zeros to the right of a decimal produces *equivalent* decimals.

So, $0.3 = 0.30$.   *The models also show that 0.3 and 0.30 are equivalent.*

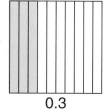

0.3

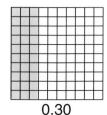

0.30

Real World APPLICATION

Libraries When Katie was doing research on whales for a science project, she found the titles of four books. She knows that the books are placed on the shelf by numbers ordered from least to greatest. In what order will she find the books?

Books About Whales

Where the Whales Are	599.5097
The World of the Arctic Whales	599.5091
Gentle Giant: The Humpback Whale	599.51
Giants of the Sea	599.5

The numbers of the books are 599.5097, 599.5091, 599.51, and 599.5. Locate the numbers on a number line.

599.5   599.5091   599.5097   599.51

599.500   599.505   599.510

The least number is 599.5.
The greatest number is 599.51.

From least to greatest, the numbers are 599.5, 599.5091, 599.5097, 599.51. So, Katie will find the books in this order from left to right.

CHECK FOR UNDERSTANDING

Communicating Mathematics

Read and study the lesson to answer each question.

1. *Write* a number sentence comparing two of the numbers shown on the number line.

0.30   0.50   0.70

2. *Draw* models of 0.2 and 0.18. Then write a number sentence that compares them.

3. *Write* a few sentences explaining why 0.4 is the same as 0.40.

Guided Practice

Replace each ● with <, >, or = to make a true sentence.

4. 1.22 ● 1.02   5. 0.97 ● 1.06

6. 7.90 ● 7.9   7. 1.3 ● 1.31

8. Order 5.13, 5.07, and 5.009 from least to greatest.

9. *Food* One serving of chocolate jimmies on your frozen yogurt adds 144 Calories and 5.9 grams of fat. One serving of peanut butter cup crumbles adds 92 Calories and 5.3 grams of fat. Which topping has more grams of fat?

Practice   Replace each ● with <, >, or = to make a true sentence.

10. 4.03 ● 4.01   **11.** 0.77 ● 0.69   **12.** 0.8 ● 0.08

13. 0.68 ● 0.680   **14.** 3.28 ● 3.279   **15.** 0.23 ● 0.32

16. 0.55 ● 0.65   **17.** 1.29 ● 1.43   **18.** 2.36 ● 2.3600

19. 0.0034 ● 0.034   **20.** 1.67 ● 0.48   **21.** 9.09 ● 9

22. Which is the greatest, 0.9, 0.088, 1.02, or 0.98?

23. Which is the least, 0.087, 0.901, 2, or 1.001?

24. Order 12.3, 12.008, 1.273, 12.54 from least to greatest.

25. Order 6.5, 6.05, 6.55, 6.505 from greatest to least.

Applications and Problem Solving

26. *Games* During the bonus round on a television word game show, the consonants R, S, T, L, N and the vowel E are automatically turned over. The contestant then gets to choose three more consonants and one more vowel. Refer to the chart on page 44. What letters should the contestant choose?

27. *Leisure Time* The chart shows how teenagers and unmarried adults spend some of their leisure time.
  a. On which activities do the teenagers spend less time than the adults?
  b. Order each list from greatest number of hours to least number of hours.

Leisure Time (hours per week)		
Activity	Teenagers age 12–17	Unmarried Adults age 18–29
organizations	1.2	0.5
visiting friends	4.4	7.8
hobbies	1.1	0.7
listening to music	0.9	0.7
television	17.7	14.2
reading	1.3	1.9

Source: Americans' Use of Time Project

  c. Make a graph to show the information.

28. *Working on the* CHAPTER Project Refer to the table on page 43.
  a. Make a chart listing the planets in order from greatest to least diameter.
  b. Make another chart listing the planets in order from greatest to least average distance from the Sun.

29. *Critical Thinking* Place decimal points in 999, 463, 208, and 175 so that the resulting decimals will be in order from least to greatest.

Mixed Review

30. *Geometry* Find the width of a rectangle with an area of 35 square inches and a length of 7 inches. *(Lesson 1-7)*

31. **Standardized Test Practice** Carla is six years old and is 12 years younger than Maria. Which equation can be used to find Maria's age? *(Lesson 1–5)*

   A $c = m + 12$ **B** $c = m - 12$ **C** $m = 12$ **D** $c + m = 12$

For **Extra Practice,** see page 570.

32. Write 2^5 as a product. *(Lesson 1–4)*

33. Evaluate $3(4 + 8) - 2 \cdot 5$. *(Lesson 1–2)*

2-2 Rounding Decimals

What you'll learn

You'll learn to round decimals.

When am I ever going to use this?

Knowing how to round can help you estimate with money.

Cultural Kaleidoscope

One of the best historical estimates for π was made by the Chinese astronomer Tsu Ch'ung-Chi (A.D. 470). He stated the value of π as 3.1415929, which is correct to six decimal places.

Are you a science fiction fan? In a TV show about a spaceship, an evil computer threatened to take over. To distract the computer, the ship's captain asked it to compute the exact value of pi.

Pi is a decimal that never ends and never has a pattern in its digits. So, the computer couldn't find the exact value, and the spaceship was saved!

Today, pi has been computed to over two billion decimal places. Here are the first fifteen. *The symbol for pi is π.*

$$\pi \approx 3.141592653589793...$$

On the number line, the graph of π is closer to 3.14 than to 3.15. To the nearest hundredth, π rounds to 3.14.

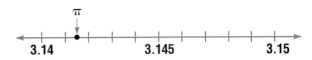

You can round to any place-value position without using a number line.

Rounding Decimals	Look at the digit to the right of the place being rounded. • The digit being rounded remains the same if the digit to the right is 0, 1, 2, 3, or 4. • Round up if the digit to the right is 5, 6, 7, 8, or 9.

Examples

1 Round 3.92 to the nearest tenth.

Look at the digit to the right of the tenths place.

3.9↑2   *Since 2 < 5, the digit in the tenths place stays the same.*
tenths place

3.92 rounded to the nearest tenth is 3.9.

2 Round 46.297 to the nearest hundredth.

Look at the digit to the right of the hundredths place.

46.29↑7   *Round up since 7 > 5.*

hundredths place

46.297 rounded to the nearest hundredth is 46.30.

Whole numbers are often expressed using a combination of decimals and words. For example, 2,500,000 is sometimes written as 2.5 million. In this case, the 2 is in the millions place.

Example

Real World APPLICATION

Computers The number of personal computers grew from 82.4 million in 1997 to 97.3 million in 1998. To the nearest million, about how many more personal computers were expected to be in use in 1998 than in 1997?

Since both numbers are expressed in millions, you can subtract as you would with decimals.

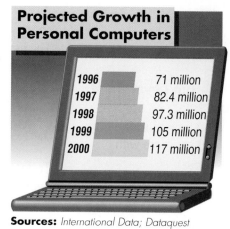

Projected Growth in Personal Computers

1996	71 million
1997	82.4 million
1998	97.3 million
1999	105 million
2000	117 million

Sources: *International Data; Dataquest*

$$\begin{array}{r} 97.3 \\ -\ 82.4 \\ \hline 14.9 \end{array}$$ *Align the decimal points.*

To the nearest million, 14.9 million rounds to 15 million. So, there were about 15 million more personal computers in 1998 than in 1997.

CHECK FOR UNDERSTANDING

Communicating Mathematics

Read and study the lesson to answer each question.

1. *Show,* using a number line, to what tenth you would round 14.37.

14.37

14.3   14.35   14.4

2. *Explain* why you can ignore all the digits to the right of 7 when rounding 4.23715 to the nearest hundredth.

3. *You Decide* Tomas rounds 11.96 to the nearest tenth and gets 12.0; Cynthia gets 12. Who is correct? Explain your reasoning.

Guided Practice

Round each number to the place indicated.

4. 0.315; tenth   5. 0.2456; hundredth   6. 17.499; tenth

Round each number to the underlined place-value position.

7. 0.7<u>8</u>9   8. 0.9<u>6</u>   9. 1.5<u>7</u>246

10. *Shopping* Calculations involving money are usually rounded to the nearest cent or hundredth. Diego uses his calculator to find the sales tax on purchases that total $15.99. His calculator display is shown at the right. How much sales tax will Diego pay?

Practice

Round each number to the place indicated.

11. 0.219; hundredth   **12.** 15.552; tenth   **13.** 9.6; unit

14. 7.0375; thousandth   **15.** 16.399; tenth   **16.** 6.95; tenth

17. 9.1283; thousandth   **18.** 0.37; tenth   **19.** 0.445; hundredth

Round each number to the underlined place-value position.

20. 2_3_.48   **21.** 1.7_0_4   **22.** 0.1_6_3

23. 15._4_51   **24.** 4.52_9_88   **25.** 0._7_87

26. _3_8.56   **27.** 5_9_.61   **28.** 0._5_55

29. Draw a number line to show how 5.67 rounds to 6.

Applications and Problem Solving

30. *Measurement Precision* After adding or subtracting measurements, the sum or difference should always be rounded to the least precise place-value position. In the triangle, 8.2 is the least precise measure because it is expressed in tenths and 9.25 and 10.73 are expressed in hundredths. Find the perimeter of the triangle to the nearest tenth.

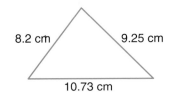

31. *Food* The graph shows how many pounds of breakfast cereal are consumed each year by the average person in several countries.

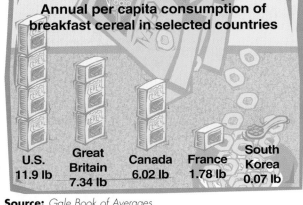

a. Which measurement is the least precise?

b. To the nearest tenth, how many more pounds of cereal are consumed by a person in the United States than in Canada?

Source: *Gale Book of Averages*

32. *Critical Thinking* Write three different decimals that round to 4.63.

Mixed Review

33. Order 8.75, 9.5, 8, and 9.15 from least to greatest. *(Lesson 2-1)*

34. *Algebra* Solve $\frac{x}{3} = 6$ mentally. *(Lesson 1-5)*

35. *Algebra* Evaluate $y^3 + 2$ if $y = 3$. *(Lesson 1-4)*

36. **Standardized Test Practice** Juanita is preparing for her birthday party. She buys 2 boxes of cookies containing 24 cookies each and 3 packages of brownies containing 15 brownies each. Which expression *cannot* be used to find the total number of dessert items she has bought? *(Lesson 1-2)*

A $3 \times 15 + 2 \times 24$ B $5 \times (24 + 15)$ C $2 \times 24 + 3 \times 15$

D $48 + 45$ E $15 + 15 + 15 + 24 + 24$

For **Extra Practice,** see page 571.

What you'll learn

You'll learn to estimate with decimals.

When am I ever going to use this?

You'll use estimation to help determine the total cost of items at the grocery store.

Word Wise

clustering

Do you like to swim but get bored swimming lap after lap? Wouldn't it be more exciting to swim the English Channel or the Atlantic Ocean? Believe it or not, you can swim that in a pool! The chart at the right shows the swimming equivalents assuming that a lap in the pool is 60 feet. If you wanted to swim all of the bodies of water, *about* how many laps would you need to swim?

Swimming Equivalents (thousands of laps)	
English Channel	1.848
Lake Michigan	10.384
Mississippi River	206.624
Atlantic Ocean	365.200

Source: *Health Scan*

To solve this problem, you can use rounding to estimate the answer. First, round each addend to the same place-value position. In this case, we'll round to tens. Then add.

$$
\begin{array}{rcr}
1.8 & \to & 0 \\
10.4 & \to & 10 \\
206.6 & \to & 210 \\
365.2 & \to & +\,370 \\
\hline
& & 590
\end{array}
$$

Change 590 to 590,000 since the answer is in thousands.

You would need to swim about 590,000 laps if you wanted to swim the English Channel, Lake Michigan, the Mississippi River, and the Atlantic Ocean.

Examples

Estimate by rounding.

1 23.485 − 9.757

$$
\begin{array}{rcr}
23.485 & \to & 23 \\
-9.757 & \to & -10 \\
\hline
& & 13
\end{array}
$$

The difference is *about* 13.

2 43.9 × 37.5

$$
\begin{array}{rcr}
43.9 & \to & 40 \\
\times 37.5 & \to & \times 40 \\
\hline
& & 1{,}600
\end{array}
$$

The product is *about* 1,600.

3 6.43 + 2.17 + 9.1 + 4.87

$$
\begin{array}{rcr}
6.43 & \to & 6 \\
2.17 & \to & 2 \\
9.1 & \to & 9 \\
+4.87 & \to & +\,5 \\
\hline
& & 22
\end{array}
$$

The sum is *about* 22.

4 432.87 ÷ 8.9

Round the divisor.

$$8.9 \to 9$$

Round the dividend to a multiple of 9.

$$432.87 \to 450$$

$$8.9\overline{)432.87} \to 9\overline{)450}\ \ \overset{50}{}$$

The quotient is *about* 50.

You can also use **clustering** to estimate sums. Clustering is used in addition situations if the numbers seem to be clustered around a common quantity.

Example ⑤ **APPLICATION**

Gymnastics Marissa watched her favorite gymnast, Shannon Miller, compete during the Summer Olympics in Atlanta. Marissa used her calculator to determine the total score before it was announced on TV.

Scores Shannon Miller	
Uneven bars	9.775
Balance beam	9.862
Floor exercise	9.475
Vault	9.724

Source: *USA TODAY*

Marissa's total read 30.0594. Use clustering to check her answer.

All of the scores were clustered around 10. There are four numbers. So, the sum is *about* 10 × 4, or 40.

30.0594 is not very close to 40. Marissa must have made an error in entering the numbers. She should add the four numbers again.

CHECK FOR UNDERSTANDING

Communicating Mathematics

Read and study the lesson to answer each question.

1. *Tell* why estimation is helpful when using a calculator to solve math problems.

2. *Write* a sentence describing when it makes sense to use the clustering method to estimate a sum.

3. *Write* about a situation where you have used estimation to solve a problem involving decimals.

Guided Practice

Estimate by rounding.

4.   8.56
   +5.34

5.   34.84
   −17.69

6.   6.8
   ×2.4

7. $6.8\overline{)40.79}$

8. $38.1\overline{)984.76}$

Estimate by clustering.

9. $18.4 + 22.5 + 20.7$

10. $56.9 + 63.2 + 59.3 + 61.1$

11. *Population* In 1996, Hispanics replaced non-Hispanic blacks as the second largest ethnic/racial group for ages 19 and under. There were 12.0 million Hispanic teens and 11.4 million non-Hispanic, black teens. About how many more Hispanic teens are there?

Practice

Estimate by rounding.

12.   23.84
      +12.13

13.   34.3
      −18.9

14.   7.5
      ×8.4

15. 9.3)‾65.48‾

16. 26.3
      ×9.7

17. 33.21
      −8.23

18. 67.86
      −24.35

19. 18.4)‾41.7‾

20. $121.5
      +487.8

21. 8.1)‾73.8‾

22. 2.6)‾8.99‾

23.   32.5
      ×81.4

24. 23.3)‾119‾

25. 11.4)‾35.7‾

26. 373.4)‾645.49‾

27. Estimate the product of 6.8 and 5.2.

28. Estimate 41.79 divided by 7.23.

Estimate by clustering.

29.   42.3
       41.5
       39.8
      +40.4

30.   77.8
       75.6
       81.2
      +79.9

31.   239.8
       242.43
       236.20
      +240.77

32. 9.9 + 10.0 + 10.3 + 11.1 + 9.8 + 11.2

33. 50.4 + 51.1 + 48.9 + 49.5 + 50.8

34. 100.5 + 97.8 + 101.6 + 100.2 + 99.3 + 99.1

Applications and Problem Solving

35. *Travel* Some of the busiest airports in the United States are listed in the chart at the right. About how many million passengers use these airports in one year?

36. *Life Science* A blue whale, the largest creature to live on Earth, can weigh up to 153.26 tons. An eighteen-wheel tractor and semitrailer, fully loaded, can weigh up to 48.5 tons. About how many times heavier is the blue whale?

Airport	Passengers (millions)
Atlanta	73.5
Chicago (O'Hare)	72.5
Dallas/Ft. Worth	60.5
Denver	36.8
Los Angeles	61.2
San Francisco	40.0

Source: *Airports Council International*

37. *Entertainment* The graph shows the number of people who visited several theme parks. About how many people visited the theme parks?

THEME PARK VISITORS

	Visitors (millions)
Theme Park A	16
Theme Park B	11.2
Theme Park C	10.3
Theme Park D	9.7
Theme Park E	8.8
Theme Park F	8

38. *Mountain Bikes* The Marion Police Department is planning to start a bicycle patrol in its city. The chart at the right shows prices of several mountain bikes. If the department is planning to buy 7 bikes from this list, about how much money will they spend?

Bike	Price
Brand A	$325
Brand B	270
Brand C	300
Brand D	270
Brand E	320
Brand F	260
Brand G	270
Brand H	290

39. *Critical Thinking* Mr. Stewart's car can travel between 12 and 18 miles on each gallon of gasoline. Gasoline costs between $1.23 and $1.31 per gallon. About how much will Mr. Stewart pay to travel 100 miles?

Mixed Review

40. Standardized Test Practice A ski resort advertises a new cross-country ski trail that is 9.673 kilometers long. To the nearest 0.1 kilometer, what is the length of the trail? *(Lesson 2-2)*

A 9.67 km   **B** 9.68 km

C 9.6 km   **D** 9.7 km

41. Replace the ● in 0.2 ● 0.214 with $<$, $>$, or $=$ to make a true sentence. *(Lesson 2-1)*

42. *Algebra* Solve $m + 18 = 33$. *(Lesson 1-5)*

43. *Algebra* Write an expression that represents a $500 donation plus $5 for every event. Let n represent the number of events. *(Lesson 1-3)*

For **Extra Practice,** see page 571.

44. Evaluate $7 \cdot 4 - 9 \div 3$. *(Lesson 1-2)*

PROBLEM SOLVING

2-3B Reasonable Answers

A Follow-Up of Lesson 2-3

Megan and her friend Molly are standing in the snack line at the movies.
Megan is trying to figure out whether she has enough money
to buy the snacks she wants. Let's listen in!

ABC THEATERS

Adult Ticket	$7.50	Candy	1.75
Student Ticket	6.50	Nachos	3.75
Popcorn, Large	3.25	Soft Drinks, Large	2.25
Popcorn, Medium	2.75	Soft Drinks, Med.	1.75
Popcorn, Small	2.00	Soft Drinks, Small	1.25

Molly

Megan

I'm starving! I hope I have enough money to buy a large popcorn, a large drink, and some candy.

Well, how much money do you have?

I have $7.00. A large popcorn costs $3.25, a large drink is $2.25, and the candy is $1.75. I think I have enough because $3 + 2 + 2 = $7.

I'm not sure. I always round up to the next highest dollar. Since $4 + 3 + 2 = $9, you probably don't have enough money.

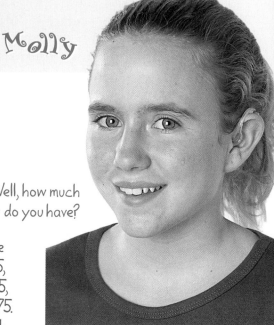

THINK ABOUT IT

Work with a partner. 1–3. See students' work.

1. ***Compare and contrast*** Megan's and Molly's thinking. Whose thinking do you like better? Why?

2. ***Think*** of another way to help Megan decide whether she has enough money.

3. ***Choose*** three items that Megan can buy with $7.00. Explain your reasoning.

4. ***Apply*** the strategy of determining **reasonable answers** to solve this problem.

   Chase earns $4.25 per hour at a sub shop. He usually works 12 hours each week. At his six-month review, he is given a $0.35 per hour raise. Chase used a calculator to determine that his weekly pay will increase by $42.00. Is this answer reasonable? No; it would increase only about $4 per week.

For **Extra Practice,** see page 571.

ON YOUR OWN

5. The last step of the 4-step plan for problem solving asks you to *examine* your solution. **Explain** how you can use estimation with decimals to help you examine a solution.

6. *Write a Problem* with an unreasonable answer and ask a classmate to explain why they think the answer is unreasonable.

7. *Explain* how you know that the answer to Exercise 37 on page 53 is reasonable.

MIXED PROBLEM SOLVING

Solve. Use any strategy.

STRATEGIES
Look for a pattern.
Solve a simpler problem.
Act it out.
Guess and check.
Draw a diagram.
Make a chart.
Work backward.

8. *Money Matters* Vinny's Video Haven is selling 3 blank tapes for $14.96. Brad says he can get 9 tapes at Vinny's for under $40. Is this reasonable?

9. *Transportation* New car carriers deliver new cars from the loading dock at the auto plant to car dealerships. Each truck can carry about 20,000 pounds. If an economy-size car weighs about 2,330 pounds, what is a reasonable number of cars that could be transported on one truck?

10. *Money Matters* At the Book Fair, Ernesto wants to buy 2 science fiction books for $3.95 each, 3 magazines for $2.95 each, and 1 bookmark for $0.39. Does he need to bring $20 or will $15 be enough? Explain your reasoning.

11. *Life Science* A photograph of a paramecium that is 0.23 millimeter long is enlarged to 60 millimeters for a science book. *About* how many times longer is the paramecium in the photograph than the actual paramecium?

12. *Money Matters* Suppose a relative matches your age with dollars each birthday. You are 13. How much money have you been given over the years by this relative?

13. *Communication* Peta places a long distance phone call to her grandparents in California and talks for 45 minutes. The phone company bills the call at a rate of $0.10 per half-minute. How much does the call cost Peta?

14. *Geography* The graph shows the lengths in miles of the longest rivers in the world. *About* how many total miles long are the three rivers?

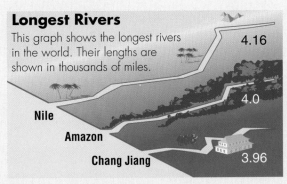

Longest Rivers
This graph shows the longest rivers in the world. Their lengths are shown in thousands of miles.

4.16

4.0

Nile

Amazon

Chang Jiang 3.96

Source: *The World Almanac*

15. *Standardized Test Practice* For lunch yesterday, Micah bought a hot dog for $1.79, potato chips for $0.89, and a milkshake for $1.15. How much did this lunch cost, not including tax?

A $2.68   **B** $2.94
C $3.83   **D** $9.51
E Not Here

2-4 Multiplying Decimals

What you'll learn

You'll learn to multiply decimals.

When am I ever going to use this?

Knowing how to multiply decimals can help you find the amount of interest earned on a savings account.

You use models in geometry to picture the concepts that you are learning. For example, the model below shows that the area of a rectangle is found by multiplying its length by its width. You can also use grid paper to represent multiplication problems.

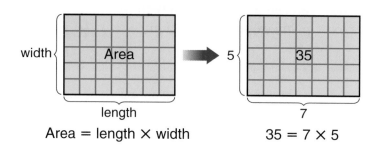

Area = length × width

35 = 7 × 5

HANDS-ON MINI-LAB

Work with a partner.   grid paper   markers

Multiplying decimals is similar to multiplying whole numbers. Here the area of the large square represents 1.00, and the area of each small square represents 1 hundredth or 0.01.

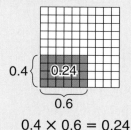

The length of a side of each small square is 1 tenth or 0.1.

0.4 × 0.6 = 0.24

Try This

1. Use grid paper to show each product.
   a. 0.2 × 0.9   b. 0.7 × 0.5
   c. 0.3 × 0.3   d. 0.6 × 2

2. Tell how many decimal places there are in each factor and in each product in Exercise 1.

Talk About It

3. How does the number of decimal places in the product relate to the number of decimal places in the factors?

Study Hint

Problem Solving
Organize your data for Exercise 2 into a table.

These and other similar models suggest the following.

Multiplying Decimals	The number of decimal places in the product of two decimals is the sum of the number of decimal places in the factors.

Examples

1 **Multiply 1.3 and 0.9.** *Estimate: 1 × 1 = 1*

$$\begin{array}{r} 1.3 \\ \times 0.9 \\ \hline 1.17 \end{array}$$ ← *one decimal place*
← *one decimal place*
← *two decimal places*

The product is 1.17. Compared to the estimate, the product is reasonable.

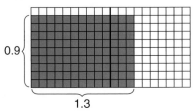
0.9
1.3

2 **Find the product of 0.054 and 1.6.** *Estimate: 0 × 2 = 0*

0.054 ← *three decimal places*
× 1.6 ← *one decimal place*
0.0864 ← *To make four decimal places, annex a zero on the left.*

The product is 0.0864. *Is the answer reasonable?*

Many real-world problems can be solved by multiplying decimals. For example, did you know that all sound, including music, is caused by vibrations? The number of vibrations per second determines the pitch of the sound. The more vibrations per second, the higher the pitch. The number of vibrations per second is called the *frequency.*

Example

CONNECTION

3 **Music** The frequency of any note multiplied by 1.06 gives the frequency of the note one-half step higher. The frequency of the A directly above middle C on a piano is 440 vibrations per second. What is the frequency of A#? *Read A# as "A sharp."*

To find the frequency of A#, multiply 440 by 1.06.

$$\begin{array}{r} 440 \\ \times 1.06 \\ \hline 466.40 \end{array}$$ *Estimate: 440 × 1 = 440*

The frequency of A# is 466.4 vibrations per second.

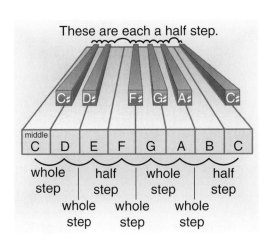
These are each a half step.

Did you know Of today's musicians, 63% learned to play before they were 12 years old.

Communicating Mathematics

Read and study the lesson to answer each question.

1. *Explain* how you can use estimation to check whether you have placed the decimal correctly in the product of two decimals.

2. *Write* a multiplication problem using decimals for the model shown at the right.

3. *Make* a model to show why $0.1 \times 0.1 = 0.01$.

Guided Practice

Place the decimal point in each product.

4. $1.32 \times 4 = 528$

5. $0.07 \times 1.1 = 77$

Multiply.

6. $\begin{array}{r} 3.4 \\ \times\ 7.8 \\ \hline \end{array}$

7. $\begin{array}{r} 0.15 \\ \times\ 1.23 \\ \hline \end{array}$

8. $\begin{array}{r} 11.5 \\ \times\ 0.47 \\ \hline \end{array}$

9. *Life Science* A giant tortoise can travel at a speed of about 0.2 kilometer per hour. At this rate, how far can it travel in 1.75 hours?

EXERCISES

Practice

Place the decimal point in each product.

10. $0.4 \times 0.7 = 28$

11. $1.9 \times 0.6 = 114$

12. $1.4 \times 0.09 = 126$

13. $5.48 \times 3.6 = 19728$

14. $4.5 \times 0.34 = 153$

15. $0.45 \times 0.02 = 9$

Multiply.

16. $\begin{array}{r} 0.2 \\ \times\ 6 \\ \hline \end{array}$

17. $\begin{array}{r} 0.3 \\ \times\ 0.9 \\ \hline \end{array}$

18. $\begin{array}{r} 0.45 \\ \times\ 0.12 \\ \hline \end{array}$

19. $\begin{array}{r} 0.0023 \\ \times\ 32 \\ \hline \end{array}$

20. $\begin{array}{r} 10.1 \\ \times\ 9 \\ \hline \end{array}$

21. $\begin{array}{r} 0.0023 \\ \times\ 0.35 \\ \hline \end{array}$

22. $\begin{array}{r} 6.78 \\ \times\ 1.3 \\ \hline \end{array}$

23. $\begin{array}{r} 1.5 \\ \times\ 2.7 \\ \hline \end{array}$

24. 5.1×4.3

25. 0.08×1.9

26. 0.25×0.004

27. 1.17×0.09

28. *Algebra* Evaluate xy if $x = 0.32$ and $y = 3.1$.

Applications and Problem Solving

29. *Travel* Karen was helping plan the Spanish Club's trip to Spain. On a recent trip, one United States dollar could be exchanged for 128.46 Spanish pesetas. How many pesetas would she receive for $50 in United States money?

30. *Geometry* To the nearest tenth, find the area of the rectangle.

0.4 m

1.5 m

31. Statistics The graph shows what part of all teenagers own or use technology.

a. Suppose you survey 35 teenagers. Predict about how many of those surveyed use each kind of technology. Round to the nearest tenth.

b. Survey your classmates. How do these results compare with the data in the graph? Explain any differences.

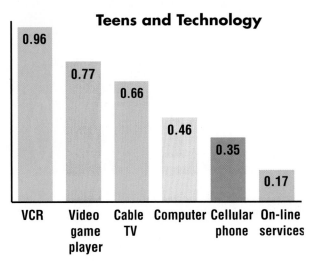

Teens and Technology

0.96 — VCR
0.77 — Video game player
0.66 — Cable TV
0.46 — Computer
0.35 — Cellular phone
0.17 — On-line services

Source: Chilton Research Services

32. Working on the CHAPTER Project Refer to the table on page 43.

a. The diameter of Earth is 12,756 kilometers. Find the diameters of all the other planets to the nearest kilometer. Include this information in your diameter chart.

b. The average distance from the Sun to Earth is 149 million kilometers. Find the actual average distances of all the other planets in millions of kilometers. Include this information in your distance chart.

33. Critical Thinking Two decimals, both less than 1, are multiplied. Is the product *always* less than 1, *sometimes* less than 1, or *never* less than 1? Explain.

Mixed Review

34. Estimate the sum of 5.82, 2.19, 8.1, and 6.05. *(Lesson 2-3)*

35. Round 0.99 to the nearest tenth. *(Lesson 2-2)*

36. Draw a number line to show which is greater, 3.77 or 3.7. *(Lesson 2-1)*

37. Standardized Test Practice Which expression could be used to find the cost of buying *b* baseball bats at $75 each and *g* baseball gloves at $98 each? *(Lesson 1-3)*

　A $75 + 98$　　**B** $b + g$　　**C** $75b + 98g$　　**D** $75b \times 98g$

For **Extra Practice**, see page 572.

CHAPTER 2 — Mid-Chapter Self Test

1. Which is greater, 0.28 or 0.028? *(Lesson 2-1)*

2. Round 0.$\underline{4}$9 to the underlined place-value position. *(Lesson 2-2)*

3. Estimate the sum of 71.28, 68.4, 70.73, 69.45, and 73.21. *(Lesson 2-3)*

4. Find the product of 0.7 and 0.9. *(Lesson 2-4)*

5. **Life Science** A snail moves at a speed of about 0.005 kilometer per hour. How far can it travel in 0.5 hour? *(Lesson 2-4)*

AEROSPACE

Dr. Mae Carol Jemison
ASTRONAUT

When Dr. Jemison flew on the space shuttle *Endeavor*, she became the first African-American woman in the space program. She took objects from Africa with her to symbolize that space exploration belongs to all nations.

To be an astronaut, you'll need at least a bachelor's degree in engineering, biological sciences, physical sciences, or mathematics and three years of professional experience related to the degree. However, there are many positions in the aerospace industry that do not require a college degree. A typical space-related company also needs mechanics, electricians, drafters, salespeople, personnel specialists, and assembly workers.

For more information:
Astronaut Selection Office
Mall Code AHX
Johnson Space Center
Houston, Texas 77058

inter NET CONNECTION
www.glencoe.com/sec/math/mac/mathnet

Someday, I'd like to fly in the space shuttle like Dr. Jemison.

Your Turn
Write and design a brochure that describes the benefits of a career in the aerospace industry.

2-5 Powers of Ten

What you'll learn

You'll learn to multiply decimals mentally by powers of ten.

When am I ever going to use this?

Mental math strategies can be used to find the cost of 100 items quickly.

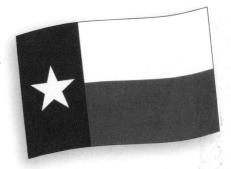

Texas is now the second most populous state according to Census Bureau estimates. For the 12-month period ending July 1, 1998, Texas' population grew to 19.7 million, passing New York's 18.2 million population.

You can write numbers like 19.7 million in standard form by multiplying by a power of ten. In this case, multiply by 1,000,000.

$$19.7 \times 1,000,000 = 19,700,000$$

In standard form, 19.7 million is 19,700,000.

How can you find the product of a power of 10 and another number without using a calculator or paper and pencil? Look for a pattern in the following products.

Study Hint

Mental Math You can also determine the number of places to move the decimal point by counting the number of zeros in the power of ten.

Decimal		Power of Ten		Product
19.7	$\times$	10^0 (or 1)	=	19.7
19.7	$\times$	10^1 (or 10)	=	197
19.7	$\times$	10^2 (or 100)	=	1,970
19.7	$\times$	10^3 (or 1,000)	=	19,700
19.7	$\times$	10^4 (or 10,000)	=	197,000

What pattern did you find? Notice that the digits in the original decimal and the product are the same. The difference is the position of the decimal point. The exponent in the power of 10 tells you the number of places to move the decimal point.

Since you are multiplying by a power of ten that is greater than one, the product will be greater than the original decimal. Thus, the decimal point moves to the right. You can use this pattern to multiply mentally.

Examples

1 Multiply 0.59 and 10^4 mentally.

$0.59 \times 10^4 = 5,900.$

Move the decimal point four places to the right.

$0.59 \times 10^4 = 5,900$

2 Solve $c = 27.2 \times 1,000$.

$c = 27.2 \times 1,000$

$c = 27.2 \times 10^3$ *$1,000 = 10^3$*

$c = 27,200.$

Move the decimal point three places to the right.

The solution is 27,200.

You can also use a similar pattern when multiplying by a power of ten that is less than 1. Look for a pattern in the following products.

Decimal		Power of Ten		Product
23.9	×	$0.1 \left(\text{or } \frac{1}{10^1}\right)$	=	2.39
23.9	×	$0.01 \left(\text{or } \frac{1}{10^2}\right)$	=	0.239
23.9	×	$0.001 \left(\text{or } \frac{1}{10^3}\right)$	=	0.0239

Since you are multiplying by a power of ten that is less than 1, the product is less than the original decimal. Therefore, the decimal point moves to the left.

Examples

3 **Multiply 1.05 and 0.01 mentally.**

$1.05 \times 0.01 = 0.0105$ $\quad 0.01 = \frac{1}{100}$

Move the decimal point two places to the left.

$1.05 \times 0.01 = 0.0105$

APPLICATION

4 **Money Matters** A department store is having a special sale in which the price of every item in the store is $\frac{1}{10}$ or 0.1 off the original price.

  a. If the original price of a sweater is $35, how much does Samuel save by buying it on sale?

  b. What is the sale price?

  a. To find how much is saved, find 0.1 × $35.

   $0.1 \times 35 = 3.5$ *Move the decimal point one place to the left.*

   Samuel saves $3.50.

  b. To find the sale price, find $35 − $3.50. *Estimate: 35 − 4 = 31*

   $35 − 3.5 = 31.5$ The sale price is $31.50.

CHECK FOR UNDERSTANDING

Communicating Mathematics

Read and study the lesson to answer each question.

1. **Tell** a classmate how you would solve $x = 2.378 \times 100$ mentally.

2. **Explain** how you can mentally find the cost of 10 items that are the same price.

3. **Write** the steps you would take to multiply 0.01 × 28.8 using paper and pencil. Compare it to solving the problem mentally.

Guided Practice

Multiply mentally.

4. 12.53×10 　　　　5. 4.6×10^3 　　　　6. 78.4×0.01

Solve each equation.

7. $n = 2.031 \times 10^4$ **8.** $a = 0.78 \times 10^2$ **9.** $x = 0.1 \times 0.45$

10. *Food* Suppose a can of soup costs $0.73. If you purchase 100 cans to donate to a food bank, what will the cost be?

EXERCISES

Practice

Multiply mentally.

11. 0.05×100 **12.** 4.527×10^0 **13.** $2.78 \times 1{,}000$

14. 13.58×0.01 **15.** 5.49×10^3 **16.** 0.1×0.8

17. 0.925×10 **18.** 99.44×10^2 **19.** 0.01×16

Solve each equation.

20. $1.32 \times 10^3 = x$ **21.** $c = 0.56 \times 1{,}000$ **22.** $m = 1.4 \times 0.1$

23. $h = 11.23 \times 10^5$ **24.** $68.94 \times 0.01 = y$ **25.** $0.8 \times 10^0 = w$

26. $28.1 \times 0.01 = z$ **27.** $t = 9.3 \times 100$ **28.** $3.76 \times 10{,}000 = b$

29. Find the product of 0.1 and 25.3 mentally.

Applications and Problem Solving

30. *Music* The chart shows the sales of selected music products for the United States.

   a. Write the number of dollars spent for percussion in standard form.

   b. How much money was spent for guitars and sound amplifiers? Write in standard form.

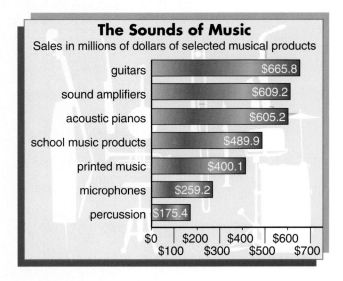

The Sounds of Music
Sales in millions of dollars of selected musical products

guitars	$665.8
sound amplifiers	$609.2
acoustic pianos	$605.2
school music products	$489.9
printed music	$400.1
microphones	$259.2
percussion	$175.4

$0 $100 $200 $300 $400 $500 $600 $700

31. *Population* In 1995, the population of Georgia was estimated at 7.2 million. It is expected to increase $\frac{1}{10}$ by the year 2000.

   a. By how many people will the population increase?

   b. Predict the population of Georgia in the year 2000.

32. *Critical Thinking* Devise a method to find the product of 0.2 and a number mentally. Use your method to find 0.2×42.

Mixed Review

33. *Algebra* Evaluate the expression xy if $x = 0.4$ and $y = 3$. *(Lesson 2-4)*

34. **Standardized Test Practice** On Manuel's trip, he drove 178.5 miles in 3.3 hours. Which is a reasonable average speed for his trip? *(Lesson 2-3)*

   A 6 mph   **B** 40 mph   **C** 60 mph   **D** 90 mph   **E** 180 mph

35. Round 0.006 to the hundredths place. *(Lesson 2-2)*

36. *Algebra* Solve $12m = 120$ mentally. *(Lesson 1-5)*

For **Extra Practice,** see page 572.

COOPERATIVE LEARNING

2-6A Division with Decimal Models

grid paper

markers

A Preview of Lesson 2-6

Since multiplication and division are inverse operations, you can also use grid paper to make decimal models to show division of decimals.

TRY THIS

Work with a partner.

1 To model 0.24 ÷ 0.6, follow these steps.

- You need to shade a rectangle with an area of 0.24. So, shade 24 small squares in a decimal model.

- There are many rectangles with an area of 0.24. You need to shade one that has a length of 0.6.

- The missing factor is 0.4.

The area of a 0.4 by 0.6 rectangle is 0.24. Therefore, 0.24 ÷ 0.6 = 0.4.

2 To model 0.2 ÷ 0.4 you need to shade a rectangle with an area of 0.2.

- Since you are using decimal models, first write 0.2 as 0.20.

- Shade a rectangle with an area of 0.20 and a length of 0.4.

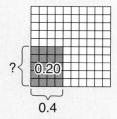

- The missing factor is 0.5.

The area of a 0.4 by 0.5 rectangle is 0.20 or 0.2. Therefore, 0.2 ÷ 0.4 is 0.5.

ON YOUR OWN

Write a division problem using decimals for each model shown below.

1.

2.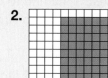

3.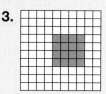

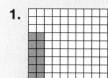

Use decimal models to show each quotient.

4. 0.35 ÷ 0.5
5. 0.36 ÷ 0.6
6. 0.4 ÷ 0.8

7. 0.9 ÷ 0.9
8. 0.64 ÷ 0.8
9. 0.48 ÷ 0.8

TRY THIS

Work with a partner.

3 To model 1 ÷ 0.5, follow these steps.

- You need to shade a rectangle with an area of 1. In a decimal model, 1 is represented by 100 small squares.

- The length of the rectangle is 0.5.

- If you shade a rectangle with a length of 0.5, you will only shade 50 squares instead of 100 squares. So, you need to use two decimal models, side by side.

- The missing factor is 2.

The area of a 0.5 by 2 rectangle is 1. Therefore, 1 ÷ 0.5 = 2.

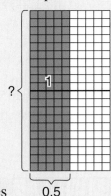

4 Model 1.2 ÷ 0.4.

- Shade a rectangle whose area is 120 small squares. The width is 0.4.

The area of a 0.4 by 3 rectangle is 1.2. Therefore, 1.2 ÷ 0.4 = 3.

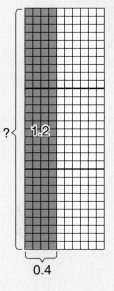

ON YOUR OWN

Write a division problem using decimals for each model shown below.

10.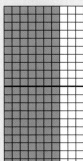

11.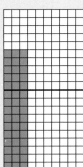

Use decimal models to show each quotient.

12. 1.2 ÷ 0.6

13. 1.6 ÷ 0.4

14. 2.25 ÷ 1.5

15. *Look Ahead* Find the quotient 0.49 ÷ 0.7 without using models.

Dioviding Decimals

Ceres, one of the largest known asteroids, is 690.4 kilometers in diameter. Jupiter, the first planet beyond the asteroid belt, has a diameter of 142,748.8 kilometers. How many times greater is Jupiter's diameter than Ceres' diameter? To find out, divide 142,748.8 by 690.4. *This problem will be solved in Exercise 38.*

As you saw in Lab 2-6A, area models are useful for simple division problems. However, as the numbers increase, you need another method for dividing. In the following Mini-Lab, you will use a calculator to discover a pattern for dividing any two decimals.

MINI-LAB

Work with a partner. calculator

Try This

1. Use a calculator to find each quotient.
   a. $0.035 \div 0.05$ b. $0.00132 \div 0.012$
        $0.35 \div 0.5$ $0.0132 \div 0.12$
        $3.5 \div 5$ $0.132 \div 1.2$
        $35 \div 50$ $1.32 \div 12$
                                $13.2 \div 120$

2. Find the similarities and differences in Exercise 1.

Talk About It

3. Which of the quotients in Exercise 1 would be easier to find *without* a calculator? Explain your reasoning.
4. Rewrite each problem so you can find the quotient without using a calculator. Then find the quotient.
   a. $0.36 \div 0.4$ b. $1.25 \div 0.5$ c. $1.68 \div 0.2$

The pattern in the Mini-Lab suggests the following.

Dividing Decimals	To divide two decimals, change the divisor to a whole number by moving the decimal point to the right. Then move the decimal point in the dividend the same number of places to the right. Then divide as with whole numbers.

Moving the decimal points in the dividend and divisor is a result of multiplying both numbers by a power of ten.

1 Find 199.68 ÷ 9.6.   *Estimate: 200 ÷ 10 = 20*

$$
\begin{array}{r}
20.8 \\
9.6\,\overline{)199.68} \\
-192 \\
\hline
7\,68 \\
-7\,68 \\
\hline
0
\end{array}
$$

Change 9.6 to 96 and 199.68 to 1,996.8 by moving each decimal point one place to the right.

So, 199.68 ÷ 9.6 = 20.8.   *Compared to the estimate, the quotient is reasonable.*

2 Solve $n = 0.9 \div 0.05$.

$$
\begin{array}{r}
18 \\
0.05\,\overline{)0.90} \\
-5 \\
\hline
40 \\
-40 \\
\hline
0
\end{array}
$$

Annex a zero.

The solution is 18.

You can use decimal division to compare the costs of items that are different sizes. It may be necessary to round quotients.

3 **Shopping** The cost of three different sizes of peanut butter are shown at the right. Which size jar costs the least per ounce?

To find the price per ounce, divide the price by the number of ounces of peanut butter in each jar. The quotient tells you the price of one ounce of peanut butter, which is also called the *unit price*. Round the unit price to the nearest cent.

PEANUT BUTTER	
SIZE	PRICE
12 oz	$1.99
40 oz	$4.99
80 oz	$8.19

> **Study Hint**
>
> **Reading Math** The symbol ≈ is read as *is approximately equal to.*

12-ounce jar

1.99 ⟦÷⟧ 12 ⟦=⟧ *0.1658333*

   ≈ 0.17

The unit price is $0.17.

40-ounce jar

4.99 ⟦÷⟧ 40 ⟦=⟧ *0.12475*

   ≈ 0.12

The unit price is $0.12.

80-ounce jar

8.19 ⟦÷⟧ 80 ⟦=⟧ *0.102375*

   ≈ 0.10

The unit price is $0.10.

The 80-ounce jar of peanut butter costs the least per ounce.

Communicating
Mathematics

Read and study the lesson to answer each question.

1. *Tell* whether $35 \div 0.5$ is the same as $3.5 \div 0.05$. Explain your reasoning.

2. *Give an example* of a division problem in which it is necessary to annex one or more zeros to the dividend.

Guided Practice

Without finding or changing each quotient, change each problem so that the divisor is a whole number.

3. $0.36 \div 0.4$ 4. $4.4 \div 1.1$ 5. $50.4 \div 0.56$

Divide.

6. $3 \div 0.6$ 7. $0.056\overline{)0.084}$ 8. $51 \div 0.8$

Solve each equation.

9. $0.42 \div 3.5 = w$

10. $1.35 \div 0.5 = s$

11. *Sports* A table tennis table has an area of 4.165 square meters, while a tennis court has an area of 260.76 square meters. To the nearest hundredth, how many times larger is the tennis court than the table tennis table?

Practice

Without finding or changing each quotient, change each problem so that the divisor is a whole number.

12. $1.05 \div 0.7$ 13. $2.94 \div 0.084$ 14. $1.89 \div 0.9$

15. $0.82 \div 0.4$ 16. $68.13 \div 0.003$ 17. $2.6 \div 1.3$

18. $0.00945 \div 0.21$ 19. $1.488 \div 3.1$ 20. $14.42 \div 0.206$

Divide.

21. $4.2 \div 1.2$ 22. $0.287 \div 0.035$ 23. $0.245 \div 0.7$

24. $0.6\overline{)4.8}$ 25. $0.7\overline{)0.21}$ 26. $0.5\overline{)35}$

27. $1.6\overline{)0.768}$ 28. $9\overline{)8.19}$ 29. $0.075\overline{)0.345}$

Solve each equation.

30. $3.68 \div 0.92 = x$ 31. $g = 0.4664 \div 5.3$ 32. $a = 7.56 \div 0.63$

33. $74.2 \div 0.53 = y$ 34. $17.94 \div 2.3 = m$ 35. $c = 2.665 \div 4.1$

36. Round the quotient of 56.38 and 2.6 to the nearest tenth.

37. What is $7.69 divided by 5, rounded to the nearest cent?

For **Extra Practice,**
see page 572.

38. *Earth Science* Refer to the beginning of the lesson. To the nearest tenth, how many times greater is Jupiter's diameter than Ceres' diameter?

39. *Shopping* A snack-sized box of microwave popcorn has five 1.75-ounce bags and costs $2.29. A large box has six 3.5-ounce bags and costs $4.99.

Microwave Popcorn			
Kind of Box	Number of Bags	Size of Bags	Cost
Snack Size	5	1.75 oz	$2.29
Large	6	3.5 oz	$4.99

   a. Find the unit price of each box.

   b. Which box costs less per ounce?

   c. Describe a situation when it would make sense to buy the box that is *not* less per ounce.

   d. A regular box of popcorn has three 3.5-ounce servings and costs $2.29. Would it make sense to buy one large box or two regular boxes? Explain your reasoning.

40. *Geometry* One rectangle has an area of 2.04 square meters, and a second has an area of 1.86 square meters. If the length of both rectangles is 0.6 meter, which rectangle is wider? Explain your reasoning.

41. *Track* At the Atlanta Olympics, U.S. track star Michael Johnson set an Olympic record of 43.49 seconds in the 400-meter event. To the nearest tenth, find his speed in meters per second.

42. *Critical Thinking* Without actually dividing, choose the division problem that has the greatest quotient.

   a. $2.4 \div 6$ **b.** $2.4 \div 0.6$

   c. $2.4 \div 0.06$ **d.** $2.4 \div 0.006$

Family Activity

Search newspapers or magazines for examples of how decimals are used in sports. Then explain it to a family member.

Mixed Review

43. Solve $x = 2.83 \times 100$ mentally. *(Lesson 2-5)*

44. Standardized Test Practice Terrence walks to and from school every day. The round trip is 3.21 kilometers. If he walks every day for 5 days, how far does he walk? *(Lesson 2-4)*

   A 16.25 kilometers

   B 16.07 kilometers

   C 12.64 kilometers

   D 1.605 kilometers

   E Not Here

45. Estimate 21.7×6.3. *(Lesson 2-3)*

46. *Money Matters* A survey of the weekly average amount spent on groceries for a family of four is $147.2653. Find the weekly average to the nearest dollar. *(Lesson 2-2)*

47. *Algebra* Evaluate $3x - y \div 6$ if $x = 4$ and $y = 12$. *(Lesson 1-3)*

What you'll learn

You'll learn to express fractions as terminating or repeating decimals.

When am I ever going to use this?

Knowing how to express a fraction as a decimal can help you calculate a batting average.

Word Wise

terminating decimal
repeating decimal
bar notation

Are you a "lefty?" President Clinton, Oprah Winfrey, and Tom Cruise are all left-handed. Actually, about 3 out of every 25 people or $\frac{3}{25}$ of the population are left-handed. Any fraction can be written as a decimal by dividing.

Method 1

Use paper and pencil.
$\frac{3}{25}$ indicates $3 \div 25$.

$$
\begin{array}{r}
0.12 \\
25\overline{)3.00} \\
\underline{25} \\
50 \\
\underline{50} \\
0
\end{array}
$$

Write 3 as 3.00. Place the decimal point in the quotient. Divide as with whole numbers.

Method 2

Use a calculator.

$3 \boxed{\div} 25 \boxed{=} \; 0.12$

The fraction $\frac{3}{25}$ can be written as the decimal 0.12. A decimal like 0.12 is called a **terminating decimal** because the division ends, or terminates, when the remainder is zero.

Study Hint

Technology
Calculators may round or truncate answers. *Truncate* means to cut off at a certain place-value position, ignoring the digits that follow.

However, not all decimals are terminating decimals. Decimals like 0.44444444. . . are called **repeating decimals** because there is a pattern in the digits that repeats forever. You can use the **bar notation** $0.\overline{4}$ to indicate that the 4 repeats forever. Study the pattern below.

$0.131313131313. . . = 0.\overline{13}$	*The digits 13 repeat.*
$5.8666666666. . . = 5.8\overline{6}$	*The digit 6 repeats.*
$72.0831831831. . . = 72.0\overline{831}$	*The digits 831 repeat.*

Example 1 Express $\frac{2}{3}$ as a decimal using division.

Method 1 Use paper and pencil.

$$
\begin{array}{r}
0.666. . . \\
3\overline{)2.000} \\
\underline{18} \\
20 \\
\underline{18} \\
20 \\
\underline{18} \\
2
\end{array}
$$

The digit 6 will repeat since 2 will continue to be the remainder.

Method 2 Use a calculator.

$2 \boxed{\div} 3 \boxed{=} \; 0.666666667$
This calculator rounds.

$2 \boxed{\div} 3 \boxed{=} \; 0.666666666$
This calculator truncates.

Use bar notation to indicate that the digit 6 repeats. So, $\frac{2}{3} = 0.\overline{6}$.

Example ▶ **2** Express $4\frac{5}{6}$ as a decimal.

Method 1 Paper and pencil

$$\begin{array}{r} 0.833\ldots \\ 6\overline{)5.000} \\ \underline{48} \\ 20 \\ \underline{18} \\ 20 \\ \underline{18} \\ 2 \end{array}$$

Divide 5 by 6.
The digit 3 will
repeat since 2 will
continue to be
the remainder.

Method 2 Calculator

5 ÷ 6 + 4 = *4.833333333*

Therefore, $4\frac{5}{6} = 4.8333\ldots$ or $4.8\overline{3}$.

Repeating decimals often occur in real-world situations. However, they are usually rounded to a certain place-value position.

Example  **3** **Sports** In the Summer Olympics in Atlanta, team captain Dr. Dot Richardson led the United States team to the first-ever gold medal in softball. During the Olympics, she had 6 hits in 22 at-bats. To the nearest thousandth, find her batting average in the Olympics.

To find a batting average, divide the number of hits, 6, by the times at bat, 22.

6 ÷ 22 = *0.272727273*

Look at the digit to the right of the thousandths place. Round up since $7 > 5$.

Dr. Dot Richardson's batting average in the Olympics was 0.273.

Did you know The United States women's soccer team also won the first gold medal awarded in that event at the Summer Olympics in Atlanta.

CHECK FOR UNDERSTANDING

Communicating Mathematics

Read and study the lesson to answer each question.

1. *Explain* how to express a fraction as a decimal.

2. *Describe* the difference between a terminating and a repeating decimal. Give an example of each.

3. *You Decide* Mika thinks that 0.5 and $0.\overline{5}$ are equal. Kim thinks they are not. Who is correct? Explain your reasoning.

Guided Practice

Write each repeating decimal using bar notation.

4. 0.7555555...   5. 6.34343434...   6. 10.123123123...

Express each fraction or mixed number as a decimal. If the decimal is a repeating decimal, use bar notation.

7. $\frac{4}{5}$ **8.** $1\frac{7}{8}$ **9.** $\frac{7}{9}$

10. *Weather* Dallas, Texas, averages $32\frac{1}{4}$ inches of precipitation each year. Express the mixed number as a decimal.

EXERCISES

Practice   Write each repeating decimal using bar notation.

11. 0.4444444. . .   **12.** 0.6666666. . .   **13.** 1.12121212. . .

14. 4.67676767. . .   **15.** 13.245245245. . .   **16.** 0.989989989. . .

17. 0.833333. . .   **18.** 2.03454545. . .   **19.** 3.01523523. . .

Express each fraction or mixed number as a decimal. If the decimal is a repeating decimal, use bar notation.

20. $\frac{8}{25}$ **21.** $\frac{11}{20}$ **22.** $\frac{1}{11}$ **23.** $\frac{5}{6}$

24. $1\frac{5}{8}$ **25.** $\frac{8}{250}$ **26.** $\frac{15}{9}$ **27.** $3\frac{14}{16}$

Replace each ● with <, >, or = to make a true sentence.

28. $5\frac{1}{5}$ ● 5.18   **29.** 23.25 ● $23\frac{1}{4}$ **30.** 0.8 ● $\frac{8}{9}$

31. $\frac{1}{12}$ ● $0.08\overline{3}$ **32.** $\frac{1}{2}$ ● $\frac{1}{6}$ **33.** $\frac{4}{5}$ ● $\frac{3}{4}$

34. Express $\frac{7}{12}$ as a decimal using bar notation.

35. Express $\frac{34}{125}$ as a decimal rounded to the nearest hundredth.

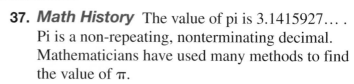

Applications and Problem Solving

36. *Life Science* Monarch butterflies migrate up to 2,000 miles from the northern United States to the warmer climates of Mexico, California, and Florida. The fastest monarch butterfly can fly $\frac{1}{3}$ mile per minute. Express $\frac{1}{3}$ as a decimal rounded to the nearest hundredth.

37. *Math History* The value of pi is 3.1415927... . Pi is a non-repeating, nonterminating decimal. Mathematicians have used many methods to find the value of π.

   a. Archimedes believed that π was between $3\frac{1}{7}$ and $3\frac{10}{71}$. Express each fraction as a decimal rounded to the nearest thousandth. Was Archimedes correct?

   b. The Rhind Papyrus records that the Egyptians used $\frac{256}{81}$ for π. Express the fraction as a decimal rounded to the nearest thousandth. Which value is closer to the actual value of π, Archimedes' or the Egyptians' value?

38. *Find a Pattern* If $\frac{1}{8} = 0.125$, find the decimal value of each fraction.

    a. $\frac{2}{8}$ **b.** $\frac{3}{8}$ **c.** $\frac{4}{8}$ **d.** $\frac{5}{8}$

39. *Critical Thinking* Find one terminating decimal and one repeating decimal between $\frac{2}{3}$ and $\frac{3}{4}$.

Mixed Review

40. Divide 0.108 by 0.2. *(Lesson 2-6)*

41. Multiply 0.54 by 0.2. *(Lesson 2-4)*

For the latest statistics, visit: www.glencoe.com/ sec/math/mac/mathnet

42. **Standardized Test Practice** Using the chart, Jeremy will write a report on the American League Batting Champions. If he wants to list the 4 years with the greatest batting averages in order from greatest to least, which should he choose? *(Lesson 2-1)*

Year	Name	Average
1990	George Brett	0.329
1991	Julio Franco	0.341
1992	Edgar Martinez	0.343
1993	John Olerud	0.363
1994	Paul O'Neill	0.359
1995	Edgar Martinez	0.356
1996	Alex Rodriguez	0.358
1997	Frank Thomas	0.347

  A 1990, 1991, 1992, 1997   **B** 1997, 1996, 1995, 1994

  C 1993, 1996, 1992, 1990   **D** 1993, 1994, 1996, 1995

43. *Algebra* Evaluate a^4, if $a = 3$. *(Lesson 1-4)*

For **Extra Practice**, see page 573.

44. *Sports* Four hundred sixty people are scheduled to attend a banquet. If each table seats 8 people, how many tables are needed? *(Lesson 1-1)*

Let the Games Begin

Match-Up

Math Skill

Expressing Fractions as Decimals

Get Ready   This game is for two players.

      10 index cards   scissors

Get Set   Cut an index card in half. On one part, write a fraction. On the other part, write its decimal equivalent. Continue until you have 10 fraction-decimal pairs.

Go
- Mix the cards and arrange them facedown into a rectangle.
- The first player turns over two cards. If they match, the player scores one point and turns over two more cards. If they do not match, the player turns the cards facedown again, no points are scored, and it becomes the next player's turn.
- Players take turns until all cards are matched. The player with the most points wins.

 Visit www.glencoe.com/sec/math/mac/mathnet for more games.

Integration: Measurement
The Metric System

What you'll learn

You'll learn to change metric units of length, capacity, and mass.

When am I ever going to use this?

You will use the metric system when working on an automobile engine.

Word Wise

meter
metric system
gram
liter

Most eagles have a wingspan of about 180 centimeters. The harpy eagle has a wingspan of 2.4 meters. Which is greater, 180 centimeters or 2.4 meters? Both measurements above are based on the **meter (m)**, which is the basic unit of length in the **metric system**. A meter is about the distance from the floor to a doorknob. All units of length in the metric system are defined in terms of the meter. A prefix is added to indicate the decimal place-value position of the measurement. The metric prefixes are shown in the chart below.

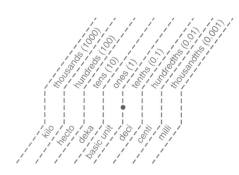

Notice that each place value is 10 times the place value to its right.

Notice that the value of each metric prefix is 10 times the value of the prefix to its right.

One way to solve the problem above is to change 2.4 meters to centimeters. Since 1 meter = 100 centimeters, multiply by 100.

$$2.4 \times 100 = 240$$

The harpy eagle's wingspan is 240 centimeters. Since 240 > 180, the harpy eagle's wingspan is greater than most other eagles.

This diagram can help you change metric units.

Study Hint

Mental Math To multiply or divide by a power of ten, you can move the decimal point.

MULTIPLY to change from larger units to smaller units.

$\times 1,000 \quad \times 100 \quad \times 10$

km   m   cm   mm

$\div 1,000 \quad \div 100 \quad \div 10$

DIVIDE to change from smaller units to larger units.

Examples

① 0.7 cm = _?_ mm

To change from centimeters to millimeters, multiply by 10 since 1 cm = 10 mm.

$0.7 \times 10 = 7$
0.7 cm = 7 mm

② 3,850 m = _?_ km

To change from meters to kilometers, divide by 1,000 since 1 km = 1,000 m.

$3,850 \div 1,000 = 3.85$
3,850 m = 3.85 km

The **kilogram (kg)** is the basic unit of mass in the metric system. *Mass* is the amount of matter that an object contains. Your math textbook has a mass of about one kilogram. Kilogram, gram, and milligram are related in a manner similar to kilometer, meter, and millimeter.

Examples

③ 8,249 g = _?_ kg

To change from grams to kilograms, divide by 1,000 since 1 kg = 1,000 g.

$8,249 \div 1,000 = 8.249$
8,249 g = 8.249 kg

④ 2 g = _?_ mg

To change from grams to milligrams, multiply by 1,000 since 1 g = 1,000 mg.

$2 \times 1,000 = 2,000$
2 g = 2,000 mg

The **liter (L)** is the basic unit of capacity in the metric system. *Capacity* is the amount of dry or liquid material an object can hold. Soft drinks often come in a 2-liter plastic container. Kiloliter, liter, and milliliter are also related in a manner similar to kilometer, meter, and millimeter.

Examples

⑤ 1.5 L = _?_ mL

Multiply by 1,000 since 1 L = 1,000 mL.

$1.5 \times 1,000 = 1,500$
1.5 L = 1,500 mL

⑥ 483 L = _?_ kL

Divide by 1,000 since 1 kL = 1,000 L.

$483 \div 1,000 = 0.483$
483 L = 0.483 kL

Scientists around the world use the metric system. Using the same system gives them a common language and makes it easy to understand each other's research.

Example

CONNECTION

⑦ **Life Science** The bacterium *E. coli* has a diameter of 0.001 millimeter. The head of a pin has a diameter of 1 millimeter. How many *E. coli* bacteria could fit across the head of a pin?

Both measures are expressed in millimeters. So you divide 1 by 0.001.

$1 \div 0.001 = 1,000$

Therefore, 1,000 *E. coli* bacteria could fit across the head of a pin.

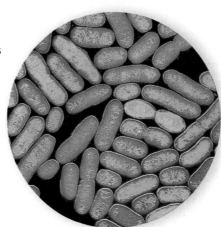

Communicating Mathematics

Read and study the lesson to answer each question.

1. *Tell* why you should divide when changing from grams to kilograms.

2. *Explain* how the metric system and decimals are similar.

Guided Practice

Complete.

3. 750 m = _?_ cm
4. 52.8 mm = _?_ cm
5. 923 g = _?_ kg
6. 16.5 g = _?_ mg
7. 8,200 mL = _?_ L
8. 76 L = _?_ mL

9. *Earth Science* The mass of a sample of rocks is 1.24 kilograms. How many grams are in 1.24 kilograms?

EXERCISES

Practice

Complete.

10. 89 km = _?_ m
11. 67.1 kg = _?_ g
12. 0.6 L = _?_ mL
13. 234 mm = _?_ cm
14. 5.8 m = _?_ cm
15. 13.2 cm = _?_ m
16. 0.9 cm = _?_ mm
17. 46 km = _?_ m
18. 6,700 m = _?_ km
19. 567 mg = _?_ g
20. 80 g = _?_ mg
21. 8.1 L = _?_ mL
22. 329 mL = _?_ L
23. 47 L = _?_ kL
24. 0.52 km = _?_ m

25. How many milliliters are in 0.09 liter?

Applications and Problem Solving

26. *Conservation* The average shower uses 19 liters of water per minute. If you take a five-minute shower each day, how many kiloliters of water do you use in one year by showering?

27. *Working on the* **CHAPTER Project** Refer to the table on page 43.
   a. Let one centimeter represent the diameter of Earth. Use this scale to find the diameter of each planet.
   b. Let one centimeter represent the average distance from Earth to the Sun. Determine the distances for all the other planets.
   c. Decide whether you want to make a model of the planets showing their diameters or their distances from the Sun. Make your model.

28. *Critical Thinking* Find the meaning of the prefixes *giga-* and *nano-*. How many nanometers are in 0.005 meter? How many meters are in 8.3 gigameters?

Mixed Review

29. Write 0.59595959. . . using bar notation. *(Lesson 2-7)*

30. **Standardized Test Practice** A 30-second advertisement on a local television station during prime time costs the advertiser $1,280. To the nearest cent, how much does the ad cost per second? *(Lesson 2-6)*

   A $4.27   B $42.67   C $43   D $426.67   E Not Here

For **Extra Practice**, see page 573.

31. Solve $n = 11.45 \times 0.01$. *(Lesson 2-5)*

32. Evaluate 5^4. *(Lesson 1-4)*

2-9 Scientific Notation

What you'll learn

You'll learn to express numbers greater than 100 in scientific notation.

When am I ever going to use this?

You can use scientific notation to express the distances between planets.

Word Wise

scientific notation

Long ago, people thought that they were the center of the universe and that everything revolved around them. Today, you know that isn't true. Earth revolves around the sun, and the sun is part of the Milky Way Galaxy. But did you know that there are other galaxies beside the Milky Way in the universe? One such galaxy, the Andromeda galaxy, is 2.2 million light-years away from Earth. *A light-year is the distance light travels in one year.*

You can write numbers like 2.2 million in **scientific notation** by using a power of ten.

$$2.2 \text{ million} = 2.2 \times 1,000,000$$
$$= 2.2 \times 10^6 \qquad 10^6 = 1,000,000$$

In scientific notation, 2.2 million is 2.2×10^6.

Scientific Notation	Numbers expressed in scientific notation are written as the product of a number that is at least one but less than 10 and a power of ten. The power of ten is written with an exponent.

To write a number in scientific notation, move the decimal point to the right of the first nonzero digit, and multiply this number by a power of ten. To find the power of ten, count the number of places you moved the decimal point.

Example **1** Write 352,000 in scientific notation.

3.52000   *Move the decimal point 5 places to get a number between 1 and 10.*

3.52×10^5

In scientific notation, 352,000 is 3.52×10^5.

The decimal part of a number written in scientific notation is often rounded to the hundredths place.

Example ②

Write 141,710,000 in scientific notation.

1.41710000   *Move the decimal point 8 places to get*
              a number between 1 and 10.

1.4171×10^8

1.42×10^8 *Round 1.4171 to the nearest hundredth.*

In scientific notation, 141,710,000 is 1.42×10^8.

When you use a calculator to compute with large numbers, the numbers are often displayed in scientific notation.

Example ③

CONNECTION

Study Hint

Technology Use the [EE] key to enter numbers in scientific notation. Enter the factor, press [EE], then enter the exponent.

Earth Science The average distance from Earth to the Sun is 93,000,000 miles. Neptune is about 30 times as far away from the Sun. Find the average distance from Neptune to the Sun.

Multiply 93,000,000 by 30.

93000000 [×] 30 [=] *2.79 09*

Some calculators display the factor 2.79 and the exponent 9. This represents the number 2.79×10^9.

Neptune is about 2.79×10^9 or 2.79 billion miles from the Sun.

CHECK FOR UNDERSTANDING

Communicating Mathematics

Read and study the lesson to answer each question.

1. **Tell** why scientific notation is used with large numbers.

2. **Show** a classmate how to write 5,280, the number of feet in a mile, in scientific notation.

3. **You Decide** Hiroshi thinks that 24.59×10^3 is written in scientific notation. Alma thinks it is not. Who is correct? Explain your reasoning.

Guided Practice

Write each number in scientific notation.

4. 890

5. 8,300

6. 6,235

7. 52,000

8. 820,000

9. 126,400,000

10. **Finance** The Social Security Administration estimates that there will be 54 million people receiving benefits in 2010. Write 54 million in scientific notation.

Practice

Write each number in scientific notation.

11. 7,500   **12.** 8,450   **13.** 40,700

14. 630,000   **15.** 600   **16.** 17,500

17. 23,000   **18.** 400,000   **19.** 32,000,000

20. 495,000,000   **21.** 570,000   **22.** 9,500,000

23. 8,080   **24.** 602,400,000   **25.** 27,100,000

26. 7,900,000   **27.** 558,000   **28.** 160,000,000

Replace each ● with $<$, $>$, or $=$ to make a true sentence.

29. $3,000 ● 3.0 \times 10^3$ **30.** $200 ● 2.0 \times 10^1$

31. $72,500 ● 7.25 \times 10^5$ **32.** 5 million $● 5.0 \times 10^6$

33. 9.3 billion $● 9.3 \times 10^8$ **34.** $5.56 \times 10^9 ●$ 5.56 billion

35. A calculator displays $3.4\ 06$. Write the number in scientific notation and in standard form.

36. Order 9.05×10^3, 5.29×10^3, and 5.29×10^4 from least to greatest.

Applications and Problem Solving

37. *Geography* Which number describes the population of Chicago, 3.0×10^6 or 3.0×10^3? Explain your reasoning.

38. *Earth Science* Scientists divide Earth's history into small units based on the types of life-forms living then. In the Jurassic Period, which occurred about 208,000,000 years ago, dinosaurs ruled. Express 208,000,000 in scientific notation.

39. *Recycling* According to the Aluminum Association, 2,031,000,000 pounds of aluminum cans were recycled in a recent year. At an average of 29.29 cans per pound, how many aluminum cans were recycled that year? Express your answer in scientific notation.

40. *Critical Thinking* One *light-year* is the distance light travels in one year. If the speed of light is 3×10^5 kilometers per second, about how many kilometers does light travel in one year?

Mixed Review

41. *Measurement* How many grams are in 1.01 kilograms? *(Lesson 2-8)*

42. Solve $16.2 \div 2.5 = n$. *(Lesson 2-6)*

43. **Standardized Test Practice** Enrique earns $4.75 per hour for babysitting. Which is the best estimate of his earnings if he babysits 12 hours during spring break? *(Lesson 2-3)*

    A $20   **B** $48   **C** $60   **D** $80   **E** $100

44. *Algebra* Solve $x \div 8 = 14$. *(Lesson 1-5)*

45. Evaluate $100 \div 10 + 2 \cdot 6 \div 4$. *(Lesson 1-2)*

For **Extra Practice,** see page 573.

 **inter**NET **CONNECTION** Chapter Review For additional lesson-by-lesson review, visit:
www.glencoe.com/sec/math/mac/mathnet

Vocabulary

After completing this chapter, you should be able to define each term, concept, or phrase and give an example or two of each.

Number and Operations
bar notation (p. 70)
clustering (p. 51)
repeating decimal (p. 70)
scientific notation (p. 77)
terminating decimal (p. 70)

Measurement
kilogram (p. 75)
liter (p. 75)
meter (p. 74)
metric system (p. 74)

Problem Solving
reasonable answers (p. 54)

Understanding and Using the Vocabulary

Choose the correct term or number to complete each sentence.

1. The number 0.04 is (<u>less</u>, greater) than 0.041.
2. When rounding decimals, the digit in the place being rounded should be rounded up if the digit to the right is a (4, <u>7</u>).
3. The number of decimal places in the product when multiplying decimals is the (<u>sum</u>, product) of the number of places in the factors.
4. In scientific notation, a number is written as a (sum, <u>product</u>) of a decimal number and a power of ten.
5. The basic unit of mass in the metric system is the (<u>kilogram</u>, meter).
6. A model that can be used to represent multiplication problems is called a(n) (<u>area</u>, perimeter) model.
7. The model at the right represents ($\underline{0.5 \times 0.3 = 0.15}$, $0.05 \times 0.03 = 0.15$).
8. The fraction $\frac{3}{4}$ can be expressed as a (<u>terminating</u>, repeating) decimal.
9. Using bar notation, 0.23333 . . . is expressed as ($0.2\overline{3}$, $0.\overline{23}$).
10. Using scientific notation, 2.4 million is written as ($\underline{2.4 \times 10^6}$, 2,400,000).

In Your Own Words   11. Sample answer: One centimeter is one-hundredth of a meter.
11. *Explain* the relationship between a meter and a centimeter.

Objectives & Examples

Upon completing this chapter, you should be able to:

● compare and order decimals *(Lesson 2-1)*

Order the following decimals from least to greatest: 3.2, 0.4, 0.43, 3.5, 4.

0.4, 0.43, 3.2, 3.5, 4

● round decimals *(Lesson 2-2)*

Round 247.359 to the nearest tenth.

The digit to the right of the 3 in the tenths place is 5, so round up.

247.359 → 247.4

● estimate with decimals *(Lesson 2-3)*

Estimate 4.7 + 5.2 + 5.1 + 4.9.

All of the numbers are clustered around 5. There are four numbers. So, the sum is about 5 × 4 or 20.

You can also estimate using rounding.

5 + 5 + 5 + 5 = 20

● multiply decimals *(Lesson 2-4)*

$\quad$ 3.2 $\quad$ ← $\quad$ *1 decimal place*

$\underline{\times 0.6}$ $\quad$ ← $\quad$ *1 decimal place*

$\quad$ 1.92 $\quad$ ← $\quad$ *Count 2 decimal places from the right.*

Review Exercises

Use these exercises to review and prepare for the chapter test.

Order each set of decimals from least to greatest.

12. 4.2, 3.9, 3.15, 3.04, 3.7

13. 15.91, 1.59, 0.159, 0.06, 1.4

14. 0.15, 0.149, 0.105, 0.015, 0.501

15. 16.3, 16.03, 16, 15.99, 15.0

Round each number to the underlined place-value position.

16. 5.75   **17.** 13.274

18. 129,342   **19.** 0.076

20. 81.349   **21.** 57.196

Estimate.

22. 13.72 + 12.07

23. 36.8 + 39.2 + 41.3

24. 25.73 − 2.19

25. 11.75 × 3.13

26. 72.4 ÷ 9.3

27. 150.96 ÷ 4.76

Multiply.

28. 2.6 × 3.7

29. 0.13 × 2

30. 12.5 × 0.0017

31. 7.5 × 3.03

32. 1.001 × 0.4

Chapter 2   Study Guide and Assessment

Objectives & Examples

multiply decimals mentally by powers of ten *(Lesson 2-5)*

$100 \times 2.3 = 230$

$0.1 \times 25.16 = 2.516$

$8.37 \times 10^2 = 837$

Review Exercises

Multiply.

33. 13.7×10^3

34. $0.0065 \times 10,000$

35. 6.37×0.01

36. 128.63×10^4

---

divide decimals *(Lesson 2-6)*

$$
\begin{array}{r}
12.6 \\
3.6\overline{)45.36} \\
-36 \\
\hline
9\,3 \\
-7\,2 \\
\hline
2\,1\,6 \\
-2\,1\,6 \\
\hline
0
\end{array}
$$

Divide.

37. $12 \div 1.2$

38. $8.4 \div 0.2$

39. $0.0036 \div 0.9$

40. $5 \div 0.005$

Divide. Round to the indicated place-value position.

41. $3.5 \div 1.3$; tenth

42. $14.78 \div 2.6$; whole number

---

express fractions as terminating or repeating decimals *(Lesson 2-7)*

$$
\frac{1}{6} \;\to\; 6\overline{)1.000} \;\to\; 0.1\overline{6}
$$

$$
\begin{array}{r}
0.166... \\
6\overline{)1.000} \\
-6 \\
\hline
40 \\
-36 \\
\hline
40
\end{array}
$$

Express each fraction or mixed number as a decimal. If the decimal is a repeating decimal, use bar notation.

43. $\dfrac{10}{25}$ 44. $1\dfrac{2}{3}$

45. $\dfrac{3}{8}$ 46. $10\dfrac{1}{4}$

47. $\dfrac{5}{9}$ 48. $\dfrac{5}{12}$

---

change metric units of length, capacity, and mass *(Lesson 2-8)*

$1.39 \text{ kg} = \underline{\ ?\ } \text{ g}$

Multiply by 1,000 since 1 kg = 1,000 g.

$1.39 \text{ kg} = 1,390 \text{ g}$

Complete.

49. $27 \text{ mm} = \underline{\ ?\ } \text{ m}$ 50. $3.9 \text{ mg} = \underline{\ ?\ } \text{ g}$

51. $3.3 \text{ mL} = \underline{\ ?\ } \text{ L}$ 52. $6.85 \text{ km} = \underline{\ ?\ } \text{ m}$

53. $16 \text{ cm} = \underline{\ ?\ } \text{ mm}$ 54. $0.04 \text{ kL} = \underline{\ ?\ } \text{ L}$

55. $43 \text{ g} = \underline{\ ?\ } \text{ kg}$ 56. $3.9 \text{ kL} = \underline{\ ?\ } \text{ mL}$

---

express numbers greater than 100 in scientific notation *(Lesson 2-9)*

$256,000 = 2.56 \times 10^5$

Write each number in scientific notation.

57. $6,000$

58. $459,000,000$

Applications & Problem Solving

59. Weather A barometer is an instrument that measures atmospheric pressure in terms of millimeters of mercury. The higher the mercury rises in the tube, the higher the atmospheric pressure. Order the following barometer readings from least to greatest: 29.97 mm, 30.22 mm, 29.13 mm, 30.53 mm, 31.01 mm. *(Lesson 2-1)*

60. Reasonable Answers Adam bought a pair of sunglasses for $15.79, two rolls of film at $2.29 per roll, and a bottle of sunscreen for $3.69. The cashier asked for $24.06. Is that total reasonable? *(Lesson 2-3B)*

61. Earth Science In 1872, Yellowstone National Park became the first national park in the United States. The park has about 2,000 hot springs. If 0.1 of these are geysers, how many geysers are there in the park? *(Lesson 2-5)*

62. Buildings Each story in an office building is about 3.66 meters tall. Use the graph below to find the number of stories for each structure. Round to the nearest whole number. *(Lesson 2-6)*
   a. Great Pyramid of Cheops
   b. Gateway Arch
   c. Statue of Liberty

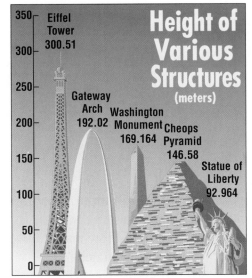

Source: *World Book Encyclopedia*

Alternative Assessment

● **Open Ended**

Suppose you are in charge of supplying a certain brand of soft drink for a picnic. At the store, you find two possible ways to buy the soft drink: 2-L bottles costing $1.30 each, and six-packs of 355-mL cans costing $2.50 per pack. Assuming cups are already provided, how will you decide which type of package is the more economical way to purchase the soft drink?

Suppose you have $40 to purchase this brand of soft drink. What is the maximum number of liters of soda you can purchase?

● **Completing the CHAPTER Project**

Use the following checklist to make sure your solar system model is complete.
☑ The planets are the correct size.
☑ The planets are the correct distance from the Sun.
☑ The charts of the planets' actual diameters and average distances from the Sun are included.

● **PORTFOLIO** Select one of the assignments from this chapter and place it in your portfolio. Attach a note to it explaining why you selected it.

A practice test for Chapter 2 is provided on page 608.

Section One: Multiple Choice

There are twelve multiple-choice questions in this section. Choose the best answer. If a correct answer is *not here,* choose the letter for Not Here.

1. Which is equivalent to 3^6?

  A 36

  B 18

  C $6 \cdot 6 \cdot 6$

  D $3 \cdot 3 \cdot 3 \cdot 3 \cdot 3 \cdot 3$

2. What is the value of $x + y + 5$ if $x = 6$ and $y = 15$?

  F 21

  G 25

  H 26

  J 28

3. Suppose you need 0.65 liter of water for a science experiment, but the container is measured in milliliters. How many milliliters of water do you need?

  A 0.00065 milliliter

  B 0.65 milliliter

  C 6.5 milliliters

  D 650 milliliters

4. Which is correct for rounding to the nearest tenth?

  F 0.56 rounds to 0.5

  G 0.95 rounds to 1.0

  H 1.205 rounds to 1.3

  J 0.4173 rounds to 0.42

5. What is the solution of the equation $x + 4 = 15$?

  A 10

  B 11

  C 12

  D 13

Please note that Questions 6–12 have five answer choices.

6. At age 16, about how much taller is the average boy than the average girl?

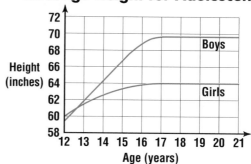

Average Height for Adolescents

  F 2 in.

  G 5 in.

  H 6 in.

  J 63.5 in.

  K 67.5 in.

7. Brookside Middle School's 72 choir students and 10 adult chaperones are planning a trip to see a musical. Each school bus will carry at most 48 people. All tickets to the musical cost $12.95, but schools get a $3.00 discount per ticket. Which piece of information is *not* needed for the school principal to determine the amount of money required for the musical tickets?

  A There are 72 students in the choir.

  B There are 10 adult chaperones.

  C Each school bus will carry at most 48 people.

  D The price of each ticket is $12.95.

  E The discount per ticket is $3.00.

8. Earth is about 93,000,000 miles from the Sun. How is this written in scientific notation?

   F 93×10^6

   G 9.3×10^7

   H 9.3×10^8

   J 0.93×10^8

   K 93 million

9. Alexis charges $5.25 per hour to mow lawns. Which is the best estimate of her earnings if she mows lawns 15 hours during the week?

   A $3   **B** $45

   C $55   **D** $75

   E $95

10. Tomato juice is priced at three cans for $2.39. To the nearest cent, what is the cost of one can?

    F $0.79   **G** $0.80

    H $0.89   **J** $7.17

    K Not Here

11. Three tablecloths are sewn together end-to-end to make one long tablecloth. The tablecloths are about 48 inches, 64 inches, and 54 inches long. What is the combined length?

    A 102 in.   **B** 112 in.

    C 118 in.   **D** 166 in.

    E 226 in.

12. Marcus had $65.72 in his pocket. He spent $32 at a clothing store. How much money did he have left?

    F $31.72   **G** $34.72

    H $65.65   **J** $72.72

    K Not Here

Test-Taking Tip   THE PRINCETON REVIEW

Most standardized tests have a time limit, so you must budget your time carefully. If you cannot answer a question within a few minutes, go on to the next one. If there is still time left when you get to the end of the test, go back to the questions that you skipped.

Section Two: Free Response

This section contains five questions for which you will provide short answers. Write your answers on your paper.

13. Felicia's vacation lasted 8 days and 7 nights. She spent $95 per night for the hotel and $30 per day for food. How much did she spend on food and lodging?

14. Evaluate $2a + 5b$ if $a = 15$ and $b = 4$.

15. To the nearest tenth of a centimeter, what is the length of the ribbon shown below?

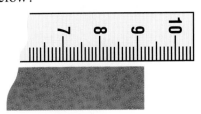

16. Write the product $4 \cdot 4 \cdot 4 \cdot 4 \cdot 4$ using exponents.

17. One of the fastest roller coasters in the world is The Beast at King's Island in Ohio. Its top speed is 64.77 miles per hour. What is its top speed rounded to the nearest whole number?

Test Practice For additional test practice questions, visit:

www.glencoe.com/sec/math/mac/mathnet

What you'll learn in Chapter 3

- to organize data in a frequency table,
- to solve problems and make predictions by using a graph,
- to find the mean, median, and mode of a set of data,
- to construct line plots, stem-and-leaf plots, and box-and-whisker plots, and
- to recognize when statistics and graphs are misleading.

CHAPTER Project

LIGHTS! CAMERA! ACTION!

In this project, you will use statistics to investigate the most popular movies, according to ticket sales. You will also conduct your own survey about favorite movies and display your data using graphs.

Getting Started

Survey 15 to 20 people. Ask each person to pick a movie from the list below. Record the results.

Movie	Ticket Sales (millions $)
A	460.9
B	399.8
C	356.8
D	329.5
E	312.8
F	309.0
G	305.4
H	290.2
I	285.8
J	260.0

Technology Tips

- Use a **spreadsheet** to record the results of your survey.
- Use **computer software** to make graphs.
- Use a **word processor** to summarize your survey results.

 **inter NET** **CONNECTION** Data Update **For up-to-date information on movie statistics, visit:**

www.glencoe.com/sec/math/mac/mathnet

Working on the Project

You can use what you'll learn in Chapter 3 to help you conduct your survey.

Page	Exercise
91	22
105	19
121	10
125	Alternative Assessment

3-1 Frequency Tables

What you'll learn

You'll learn to choose appropriate scales and intervals for data, and organize data in a frequency table.

When am I ever going to use this?

Frequency tables are useful when you take surveys.

Word Wise

range
frequency table
scale
interval

There is only one range for a set of data. However, there is more than one way to choose the scale and the interval for a set of data.

Single-use cameras are convenient to use, and they come in panoramic, waterproof, and 3-D models. The chart shows the prices of 21 single-use cameras. What could you conclude about the prices? *This problem will be solved in Example 1.*

One way to summarize data is to use the **range**. The range is the difference between the greatest number and the least number in a set of data.

Prices of Cameras ($)		
10	6	10
5	10	9
7	8	15
10	16	10
11	14	8
15	10	9
18	6	7

The range for the data above is 18 − 5 or 13.

greatest number ⎤ ⎡ least number

So, the difference between the least expensive and most expensive camera is $13.

A useful way to organize large amounts of data is in a **frequency table**. This kind of table shows the number of times each item of data appears.

First, choose a **scale** for the data. The scale must include all of the numbers from 5 to 18. One scale that will allow you to record all of the numbers is 1 to 20.

You must also decide on the **interval**. The interval separates the scale into equal parts. One possible interval is 5.

Example 1 — Real World APPLICATION

About four to six categories is a good number, though more could be used.

Cameras Refer to the application above. Make a frequency table of the data. What could you conclude about the prices of single-use cameras?

The scale is 1 to 20, and the interval is 5. Therefore, the categories are 1-5, 6-10, 11-15, and 16-20.

In the "Tally" column, record the number of cameras in each category.

Cost ($)	Tally	Frequency
1-5	I	1
6-10	IIII IIII IIII	14
11-15	IIII	4
16-20	II	2

Write the number of tallies or frequency in the "Frequency" column. There should be at least one number in the data set in the highest category, and one in the lowest category. If there are not, choose the scale and interval again.

The frequency table shows that most of these single-use cameras cost from $6 to $10.

Not all frequency tables have categories with scales or intervals.

In-Line Skating Recently, 26 types of in-line skates were tested. The chart shows the kind of brakes that the skates had: toe-stop (*T*), heel-stop (*H*), cuff-activated (*C*), rear-wheel (*R*) or heel-stop/rear-wheel (*H/R*).

H	H	R	H	H
C	C	H/R	H	C
H	H	H	C	R
H	H	H	H/R	H
H	H	H	H	T
H				

Source: *Zillions*

a. Make a frequency table of the data.

b. What advantage is there to using a frequency table instead of a chart?

a. Draw a table with three columns.

In the first column, list the types of brakes. In the second column, tally the data. In the third column, add the number of tallies.

Brakes	Tally	Frequency
toe-stop	|	1
heel-stop	ℍℍ ℍℍ ℍℍ ||	17
cuff-activated	||||	4
rear-wheel	||	2
heel-stop/ rear-wheel	||	2

b. The frequency table makes it easier to see quickly the number of skates with each type of brakes.

CHECK FOR UNDERSTANDING

Communicating Mathematics

Read and study the lesson to answer each question.

1. *Explain* how to find the range, scale, and interval for a set of data.

2. *Make* a frequency table for the data in Example 1 using a different scale and interval. Summarize the data in the new table.

3. *You Decide* Tatanka says that a frequency table includes the least number and the greatest number in a set of data. Stephen argues that only the numbers appearing most frequently are included in a frequency table. Who is correct? Explain.

Guided Practice

4a. Copy and complete the frequency table.

b. Name the scale and the interval.

Fitness Test Scores		
Score	Tally	Frequency
61-70	|	1
71-80		2
81-90	ℍℍ	
91-100	ℍℍ |	

Find the range for each set of data. Choose an appropriate scale and interval. Then make a frequency table.

5. 9, 0, 18, 19, 2, 9, 8, 13, 4

6. 4.5, 2.3, 4.5, 7.8, 5.5, 5.1, 3.9

7. History The table shows the length of reign of the 11 most recent rulers of England.

a. What is the range of the data?

b. Choose an appropriate scale and an interval and make a frequency table.

Ruler	Reign (years)
George I	13
George II	33
George III	59
George IV	10
William IV	7
Victoria	63
Edward VII	9
George V	25
Edward VIII	1
George VI	15
Elizabeth II	45*

*as of 1997
Source: *Academic American Encyclopedia*

EXERCISES

Practice

Find the range for each set of data. Choose an appropriate scale and interval for a frequency table.

8. 6, 2, 8, 9, 12, 4

9. 30, 20, 60, 80, 90, 120, 40

10. 456, 900, 785, 832, 678

11. 14, 19, 4, 0, 13, 8, 2

12. 13, 15, 17, 21, 28, 25, 26, 29, 31, 32, 26, 23, 34, 29

13. 18.2, 14.5, 21.6, 18.8, 17.3, 14.1, 14.6, 15.0

Choose an appropriate scale and interval for each set of data. Then make a frequency table.

14.

Lengths of TV Commercial (s)						
25	30	10	20	60	10	10
30	60	15	20	20	30	45
20	10	60	20	35	30	30

15.

Number of Books Read During the Summer							
0	2	6	2	5	1	3	0
1	2	5	1	3	1	3	1
2	0	2	2	4	1	2	2

16.

Favorite Fast Food Restaurant						
M	M	T	B	T	M	T
B	T	B	M	M	T	M
B	M	T	M	T	B	B

B=Burger Barn   M=Murray's   T=Terry's Tacos

17.

Average Rainfall in Little Rock, Arkansas (in.)			
Jan.	10	July	8
Feb.	9	Aug.	7
Mar.	10	Sept.	7
Apr.	10	Oct.	7
May	10	Nov.	8
June	8	Dec.	9

Source: *Statistical Abstract of the United States, 1996*

18. Name the scale and the interval of the number line.

2 4 6 8 10

19. Draw a number line that shows a scale of 0 to 50 and an interval of 5.

Applications and Problem Solving

Real World

20. *Baseball* The chart shows recent prices of field box seats for baseball teams.

    a. Find the range of the prices.

    b. Choose an appropriate scale and an interval and make a frequency table of the data.

    c. In which interval do the greatest number of prices fall?

Team	Price	Team	Price
A's	$17.50	Mets	$17
Angels	14.50	Orioles	20
Astros	21	Padres	16
Braves	30	Phillies	16
Brewers	20	Pirates	15
Blue Jays	25	Rangers	20
Cardinals	19	Red Sox	23
Cubs	21	Reds	14
Dodgers	19	Rockies	22
Expos	20	Royals	13
Giants	21	Tigers	15
Indians	23	Twins	18
Mariners	22	White Sox	20
Marlins	40	Yankees	23

Source: *USA TODAY*

21. *Earth Science* The chart shows years in the twentieth century in which major hurricanes occurred in the United States.

    a. Make a frequency table of the data. Use the intervals 1900-1919, 1920-1939, 1940-1959, 1960-1979, and 1980-1999.

    b. In which time interval did the greatest number of major hurricanes occur?

Years of Major Hurricanes				
1900	1909	1957	1965	1915
1938	1935	1989	1980	1926
1961	1969	1979	1972	1992
1955	1947	1954	1992	1928
1944	1960			

Source: *The World Almanac*

22. *Working on the* **CHAPTER** *Project* Make a frequency table of the data from your survey.

23. *Critical Thinking* Refer to Exercise 2. Tell the advantages and disadvantages of the two different scales and intervals used for the data in Example 1.

Mixed Review

24. *Entertainment* The pool for Shamu the whale contains 6 million gallons of water. Write 6 million in scientific notation. *(Lesson 2-9)*

25. *Standardized Test Practice* Kalinda bought a skirt for $21.95, a shirt for $19.30, and a pair of shoes for $39.60. All of the prices included tax. Which is the best estimate of the total cost of the three items? *(Lesson 2-3)*

    A less than $65   **B** between $65 and $75

    C between $75 and $85   **D** between $85 and $95

    E more than $95

26. *Real Estate* A house advertised in the real estate section of the newspaper claims to have a rectangular lot with a length of 250 feet and a width of 120 feet. What is the area of the lot? *(Lesson 1-7)*

For **Extra Practice,** see page 574.

PROBLEM SOLVING

3-2A Use a Graph

A Preview of Lesson 3-2

Robert

Marian received information on 8 mountain bikes. She's talking to her friend Robert about which one to buy. Let's listen in!

Okay, I have $350 to spend on a new bike. And I want the best bike for my money.

Well, I graphed the data, and the graph shows that the most expensive bike isn't the best.

Brand	Rating	Price
A	30	$225
B	48	$300
C	48	$370
D	46	$290
E	45	$320
F	37	$240
G	54	$325
H	43	$250

Higher ratings represent better bikes.

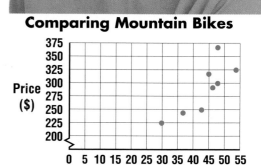
Comparing Mountain Bikes

Marian

The *scatter plot* shows the relationship between rating and price.

THINK ABOUT IT

Work with a partner.

1. **List** different advantages in using the table and the graph to get information about mountain bikes.

2. **Analyze** the data in the graph to see which bike you would recommend for Marian to purchase and why.

3. **Apply** the **use-a-graph** strategy to solve the following problem.

   Kids ages 11-13 tested 18 computer games and rated them based on graphics, sound, action, challenge, and fun. Ratings of 41-60 were Good, 61-80 were Very Good, and 81-100 were Excellent. Use the graph at the right to find how many games were rated Excellent.

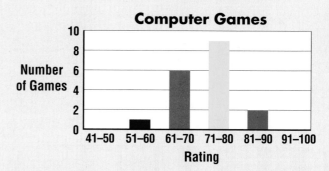

92 Chapter 3 Statistics: Analyzing Data

For **Extra Practice,** see page 574.

ON YOUR OWN

4. *Explain* why the use-a-graph strategy can make solving problems easier.

5. The last step of the 4-step plan for problem solving asks you to *examine* the solution. *Explain* how you can use a graph to help you examine a solution.

MIXED PROBLEM SOLVING

STRATEGIES

Look for a pattern.
Solve a simpler problem.
Act it out.
Guess and check.
Draw a diagram.
Make a chart.
Work backward.

Solve. Use any strategy.

6. *Safety* An elevator sign reads *Do not exceed 2,500 pounds.* How many people, each weighing about 150 pounds, can be in the elevator at the same time?

7. *Telecommunications* Fiber optic cables are made up of bundled strands of glass. If each fiber is 0.0005 inch thick, how thick is a bundle that is 3,000 fibers wide?

8. *Sports* The graph shows the percent of kids ages 9-13 who say they know "a lot" about the rules for playing sports.

PLAYING BY THE RULES

Boys	Girls						
59%	45%	58%	21%	34%	36%	28%	14%
Basketball		Football		Soccer		Hockey	

Source: *Sports Illustrated for Kids Omnibus*

According to the survey, do a greater percent of girls understand the rules of basketball or soccer?

9. *Food* A soup can display has 66 cans. There is one less can in each row than in the row below, with a single can in the top row. How many cans are in the bottom row?

10. *Geography* Alaska, the largest U.S. state, has an area of 1,478,458 square kilometers. Rhode Island, the smallest state, has an area of 2,732 square kilometers. How many times larger is Alaska?

11. *Earth Science* In 1997, for the first time on record, Los Angeles had no rain during March or April.

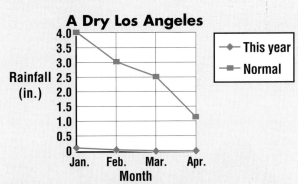

A Dry Los Angeles

— This year
— Normal

Rainfall (in.)

Jan.   Feb.   Mar.   Apr.
Month

Source: National Weather Service

a. About how many inches below normal was the rainfall in February?

b. Use the graph to predict what month this year's rainfall will equal the normal rainfall.

12. **Standardized Test Practice** Francisca bought 6 tickets to the circus. She gave the cashier $170 and received $8 in change. How much did one ticket cost?

  A $20.33   **B** $48.00

  C $27.00   **D** $28.33

  E $15.25

Making Predictions

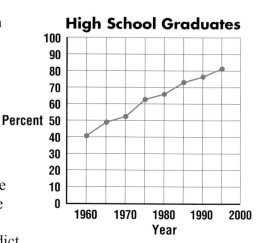

High School Graduates

What you'll learn

You'll learn to make predictions from graphs.

When am I ever going to use this?

You'll be able to make predictions from graphs in newspapers and in magazines.

Word Wise

line graph
bar graph
scatter plot

More students are graduating from high school than did in the past. The graph shows the percent of adults that have a high school diploma from 1960 to 1995. Can you predict what percent of the population will have a high school diploma in 2000?

Line graphs such as the one at the right are useful in predicting future events since they show trends over time. You can use the graph to predict that in the year 2000, about 90% of adults will have high school diplomas.

Example

CONNECTION ①

Life Science In 1995, biologists began releasing gray wolves into different areas of the wild in an effort to increase the species. The table shows the projected population of wolves in one area. Make a line graph of the data and predict the number of wolves in that area in 2002.

Year	Number of Wolves
1995	8
1996	14
1997	27
1998	45
1999	56
2000	68
2001	83

Source: *USA Today*

Step 1 Draw a horizontal and a vertical axis. Label the axes and include a title of the graph.

Step 2 Since the data values go from 8 to 83, an appropriate scale for the vertical axis is 0-100 with an interval of 10.

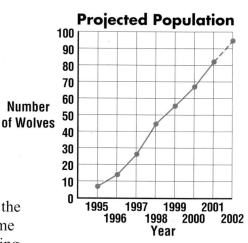

Projected Population

Step 3 Graph the data.

Step 4 Connect the points.

Step 5 Draw a continuation of the graph, dotted, in the same direction it has been going in, to take it one more year into the future.

In 2002, the projected wolf population in that area would be about 95.

A **bar graph** can also be used to make predictions. A bar graph uses bars to make comparisons.

Example 2

Sales The manager of Sweatshirts Unlimited kept a record of how many sweatshirts of each color she sold last month. Use the graph to predict which color will sell the most in the next three months.

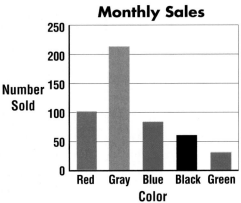

Monthly Sales

The graph shows that gray sweatshirts sold the most. You could predict that gray would continue to be the greatest number sold during the next three months.

A **scatter plot** shows two sets of related data on the same graph.

Example 3

CONNECTION

Life Science The graph shows the average heights and weights of different species of buffalo. Could you conclude that the tallest buffalo are the heaviest?

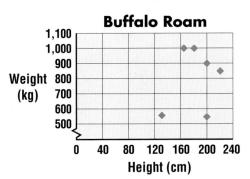

Buffalo Roam

The tallest species of buffalo measure 220 centimeters, but weigh only 850 kilograms. Therefore, the tallest buffalo are not the heaviest.

 HANDS-ON

MINI-LAB

Work with a partner.  ruler marbles drinking glass

Try This

- Make a table. Label the first column "Number of Marbles" and the second column "Height."
- Put one cup of water in a glass. Measure the height of the water.
- Place two marbles in the glass and measure the height of the water. Record the data in your table. Repeat this procedure with 4, 6, and 8 marbles.

Talk About It

1. Draw a graph of the data.
2. Predict the height of the water if a total of ten marbles were in the glass.

Communicating Mathematics

Read and study the lesson to answer each question.

1. *Tell* how a graph can be used to make predictions.

2. Refer to the beginning of the lesson. *Predict* what percent of the population will have high school diplomas when you graduate from high school.

HANDS-ON MATH

3. Take a balloon and blow it up slightly. Record the number of breaths you took and use a ruler to measure the width of the balloon. Repeat, taking measurements every couple of breaths. Make a table of your data and use the data to draw a graph. Summarize your results.

Guided Practice

4. *Food* Extra help is needed at Ralphy's Burgers whenever the number of orders in one hour exceeds 60. During which hours would extra help be needed?

Orders per Hour

5. *Snowboarding* The graph shows the millions of people who have snowboarded more than once. Predict how many people will snowboard more than once in the year 2001.

Snowboarding

Source: National Sporting Goods Association

Practice

The 1940 and 1944 Olympics were cancelled due to World War II.

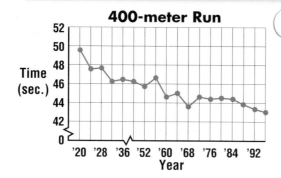

400-meter Run

6. *Sports* The line graph shows the winning times in seconds for the 400-meter run in the Olympic Games from 1920 to 1996. Predict the winning time for the 400-meter run in the 2000 Olympics.

7. *Computers* A survey showed the number of hours each week that students use a computer.

   a. Draw a bar graph of the data.

   b. Determine what grades computer companies should design new software for, according to this data. Explain.

Grade	Hours Per Week
Pre K–K	3.9
1st–3rd	4.9
4th–6th	4.2
7th–8th	6.9
9th–12th	6.7

Source: Find/SVP American Learning Household Survey

8. **Business** Montez started his own lawn care company. The graph shows how much money he made over 15 weeks last summer.

My Profits

His friends are thinking about starting a similar company. Based on the graph of Montez's profits, what would you recommend to his friends?

9. **Entertainment** Although attendance at Wally World is growing, it still attracts fewer people than Valley World. Use the graph to make predictions about attendance at the two parks. Do you think that Wally World will catch up to Valley World in attendance? Explain.

That's Entertainment

Valley World

Attendance (millions)

Wally World

'94   '96   '98   2000
Year

10. **Critical Thinking** Refer to the beginning of the lesson.

   a. Draw a bar graph of the data. Which graph seems easier to understand, the bar graph or the line graph? Explain.

   b. Which types of graphs are more useful to use in making predictions, line graphs, bar graphs, or scatter plots? Explain.

Mixed Review

11. **Statistics** Find the range for the following set of data. Choose an appropriate scale and an interval: 35, 42, 18, 25, 32, 47, 34. *(Lesson 3-1)*

12. **Money Matters** Suzanne bought 3.5 yards of fabric for $7.52. Find the price per yard rounded to the nearest cent. *(Lesson 2-6)*

13. **Standardized Test Practice** The seventh graders at McKinley Middle School are going to an amusement park for a class trip. There are 90 students, and the admission price is $10.75 per person. What is the total cost for all of the students to enter the amusement park? *(Lesson 2-4)*

   A $1,350   B $1,100
   C $967.50   D $875.50
   E Not Here

For **Extra Practice**, see page 574.

What you'll learn

You'll learn to construct line plots.

When am I ever going to use this?

In line plots, you can see useful characteristics of data that can't be seen in tables.

Word Wise

line plot
cluster

Cultural Kaleidoscope

Phiops II became pharaoh of Egypt in 2281 B.C., when he was six years old. It is believed that he ruled until the age of 100.

At 42 years old, Theodore Roosevelt became the youngest President of the United States.

The chart lists the Presidents from Grover Cleveland to William Clinton and their age when they first took office.

One way to organize this data is to use a **line plot**. A line plot is a picture of information on a number line. To make a line plot, you must first determine the scale and the interval.

Grover Cleveland	47	Harry Truman	60
Benjamin Harrison	55	Dwight Eisenhower	62
William McKinley	54	John Kennedy	43
Theodore Roosevelt	42	Lyndon Johnson	55
William Taft	51	Richard Nixon	56
Woodrow Wilson	56	Gerald Ford	61
Warren Harding	55	James Carter	52
Calvin Coolidge	51	Ronald Reagan	69
Herbert Hoover	54	George Bush	64
Franklin Roosevelt	51	William Clinton	46

Step 1 Draw a number line. Since the least age is 42 and the greatest is 69, you can use a scale of 40 to 70 and an interval of 2. *Other scales and intervals could also be used.*

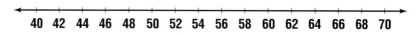

A line plot does not need to start at 0.

Step 2 Put an "✕" above the number that represents the age of each President. If the number is odd, place the "✕" halfway between the appropriate notches.

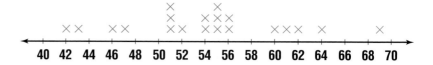

You can make some observations about the data from the line plot.

• The age that occurred most often is 55.

• There seems to be a **cluster** of data between 50 and 55. Data that are grouped closely together are called a cluster.

1 **Food** Restaurants often serve meals that are much larger than the "official" serving sizes shown on labels of similar store-bought items. The chart shows the fat content of typical portions of selected restaurant foods.

Food	Fat (grams)
Pancakes	16
Blueberry muffin	18
Spaghetti with meatballs	39
French fries	26
Ranch salad dressing	21
Tuna salad sandwich	43
Ham sandwich	27
Hamburger with bun	36
Pepperoni pizza	28
Small movie popcorn	27
Chocolate chip cookie	13
Sirloin steak	20
Chicken pot pie	42

Source: *Nutrition Action Newsletter*

a. Draw a line plot of these data.

b. Are there any clusters?

a. The least number of fat grams is 13, and the greatest is 43. An appropriate scale for this graph is 12 to 44 with an interval of 2.

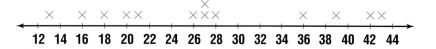

12 14 16 18 20 22 24 26 28 30 32 34 36 38 40 42 44

b. There is a small cluster of data from 26–28.

CONNECTION **2** **Geography** Every 10 years, the Census Bureau conducts a census of the United States population. The top 15 ancestry groups reported in the 1990 census are shown in the table.

Ancestry	% of Population
Afro-American	10
American	5
American Indian	4
Dutch	3
English	13
French	4
German	23
Irish	16
Italian	6
Mexican	5
Norwegian	2
Polish	4
Scotch-Irish	2
Scottish	2
Swedish	2

Source: Bureau of the Census

a. Draw a line plot of the data.

b. Is there a cluster? What does this represent?

a. The least number is 2, and the greatest is 23. An appropriate scale for this graph is 2 to 24 with an interval of 2.

There are 18 other ancestry groups not shown in the table. People could choose more than one ancestry.

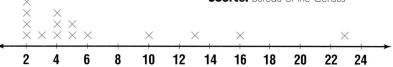

2   4   6   8   10   12   14   16   18   20   22   24

b. There is a cluster from 2 to 6, so most of the ancestries make up between 2% and 6% of the population.

Communicating Mathematics

Read and study the lesson to answer each question.

1. *Tell* the advantage of using a line plot rather than a table to display data.

2. *Draw* a line plot that displays ten pieces of data. Use a range of at least 15. Arrange the data so that there is a cluster. Then describe what the data represent.

Guided Practice

Make a line plot for each set of data.

3. 32, 41, 45, 35, 45, 35, 15, 41, 38, 30, 33

4. 110, 112, 106, 104, 110, 112, 112, 110, 115, 107

5. *Food* The table shows various beverages and their caffeine content.

Beverage	Caffeine (mg)	Beverage	Caffeine (mg)
Caffè Americano (8 oz)	35	Coffee, decaf (16 oz)	15
Caffè Latte (8 oz)	35	Coffee, decaf (8 oz)	10
Caffè Mocha (8 oz)	35	Espresso (1 oz)	35
Cappuccino (8 oz)	35	Lemon Lime (12 oz)	55
Cola (12 oz)	35	Tea, bottled (12 oz)	15
Cola (16 oz)	50	Tea, instant (8 oz)	30
Cola (20 oz)	60	Tea, bag (8 oz)	50

Source: *Nutrition Action Healthletter, 1996*

a. Make a line plot of the data.

b. Name the cluster and describe what it tells you about the caffeine content of the items in the table.

Practice

Make a line plot for each set of data.

6. 45, 35, 50, 40, 40, 55, 30, 35, 45, 35

7. 500, 640, 600, 340, 730, 600, 520, 560, 490, 670

8. 17, 12, 30, 22, 36, 18, 18, 4, 20, 12

9. 1990, 1993, 1995, 1995, 1994, 1995, 1990, 1999, 1988, 1992, 1999

10. 3.8, 4.0, 3.2, 3.6, 3.7, 3.3, 3.2, 3.0, 4.0, 3.6, 3.2

11. 101, 110, 103, 111, 111, 102, 110, 101, 105, 107, 110, 108

12. Make a line plot of the test scores: 84, 100, 89, 88, 83, 90, 97, 100, 89, 90, 90, 80, 91, 95. Name any clusters.

13. Make a line plot of shampoo prices: $2.40, $2.80, $2.50, $2.35, $3.25, $2.75, $2.50, $3.00, $3.25.

14. Geography

The table shows the approximate square miles of water in fifteen states.

State	Water (sq mi)	State	Water (sq mi)
Colorado	350	New Hampshire	400
Georgia	1,500	South Dakota	1,200
Illinois	2,300	Tennessee	900
Iowa	400	Texas	6,700
Kansas	450	Utah	2,700
Maryland	2,600	Virginia	3,200
Montana	150	Wyoming	700
Nevada	750		

Source: *Statistical Abstract*

a. Find the range and determine the scale and an interval.

b. Make a line plot of the data.

c. Do the data cluster in one area?

15. Health Professional basketball players can lose as much as 12 pounds of water each game. The list shows the number of pounds lost by twelve players during one game.

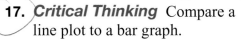

3, 4, 2, 11, 10.5, 8, 8.5, 7.5, 9, 10, 8.5, 9

a. Make a line plot of the data.

b. What conclusions can you draw from your graph?

16. Food Refer to Example 1. The chart shows the fat content of the "official" serving size of each store-bought food item.

Food	Fat (grams)
Pancakes	6
Blueberry muffin	8
Spaghetti with meatballs	5
French fries	10
Ranch salad dressing	11
Tuna salad sandwich	17
Ham sandwich	8
Hamburger with bun	19
Pepperoni pizza	11
Small movie popcorn	11
Chocolate chip cookie	8
Sirloin steak	8
Chicken pot pie	16

a. Draw a line plot of the data.

b. Compare this line plot to the line plot in Example 1. What conclusions could you make, based on these line plots?

17. Critical Thinking Compare a line plot to a bar graph.

a. How is a line plot similar to a bar graph?

b. How is it different?

c. Which do you think is easier to construct? Explain.

Mixed Review

18. Elections On the day before the class elections, 100 students were asked who they would choose for class president. The graph shows the results. Who do you think will win? *(Lesson 3-2)*

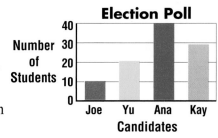

19. Standardized Test Practice The length of a pencil is about — *(Lesson 2-8)*

A 6 millimeters.

B 6 centimeters.

C 6 meters.

D 6 kilometers.

For **Extra Practice,** see page 575.

20. Write 4.23232323… using bar notation. *(Lesson 2-7)*

Mean, Median, and Mode

What you'll learn

You'll learn to find the mean, median, and mode of a set of data.

When am I ever going to use this?

You can use mean to find the average of your test scores.

Word Wise

mean
arithmetic average
mode
median

Priscilla's allowance is $6 per week, and she wants to ask her parents to increase it. She surveyed other students in her class to see how much they get for an allowance. The results are shown below.

$15, $0, $5, $10, $3, $4, $10, $6, $5, $10, $0, $6, $8, $8, $15

In mathematics, there are three common ways to summarize the data with a single number: the mean, the mode, and the median. These are all types of averages.

Mean	The mean of a set of data is the arithmetic average.

The **mean**, or **arithmetic average**, is found by adding the numbers in the data set and dividing by the number of items in the set.

$$\text{mean} = \frac{15 + 0 + 5 + 10 + 3 + 4 + 10 + 6 + 5 + 10 + 0 + 6 + 8 + 8 + 15}{15}$$

$$= \frac{105}{15} \text{ or } 7 \quad \text{The mean allowance is \$7 per week.}$$

Mode	The mode of a set of data is the number(s) or item(s) that appear most often.

Study Hint

Technology You can use a graphing calculator to find the mean and median. Press [STAT] 1 and enter the data into list L1. Then press [STAT] [▶] 1 [ENTER]. $\bar{x}$ is the mean, and Med is the median.

A line plot of the survey results can quickly give you the mode.

The **mode** is $10 because $10 occurs most often.

Median	The median is the middle number in a set of data when the data are arranged in numerical order.

The **median** can also be found by using the line plot. Since there are 15 numbers, the eighth number is the median. If you count from either end of the plot, you will find that the eighth data point is 6.

Note that the mean, the median, and the mode for this set of data are all different. Priscilla could use the mode to argue that more kids get $10 for an allowance than any other amount. She could also say that the average student's allowance is $7.

1 **Meteorology** The high temperatures (°F) in Nashville, Tennessee, for the first week in April were 68°, 65°, 60°, 62°, 67°, 72°, and 71°. Find the mean, mode, and median.

mean   Calculate the arithmetic average.

$\boxed{(}$ 68 $\boxed{+}$ 65 $\boxed{+}$ 60 $\boxed{+}$ 62 $\boxed{+}$ 67 $\boxed{+}$ 72 $\boxed{+}$ 71 $\boxed{)}$

$\boxed{\div}$ 7 $\boxed{)}$ 66.42857143 The mean is about 66.

mode   None, since each temperature occurs only once.

median   60, 62, 65, 67, 68, 71, 72   *Arrange the numbers in order.*

Since there are 7 numbers, the median is the fourth number, or 67.

APPLICATION   **2**   **School** Darrell's test scores for the first grading period are graphed on the line plot. How might he use the mean, mode, or median of the data to describe his scores to his parents? Is this statistic an accurate description?

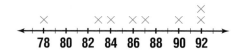

Explore   Which statistic would make the test scores appear better? This would be the greatest number.

Plan   Find the mean, mode, and median of the test scores and find which is the greatest.

Solve $\text{mean} = \dfrac{78 + 83 + 84 + 86 + 87 + 90 + 92 + 92}{8}$ or $\dfrac{692}{8}$ or 86.5

mode = 92, since it occurred most often

Since there is an even number of data points, the median is the mean of the two middle numbers, 86 and 87.

78, 83, 84, 86, 87, 90, 92, 92

The mean of 86 and 87 is 86.5. Therefore, the median is 86.5. Since the mode is the greatest, Darrell might use this statistic to describe his test scores.

Examine   The mode, 92, is not an accurate description of Darrell's test scores since all his other scores were less than 92.

For this particular data set, the mean and the median are the same. Thus, 86.5 might be a better indicator of his score.

Communicating Mathematics

Read and study the lesson to answer each question.

1. *Tell* how you would find each of the following from a set of data.
   a. the median, if there is an odd number of items
   b. the median, if there is an even number of items

2. *Write a Problem* in which a set of data meets each condition.
   a. one mode   b. two modes   c. no modes

3. *You Decide* Cynthia says that if the mean, median, and mode of a set of data are equal, then all the numbers in the set must be the same. Erica says that this is not always true. Who is correct? Explain.

Guided Practice

Find the mean, mode(s), and median for each set of data.

4. 2, 6, 7, 4, 3, 5, 7, 8   5. 8.0, 9.1, 8.9, 9.0, 9.3, 9.4

6.

Length of Stoplights (s)	Tally	Number of Stoplights
10					3
20	₩₩₩	5
30				2
40					3
50			1
60			1

7. *Shoes* The manager of a shoe store keeps a record of the sizes of each athletic shoe sold. When she is ready to place an order, she uses the information to decide what sizes she needs. Which number is probably most useful to her, the mean, mode, or median? Explain.

Practice

Find the mean, mode(s), and median for each set of data.

8. 17, 13, 18, 20, 17, 15, 12   9. 90, 92, 94, 91, 90, 94, 95, 98

10. 2, 7, 1, 5, 8, 3, 5, 4, 6, 3   11. 56, 65, 57, 75, 76, 66, 65, 64

12. 14, 80, 78, 25, 30, 59, 69, 55, 25, 59, 50, 59

13. 1,780; 1,755; 1,755; 1,805; 1,805

14. number of minutes spent on homework

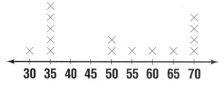

15.

Quiz Score	Tally	Number
100						4
95	₩₩ ||	7
90					3
85						4
80				2
75			1

16. Find the mean, mode, and median for 16, 4, 14, 5, 15, 9, 7, 7, 14, 2.

Applications and Problem Solving

17. *Literature* In *The Phantom Tollbooth,* a boy visits a place called Digitopolis, where each family has 2.58 children. National statistics often report American family size using a "fractional size" family. Do you think they are using the mode, median, or mean? Is this the most informative way to report family size? Explain why or why not.

18. *Space* Find the mean, mode, and median for the length of the space shuttle flights (in days) from February, 1994 to March, 1996.

8, 14, 11, 14, 10, 11, 10, 8, 16, 9, 8, 10, 15, 8, 8, 15, 9

19. *Working on the* **CHAPTER Project** Refer to the table on page 87. Find the mean and median of the ticket sales.

20. *Critical Thinking* A data set contains 50, 100, 75, 60, 75, 1,000, 90, 100, 125, and 75. Without calculating, would the mean, median, or mode be most affected by eliminating 1,000 from the list? Which would be the least affected? Explain.

Mixed Review

21. *Nutrition* The grams of fiber in 15 different cereals are 5, 5, 4, 3, 3, 3, 1, 1, 1, 2, 1, 1, 1, 1 and 0. Make a line plot of the data. *(Lesson 3-3)*

22. Solve $a = 0.8 \div 0.04$. *(Lesson 2-6)*

23. *Standardized Test Practice* Which expression is equivalent to 6×2^3? *(Lesson 1-4)*

A 6×4 **B** $12 \times 12 \times 12$

C $6 \times 2 \times 2 \times 2$ **D** $6 \times 6 \times 6 \times 2 \times 2 \times 2$

For **Extra Practice,** see page 575.

CHAPTER 3

Mid-Chapter Self Test ✓

1. Find the range for the following set of data. Choose an appropriate scale and interval. 12, 15, 23, 20, 18, 19, 10, 15, 20, 11 *(Lesson 3-1)*

2. *Use a Graph* How much more money is spent on photography than design? *(Lesson 3-2A)*

Publishing Budget for School Newspaper

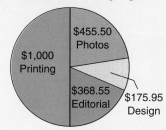

$455.50 Photos
$1,000 Printing
$368.55 Editorial
$175.95 Design

3. *Agriculture* The production of cotton in the United States is shown in the graph. Predict the number of bales that will be produced in the year 2000. *(Lesson 3-2)*

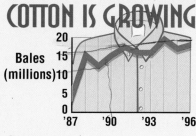

COTTON IS GROWING

Bales (millions)

Source: Agriculture Department

4. Make a line plot of the data: 48, 50, 44, 52, 46, 45, 45, 48, 45, 46. *(Lesson 3-3)*

5. Find the mean, mode(s), and median for the data set in Exercise 4. *(Lesson 3-4)*

COOPERATIVE LEARNING

3-4B Are You Average?

A Follow-Up of Lesson 3-4

markers

ruler

How would you describe an "average" student in your school? The average student in your school may vary quite a bit from the average student in another school. In this lab, you will find out what the average student in your math class is like.

TRY THIS

Work together as a class.

Step 1 List at least ten questions you would like to ask to help you describe what the "average" student is like. For example:

What is your height in inches?

What is your age in months?

How many children are in your family?

What is your favorite TV program?

What is your favorite extracurricular activity?

Step 2 Prepare a survey with your ten questions. Each student in the class should complete the survey.

Work in groups of three.

Step 3 Each group should take at least two of the questions and compile the data in a frequency table or on a line plot.

Step 4 Find the mean, mode, median, and range of the data for each question. Decide which one best describes each set of data and justify your choices.

Step 5 Compile the results of all the groups. Choose an appropriate graph to display your data.

Step 6 Make a poster that describes the "average" student in the classroom.

ON YOUR OWN

1. If you surveyed students in another class in your school, would you expect the same results? Why or why not?

2. If you designed another survey, which questions would you change and why?

3. *Reflect Back*

a. Which did you use to represent your data, the mean, mode, median, or range? Explain why.

b. Which graph did you use to display your data? Explain why you chose that graph.

Can You Guess?

Get Ready This game is for the entire class divided into 3-person teams.

 self adhesive notes   coins

cup   water droppers

Get Set Fill the cup with water. Draw a number line on the chalkboard with a scale from 0 to 15.

Go
- Each team tries to see how many drops of water they can get to stay on the head of a dime. Write the results on a self-adhesive note. Post it on the chalkboard above the appropriate place on a number line.

- Use the information on the chalkboard to predict how many drops of water will fit on the head of a penny. Record your prediction.

- Find how many drops of water you can get to stay on the head of a penny. Post the results and find the mean. If your prediction is within five drops of the mean, your team gets to stay in the game.

- Repeat using a nickel. Continue predicting and testing using different coins. The teams left at the end of the game win.

interNET CONNECTION Visit www.glencoe.com/sec/math/mac/mathnet for more games.

Stem-and-Leaf Plots

What you'll learn

You'll learn to construct and interpret stem-and-leaf plots.

When am I ever going to use this?

You can use a stem-and-leaf plot to record data in science experiments.

Word Wise

stem-and-leaf plot
stem
leaf
back-to-back
  stem-and-leaf plot

Which countries do you think consume the most pasta each year? Many people would guess that Italy tops the list, but would you have guessed that Venezuela is second? The data can be displayed in a **stem-and-leaf-plot**.

In a stem-and-leaf plot, the last digit can be used for the **leaves**, and the digits in front of it can be used as the **stems**.

Follow these steps to make a stem-and-leaf plot of the data.

Country (EU = European Union)	Pasta per Person Each Year (lb)
Argentina	15
Australia	5
Canada	14
Egypt	15
France (EU)	15
Germany (EU)	10
Greece (EU)	19
Italy (EU)	59
Japan	3
Mexico	8
Netherlands (EU)	9
Portugal (EU)	15
Russia	15
Spain (EU)	9
Sweden (EU)	8
Switzerland (EU)	20
Turkey	11
United Kingdom (EU)	4
United States	20
Venezuela	28

Source: National Pasta & National Restaurant Associations

Step 1 Find the least and greatest data values. In the data on pasta consumption, 3 is the least value, and 59 is the greatest. So the tens digits will form the stems, and the ones digits will form the leaves.

Stem	Leaf
0	
1	
2	
3	
4	
5	

Step 2 The stems will be the digits in front of the final digit: that is, the digit in the tens place. List the digits 0 to 5 in order from least to greatest.

Step 3 The leaves are the digits in the ones place for each stem. For example, there are three numbers that have a 2 in the tens place. They are 28, 20, and 20. The 8, 0, and 0 are the leaves for the stem 2. Always write every leaf, even if it is a repeat of another leaf. The leaves are written in order from least to greatest.

Step 4 Include a key to the data.

On the stem-and-leaf plot, it is easy to see that 59 pounds is by far the greatest amount.

Stem	Leaf
0	3 4 5 8 8 9 9
1	0 1 4 5 5 5 5 5 9
2	0 0 8
3	
4	
5	9

$5\,|\,9 = 59$ pounds

You can use a **back-to-back stem-and-leaf plot** to compare two sets of data. In this type of plot, the leaves for one set of data are on one side of the stems, and the leaves for the other set of data are on the other side of the stems. Two keys to the data are needed.

Example
APPLICATION

Food Refer to the beginning of the lesson. Make a back-to-back stem-and-leaf plot to compare pasta consumption in European Union (EU) countries and nonmember countries.

European Union	Stem	Nonmembers
9 9 8 4	0	3 5 8
9 5 5 0	1	1 4 5 5 5
0	2	0 8
	3	
	4	
9	5	

$9 \mid 5 = 59$ pounds $2 \mid 0 = 20$ pounds

Notice that the greatest data for each stem are always the outermost leaves.

The back-to-back stem-and-leaf plot shows that pasta consumption in most European Union and nonmember countries is between 0 and 20 pounds per person each year. The greatest number of countries fall in the interval 10-19 under nonmembers.

CHECK FOR UNDERSTANDING

Communicating Mathematics

Read and study the lesson to answer each question.

1. *Compare and contrast* a stem-and-leaf plot and a bar graph.

2. *Give two examples* of data that could be organized in a back-to-back stem-and-leaf plot. Why would you choose a back-to-back stem-and-leaf plot?

3. *Find* some data that you find interesting in a magazine, newspaper, or on the Internet. Make a stem-and-leaf plot of the data. Write a few sentences in your journal about the results.

Guided Practice

4. Refer to the stem-and-leaf plot in the Example.
   a. What is the smallest average amount of pasta consumed by each person in the European Union?
   b. Would you say that, in general, European Union countries consume more or less pasta than nonmember countries? Explain.

Write the stems that would be used in a stem-and-leaf plot for each set of data. Then make the stem-and-leaf plot.

5. 28, 32, 38, 30, 31, 13, 36, 35, 38, 32, 38, 15, 13, 24

6. 80, 80, 69, 93, 66, 55, 95, 63, 90, 93, 60, 91, 67, 60, 56, 70, 96, 62

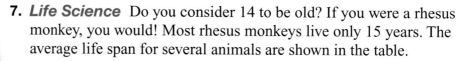

7. *Life Science* Do you consider 14 to be old? If you were a rhesus monkey, you would! Most rhesus monkeys live only 15 years. The average life span for several animals are shown in the table.

a. What numbers would be used as the stems in a stem-and-leaf plot of these data?

b. Make a stem-and-leaf plot of the life spans.

c. What are the shortest and longest life spans?

d. What interval is most representative of these life spans?

Animal	Span (yr)	Animal	Span (yr)
Asian elephant	40	Kangaroo	7
Black bear	18	Leopard	12
Box turtle	100	Lion	15
Camel	12	Mouse	3
Cat	12	Opossum	1
Chipmunk	6	Pig	10
Dog	12	Polar bear	20
Giraffe	10	Rabbit	5
Gorilla	20	Red fox	7
Grizzly bear	25	Rhesus monkey	15
Guinea pig	4	Squirrel	10
Horse	20	White rhinoceros	20

Source: *World Almanac for Kids*

EXERCISES

Practice Write the stems that would be used in a stem-and-leaf plot for each set of data. Then make the stem-and-leaf plot.

8. 16, 19, 21, 23, 25, 25, 29, 31, 33, 34, 35, 39, 41, 47, 49

9. 14, 19, 11, 2, 21, 8, 12, 7, 18, 9, 22, 18, 31, 1

10. 498, 472, 459, 443, 491, 481, 469, 403, 439, 444, 411, 492

11. 56, 49, 15, 18, 36, 39, 27, 75, 44, 90, 37, 26, 68, 61, 58

12. 9.3, 8.2, 9.9, 10, 8, 9.2, 8.7, 8, 8.2, 9, 9.9, 8.7, 8.5, 8.1, 8.8, 9.3

Applications and Problem Solving

13. ***Money Matters*** You may have heard it said, "You get what you pay for." Do you think it's true? Study the table of jeans prices and quality ratings below.

a. Make a back-to-back stem-and-leaf plot of the prices of top-ranked and lower-ranked jeans. Do better jeans cost more? Explain.

Type	Top Ranked Price ($)	Lower Ranked Price ($)
Girls'	20, 28, 18	18, 26, 31
Boys'	17, 25, 25	13, 15, 20
Women's	30, 29, 25	38, 17, 30
Men's	30, 29, 30	30, 22, 22

Source: *Zillions*

b. Make a back-to-back stem-and-leaf plot of the prices of jeans for males and females. Do males or females pay more for jeans? Explain.

14. *Marketing* Advertisers decide when to advertise their products on television based on when the people who are likely to buy will be watching. The table shows the percents of boys and girls ages 6 to 14 who watch television at different times of day. (Values are rounded to the nearest percent.)

Time	Boys	Girls
Monday-Friday, 6 A.M.-9 A.M.	11	9
Monday-Friday, 3 P.M.-5 P.M.	21	22
Monday-Friday, 5 P.M.-8 P.M.	30	29
Monday-Saturday, 8 P.M.-10 P.M., and Sunday, 7 P.M.-10 P.M.	29	27
Saturday, 6 A.M.-8 A.M.	7	4
Saturday, 8 A.M.-1 P.M.	26	23
Saturday, 1 P.M.-5 P.M.	12	8
Saturday, 5 P.M.-8 P.M.	18	12
Sunday, 6 A.M.-8 A.M.	3	3
Sunday, 8 A.M.-1 P.M.	10	9
Sunday, 1 P.M.-5 P.M.	12	7
Sunday, 5 P.M.-7 P.M.	15	9

Source: CMR KIDTRENDS REPORT

a. Make a stem-and-leaf plot of the data for percents of boys and girls. Who watches television more often, boys or girls?

b. If you were scheduling advertising for a product aimed at pre-teen girls, when would you advertise? Explain your reasoning.

15. *Critical Thinking* Refer to the stem-and-leaf plot on page 109.

a. Make a stem-and-leaf plot of the data, replacing each leaf with either E for a European Union (EU) country or N for a nonmember country.

b. What kind of information is gained and what type of information is lost in this type of plot?

c. How does this plot compare to a line plot with two symbols for different sets of data?

Mixed Review

16. Standardized Test Practice The Cardinals baseball team played 8 games. They scored a total of 96 runs. What was the mean number of runs scored per game? *(Lesson 3-4)*

A 12

B 13

C 88

D 104

17. Solve $n = 0.8 \div 0.05$. *(Lesson 2-6)*

18. *Life Science* Refer to the data in Exercise 7. Suppose the life span of a white rhinoceros is r years. Find which animal has an average life span of $r - 16$ years. *(Lesson 1-5)*

Family Activity

Survey at least 20 relatives and friends about their birthdays. Make a stem-and-leaf plot of the day of the month for each birthday. Write about any patterns you see in the plot.

For **Extra Practice**, see page 575.

COOPERATIVE LEARNING

3-6A Quartiles

A Preview of Lesson 3-6

In a large set of data, it is helpful to separate the data into four equal parts called *quartiles*. In this lab, you will find and graph the quartiles. The *interquartile range* is the range of the middle half of the data.

TRY THIS

Work with a partner.

The table shows various waterfalls in the United States and their heights.

Name, Location	Height (ft)	Name, Location	Height (ft)
Akaka, Hawaii	442	Niagara, New York	182
Big Manitou, Wisconsin	165	Passaic, New Jersey	70
Cumberland, Kentucky	68	Seven, Colorado	300
Fall Creek, Tennessee	256	Shoshone, Idaho	212
Feather, California	640	Sluiskin, Washington	300
Great, Maryland	71	Snoqualmie, Washington	268
Illilouette, California	370	Taughannock, New York	215
Minnehaha, Minnesota	53	Yellowstone, Wyoming	308
Multnomah, Oregon	620		

Source: *The World Almanac, 1996*

Step 1 List the data in order from least to greatest.

53 68 70 71 165 182 212 215 256 268 300 300 308 370 442 620 640

Step 2 Find the median.

53 68 70 71 165 182 212 215 (256) 268 300 300 308 370 442 620 640

       8   *8*

The median separates the data into two equal groups.

Step 3 Find the median of the lower group and the median of the upper group.

When the data set has an odd number of members, don't include the original median in either group.

53 68 70 71 ↓ 165 182 212 215 (256) 268 300 300 308 ↓ 370 442 620 640

$$\frac{71 + 165}{2} = 118 \quad \textit{Lower quartile} \qquad \frac{308 + 370}{2} = 339 \quad \textit{Upper quartile}$$

Half of the data numbers (the middle half) lie between the lower and upper quartiles, 118 and 339.

1. What is the interquartile range?

2. Copy the number line below.

•

+←+―+―+―+―+―+―+―+―+―+―+―+―+→+
0   100 200 300 400 500 600

The median of the waterfall data is graphed above the number line. Graph the least value, the greatest value, the upper quartile, and the lower quartile above the number line.

3. The five numbers that are graphed on the number line divide the data into four groups. How many of the numbers in the data set fall in each group?

4. **Look Ahead** The table shows various waterfalls in Europe and their heights.

Name, Location	Height (ft)	Name, Location	Height (ft)
Frua, Italy	470	Reichenbach, Switzerland	656
Gastein, Austria	492	Rhaiadr, Wales	240
Gavarnie, France	1,385	Simmen, Switzerland	459
Giessbach, Switzerland	984	Skjeggedal, Norway	1,378
Glomach, Scotland	370	Skykje, Norway	984
Handol, Sweden	427	Staubbach, Switzerland	984
Krimml, Austria	1,312	Trummelbach, Switzerland	1,312
Mardalsfossen (N), Norway	1,535	Vetti, Norway	900
Mardalsfossen (S), Norway	2,149		

Source: *The World Almanac, 1996*

a. What is the median?
b. What is the upper quartile?
c. What is the lower quartile?
d. What is the least value?
e. What is the greatest value?
f. Draw a number line and graph the median, upper and lower quartiles, and least and greatest values.
g. Compare this number line with the number line in the activity above. What can you conclude about the heights of the waterfalls in the United States compared to the heights of the waterfalls in Europe?

Box-and-Whisker Plots

You'll learn to construct and interpret box-and-whisker plots.

When am I ever going to use this?

Business executives use box-and-whisker plots to analyze their employees' salaries.

Word Wise

box-and-whisker plot
upper quartile
lower quartile
upper extreme
lower extreme
interquartile range
outlier

Study Hint

Technology You can use a graphing calculator to find the quartiles and extreme values. Refer to the Study Hint on page 102. minX = lower extreme, Q1 = LQ, Q3 = UQ, and maxX = upper extreme.

Scientists estimate that there are more than 40,000 earthquakes each year. The table gives information on recent major earthquakes.

Date	Location	Magnitude	Date	Location	Magnitude
3/13/92	Turkey	6.2	6/6/94	Colombia	6.8
3/15/92	Turkey	6.0	8/19/94	Algeria	6.0
6/28/92	California	7.5	1/17/95	Japan	7.2
12/12/92	Indonesia	7.5	5/27/95	Russia	7.6
7/12/93	Japan	7.7	10/1/95	Turkey	6.0
9/29/93	India	6.4	10/9/95	Mexico	7.6
1/17/94	California	6.8	2/3/96	China	7.0
2/15/94	Indonesia	7.0	2/17/96	Indonesia	7.5

Source: Global Volcanism Network, Smithsonian Institution, U.S.

You can use a **box-and-whisker plot** to display and summarize data. A box-and-whisker plot summarizes data using the median, the **upper quartile (UQ)**, the **lower quartile (LQ)**, the **upper extreme** and the **lower extreme**.

Step 1 First, find the median, the quartiles, and the extreme values. Write the data in order from least to greatest.

$$6.0 \quad 6.0 \quad 6.0 \quad 6.2 \quad 6.4 \quad 6.8 \quad 6.8 \quad 7.0$$
$$7.0 \quad 7.2 \quad 7.5 \quad 7.5 \quad 7.5 \quad 7.6 \quad 7.6 \quad 7.7$$

$$\text{median} = \frac{7.0 + 7.0}{2} \text{ or } 7.0$$

The lower quartile is the median of the lower half of the data. The upper quartile is the median of the upper half of the data.

$$LQ = \frac{6.2 + 6.4}{2} \text{ or } 6.3 \qquad UQ = \frac{7.5 + 7.5}{2} \text{ or } 7.5$$

The lower extreme is the least value, 6.0. The upper extreme is the greatest value, 7.7.

Step 2 Graph each value above a number line.

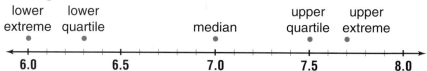

Step 3 Draw a box around the upper quartile and lower quartile and draw a vertical line through the median value.

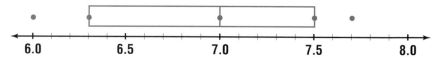

Step 4 Extend *whiskers* from each quartile to the extreme points.

A box-and-whisker plot divides the data into four parts using the lower extreme, LQ, median, UQ, and upper extreme. One-fourth of the data fall between each of these adjacent numbers.

Example 1

APPLICATION

Architecture Make a box-and-whisker plot of the data on the heights of the tallest buildings in St. Louis, Missouri.

Find the median, the quartiles, and the extremes. Then construct the plot.

$$\text{median} = \frac{434 + 420}{2} = 427$$

$\text{LQ} = 394 \quad \text{UQ} = 564$

upper extreme $= 593$
lower extreme $= 375$

Building	Height (ft)
Metropolitan Square Tower	593
One Bell Center	588
Mercantile Center Tower	540
Laclede Gas Building	434
Boatmen's Plaza	420
SW Bell Telephone Building	398
Civil Courts Building	390
One City Center	375

Source: *World Almanac*

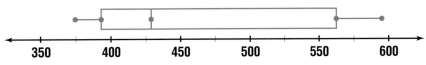

St. Louis Metropolitan Square Tower

The graph shows how the data are spread. Half of the tallest buildings are between 394 and 564 feet high. The largest range of the four quartiles is from 427 to 564 feet. One-fourth of the tallest buildings are within these heights.

You can locate the quartiles by the left and right sides of the box and then subtract to find the **interquartile range**. *The interquartile range is 170.*

Some values in a set of data may be much greater or much less than the other data. Data that are more than 1.5 times the interquartile range from the quartiles are called **outliers**.

Example 2

APPLICATION

Food Snacks sold at movie theaters often have as many Calories as fast-food meals. Make a box-and-whisker plot of the data.

Study Hint

Technology You can use a graphing calculator to draw a box-and-whisker plot.

Snack (oz)	Calories	Snack (oz)	Calories
Peanut butter cups, 3.2	380	Popcorn, medium unbuttered	901
Fruit-flavored candy, 2.6	286	Raisins, 2.3	270
Chocolate wafer candy bar, 4	588	Candy bar, 4	492
Colored peanut candy, 2.6	363	Mints, 3	360
Colored chocolate candy, 2.6	350	Licorice, 5	500
Popcorn, medium buttered	1,221		

270   286   350   360   363   380   492   500   588   901   1,221
            ↑   ↑   ↑
            LQ   Median   UQ

(continued on next page)

Draw a box to show the median and the quartiles.

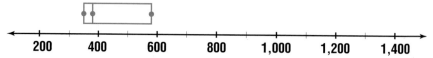

The interquartile range is 588 – 350 or 238. So, data more than 1.5 · 238, or 357, from the quartiles are outliers.

Subtract 357 from the lower quartile. 350 – 357 = –7
Add 357 to the upper quartile. 588 + 357 = 945

So, –7 and 945 are the limits for the outliers. There is one outlier in the data, 1,221. Plot the outlier with an asterisk. Then draw the lower whisker to the lower extreme, 270, and the upper whisker to the last value that is not an outlier, 901.

Outliers may be noted with asterisks in a box-and-whisker plot.

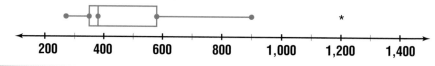

CHECK FOR UNDERSTANDING

Communicating Mathematics

Read and study the lesson to answer each question.

1. *Describe* what information you can get about a data set by looking at a box-and-whisker plot.

2. *Explain* how to determine whether a number is an outlier.

3. *Write* a few sentences to tell why a box-and-whisker plot might be used to display a large number of data. What drawbacks does a box-and-whisker plot have, if any?

Guided Practice

4. *Life Science* The box-and-whisker plot represents data on the average life spans of different animals found in Exercise 7 on page 110.

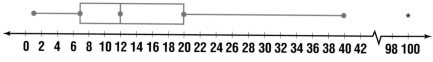

   a. Find the median. b. Find the interquartile range.

5. *History* Refer to the chart of Presidents' ages on page 98.

   a. What is the median? b. What is the upper quartile?
   c. What is the lower quartile? d. What is the upper extreme?
   e. What is the lower extreme? f. What is the interquartile range?
   g. What are the limits on outliers?
   h. Draw a box-and-whisker plot of the ages.
   i. Compare the box-and-whisker plot you drew in part h to the line plot on page 98. What similarities and differences do you see?

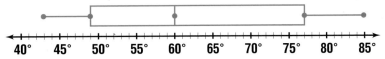

EXERCISES

Practice

6. Use the box-and-whisker plot of high temperatures (°F) below.

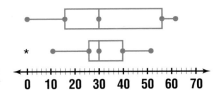

40° 45° 50° 55° 60° 65° 70° 75° 80° 85°

a. What is the median?   **b.** What is the upper quartile?

c. What is the lower quartile?   **d.** What is the upper extreme?

e. What is the lower extreme?   **f.** What is the interquartile range?

g. What are the limits on outliers? Are there any outliers?

h. What fraction of the temperatures are less than 49°?

7. Make a box-and-whisker plot of the data shown in the stem-and-leaf plot.

Stem	Leaf
1	2 7 7
2	0 4 5 5 6 7 9 9
3	1 3 5 8 1|7 = 17

8. Compare the box-and-whisker plots.

a. What is similar about the data in the two plots?

b. What is different about the data in the two plots?

c. Which set of data is more concentrated around the median? Explain.

```
0  10  20  30  40  50  60  70
```

Applications and Problem Solving

9. *Sports* The table shows the games won by each men's professional basketball team in the 1995–1996 season. Make a box-and-whisker plot of the data.

Team	Wins	Team	Wins	Team	Wins
Atlanta	46	Indiana	52	Phoenix	41
Boston	33	L.A Clippers	29	Portland	44
Charlotte	41	L.A. Lakers	53	Sacramento	39
Chicago	72	Miami	42	San Antonio	59
Cleveland	47	Milwaukee	25	Seattle	64
Dallas	26	Minnesota	26	Toronto	21
Denver	35	New Jersey	30	Utah	55
Detroit	46	New York	47	Vancouver	15
Golden State	36	Orlando	60	Washington	39
Houston	48	Philadelphia	18		

For the latest statistics, visit:

www.glencoe.com/sec/math/mac/mathnet

For **Extra Practice**, see page 576.

10. *Critical Thinking* Describe a set of data in which there is only one whisker in its box-and-whisker plot.

Mixed Review

11. *Statistics* Make a stem-and-leaf plot for 20, 21, 35, 34, 18, 56, 11, 10, 12, 22, and 38. *(Lesson 3-5)*

12. *Standardized Test Practice* Choose the solution to the equation
$\frac{48}{c} = 4 + 2$. *(Lesson 1-5)*

A 2   **B** 4   **C** 6   **D** 8

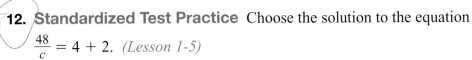

Lesson 3-6 Box-and-Whisker Plots **117**

COOPERATIVE LEARNING

3-6B How Much is a Handful?

A Follow-Up of Lesson 3-6

🍿 popped
popcorn

Leon and his older brother ask their mother if they can have some popcorn. She replies, "Only one handful each." How much is a handful? Will Leon and his older brother get the same amount of popcorn? You can use data to make predictions.

TRY THIS

Work together as a class.

Step 1 Each student should grab a handful of popped popcorn.

Step 2 Count the number of popped kernels in each handful and record the information on the chalkboard.

Work in groups of three.

Step 3 Choose one of the following ways to represent the data: frequency table, bar graph, line plot, stem-and-leaf plot, or box-and-whisker plot.

Step 4 Find the mode, median, and mean of the data.

Step 5 Display your graphical representation and explain it to the rest of the class.

ON YOUR OWN

1. Which value, the mode, median, or mean, best describes the set of data? Why?

2. Which graphical representation is the best representation of the data? Why?

3. *Reflect Back* Examine the data you collected.

  a. Predict how many popped kernels other students your age might get in one handful.

  b. Randomly choose 10 students who are not in your math class and have them grab a handful of popcorn. Compare the number of popped kernels in their handfuls with your prediction.

  c. Ask your teacher to grab a handful of popcorn. How does the number of kernels that he or she grabbed compare with your data? Could you use this information to predict about how many kernels other adults in your school might get in one handful? Why or why not?

Misleading Statistics

6 days, 7 nights for only $299

Orlando

What you'll learn

You'll learn to recognize when statistics and graphs are misleading.

When am I ever going to use this?

Being able to recognize misleading statistics is useful when you're reading advertisements.

Have you ever seen a vacation advertisement that appeared too good to be true? Was the ad misleading, or did it not give all the information? As a selling strategy, advertisements sometimes distort the facts by taking images and words out of context or by leaving out important information. Similarly, when you are given insufficient background information or an incomplete picture of the data, you may be looking at misleading statistics.

Whenever there are outliers in the data, the mean is not a good way to describe the data.

Examples

APPLICATION ① **Salaries** Amara is interviewing for a technology company. She is told that the average salary of the 37 employees is more than $40,000. Using the information at the right, should Amara expect a salary of more than $40,000 if she gets the job? Explain.

Employee	Salary
President	$375,000
Vice President	$325,000
Sales Staff (15)	$35,000
Secretaries (10)	$16,000
Phone Order Staff (10)	$12,000

The company has 37 employees, but only two have a salary over $40,000. The two very high salaries result in a misleading mean. Amara should not expect a salary of more than $40,000.

APPLICATION ② **Television** Both bar graphs show the percent of viewers that watch network television. Which graph could be misleading?

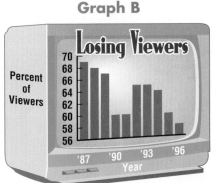

Graph A

Graph B

Losing Viewers

Percent of Viewers

70 60 50 40 30 20 10 0

'87 '90 '93 '96
Year

Losing Viewers

Percent of Viewers

70 68 66 64 62 60 58 56

'87 '90 '93 '96
Year

Although both graphs show a decrease in network viewers, a change in the vertical scale makes Graph B look misleading. The loss of viewer looks more drastic than in Graph A.

Communicating Mathematics

Read and study the lesson to answer each question.

1. *Tell* three ways data can be misleading.

2. *Tell* which graph in Example 2 could be better used to convince advertisers not to air their commercials on network television.

3. *Write* about at least two newspaper or magazine advertisements that you find that are misleading.

Guided Practice

4. The line graphs both show monthly CD sales for one year at the Music Barn. Which graph could be misleading? Explain.

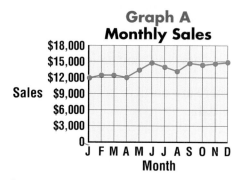

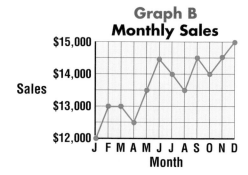

5. *Recycling* Both graphs show the number of pounds of aluminum cans that students at Walker Middle School recycled in eight weeks.

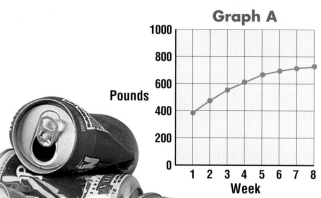

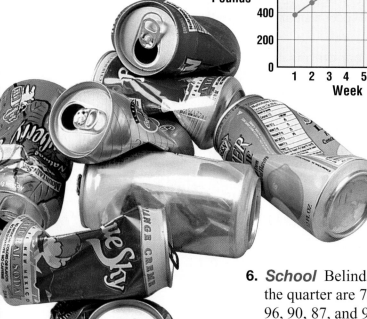

a. Which graph might be used to show that recycling efforts are greatly improving?

b. Which graph could be considered misleading? Explain.

6. *School* Belinda's math scores for the quarter are 72, 89, 92, 96, 82, 96, 90, 87, and 91. What misleading statistic could Belinda use to describe how well she's done in class? Explain.

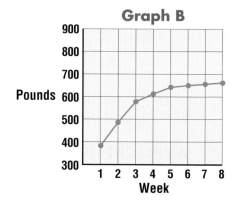

Practice

7. Both line graphs show ship sales at Marvin's Marina in thousands of dollars. Which graph could be misleading? Explain.

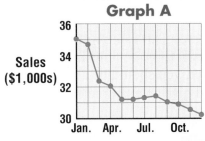
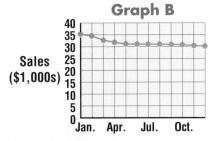

8. *Geography* The table shows the areas of 10 Caribbean islands.

a. Find the mean, mode, and median.

b. Which average is misleading?

Island	Area (sq mi)	Island	Area (sq mi)
Antiqua	108	Martinique	425
Aruba	75	Puerto Rico	3,339
Barbados	166	Tobago	116
Curacao	171	Virgin Islands, UK	59
Dominica	290	Virgin Islands, U.S.	134

Source: Bureau of the Census

c. Which average would most accurately describe the data?

Applications and Problem Solving

9. *Marketing* *Best Bikes* is running an ad claiming that their prices are comparable with *Deals on Wheels,* but their bikes are a much better quality.

Draw graphs of the same data so that Deals on Wheels' lower prices are better shown.

10. *Working on the* CHAPTER Project Refer to the table on page 87.

a. Draw a bar graph of the data that could be misleading. Explain how the graph is misleading.

b. Draw a bar graph that better represents the data.

For **Extra Practice,** see page 576.

11. *Critical Thinking* Do great or small values affect the median of a set of data? Give several examples to support your answer.

Mixed Review

12. *Statistics* Refer to the data on countries consuming pasta on page 108. Find the interquartile range. *(Lesson 3-6)*

13. **Standardized Test Practice** The rental charge for a car is $78 per day. The 7% sales tax is $5.46. What is the total cost of renting a car for one day? *(Lesson 2-3)*

A $73.30   B $73.46   C $83.54   D $90.46   E Not Here

Chapter Review For additional lesson-by-lesson review, visit:
www.glencoe.com/sec/math/mac/mathnet

Vocabulary

After completing this chapter, you should be able to define each term, concept, or phrase and give an example or two of each.

Statistics and Probability
arithmetic average (p. 102)
back-to-back stem-and-leaf plot (p. 109)
bar graph (p. 94)
box-and-whisker plot (p. 114)
cluster (p. 98)
frequency table (p. 88)
interquartile range (pp. 112, 115)
interval (p. 88)
leaf (p. 108)
line graph (p. 94)
line plot (p. 98)
lower extreme (p. 114)
lower quartile (p. 114)
mean (p. 102)

median (p. 102)
mode (p. 102)
outlier (p. 115)
quartile (p. 112)
range (p. 88)
scale (p. 88)
scatter plot (p. 92, 95)
stem (p. 108)
stem-and-leaf plot (p. 108)
upper extreme (p. 114)
upper quartile (p. 114)

Problem Solving
use a graph (p. 92)

Understanding and Using the Vocabulary

Choose the letter of the term that best matches each phrase.

1. the difference between the greatest number and the least number in a set of data
2. separates the scale into equal parts
3. graphs that are useful in predicting future events since they show trends over time
4. graphs that use bars to make comparisons
5. plots showing two sets of related data on the same graph
6. data that are grouped closely together
7. the arithmetic average of a set of data
8. the number or item that appears most often in a set of data
9. the middle number in a set of data when the data are arranged in numerical order
10. a plot used to display a large data set to make it easier to read

a. interval
b. scatter plot
c. outlier
d. median
e. range
f. bar graph
g. line graph
h. mode
i. stem-and-leaf plot
j. cluster
k. mean

In Your Own Words

11. *Explain* how to make a stem-and-leaf plot.

Objectives & Examples

Upon completing this chapter, you should be able to:

● choose appropriate scales and intervals for data and organize data in a table *(Lesson 3-1)*

Find the range and an appropriate scale and interval for 11, 2, 22, 13, 15, 14, 18, 7, 20, 10, 19.

range = 22 − 2 = 20
scale: 0 to 24
interval of 6

Make a frequency table of the data.

Interval	Tally	Frequency
0-6	I	1
7-12	III	3
13-18	IIII	4
19-24	III	3

● make predictions from graphs *(Lesson 3-2)*

What is the most common response to the question "How many children are in your family?"

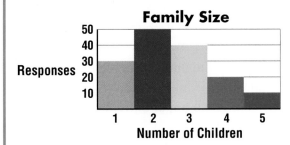

Family Size

The most common response is 2.

● construct line plots *(Lesson 3-3)*

Make a line plot for 5, 7, 3, 8, 3, 2, 8, 4, 15, 12.

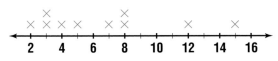

Review Exercises

Use these exercises to review and prepare for the chapter test.

Find the range for each set of data. Choose an appropriate scale and interval.

12. 3, 12, 1, 43, 25, 16

13. 75, 150, 100, 400, 550

14. 2.3, 11.9, 7.6, 1.3, 4.8

15. Choose the appropriate scale and an interval for the data. Then make a frequency table.

People in Your Family					
3	2	4	4	3	5
6	4	5	6	4	4
5	3	4	5	4	5
3	4	2	4	3	4

16. *Finance* Isabel displayed her salary for the last five years in the line graph. Predict her salary after three more years.

Salary for Five Years

Make a line plot for each set of data.

17. 10, 12, 10, 8, 13, 10, 8, 12

18. 7.9, 8.3, 8.1, 8.3, 8.5, 8.9, 8.3, 8.5

19. 550, 554, 545, 553, 550, 554, 548, 553, 554

20. 43, 41, 42, 45, 43, 42, 43, 46, 44, 44

Objectives & Examples

find the mean, median, and mode of a set of data *(Lesson 3-4)*

Find the mean, mode(s), and median for 23, 21, 18, 19, 20, 19, 20, and 19.

mean: $\dfrac{23 + 21 + 18 + 19 + 20 + 19 + 20 + 19}{8} = 19.875$

mode: 19　　median: 19.5

Review Exercises

Find the mean, mode(s), and median for each set of data.

21. 2, 3, 4, 3, 4, 3, 8, 7, 2

22. 31, 24, 26, 18, 23, 31, 18

23. 89, 76, 93, 100, 72, 86, 74

24. 54,000, 49,000, 112,000, 89,000, 76,000, 65,000

---

construct and interpret stem-and-leaf plots *(Lesson 3-5)*

Construct a stem-and-leaf plot for 12, 15, 17, 20, 22, 22, 23, 25, 27, 35, 52.

Stem	Leaf
1	2 5 7
2	0 2 2 3 5 7
3	5
4	
5	2　　$5|2 = 52$

Write the stems that would be used in a stem-and-leaf plot for each set of data. Then make the stem-and-leaf plot.

25. 75, 61, 83, 99, 78, 85, 87, 92, 77, 78, 60, 53, 87, 89, 91, 90

26. 29¢, 54¢, 31¢, 26¢, 38¢, 46¢, 23¢, 21¢, 32¢, 37¢

27. Make a back-to-back stem-and-leaf plot for the following temperatures (°F).

Seattle: 52, 60, 61, 46, 80, 62, 70, 77, 53, 85, 54, 62, 72, 68, 78, 69

Olympia: 60, 53, 68, 72, 66, 80, 73, 51, 62, 48, 56, 84, 77, 45, 79, 65

---

construct and interpret box-and-whisker plots *(Lesson 3-6)*

Construct a box-and-whisker plot of the English quiz scores: 12, 12, 13, 14, 14, 15, 15, 15, 16, 16, 17, 18, 20.

median: 15　　UQ: 16.5　　LQ: 13.5
upper extreme: 20　　lower extreme: 12

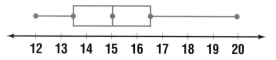

28. The points scored in a game by each player on the girls basketball team are 6, 4, 15, 3, 11, 8, 2, 9, 4, and 2.

　a. What is the median?

　b. What is the upper quartile and lower quartile?

　c. What is the upper extreme and the lower extreme?

　d. Draw a box-and-whisker plot of the points scored.

---

recognize when statistics and graphs are misleading *(Lesson 3-7)*

When you are given insufficient background information or an incomplete picture of the data, the data may be misleading.

29. Gina's test scores are 89, 92, 87, 86, 95, 93, and 29.

　a. Find the mean, mode, and median of the scores.

　b. Which number is misleading? Explain.

Applications & Problem Solving

30. Use a Graph Francine is keeping track of her savings account balance each month. How much money do you think she will have in her savings account in the 7th month? *(Lesson 3-2A)*

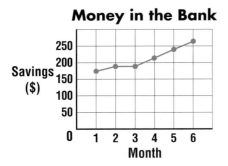

Money in the Bank

31. Jobs The hourly wage of eight students at their after-school jobs is $4.65, $5.15, $4.90, $5.25, $4.90, $5.00, $5.50, $4.80. Find the mean, mode, and median. *(Lesson 3-4)*

32. Earth Science The table shows the average water speeds of various ocean currents, rivers, and bores (tidal waves). Make a stem-and-leaf plot of the data. *(Lesson 3-5)*

Water	Speed (mph)
Amazon River	2
Antarctic Circumpolar Current	1
Ganges Bore	17
Gulf Stream	5
Lava Falls, Colorado R.	30
Mississippi River	2
Pentland Firth	12
Severn Bore	13
Saltstraumen Current	18

Source: *Encyclopedia Britannica*

Alternative Assessment

● **Open Ended**

Suppose you took a survey, asking your classmates how much they spent for lunch yesterday. The responses are $1.50, $2.30, $0.99, $1.50, $2.00, $1.50, $1.75, $2.30, $1.75, $2.00, $3.15, $1.75, $2.00, $3.15, $1.75, $2.00, $1.75, $2.30, $1.75, and $1.50. How can you organize the data to determine the most common response?

● **PORTFOLIO** Select your favorite application problem from this chapter and place it in your portfolio. Attach a note explaining why it is your favorite.

● **Completing the CHAPTER Project**

Use your survey results and make a graph of the data. Prepare a poster or display comparing the results of your survey with the data about ticket sales for movies. Use the following checklist to make sure your project is complete.

☑ The frequency table of your survey results is included.

☑ A graph of your survey results is clear and easy to read.

☑ A paragraph comparing your survey results to the rating in the table on page 87 includes which method you think is best for determining the most popular movies— using ticket sales or surveying people.

A practice test for Chapter 3 is provided on page 609.

Section One: Multiple Choice

There are eight multiple choice questions in this section. Choose the best answer. If a correct answer is *not here,* choose the letter for Not Here.

1. Armando has 3 fewer hats than Denise. Tien has twice as many hats as Armando. If Denise has x hats, which number sentence represents the number of hats that Tien has?

   A $x - 3$

   B $3x$

   C $2(x - 3)$

   D $3(x - 3)$

2. Which expression is the same as v^6?

   F $v + v + v + v + v + v$

   G $6v$

   H $v^3 + v^3$

   J $v \times v \times v \times v \times v \times v$

3. The line plot shows how far in kilometers some students live from the school. How many students are represented in the plot?

   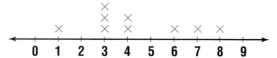

   A 5

   B 7

   C 9

   D 8

Please note that Questions 4–8 have five answer choices.

4. Students at Karlon High School have collected money this year for a charity.

   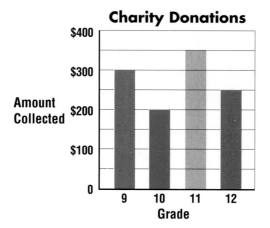

   Charity Donations

   According to the graph, how much did all four classes collect this year?

   F $950

   G $1,100

   H $1,050

   J $1,000

   K $1,150

5. There are 22 tables in the cafeteria at Evan Middle School. If there are 10 chairs at each table and 15 chairs in the cafeteria are empty, which number sentence can be used to find the number of students seated, N?

   A $N = (22 \times 10) - 15$

   B $N = (22 + 10) + 15$

   C $N = (22 - 10) + 15$

   D $N = (22 \times 10) + 15$

   E $N = (22 - 10) \times 15$

6. The table shows the cost of three items that Ricci bought, all including tax.

Item	Cost
purse	$12.45
pair of earrings	$3.95
wallet	$5.85

Which is the best estimate of the total cost of the three items?

F less than $10

G between $10 and $20

H between $20 and $30

J between $30 and $40

K more than $40

7. Franklin earned $13.75 in one week by baby-sitting 5.5 hours. How much was Franklin paid per hour?

A $2.10

B $2.50

C $3.05

D $3.50

E Not Here

8. A soybean harvesting company collects soybeans to be used in the production of baby formula. If the company sold 158.6 tons, 200 tons, and 82.04 tons of soybeans in the last three months, what were the company's total soybean sales for that period?

F 440.64 tons

G 438.84 tons

H 441.0 tons

J 438.0 tons

K Not Here

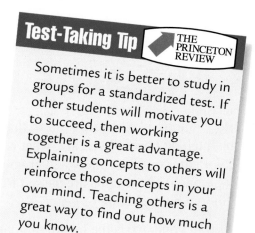

Test-Taking Tip THE PRINCETON REVIEW

Sometimes it is better to study in groups for a standardized test. If other students will motivate you to succeed, then working together is a great advantage. Explaining concepts to others will reinforce those concepts in your own mind. Teaching others is a great way to find out how much you know.

Section Two: Free Response

This section contains four questions for which you will provide short answers. Write your answers on your paper.

9. What is the value of $3[2(20-18)-4]$?

10. Use the data in Exercise 3. What is the median distance? If necessary, round to the nearest tenth.

11. A business had weekly profits of $5,000, $3,000, $2,000, $2,500, and $5,000. Which average might be misleading: the mode, the median, or the mean?

12. To the nearest tenth, find the area of the rectangle shown below.

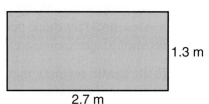

1.3 m

2.7 m

 **inter NET CONNECTION** Test Practice **For additional test practice questions, visit:**

www.glencoe.com/sec/math/mac/mathnet

Interdisciplinary Investigation

IF THE SHOE FITS . . .

What do you look like? The answer to this question would probably include how tall you are, what color eyes you have, what color hair you have, and whether you are light-skinned or dark-skinned. These traits are passed from parents to their children by the action of *genes*.

Another trait you have is the length of your foot. There may be other students in your class who have the same foot length as you do. However, most probably have a different foot length. There are many different foot lengths because this trait is influenced by genetic and other factors.

What You'll Do

In this investigation, you will collect data about the human foot, analyze it, and design a method to display the results.

Materials tape measure

calculator

Procedure

1. Work in a group. Each group should measure the feet of at least ten people who are between the ages of 10 and 18. Include both boys and girls. Measure only those people who are willing to let you. Your class should agree on exactly how to measure the length.

2. Calculate the mean, median, mode, and range of foot length for boys, for girls, and for boys and girls together.

3. Display your results graphically.

4. Write a concluding statement that describes the data.

Technology Tips

- Use a **spreadsheet** or **database** to determine the mean.

- Use **graphing software** to display the data.

- Use **publishing software** to develop ads and brochures.

Making the Connection

Use the data collected about foot lengths as needed to help in these investigations.

Language Arts

Suppose you are the manager of a shoe store and are planning to sell some hot new shoes for the teenage market. Predict how many shoes of each size you expect to sell. Plan a newspaper advertisement to market the shoes.

Social Studies

Research the contribution Jan Earnst Matzeliger made to the shoe industry. Include some statistics that tell the economic effect of his invention.

Science

Show how a Punnett Square can be used to make predictions about the traits of a population.

Go Further

- Collect data from different age groups about foot length. Make a graph for each age group.

- Ask a local podiatrist for charts that list the average foot size for various age groups. See how close your results come to these published averages.

 **Research For current information on shoes, visit the following website.**

Data Collection and Comparison To share and compare your data with other students in the U.S., visit:

www.glencoe.com/sec/math/mac/mathnet

 You may want to place your work on this investigation in your portfolio.

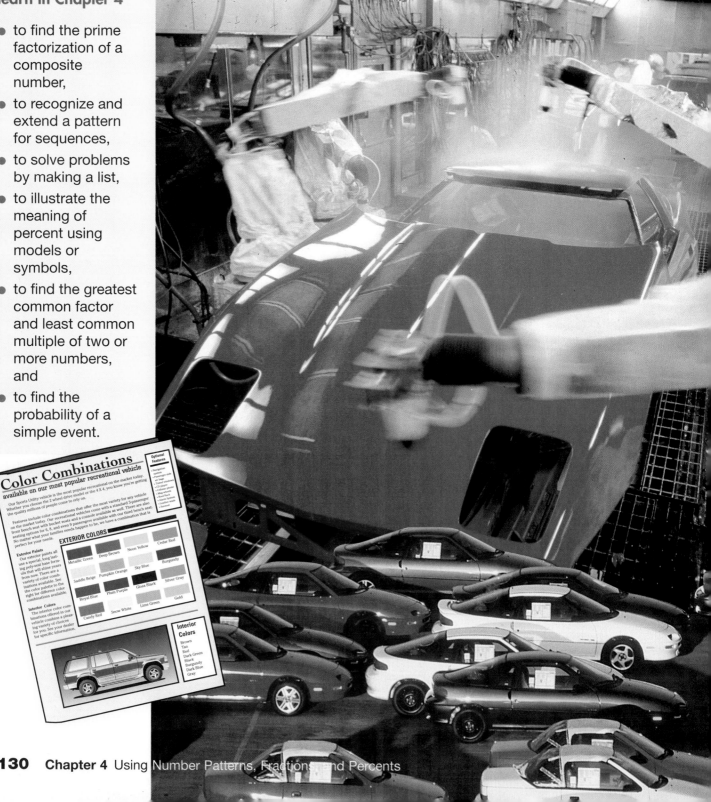

Using Number Patterns, Fractions, and Percents

What you'll learn in Chapter 4

- to find the prime factorization of a composite number,
- to recognize and extend a pattern for sequences,
- to solve problems by making a list,
- to illustrate the meaning of percent using models or symbols,
- to find the greatest common factor and least common multiple of two or more numbers, and
- to find the probability of a simple event.

Color Combinations
available on our most popular recreational vehicle

Our Sports Utility vehicle is the most popular recreational on the market today. Whether you choose the 2 wheel drive model or the 4 X 4, you know you're getting the quality millions of people come to rely on.

Features include color combinations that offer the most variety for any vehicle on the market today. Our recreational vehicles come with a standard 3-passenger front bench seat with bucket seats and a console available as well. There are also seating options for 6, 8, and even 9 passengers available with our third bench seat. No matter what your families needs happen to be, we have a combination that is perfect for your needs.

Exterior Paints
Our exterior paints all use a special, long lasting poly-seal base formula that will shine years from now. There are a variety of color combinations available. See the color palette to the right for different color combinations available.

Interior Colors
The interior color combinations offered in our vehicle combine a pleasing variety of choices for you. See your dealer for specific information.

Optional Features
- Navigation system
- Passenger side air bags
- Tinted windows
- CD player
- Cellular phones
- Mag wheels
- Roof Rack
- Alarm System
- Leather seats
- Sunroof

EXTERIOR COLORS

Metallic Green	Deep Brown	Neon Yellow	Cedar Red
Saddle Beige	Pumpkin Orange	Sky Blue	Burgundy
Royal Blue	Plum Purple	Gloss Black	Silver Gray
Candy Red	Snow White	Lime Green	Gold

Interior Colors
Brown
Tan
Red
Dark Green
Black
Burgundy
Dark Blue
Gray

CHAPTER Project

WHAT COLOR WAS THAT CAR?

In this project, you will design and conduct a survey about favorite car colors. You will use fractions, decimals, and percents to display the results of your survey, and you will find probabilities based on your survey. You will also investigate data about the production of cars in the United States.

Getting Started

- Look at the car production table. What company produced the greatest number of cars? How many total cars were produced?

- Select six common car colors and include a category labeled, "other colors." Survey 20 people and record their favorite car color. Try to survey both students and adults.

U.S. Car Production	
Company	Number of Cars
A	149,562
B	576,846
C	1,395,710
D	2,515,136
E	552,995
F	218,161
G	333,234
H	80,660
I	516,557
J	11,872
Total Production	**6,350,733**

Technology Tips

- Use a **spreadsheet** to convert fractions to decimals and percents.
- Use **computer software** to make graphs.
- Use a **word processor** to summarize your survey results.

inter**NET** CONNECTION Data Update **For up-to-date information on car production, visit:**

www.glencoe.com/sec/math/mac/mathnet

Working on the Project

You can use what you'll learn in Chapter 4 to help you conduct your survey.

Page	Exercise
157	34
164	43
168	26
179	Alternative Assessment

COOPERATIVE LEARNING

4-1A Exploring Factors

30 index cards

A Preview of Lesson 4-1

You can use number cards to discover the factors of whole numbers.

TRY THIS

Step 1 Number the index cards 1 through 30.

Step 2 In order around the classroom, give thirty students one of the index cards. Have each of them stand up and write the number 1 on the back of his or her card. (The front of each index card will have a number from 1 to 30.)

Step 3 Start with the student holding the "2" card. Have every second student sit down and write the number 2 on the back of his or her card.

Step 4 Start with the student holding the "3" card. Have every third student stand up or sit down (depending on whether the student is already sitting or standing) and write the number 3 on the back of his or her card.

Step 5 Continue this process for each of the remaining numbers until the thirtieth student has stood up or sat down and has written the number 30 on the back of his or her card.

ON YOUR OWN

1. Make a conjecture about the numbers written on the back of each number card.
2. Use your conjecture to predict the numbers that would be written on the back of number cards from 31 to 35.
3. Which number cards have exactly two numbers on the back?
4. Which number cards have the fewest numbers on the back?
5. Which number cards have the most numbers on the back?
6. Which number cards are held by those students standing at the end of the activity?
7. **Look Ahead** Suppose there were 50 students holding index cards. Which numbers do you predict would be held by students standing at the end of the activity? Why?

4-1 Divisibility Patterns

What you'll learn

You'll learn to use divisibility rules.

When am I ever going to use this?

Divisibility rules can be used to find factors of numbers.

Word Wise

factor
divisible

LOOK BACK
Refer to Lesson 1-7 to review the area of a rectangle.

How many different ways can you form a rectangle using 24 squares?

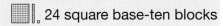

 **HANDS-ON**

MINI-LAB

Work with a partner. 24 square base-ten blocks

One possible rectangle with side lengths of 1 and 24 is shown below.

Try This

1. One partner should build a different rectangle that has 24 square units with the base-ten blocks.
2. The other partner should build a third rectangle with the 24 blocks.
3. Continue to take turns building rectangles until both partners agree that no more rectangles can be built.

Talk About It

4. What patterns do you see? How do the length and width of the rectangles you built relate to the number 24?
5. What is the area of each rectangle that you built?
6. Did you build a rectangle with a width of 5 units? Explain.

The length and width of the rectangles are the **factors** of 24.

In $24 \div 6 = 4$, the quotient, 4, is a whole number. So, we say 24 is **divisible** by 6, or 6 is a factor of 24. Study the division sentences below. What other numbers are factors of 24?

$$24 \div 1 = 24 \qquad 24 \div 6 = 4$$
$$24 \div 2 = 12 \qquad 24 \div 8 = 3$$
$$24 \div 3 = 8 \qquad 24 \div 12 = 2$$
$$24 \div 4 = 6 \qquad 24 \div 24 = 1$$

You can use the following rules to check the divisibility of numbers.

A number is divisible by:

- 2 if the digit in the ones place is even.
- 3 if the sum of the digits is divisible by 3.
- 4 if the number formed by the last two digits is divisible by 4.
- 5 if the digit in the ones place is 5 or 0.
- 6 if the number is divisible by both 2 and 3.
- 9 if the sum of the digits is divisible by 9.
- 10 if the digit in the ones place is 0.

1 Determine whether 156 is divisible by 2, 3, 4, 5, 6, 9, or 10.

   2: Yes; the ones digit, 6, is even.

   3: Yes; the sum of the digits, 12, is divisible by 3.

   4: Yes; the number formed by the last two digits, 56, is divisible by 4.

   5: No; the ones digit is not 5 or 0.

   6: Yes; the number is divisible by 2 and 3.

   9: No; the sum of the digits, 12, is not divisible by 9.

   10: No; the ones digit, 6, is not 0.

APPLICATION

Real World

2 **Money Matters** Alonso has $249 in spending money to use during his 7-day vacation. Can he spend the same amount of money each day? Use your calculator to determine whether 249 is divisible by 7.

   249 ÷ 7 = *35.57142857*

Since the quotient is not a whole number, 249 is not divisible by 7. Alonso cannot spend exactly the same amount of money each day. He will be able to spend *about* $35.57 each day.

3 Find a number that is divisible by 2, 4, 5, and 10.

The ones digit must be 0 in order for the number to be divisible by 10, which means the number will also be divisible by 2 and by 5. The number formed by the last two digits must be divisible by 4.

The numbers 120, 680, and 1,340 are just a few of the numbers that meet these requirements.

CHECK FOR UNDERSTANDING

Communicating Mathematics

Read and study the lesson to answer each question.

1. *Tell* what the figure at the right suggests about the number 15 and whether it is divisible by 4.

   · · · ·
   · · · ·
   · · · ·
   · · ·

2. *Draw diagrams* showing how 36 dots can be equally divided into 4 rows or 6 rows or 12 rows.

HANDS-ON MATH

3. *Make a model* to show the factors of 18. Use square base-ten blocks or draw rectangles on grid paper. Use rectangular arrangements to determine whether 7 is a factor of 18.

Guided Practice

Determine whether the first number is divisible by the second number.

4. 447; 3   5. 9,015; 6   6. 1,287; 9

Determine whether each number is divisible by 2, 3, 4, 5, 6, 9, or 10.

7. 712 **8.** 1,035 **9.** 8,928

10. *Recreation*
A magazine had kids test several kinds of sleds. The table shows the weights of some of the sleds. If a shipment of sleds weighs 351 pounds and contains only one kind of sled, which sleds could it contain? (*Hint:* Determine whether 351 is divisible by 2, 3, 6, or 9.)

Sled	Weight (lb)
Mega Saucer	2
Paris Glad-A-Boggan	3
Arctic Circle Steel Saucer	6
Laserluge	9

EXERCISES

Practice

Determine whether the first number is divisible by the second number.

11. 419; 3 **12.** 7,110; 5 **13.** 4,408; 4

14. 831; 3 **15.** 6,670; 4 **16.** 2,984; 9

17. 7,026; 6 **18.** 1,260; 10 **19.** 8,903; 6

Determine whether each number is divisible by 2, 3, 4, 5, 6, 9, or 10.

20. 462 **21.** 270 **22.** 1,005

23. 32,221 **24.** 8,340 **25.** 920

26. 50,319 **27.** 64,042 **28.** 75,396

29. Use your calculator to determine whether 17,136 is divisible by 8.

30. Find two numbers that are divisible by 2, 9, and 10.

Applications and Problem Solving

31. *Math History* Ancient Babylonians (who lived in what is now Iraq) used the number 60 as a base for computation and commerce, just as we use 10 as a base. There were 60 bushels in a mana and 60 mana in a talent. Some mathematicians think that the Babylonians used 60 instead of 10 because 60 is divisible by many numbers. Find all factors of 60 that are not factors of 10.

32. *History* George Washington was elected president in 1789 and re-elected in 1792. Since that time, elections are held every four years.

a. Which presidential election years since George Washington were divisible by 10?

b. Describe the pattern in the election years in part a.

c. Name the next two times in which the presidential election years will be divisible by 10.

33. *Number Puzzle* Find the mystery number from the following clues.

• This whole number is divisible by 4, 7, and 9.

• Two of its digits are the same.

• The tens digit is greater than the sum of the other two digits.

• It is less than 300, divisible by 12, and has only one odd digit.

34. *Critical Thinking* 3,075 is not divisible by 6. Rearrange the digits to form as many different numbers as possible that are divisible by 6. Use all the digits of 3,075 exactly once as you form each number.

Mixed Review

35. *Statistics* The average income for five people in an accounting department is $29,080. The individual salaries are $26,700, $23,500, $24,800, $28,300, and $42,100. Why could the average income be considered misleading to a person applying for a job? *(Lesson 3-7)*

36. *Surveys* The bar graph shows the responses to the question, "What is your favorite soft drink?" Which two soft drinks should the manager of a local grocery always keep in stock? *(Lesson 3-2)*

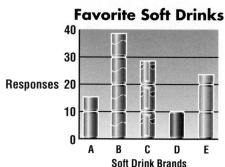

Favorite Soft Drinks

37. Write 635,000 in scientific notation. *(Lesson 2-9)*

38. *Standardized Test Practice* Which of the following expressions is equivalent to $7 \cdot 2^4$? *(Lesson 1-4)*

   A $7 \cdot 8$

   B $14 \cdot 14 \cdot 14$

   C $7 \cdot 2 \cdot 2 \cdot 2 \cdot 2$

   D $7 \cdot 7 \cdot 7 \cdot 7 \cdot 2 \cdot 2 \cdot 2 \cdot 2$

For **Extra Practice**, see page 576.

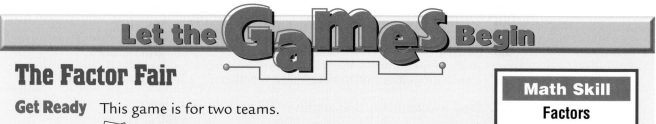

Let the Games Begin

The Factor Fair

	Math Skill
	Factors

Get Ready   This game is for two teams.

   36 index cards   tape

Get Set   Number the index cards 1 through 36. Tape the index cards in order on the chalkboard. Divide the class into two teams, Team A and Team B.

Go

- Team A chooses one of the index cards, such as the 8 card, and takes it off the chalkboard. Team A receives 8 points. Team B gets all the cards that are factors of 8 that have not yet been taken. The factors are the points they receive. In this case, Team B would receive $1 + 2 + 4$ or 7 points.

Team A	Team B
8	1
	2
	4
	7

- Team B then chooses a card and gets that many points.

- A team loses a turn if it selects an illegal number. A number is considered illegal if the opposing team does not have at least one factor available.

- Teams continue to take turns until there are no legal plays remaining. The team with the most points wins.

 Visit www.glencoe.com/sec/math/mac/mathnet for more games.

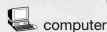
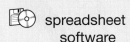

SPREADSHEETS

4-1B Divisibility

A Follow-Up of Lesson 4-1

💻 computer

💿 spreadsheet
software

You can test for divisibility by using a *spreadsheet*. A spreadsheet is made up of *cells*. A cell can contain data, labels, or formulas. In the cells in column A, enter the numbers to be tested. Then divide each number in column A by the numbers at the top of columns B, C, and D. Divisibility is indicated by whole number quotients.

TRY THIS

Work with a partner.

Use the spreadsheet to determine whether 96, 57, 108, 36, and 154 are divisible by 12, 14, or 19 by making the following substitutions.

A2 = 96, A3 = 57, A4 = 108,

A5 = 36, A6 = 154, B1 = 12,

C1 = 14, D1 = 19

	A	B	C	D
1		B1	C1	D1
2	A2	=A2/B1	=A2/C1	=A2/D1
3	A3	=A3/B1	=A3/C1	=A3/D1
4	A4	=A4/B1	=A4/C1	=A4/D1
5	A5	=A5/B1	=A5/C1	=A5/D1
6	A6	=A6/B1	=A6/C1	=A6/D1

The screen at the right shows the results of making the substitutions and running the spreadsheet. Since cell B2 contains a whole number, the number in cell A2, 96, is divisible by the number in cell B1, 12.

	A	B	C	D
1		12	14	19
2	96	8	6.857143	5.052632
3	57	4.75	4.071429	3
4	108	9	7.714286	5.684211
5	36	3	2.571429	1.894737
6	154	12.83333	11	8.105263

ON YOUR OWN

1. Which numbers are divisible by 12?

2. Which number is not divisible by 12 or 19?

3. How could you use the spreadsheet to determine whether a number is divisible by 228?

4. Change row 1 of the spreadsheet to test divisibility by 15, 18, and 23. Then change column A to test numbers 90, 253, and 574. Which numbers are divisible by 15, 18, or 23?

5. Create a spreadsheet that determines whether the numbers 1 through 100 are divisible by 2, 3, 4, 5, 6, 7, 8, 9, or 10.

Prime Factorization

What you'll learn

You'll learn to find the prime factorization of a composite number.

When am I ever going to use this?

You'll use prime factorization to determine whether numbers are composite or prime and to help simplify fractions.

Word Wise

prime number
composite number
prime factorization
factor tree

The world's most popular card game has 108, or 9 dozen, numbered cards. The numbers 9 and 12 (one dozen) are factors of 108 because $9 \times 12 = 108$.

For some numbers, such as 2, 3, 7, and 13, the only product that can be written is 1 times the number itself. These numbers are called **prime numbers**.

Prime Number	A prime number is a whole number greater than 1 that has exactly two factors, 1 and itself.

Numbers that have more than two factors, such as 9 and 12, are called **composite numbers**.

Composite Number	A composite number is a whole number greater than 1 that has more than two factors.

Notice that 0 and 1 are neither prime nor composite numbers.

Every composite number can be written as the product of prime numbers in exactly one way if you ignore the order of the factors. This product is called the **prime factorization** of the number.

Example 1

The figure formed by the steps of the factorization of 108 is called a factor tree.

Write the prime factorization of 108.

Factor Tree A

```
        108
        / \
      9 × 12
     /|   |\
   3 × 3 2 × 6
         /|
3 × 3 × 2 × 2 × 3
```

Factor Tree B

```
        108
        / \
      4 × 27
     /|   |\
   2 × 2 3 × 9
             /|
2 × 2 × 3 × 3 × 3
```

Notice that both trees give the same prime factors, except in different orders. Since 2 and 3 are prime numbers, $2 \times 2 \times 3 \times 3 \times 3$ is the prime factorization of 108. Using exponents, $2 \times 2 \times 3 \times 3 \times 3 = 2^2 \times 3^3$.

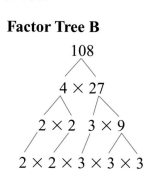

2 **Find the prime factors of 420. Then write the prime factorization.**

Method 1 Use a factor tree.

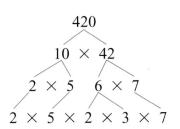

$$2 \times 5 \times 2 \times 3 \times 7$$

Method 2 Use division.

$$420 \div 2 = 210$$
$$\div 2 = 105$$
$$\div 3 = 35$$
$$\div 5 = 7$$
$$\div 7 = 1$$

The divisors are 2, 2, 3, 5, and 7.

The prime factors of 420 are 2, 3, 5, and 7. The prime factorization is $2 \times 2 \times 3 \times 5 \times 7$ or, using exponents, $2^2 \times 3 \times 5 \times 7$.

INTEGRATION **3** **Algebra** The algebraic expression $n^2 + n + 11$ may result in a prime number when n is replaced with a whole number.

a. Evaluate the expression $n^2 + n + 11$ for $n = 0, 1, 2,$ and 3. Are the resulting numbers prime or composite?

$n = 0$	$n = 1$	$n = 2$	$n = 3$
$0^2 + 0 + 11$ $= 0 + 0 + 11$ $= 11$	$1^2 + 1 + 11$ $= 1 + 1 + 11$ $= 13$	$2^2 + 2 + 11$ $= 4 + 2 + 11$ $= 17$	$3^2 + 3 + 11$ $= 9 + 3 + 11$ $= 23$

The numbers 11, 13, 17, and 23 are all prime.

b. Find the least whole number n for which the expression $n^2 + n + 11$ is not prime.

Since the expression is prime for $n = 0, 1, 2,$ and 3, we can start evaluating for $n = 4$.

- $n = 4$
  $4^2 + 4 + 11 = 31$ (prime)

- $n = 5$
  $5^2 + 5 + 11 = 41$ (prime)

- $n = 6$
  $6^2 + 6 + 11 = 53$ (prime)

- $n = 7$
  $7^2 + 7 + 11 = 67$ (prime)

- $n = 8$
  $8^2 + 8 + 11 = 83$ (prime)

- $n = 9$
  $9^2 + 9 + 11 = 101$ (prime)

- $n = 10$
  $10^2 + 10 + 11 = 121$ (composite)

So, 10 is the least whole number for which the expression is not prime.

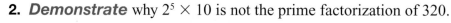

Communicating Mathematics

Read and study the lesson to answer each question.

1. *Explain* how you know that the number 387 is not a prime number.

2. *Demonstrate* why $2^5 \times 10$ is not the prime factorization of 320.

3. *Write* a paragraph explaining how you can use divisibility rules to help find prime factors.

Guided Practice

Determine whether each number is *composite* or *prime*.

4. 24   5. 19   6. 447

Use a factor tree to find the prime factorization of each number.

7. 132   8. 45   9. 288

Use your calculator to find the prime factors of each number. Then write the prime factorization of each number.

10. 66   11. 375   12. 400

13. *Algebra* Is the value of $6a + 3b$ prime or composite if $a = 5$ and $b = 1$?

EXERCISES

Practice

Determine whether each number is *composite* or *prime*.

14. 85	15. 423	16. 17	17. 333
18. 642	19. 139	20. 739	21. 1,492

Use a factor tree to find the prime factorization of each number.

22. 78	23. 96	24. 560	25. 144
26. 300	27. 990	28. 222	29. 1,700

Use your calculator to find the prime factors of each number. Then write the prime factorization of each number.

30. 166	31. 64	32. 221	33. 270
34. 146	35. 880	36. 1,740	37. 1,475

38. Find numbers that are factors of both 8 and 12.

39. Find the missing factor: $2 \times \underline{\ ?\ } \times 5^2 = 450$.

Applications and Problem Solving

40. *Contests* Sandcastle Day has been an annual event in Cannon Beach, Oregon, since 1964. Entrants from each age division are given plots measuring 81 square feet, 225 square feet, or 441 square feet, on which to build their sandcastles.

a. Find the prime factorization of 81, 225, and 441.

b. Use the prime factors to determine two possible dimensions for each plot.

41. Packaging Kien is ordering cartons to ship games to retail stores. The height, width, and length of the cartons must be measured in whole game packages. For example, a carton could be 2 packages high, 3 packages wide, and 4 packages long. Such a carton would contain 2 × 3 × 4 or 24 packages of games.

  a. List two other ways to arrange 24 packages.
  b. List three ways to arrange 36 packages in a carton.
  c. How many different ways could you arrange 30 packages?
  d. How would you arrange 17 game packages?
  e. After researching costs, Kien finds that shipping from 30-35 packages is cost effective. What number and arrangement of game packages would you recommend be shipped in each carton? Why?

42. Number Puzzle Find the mystery number from the following clues.
  • This whole number is between 30 and 40.
  • It has only two prime factors.
  • The sum of its prime factors is 5.

43. Critical Thinking Numbers that can be represented by a triangular arrangement of dots are called *triangular numbers*.

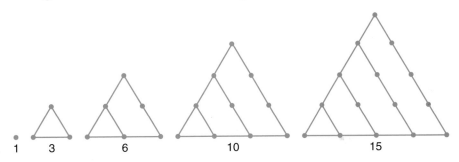

1   3   6   10   15

  a. Find the sum of any two consecutive triangular numbers.
  b. What kind of number is the sum?
  c. Determine the next two triangular numbers after 15. Verify your result in part b for these two numbers.

Mixed Review

44. Name two numbers that are divisible by both 3 and 8. *(Lesson 4-1)*

45. Standardized Test Practice Which number is an outlier in the data shown on the line plot? *(Lesson 3-6)*
  A 6
  B 9
  C 10
  D 12

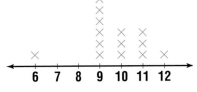

46. Measurement Express 36 milliliters as liters. *(Lesson 2-8)*

47. Racing A trainer recorded a racehorse running at 47.54 miles per hour. What is this speed rounded to the nearest mile per hour? *(Lesson 2-2)*

For **Extra Practice,** see page 577.

48. Geometry Find the area of a rectangle having a length of 8.5 inches and a width of 4 inches. *(Lesson 1-7)*

Integration: Patterns and Functions
Sequences

What you'll learn

You'll learn to recognize and extend a pattern for sequences.

When am I ever going to use this?

Sequences can help you use a simple problem to solve a more complicated one.

Word Wise

sequence
terms
arithmetic sequence
geometric sequence

Did you know Another type of cicada comes up every 13 years. How they know when to surface is still a mystery to biologists.

From North Carolina to New Hampshire, billions of cicadas (suh KAY duhz) emerged from hibernation in 1996. The last time these insects surfaced was in 1979, and biologists say they will surface again in 2013. If this pattern continues, find the next three years in which they will reappear.

Making a table or organized list can help you discover a pattern. Write the years in chronological order.

$$1979, \quad 1996, \quad 2013, \ldots \qquad 1996 - 1979 = 17$$
$$\underbrace{}_{+17} \underbrace{}_{+17} \qquad\qquad 2013 - 1996 = 17$$

The cicadas surface every 17 years, so add 17 to each year.

$$2013 + 17 = 2030 \quad\quad 2030 + 17 = 2047 \quad\quad 2047 + 17 = 2064$$

The cicadas will surface in 2030, 2047, and 2064.

A **sequence** of numbers is a list in a specific order. The numbers in the sequence are called **terms**. If you can always find the next term in the sequence by adding the same number to the previous term, the sequence is called an **arithmetic sequence**. So the years in which the cicadas surface are an arithmetic sequence.

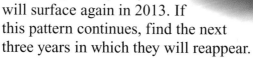

Example ❶ Identify the pattern in the sequence 4, 12, 20, 28, 36, . . . and describe how the terms are created. Then find the next three terms.

$$4, \quad 12, \quad 20, \quad 28, \quad 36, \ldots \quad 12 - 4 = 8$$
$$\underset{+8}{} \quad \underset{+8}{} \quad \underset{+8}{} \quad \underset{+8}{} \qquad 20 - 12 = 8$$
$$28 - 20 = 8$$
$$36 - 28 = 8$$

This is an arithmetic sequence in which each term after the first is created by adding 8 to the previous term.

$$36 + 8 = 44 \qquad 44 + 8 = 52 \qquad 52 + 8 = 60$$

The next three terms are 44, 52, and 60.

If you can always find the next term in the sequence by multiplying the previous term by the same number, the sequence is called a **geometric sequence**.

$$3, \quad 12, \quad 48, \quad 192 \qquad 12 \div 3 = 4$$
$$\times 4 \quad \times 4 \quad \times 4 \qquad 48 \div 12 = 4$$
$$192 \div 48 = 4$$

Example 2 Identify the pattern in the sequence 128, 64, 32, 16, . . . and describe how the terms are created. Then find the next three terms.

$$128, \quad 64, \quad 32, \quad 16, \ldots \qquad 64 \div 128 = 0.5$$
$$\times 0.5 \quad \times 0.5 \quad \times 0.5 \qquad 32 \div 64 = 0.5$$
$$16 \div 32 = 0.5$$

This is a geometric sequence in which each term after the first is created by multiplying the previous term by 0.5.

$$16 \times 0.5 = 8 \qquad 8 \times 0.5 = 4 \qquad 4 \times 0.5 = 2$$

The next three terms are 8, 4, and 2.

There are many sequences that are neither arithmetic nor geometric.

Example 3

INTEGRATION

Geometry A *diagonal* connects two nonconsecutive vertices in a figure. The number of diagonals in each figure is shown below.

| 3 sides | 4 sides | 5 sides | 6 sides |
| 0 diagonals | 2 diagonals | 5 diagonals | 9 diagonals |

a. Identify the pattern in the sequence 0, 2, 5, 9, ...
b. Find how many diagonals the next three figures would have.

a. $\quad 0, \quad 2, \quad 5, \quad 9, \ldots \qquad 2 - 0 = 2$
$$+ 2 \quad + 3 \quad + 4 \qquad 5 - 2 = 3$$
$$9 - 5 = 4$$

This sequence is neither arithmetic nor geometric. Each term after the first is created by adding 1 more than was added to the previous term.

b. $9 + 5 = 14 \qquad 14 + 6 = 20 \qquad 20 + 7 = 27$

The next three figures would have 14, 20, and 27 diagonals.

Communicating Mathematics

Read and study the lesson to answer each question.

1. *Explain* what pattern you would use to find the next term in the sequence 1, 3, 7, 13, 21, … .

2. *Write* a rule to create your own sequence. Trade your sequence with a classmate. See if you can find the rule and the next three numbers in his or her sequence.

Guided Practice

Describe the pattern in each sequence. Identify the sequence as *arithmetic*, *geometric*, or *neither*. Then find the next three terms.

3. 8, 13, 18, 23, . . .   4. 1, 2, 4, 7, 11, . . .   5. 4, 2, 1, 0.5, . . .

Create a sequence using each rule. Provide four terms for the sequence beginning with the given number. State whether the sequence is *arithmetic*, *geometric*, or *neither*.

6. Add 4 to each term; 12.

7. Multiply each term by 3; 35.

8. Add 0.2 to the first term, add 0.3 to the 2nd term, add 0.4 to the 3rd term, and so on; 17.1.

9. *Contests* Each year, a company sponsors a flying disc contest. First prize is a $1,000 U.S. savings bond. Second prize is a $500 bond, and third prize is a $250 bond. If this pattern continues, what would fourth, fifth, and sixth prizes be?

Practice

Describe the pattern in each sequence. Identify the sequence as *arithmetic*, *geometric*, or *neither*. Then find the next three terms.

10. 0, 13, 26, 39, . . .   11. 4, 12, 36, 108, . . .

12. 0.1, 0.4, 0.7, 1.0, . . .   13. 200, 202, 206, 212, . . .

14. 2.0, 3.1, 4.2, 5.3, . . .   15. 5, 15, 45, 135, . . .

16. 125, 25, 5, 1, 0.2, . . .   17. 1, 22, 333, 4,444, . . .

18. 55, 66, 77, 88, 99, . . .   19. 1, 8, 27, 64, . . .

Create a sequence using each rule. Provide four terms for the sequence beginning with the given number. State whether the sequence is *arithmetic*, *geometric*, or *neither*.

20. Add 6 to each term; 23.   21. Multiply each term by 5; 7.

22. Multiply each term by 4; 36.   23. Add 0.4 to each term; 5.

24. Add 14 to each term; 58.   25. Multiply each term by 0.1; 70.

26. Add 2 to the 1st term, add 3 to the 2nd term, add 4 to the 3rd term, and so on; 6.

27. Multiply the 1st term by 1, multiply the 2nd term by 2, multiply the 3rd term by 3, and so on; 3.

28. Find the missing terms in the sequence 12, 19, 26, ___?___, 40, 47, ___?___, ...

29. Find the missing terms in the sequence 6, 18, ___?___, 162, ___?___, 1,458, ...

30. *Earth Science* During an earthquake, about 32 times more energy is released for every increase of 1.0 on the Richter scale. (Not 10 times more energy, as was previously thought.) So, a magnitude-6.3 earthquake releases about 32 times more energy than a magnitude-5.3 earthquake. Does the energy increase represent a geometric, arithmetic, or neither type of sequence?

31. *Geometry* Refer to Exercise 43 on page 141. Write a sequence formed by the first eight triangular numbers. Then write a rule for generating the sequence.

32. *Life Science* The bamboo plant is the fastest growing plant in the world. One bamboo plant grew 36 inches in 24 hours. If the plant grew at a constant rate during the 24 hours and it was 16 inches tall at 8:00 A.M., how tall was the bamboo plant at 4:00 P.M.? (*Hint:* Find how many inches the plant grew per hour.)

33. *Critical Thinking* You can use an algebraic expression to describe the relationship between the terms in a sequence and their position in the sequence.

   a. You can find any term in an arithmetic sequence by using the expression $a + (n - 1)d$. In the expression, a is the first term, d is the common difference, and n is the position in the sequence. Find the tenth term of the sequence 3, 7, 11, 15, ...

   b. You can find any term in a geometric sequence by using the expression $a \cdot r^{n-1}$. In the expression, a is the first term, r is the common ratio, and n is the position in the sequence. Find the eighth term in the sequence 4, 12, 36, 108, ...

34. Find the prime factorization of 255. *(Lesson 4-2)*

Mixed Review

35. *Statistics* Find the range and appropriate scale and interval for the following data: 3, 17, 21, 19, 36, 15, 12, 9. Draw a number line to show the scale and interval. *(Lesson 3-1)*

36. **Standardized Test Practice** Boxes of golf balls containing 3 balls each are on sale for $3.97 per box. What is the cost of 4 boxes of golf balls before tax is added? *(Lesson 2-4)*

   A $7.97   **B** $11.91   **C** $15.88   **D** $18.88   **E** Not Here

For **Extra Practice,** see page 577.

37. *Algebra* Evaluate $3x - y \div 6$ if $x = 4$ and $y = 12$. *(Lesson 1-3)*

COOPERATIVE LEARNING

4-3B Exploring Sequences

A Follow-Up of Lesson 4-3

You can use paper folding to explore patterns in sequences.

calculator

paper

TRY THIS

Work with a partner.

1 Fold a piece of paper in half and record the number of layers of paper. (Refer to the table.)

- Shade one side of the folded paper.

- Open the piece of paper and record the fractional part of the paper that is *not* shaded. Refold the piece of paper.

- Fold your paper in half again so the unshaded side is on the outside and record the number of layers of paper.

- Shade one side of the folded paper.

- Open the piece of paper and record the fractional part of the paper that is not shaded. Completely refold the paper.

- Continue folding, shading, and recording until you can no longer fold your paper (at least five folds).

Number of Folds	Layers of Paper	Fraction of Paper Not Shaded
1	2	$\frac{1}{2}$
2	4	$\frac{1}{4}$
3	8	
4		
⋮		

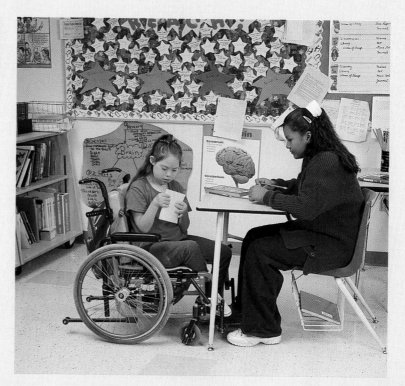

1. Examine the sequence of numbers in the "Layers" column of your table. Is this sequence arithmetic or geometric?
2. Study the sequence in the "Fraction" column. Is this sequence arithmetic or geometric?

Work with a partner.

2 Imagine continuing the paper-folding process from Activity 1 forever. Assuming that your unfolded sheet of paper is 0.002 inch thick, make a table similar to the one below for the first five folds.

Number of Folds	Thickness of the Folded Paper	
	In Layers	In Inches
1	2	0.004
2	4	0.008
3	8	
⋮		

3. How many folds would it take until the paper is as tall as you?
4. One mile is 5,280 feet. How many folds would it take until the folded paper is one mile tall?
5. How would your data change if you were folding the paper into thirds instead of halves?
6a. **Reflect Back** Write the first ten terms of the geometric sequence 3, 9, 27, … .
  b. Write the sequence of numbers formed by ones digits of successive terms in the sequence.
  c. What pattern do you see?
  d. Find the sum of the digits for each term in the sequence created in part a.
  e. Which sums are divisible by 3?
  f. Which sums are divisible by 9? Describe the pattern.

PROBLEM SOLVING

4-4A Make a List

A Preview of Lesson 4-4

Justin and Monique are on a field trip studying forest conservation. Justin is holding a pinecone, examining its scales. Let's listen in.

Monique

Have you ever noticed how the scales spiral around a pinecone?

No, I didn't think there was any kind of pattern.

Let's check it out. This pinecone has 5 spirals.

The one I have has 13 spirals!

Justin

Monique and Justin found more pinecones and counted the spirals. The chart below is an organized list of their data. You can see a pattern in each of the columns.

Pinecone	Number of Spirals	Increase from Previous Number
1	3	-
2	5	2
3	8	3
4	13	5
5	21	8

The numbers 2, 3, 5, and 8 are from the *Fibonacci sequence*. This sequence — 1, 1, 2, 3, 5, 8, 13, 21, . . . — is named after Leonardo Fibonacci, who first presented it in 1201. Notice the pattern in this sequence. Each number is the sum of the two numbers that precede it.

The 6th term in the sequence of spirals would be 13 + 21 or 34, the 7th term would be 21 + 34 or 55, and so on.

THINK ABOUT IT

Work with a partner.

1. **Write** the next three numbers in the Fibonacci sequence, 1, 1, 2, 3, 5, 8, 13, 21, 34, 55, ...

2. **Explain** in your own words how you could extend the list of Fibonacci numbers through 20 terms.

For **Extra Practice,** see page 577.

3. *Apply* what you have learned to solve the following problem.

On January 1, there was 1 pair of rabbits that lived in a pen. On February 1, a pair of bunnies was born. Now there are 2 pairs of rabbits: 1 adult pair and 1 baby pair. By March 1, the original parents have had another pair of bunnies and the original bunnies have grown into adult rabbits. Now there are 3 pairs of rabbits: 2 adult pairs and 1 baby pair. The next month the adult pairs each have a pair of babies, while the previous month's babies grow to adulthood. The pattern continues. How many total pairs of rabbits will there be by June 1? Describe the pattern.

ON YOUR OWN

4. The second step of the 4-step plan for problem solving asks you to *plan* the solution. *Explain* how you can **make a list** to help you plan a solution to a problem.

5. *Write a Problem* that you can solve by making a list. Explain your answer.

6. *Find* the greatest number by which each term of the sequence 2, 2, 4, 6, 10, 16, . . . is divisible.

MIXED PROBLEM SOLVING

STRATEGIES

Look for a pattern.
Solve a simpler problem.
Act it out.
Guess and check.
Draw a diagram.
Make a chart.
Work backward.

Solve. Use any strategy.

7. **History** During the 3rd century B.C., Greek mathematician Eratosthenes calculated that Earth's circumference was 250,000 *stades*. One mile is approximately 10 stades. If the actual circumference of Earth is about 24,901 miles, was Eratosthenes' calculation reasonable? Explain.

8. *Geography* Mount Everest, the tallest mountain on Earth, is 8,872 meters tall. Olympus Mons, the tallest mountain on Mars, is 23,775 meters tall. How many times taller is Olympus Mons than Mount Everest?

9. *Physical Science* Telephone calls travel through optical fibers at the speed of light, which is 186,000 miles per second. A millisecond is 0.001 of a second. How far can your voice travel over an optical line in 1 millisecond?

10. *Number Theory* Find two numbers that when added, equal 56, and when multiplied, equal 783.

11. *Aircraft* A Boeing 747 Jumbo Jet, the largest capacity jetliner, is 70.51 meters long. The Stits Skybaby, the smallest fully functional aircraft, is only 2.794 meters long. How many Skybabies, set end-to-end, would it take to equal the length of the 747?

12. **Standardized Test Practice** The table shows the five longest roller coasters at Paramount's Kings Island in Cincinnati, Ohio. Which is the best estimate of the total length of the five coasters?

Roller Coaster	Length (ft)
The Beast	7,400
Vortex	3,800
Racer	3,415
Adventure Express	2,963
TOP GUN	2,352

A 2,000 ft   B 12,000 ft

C 20,000 ft   D 40,000 ft

E 48,000 ft

4-4 Greatest Common Factor

What you'll learn

You'll learn to find the greatest common factor of two or more numbers.

When am I ever going to use this?

You'll use greatest common factors to simplify fractions.

Word Wise

greatest common factor (GCF)

In 1903, a company introduced its first box of 8 crayons. In 1958, the 64-box of crayons, with a built-in sharpener, was introduced. Thirty-five years later, the box of 96 crayons came out. Suppose the company is designing boxes so that all sets have the same number of crayons in each row. What would be the greatest number of crayons in each row? *This problem will be solved in Example 4.*

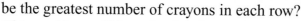

You can determine the **greatest common factor (GCF)** of two or more numbers by making an organized list. The greatest common factor of two or more numbers is the greatest number that is a factor of each number.

One of the following methods can be used to find the GCF.

Method 1 List the factors of each number. Then identify the common factors. The greatest of these common factors is the GCF.

Method 2 Write the prime factorization of each number. Then identify all common prime factors and find their product, the GCF.

Example ① **Find the GCF of 27 and 36.**

The GCF of two prime numbers is 1.

Method 1 List the factors.
factors of 27: 1, 3, 9, 27
factors of 36: 1, 2, **3**, 4, 6, 9, 12, 18, 36

common factors: 1, 3, 9

Thus, the GCF of 27 and 36 is 9.

Method 2 Write the prime factorization.

$$
\begin{array}{cc}
27 & 36 \\
3 \times 9 & 3 \times 12 \\
3 \times 3 \times 3 & 3 \times 3 \times 4 \\
 & 3 \times 3 \times 2 \times 2
\end{array}
$$

common prime factors: 3, 3

Thus, the GCF of 27 and 36 is 3×3 or 9.

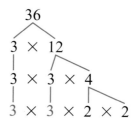

2 **Find the GCF of 105, 63, and 42.**

Method 1 List the factors.

factors of 105: 1, 3, 5, 7, 15, 21, 35, 105
factors of 63: 1, 3, 7, 9, 21, 63
factors of 42: 1, 2, 3, 6, 7, 14, 21, 42

Since the common factors are 1, 3, 7, and 21, the GCF is 21.

Method 2 Write the prime factorization.

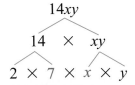

$$
\begin{array}{ccc}
105 & 63 & 42 \\
15 \times 7 & 9 \times 7 & 6 \times 7 \\
3 \times 5 \times 7 & 3 \times 3 \times 7 & 2 \times 3 \times 7
\end{array}
$$

Since the common prime factors of 105, 63, and 42 are 3 and 7, the GCF is 3×7 or 21.

INTEGRATION **3** **Algebra** **Find the GCF of $14xy$ and $7x^2$.**

Write the factors of each expression.

$$
\begin{array}{cc}
14xy & 7x^2 \\
14 \times xy & 7 \times x^2 \\
2 \times 7 \times x \times y & 7 \times x \times x
\end{array}
$$

The common prime factors are 7 and x.
Thus, the GCF of $14xy$ and $7x^2$ is $7x$.

APPLICATION **4** **Packaging** Refer to the beginning of the lesson.

a. What would be the greatest number of crayons in each row of an 8-, a 64-, and a 96-crayon box if all rows have the same number?

b. How many rows will there be in each box?

a. Find the GCF of 8, 64, and 96.

$8 = 2 \times 2 \times 2$
$64 = 2 \times 2 \times 2 \times 2 \times 2 \times 2$
$96 = 2 \times 2 \times 2 \times 2 \times 2 \times 3$

The GCF is $2 \times 2 \times 2$ or 8. Thus, there would be 8 crayons in each row.

b. Since there are 8 crayons in each row, the 8-crayon box will have $8 \div 8$ or 1 row, the 64-crayon box will have $64 \div 8$ or 8 rows, and the 96-crayon box will have $96 \div 8$ or 12 rows.

Communicating Mathematics

Read and study the lesson to answer each question.

1. *Explain* how to find the greatest common factor of 308 and 210 by using the factor trees at the right. Then find the GCF.

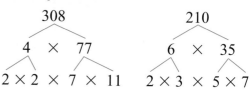

2. *Draw* factor trees for 240 and 360, and then circle common factors. What is the GCF of 240 and 360?

Guided Practice

Find the GCF of each set of numbers by listing factors.

3. 20, 24   4. 18, 30   5. 6, 8, 12

Find the GCF of each pair of numbers by listing common prime factors.

6. $80 = 2^4 \times 5$
$150 = 2 \times 3 \times 5^2$

7. $45 = 3^2 \times 5$
$60 = 2^2 \times 3 \times 5$

Find the GCF of each pair of numbers by writing prime factorizations.

8. 30, 48   9. 125, 40   10. 12, 90

11. *Algebra* Find the GCF of the terms in the expressions $15a^2$ and $10a$.

Practice

Find the GCF of each set of numbers by listing factors.

12. 33, 121   13. 28, 84   14. 96, 56

15. 6, 10, 12   16. 18, 42, 60   17. 36, 50, 130

Find the GCF of each pair of numbers by listing common prime factors.

18. $18 = 2 \times 3^2$
$54 = 2 \times 3^3$

19. $60 = 2^2 \times 3 \times 5$
$27 = 3^3$

20. $36 = 2^2 \times 3^2$
$105 = 3 \times 5 \times 7$

21. $90 = 2 \times 3^2 \times 5$
$126 = 2 \times 3^2 \times 7$

Find the GCF of each pair of numbers by writing prime factorizations.

22. 12, 78   23. 40, 50   24. 45, 75

25. 100, 30   26. 45, 54   27. 120, 72

Find the GCF of each pair of numbers. Use any method you prefer.

28. 28, 77   29. 132, 108   30. 9, 10

31. 14, 33   32. 16, 36   33. 65, 91

34. Find the GCF of 82 and 28.

35. Name two different pairs of numbers whose GCF is 26.

36. Patterns Find the GCF of all the numbers in the sequence 24, 30, 36, 42, 48, 54, ...

37. Number Sense 14 and 15 are *relatively prime* because their GCF is 1. Find the least composite numbers that are relatively prime.

38. Calendars Some ancient civilizations used the moon to calculate a lunar month of 30 days. The Maya used the Sun to calculate the solar year of 365 days.

   a. Find the GCF of these two numbers.

   b. Find how many days off the first year would be using lunar months.

Mayan astronomical observatory

39. Critical Thinking Make a sieve of Eratosthenes up to 100. You can use the Internet to help you research.

Mixed Review

40. Patterns Identify the pattern in the sequence 13, 26, 52, 104, ... and find the next three terms. *(Lesson 4-3)*

41. Standardized Test Practice The tables in the school cafeteria are arranged so there are always the same number of tables in each row. If the tables can be arranged in 4 rows, 5 rows, or 6 rows, what is the fewest number of tables in the cafeteria? *(Lesson 4-1)*

  A 15   **B** 40   **C** 60   **D** 120

42. Nutrition The following data are the grams of carbohydrates in 15 different energy bars. Find the mean, mode, and median. *(Lesson 3-4)*

    16, 15, 20, 24, 16, 16, 16, 2, 20, 26, 14, 20, 20, 16, 16

43. Statistics Round 17.9 to the nearest whole number. *(Lesson 2-2)*

44. Evaluate the expression $24 - 12 \div 3 \cdot 5 + 6$. *(Lesson 1-2)*

For **Extra Practice,** see page 578.

CHAPTER 4

Mid-Chapter Self Test

Determine whether each number is divisible by 2, 3, 4, 5, 6, 9, or 10. *(Lesson 4-1)*

  1. 370   **2.** 2,004   **3.** 135

Use a factor tree to find the prime factorization. *(Lesson 4-2)*

  4. 106   **5.** 330   **6.** 184

Describe the pattern in each sequence. Identify the sequence as *arithmetic*, *geometric*, or *neither*. Then find the next three terms. *(Lesson 4-3)*

  7. 9, 14, 19, 24, ...   **8.** 0, 1, 3, 6, 10, ...   **9.** 12, 24, 48, 96, 192, ...

10. Algebra Find the GCF of the terms in the expressions $18w$ and $12w^3$. *(Lesson 4-4)*

4-5

Simplifying Fractions and Ratios

What you'll learn

You'll learn to express fractions and ratios in simplest form.

When am I ever going to use this?

Simplest form lets you easily compare fractions and ratios.

Word Wise

ratio
simplest form

Artists have painted Martian landscapes based on actual photographs. Suppose 4 lumps of yellow paint is mixed with 16 lumps of red paint to create a red-orange color for a Martian sunset. Then the yellow-to-red ratio is 4 to 16.

The amounts of colors are compared by using a **ratio**. A ratio is a comparison of two numbers by division. The expressions below represent the ratio used.

4:16   4 to 16

Ratios can also be expressed as fractions. In this case, the fraction is $\frac{4}{16}$.

You can simplify ratios and fractions like $\frac{4}{16}$ by finding the GCF. A fraction is in **simplest form** when the GCF of the numerator and denominator is 1. To express a fraction in simplest form:

- find the GCF of the numerator and the denominator,
- divide the numerator and the denominator by the GCF, and
- write the resulting fraction.

Let's write the fraction $\frac{4}{16}$ in simplest form.

Study Hint

Reading Math The ratio 1:4 is read as 1 to 4.

factors of 4: 1, 2, 4
factors of 16: 1, 2, 4, 8, 16   The GCF of 4 and 16 is 4.

$$\frac{4}{16} = \frac{4 \div 4}{16 \div 4} = \frac{1}{4} \text{ or } 1:4$$

$$\frac{4}{16}$$ $$\frac{1}{4}$$

This means that for each lump of yellow, the artist uses 4 lumps of red.

Example

1 Express $\frac{24}{36}$ in simplest form.

$24 = 2 \times 2 \times 2 \times 3$
$36 = 2 \times 2 \times 3 \times 3$

GCF: $2 \times 2 \times 3$ or 12

Now divide the numerator and denominator by the GCF.

$$\frac{24}{36} = \frac{24 \div 12}{36 \div 12} = \frac{2}{3}$$

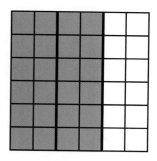

$$\frac{24}{36} = \frac{2}{3}$$

If the numerator and denominator do not have any common factors other than 1, then the fraction is in simplest form.

2 Express $\frac{54}{84}$ in simplest form.

factors of 54: 1, 2, 3, 6, 9, 18, 27, 54
factors of 84: 1, 2, 3, 4, 6, 7, 12, 14, 21, 28, 42, 84

The GCF of 54 and 84 is 6.

$$\frac{54}{84} = \frac{54 \div 6}{84 \div 6} = \frac{9}{14}$$

CONNECTION

3 **Earth Science** Australia's Great Barrier Reef is the world's largest structure of living organisms. It is home to about 410 of the 2,290 species of coral in the world. Express the ratio 410:2,290 in simplest form.

First, find the GCF.

factors of 410: 1, 2, 5, 10, 41, 82, 205, 410
factors of 2,290: 1, 2, 5, 10, 229, 458,
                    1,145, 2,290

The GCF of 410 and 2,290 is 10.

Then, write the ratio as a fraction and simplify.

$$\frac{410}{2,290} = \frac{410 \div 10}{2,290 \div 10} = \frac{41}{229}$$

In simplest form, the ratio is 41:229. So there are 41 species of coral in the Great Barrier Reef for every 229 species found in the world.

CHECK FOR UNDERSTANDING

Communicating Mathematics

Read and study the lesson to answer each question.

1. *Explain* why the method shown below does not produce an equivalent fraction in simplest form.

$$\frac{16}{36} = \frac{1\cancel{6}}{3\cancel{6}} = \frac{1}{3}$$

2. *Name* three equivalent fractions that are modeled at the right.

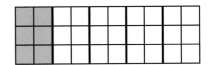

Guided Practice

Express each fraction or ratio in simplest form.

3. $\frac{20}{24}$ **4.** 18:30   **5.** $\frac{14}{35}$ **6.** 81:90

7. Write two fractions that can be expressed in simplest form as $\frac{2}{9}$.

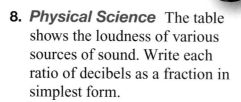

8. **Physical Science** The table shows the loudness of various sources of sound. Write each ratio of decibels as a fraction in simplest form.

Source of Sound	Number of Decibels
Niagara Falls	85
car horn	110
racing cars	125
amplified music	130
jet aircraft takeoff	140

Source: *The Sizesaurus,* Stephen Strauss

a. Niagara Falls to amplified music
b. car horn to jet aircraft takeoff
c. racing cars to amplified music

EXERCISES

Practice

Express each fraction or ratio in simplest form.

9. $\frac{35}{45}$

10. 150:350

11. $\frac{61}{102}$

12. 56:96

13. 16:32

14. $\frac{50}{75}$

15. 32:70

16. $\frac{17}{35}$

17. 44:200

18. $\frac{44}{160}$

19. 200:500

20. $\frac{80}{96}$

21. $\frac{48}{72}$

22. 117:9

23. $\frac{64}{80}$

24. 99:66

Write two different fractions that can be expressed in simplest form as each of the following.

25. $\frac{3}{4}$

26. $\frac{4}{7}$

27. $\frac{5}{9}$

28. Write two ratios that can be expressed in simplest form as 3:10.

29. Express 16:26 in simplest form.

Applications and Problem Solving

30. **Music** A *sitar* is a South Asian instrument that has two sets of strings. One type of sitar has 6 top strings and 16 bottom strings. Express the number of top strings as a fraction of the total number of strings. Write the fraction in simplest form.

31. **Cycling** A *gear ratio* of a bike is the comparison of the number of teeth on a chainwheel to the number on a freewheel. If the gear ratio for a 10-speed bike is 52:16, write this as a ratio in simplified form.

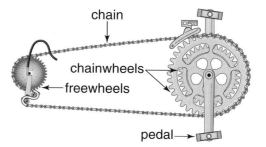

32. **Life Science** The human body has 60,000 miles of blood vessels. There are 3,000 miles across the United States from east to west. Express the miles of blood vessels to the miles across the United States as a ratio in simplest form.

33. *Exploration* A team of six explorers took three dogsleds on a 2,000-mile trip across the frozen Arctic Ocean. The table shows the weight of communication equipment needed for such a trip.

Item	Weight (lb)
radio	40
laptop computer	6
computer batteries	5
electrical repair kit	2
satellite equipment	9
satellite batteries	13
battery recharger	10

Write the weight of each item as a fraction of the total in simplest form.

a. radio

b. laptop computer

c. battery recharger

34. *Working on the* CHAPTER Project Refer to the table on page 131.

a. Round the number of cars produced by each company to the nearest hundred thousand. Make a table showing this information.

b. The total car production was approximately 6,000,000. Express the rounded numbers in part a as a fraction of 6,000,000. Write the fraction in simplest form. Include this information in your table.

35. *Critical Thinking* Explain how you could find a fraction in which the numerator and denominator are greater than 100 and they have a common factor of 13.

Mixed Review

36. Standardized Test Practice Find the GCF of 126 and 420. *(Lesson 4-4)*

    A 52,900   **B** 1,260   **C** 546   **D** 42

37. *Earth Science* The table shows the actual thermometer reading in °C and the corresponding wind chill temperatures when the wind is blowing 20 miles per hour. Find the missing temperatures. *(Lesson 4-3)*

Temperature	
Actual	**Wind Chill**
0°	−38°
5°	−31°
10°	−24°
15°	?
20°	−10°
25°	?
30°	4°

38. *Transportation* Isabel recorded the number of minutes it took her to drive to work each day for a week. Find the mean and median for the following times: 12, 23, 10, 14, and 11. *(Lesson 3-4)*

For **Extra Practice**, see page 578.

39. Multiply 1,000 and 18.7. *(Lesson 2-5)*

Ratios and Percents

What you'll learn

You'll learn to illustrate the meaning of percent using models or symbols.

When am I ever going to use this?

Percents are often used to illustrate articles in the newspaper.

Word Wise

percent

Think of a state that has the most of its area as water. You may have thought of Florida or Minnesota. But the state with the most water is Michigan! In Michigan, 43 out of every 100 square miles of territory on average is under water. You can write this ratio as a **percent**.

HANDS-ON MINI-LAB

Work with a partner. grid paper 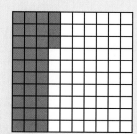 colored pencils

Try This

- Draw a 10 × 10 square on a piece of grid paper.
- Shade the small squares like the model at the right that represents 33 percent.

Talk About It

1. How many small squares are in the model?
2. How many small squares are shaded?
3. Write a ratio of shaded squares to squares in the model.
4. If the model represents 33 percent, write a conjecture about the meaning of the word *percent*.

Percent	Words:	A percent is a ratio that compares a number to 100.
	Symbols:	$\frac{n}{100} = n\%$ The symbol % means *percent*.

Examples

Write a percent to represent the shaded area of each model.

1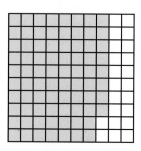

$$\frac{78}{100} = 78\%$$

2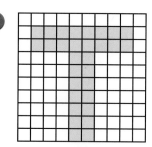

$$\frac{30}{100} = 30\%$$

Examples

Express each ratio as a percent.

3 $\dfrac{52}{100}$

$\dfrac{52}{100} = 52\%$

4 14.4 out of 100

$\dfrac{14.4}{100} = 14.4\%$

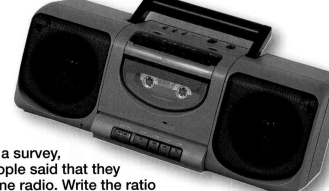

5 $88\dfrac{1}{4}$ per 100

$\dfrac{88\frac{1}{4}}{100} = 88\dfrac{1}{4}\%$

APPLICATION

Real World

6 **Technology** In a survey, 99 out of 100 people said that they owned at least one radio. Write the ratio as a percent.

99 out of 100 $= \dfrac{99}{100}$ or 99%

CHECK FOR UNDERSTANDING

Communicating Mathematics

Read and study the lesson to answer each question.

1. *Explain* in your own words the meaning of percent.

2. *Write* a percent that means 5 out of 100.

HANDS-ON MATH

3. *Draw* a model to show 25%.

Guided Practice

Write a percent to represent the shaded area.

4.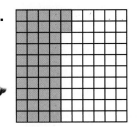

5.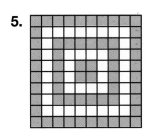

Express each ratio as a percent.

6. 18:100

7. 34 teams out of 100

8. $14.50 per $100

9. *Sports* For every 100 people who play soccer, 77 are under the age of 18. Write this ratio as a percent.

Practice

Write a percent to represent the shaded area.

10.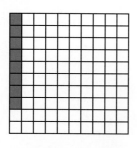

11.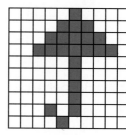

12.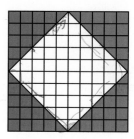

13.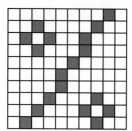

14.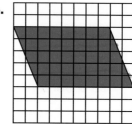

15.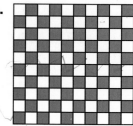

Express each ratio as a percent.

16. $\frac{44}{100}$

17. 57 out of 100

18. 33:100

19. 38.4 to 100

20. $66\frac{2}{3}$:100

21. 1 in 100

22. $89 per $100

23. 15 acres in 100

24. 4 species in 100

25. Write 36 track team members in 100 students as a percent.

26. Express a $7.50 donation from a $100 paycheck as a percent.

Applications and Problem Solving

27. *Volunteering* Do you give your time to a favorite charity? Of the 100 teens who answered a women's magazine survey, 61 volunteer more than 3 hours per week. What percent of those who were surveyed volunteer? **Source:** *Ladies Home Journal*

28. *Geometry* Square *A* measures 10 units on each side. Square *B* measures 6 units on each side. Write a percent for the ratio of the area of square *B* to the area of square *A*.

29. *Critical Thinking* What percent of the model is shaded?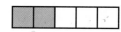

Mixed Review

30. Express $\frac{10}{24}$ in simplest form. *(Lesson 4-5)*

For **Extra Practice**, see page 578.

31. *School* Mr. Eppick likes to have his math class work in groups of at least two students. If there are 36 students in the class, list all the ways he can arrange the students in equal-sized groups. *(Lesson 4-4)*

32. Solve $b = 12.31 \times 10^4$. *(Lesson 2-5)*

33. **Standardized Test Practice** How many square feet of carpet are needed to cover a floor that measures 12 feet by 15 feet? *(Lesson 1-7)*

    A 54 ft²   **B** 90 ft²   **C** 108 ft²   **D** 180 ft²

4-7

What you'll learn

You'll learn to express fractions as percents, and percents and decimals as fractions.

When am I ever going to use this?

Fractions, decimals, and percents are used interchangeably in real-life situations such as the stock market and department store sales.

Have you ever wondered why you can't live without water? It's because your body is mostly water. Brain matter is $\frac{17}{20}$ water. What percent of brain matter is water?

Since a percent is a ratio of a number to 100, you can express $\frac{17}{20}$ as a percent by finding an equivalent fraction with a denominator of 100.

$$\frac{17}{20} \overset{\times 5}{=} \frac{85}{100} = 85\%$$

Since $100 \div 20 = 5$, multiply the numerator and denominator by 5.

So, 85% of the brain is water.

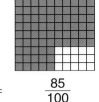

$$\frac{17}{20} = \frac{85}{100}$$

Example ① Express $\frac{3}{5}$ as a percent.

In Chapter 8, you will express fractions like $\frac{1}{3}$, $\frac{3}{8}$, and $\frac{2}{9}$ as percents.

$$\frac{3}{5} \overset{\times 20}{=} \frac{60}{100} = 60\%$$

Since $100 \div 5 = 20$, multiply the numerator and denominator by 20.

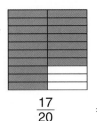

$$\frac{3}{5} = \frac{60}{100}$$

You can express a percent as a fraction by writing the number over 100 and simplifying.

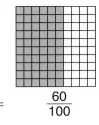

Example ②
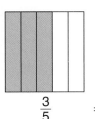
APPLICATION

Tipping Author Irene Frankel claims that the word "tip" originated in 18th-century England, where coffeehouse patrons put coins in a box labeled "To Insure Promptness." A standard tip for food service is 15%. Express this as a fraction in simplest form.

$$15\% = \frac{15}{100} \quad \text{The GCF of 15 and 100 is 5.}$$
$$= \frac{15 \div 5}{100 \div 5} \text{ or } \frac{3}{20}$$

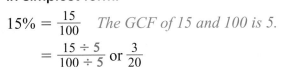

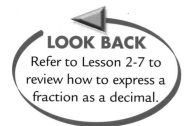

LOOK BACK

Refer to Lesson 2-7 to review how to express a fraction as a decimal.

In Chapter 2, you learned how to express a fraction as a decimal. You can also express decimals as fractions. Express the decimal as a fraction with a denominator of a power of 10. This is indicated by the place value of the final digit of the decimal. Then simplify the fraction.

$$0.14 = \frac{14}{100} = \frac{7}{50}$$ *The GCF of 14 and 100 is 2.*

Examples

Express each decimal as a fraction.

3 0.6

$$0.6 = \frac{6}{10} \text{ or } \frac{3}{5}$$

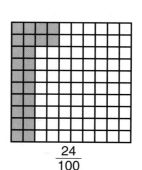

$\frac{6}{10}$

$\frac{3}{5}$

4 0.24

$$0.24 = \frac{24}{100} \text{ or } \frac{6}{25}$$

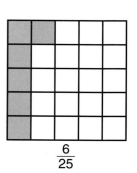

$\frac{24}{100}$ $\frac{6}{25}$

CHECK FOR UNDERSTANDING

Communicating Mathematics

Read and study the lesson to answer each question.

1. *Express* the shaded portion of the model at the right as a percent and as a fraction in simplest form.

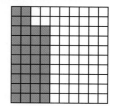

2. *Describe* two real-life situations in which you would use percents rather than fractions or decimals. (*Hint:* Recipes use fractions, such as $\frac{1}{2}$ cup, rather than percents or decimals.)

3. *You Decide* Ed thinks that 0.250 and 0.025 equal the same fraction. Juliana doesn't think they do. Who is correct? Explain your reasoning.

Guided Practice

Express each fraction as a percent.

4. $\frac{1}{2}$ **5.** $\frac{13}{20}$ **6.** $\frac{19}{25}$ **7.** $\frac{9}{10}$

Express each percent or decimal as a fraction in simplest form.

8. 45%   **9.** 0.22   **10.** 90%   **11.** 0.8

12. **Food** The graph shows the percent of family income spent on food for five different countries. Write each percent as a fraction in simplest form.

a. India **b.** China

c. Mexico **d.** Japan

e. USA

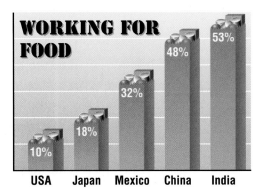

WORKING FOR FOOD

53% 48% 32% 18% 10%

USA Japan Mexico China India

Source: ACA Education Foundation, Inc.

EXERCISES

Practice   **Express each fraction as a percent.**

13. $\frac{37}{50}$ **14.** $\frac{1}{4}$ **15.** $\frac{1}{10}$ **16.** $\frac{3}{5}$

17. $\frac{91}{100}$ **18.** $\frac{1}{5}$ **19.** $\frac{3}{10}$ **20.** $\frac{23}{25}$

21. $\frac{3}{4}$ **22.** $\frac{7}{10}$ **23.** $\frac{11}{20}$ **24.** $\frac{8}{8}$

Express each percent or decimal as a fraction in simplest form.

25. 25%   **26.** 0.30   **27.** 80%   **28.** 0.15

29. 0.9   **30.** 76%   **31.** 0.32   **32.** 52%

33. 10%   **34.** 0.62   **35.** 0.02   **36.** 17%

37. Express $\frac{5}{25}$ as a percent.

38. Express *thirty-five percent* as a fraction in simplest form.

Applications and Problem Solving

39. **Life Science** Cats sleep as much as 18 hours per day. Find the percent of the day that cats spend sleeping.

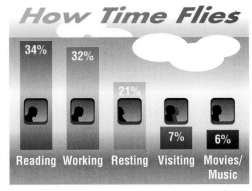

How Time Flies

34% 32% 21% 7% 6%

Reading Working Resting Visiting Movies/Music

Source: *Frequent Flyer* survey

40. **Transportation** The graph shows how people spend their time in flight. Write each percent of time as a fraction in simplest form.

a. Reading

b. Working

c. Resting

d. Visiting

e. Movies/Music

41. *History* In 1896, about 100,000 "stampeders" set out for the Klondike in Bonanza Creek, Canada, in search of gold. Only about 30,000 people made it to the Klondike.

   a. What fraction of the stampeders made it? Express the fraction in simplest form.

   b. Express the fraction as a decimal.

42. *History* Twenty-six states in America have names that come from Native American languages. What percent of the states is this?

43. *Working on the* CHAPTER Project Refer to Exercise 34 on page 157. Express each fraction in your table as a decimal rounded to the nearest hundredth and then as a percent. Add two more columns to your table labeled "Decimal" and "Percent" to display this information.

44. *Critical Thinking* Explain why 0.5% is not the same as $\frac{1}{2}$.

Mixed Review

45. Express 47 out of 100 as a percent. *(Lesson 4-6)*

46. Standardized Test Practice Which expression represents the prime factorization of 126? *(Lesson 4-2)*

   A $2^2 \times 3 \times 7$ **B** $2 \times 3^2 \times 7$ **C** $2 \times 3 \times 7^2$ **D** $2^2 \times 3^2 \times 7^2$

47. *Statistics* Construct a stem-and-leaf plot for the following data: 12, 17, 23, 5, 9, 25, 13, 16, 2, 25, 31, 10. *(Lesson 3-5)*

48. *Entertainment* The table shows the profits of the five re-released movies in millions of dollars. *(Lesson 2-9)*

Play It Again....	
Movie A	$125.0
Movie B	$70.6
Movie C	$60.8
Movie D	$46.6
Movie E	$44.6

   a. Write how much the re-release of Movie A earned in standard form.

   b. How much money did Movie C, Movie D, and Movie E earn altogether by their re-releases? Write in standard form.

49. *Sports* A triathalon competition consists of swimming 3 miles, running 10 miles, and bicycling 35 miles. How many miles does an athlete travel during the competition? *(Lesson 1-1)*

For **Extra Practice,** see page 579.

4-8

Integration: Probability
Simple Events

What you'll learn

You'll learn to find the probability of a simple event.

When am I ever going to use this?

You can use probability to find the chances of winning a raffle held by your school.

Word Wise

probability
event
random

The makers of a popular jelly bean recommend combining beans to make gourmet flavors. For example, very cherry and chocolate pudding beans make "chocolate covered cherries." If a bag of 30 jelly beans contains 18 very cherry beans, what is the **probability** that you will get a very cherry bean if you pick one from the bag at random?

The **event**, or specific outcome, we are interested in is a very cherry jelly bean.

Probability	**Words:**	The probability of an event is the ratio of the number of ways an event can occur to the number of possible outcomes.
	Symbols:	$P(\text{event}) = \dfrac{\text{number of ways event occurs}}{\text{number of possible outcomes}}$

Study Hint

Reading Math $P(A)$ is read as *the probability that A occurs.*

Outcomes occur at **random** if each outcome is equally likely to occur. Since a jelly bean will be drawn at random,

$$P(\text{very cherry}) = \frac{\text{number of very cherry jelly beans}}{\text{total number of jelly beans}}$$

$$= \frac{18}{30}$$

$$= \frac{3}{5} \quad \textit{Express the fraction in simplest form.}$$

The probability of choosing a very cherry jelly bean is $\frac{3}{5}$ or 60%.

Example

1 Find the probability of drawing a card with a factor of 12 on it from a deck of cards numbered 1 to 24. Express the fraction in simplest form.

$$P(\text{factor of 12}) = \frac{\text{number of ways of drawing a card that is a factor of 12}}{\text{number of ways a card can be drawn}}$$

$$= \frac{6}{24} \quad \begin{array}{l} \leftarrow \textit{There are 6 factors of 12: 1, 2, 3, 4, 6, and 12.} \\ \leftarrow \textit{There are 24 cards.} \end{array}$$

$$= \frac{1}{4} \quad \text{The GCF of 6 and 24 is 6.}$$

The probability of drawing a card with a factor of 12 is $\frac{1}{4}$ or 25%.

Example

Real World APPLICATION

Drawings A department store has a weekly drawing in which the winner receives a $50 gift certificate. About 300 customers register each day. If Yvonne registered 3 times this week, what is the probability that she will win this week?

number of cards: $300 \times 7 = 2{,}100$

number of cards with Yvonne's name on them: 3

$$P(\text{Yvonne}) = \frac{\text{number of cards with Yvonne's name}}{\text{total number of cards}}$$

$$= \frac{3}{2{,}100}$$

$$= \frac{1}{700}$$

The probability of Yvonne's name being drawn is $\frac{1}{700}$.

The probability that an event will happen is somewhere between 0 and 1.

- A probability of 0 means that the event is impossible.
- A probability of 1 means that the event is certain to happen.
- The closer a probability is to 1, the more likely the event is to happen.

You can express the probability of an event as a fraction, decimal, or percent. Since probability is a number, you can picture it on a number line.

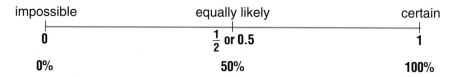

impossible   equally likely   certain

0                  $\frac{1}{2}$ or 0.5   1

0%   50%   100%

HANDS-ON

MINI-LAB

Work with a partner.    paper   ruler

Try This

1. Draw a number line like the one above.
2. Locate each of the situations on the number line.
   a. The meteorologist predicts a 75% chance for rain today.
   b. A tossed coin will come up heads.
   c. You will have math homework tonight.
   d. February will have 31 days this year.
   e. The sum of two number cubes will be greater than 1.

Talk About It

3. Order the events from least likely to most likely.
4. Write three events of your own. Add them to your number line.

Communicating Mathematics

Read and study the lesson to answer each question.

1. Refer to the beginning of the lesson. What is the probability that if you choose a jelly bean from the bag of 30 beans, that it is *not* a very cherry jelly bean?

2. **Give** an example of an event that has a probability of 0 and an event that has a probability of 1 when rolling a number cube.

3. **Draw** a spinner with three regions labeled A, B, and C. Make the spinner so that there is a 50% chance that the spinner will stop on A, a 25% chance that the spinner will stop on B, and a 25% chance that the spinner will stop on C.

Guided Practice

The spinner shown is equally likely to stop on each of its regions numbered 1 to 10. Find the probability that the spinner will stop on each of the following.

4. an even number

5. a multiple of 3

A package contains 7 bags of tortilla chips, 3 bags of cheese puffs, and 4 bags of potato chips. If you reach in the package and choose one bag at random, what is the probability that you will select each of the following? Express each ratio as a fraction in simplest form and as a percent.

6. tortilla chips

7. cheese puffs

8. **Technology** The graph shows the cost of 22 portable CD players that were tested by a magazine recently. If one of the CD players from the test is chosen at random, what is the probability that it costs between $211-260?

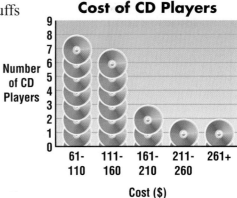

Cost of CD Players

Practice

A certain spinner is equally likely to stop on each of its regions numbered 1 to 20. Find the probability that the spinner will stop on each of the following.

9. the GCF of 12 and 18

10. a number greater than 6

11. a multiple of 2 and 3

12. an odd number

13. a prime number

14. a factor of 12

A bag of marbles contains 16 blue, 8 green, 9 red, 12 yellow, and 5 black marbles. If you reach in the bag and draw one marble at random, what is the probability that you will draw each of the following? Express each ratio as a fraction in simplest form and as a percent.

15. a green marble

16. a black marble

17. a yellow marble

18. a blue marble

19. either a green or a blue marble

20. a red or a black marble

21. All of the factors of 48 are written on separate cards. If you randomly choose a card, what is the probability of choosing a prime number?

22. There are 52 cards in a deck. Four of these are kings. What is the probability of drawing a king?

Applications and Problem Solving

23. *Reading* More than half of all books bought by Americans are fiction. A bookshelf contains the following fiction books: 34 general fiction, 15 children's books, 6 romance, 6 science fiction, 5 mystery, and 4 horror. If you randomly choose a book from the shelf, what is the probability that it is a mystery?

24. *School* Your school is raffling a television set and a total of 750 tickets have been sold. Your family bought 15 tickets.

   a. What is the probability that your family will win the television set?

   b. What is the probability that your family will *not* win the television set?

25. *Crayons* Refer to the beginning of Lesson 4-4 on page 150. The box of 96 crayons that was introduced in 1993 contained 16 unnamed crayons. If you randomly pick a crayon from such a box, what is the probability that it doesn't have a name?

Play a card game or board game with a family member. Then discuss whether chance, skill, or both chance and skill determine who wins the game. Explain the role of each in the game.

26. *Working on the* **CHAPTER Project** Refer to the results of your survey in the Chapter Project on page 131. Based on your data, what is the probability that if a new student joins the class, her or his favorite car color will be white?

27. *Critical Thinking* Marvin and Naomi are playing a game by rolling number cubes. Naomi gets a point each time the sum of the number cubes is 2, 3, 4, 9, 10, 11, or 12. Marvin gets a point when the sum is 5, 6, 7, or 8. Is this a fair game? That is, does each player have an equal chance to win? Explain.

Mixed Review

28. *Diamonds* The graph shows the percent of diamond engagement rings that are different shapes. Express each percent as a fraction in simplest form. *(Lesson 4-7)*

   a. round

   b. marquise

   c. others

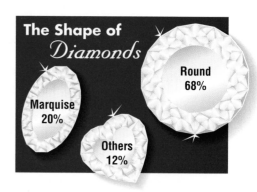

The Shape of *Diamonds*

Round 68%

Marquise 20%

Others 12%

Source: Diamond Information Center

29. **Standardized Test Practice** The Reds baseball team played 12 games. They scored a total of 168 runs. What was the mean number of runs scored per game? *(Lesson 3-4)*

   A 14   **B** 15   **C** 156   **D** 180

For **Extra Practice,** see page 579.

30. Estimate 23.69 divided by 4.05. *(Lesson 2-3)*

4-9 Least Common Multiple

What you'll learn

You'll learn to find the least common multiple of two or more numbers.

When am I ever going to use this?

You will use least common multiples to write equivalent fractions.

Word Wise

multiple

least common multiple (LCM)

Cultural Kaleidoscope

In 1925, Garrett Morgan invented the first three-way traffic signal.

Traffic engineers determine the best traffic-light sequences to ensure orderly traffic flow. Suppose a traffic light on one street turns red every 40 seconds and a second light turns red every 60 seconds. Both lights just turned red. In how many seconds will both lights turn red at the same time again?

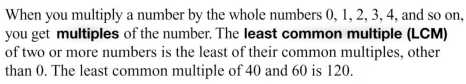

$$+ 40 \quad + 40 \quad + 40$$

first light: 0, 40, 80, 120

second light: 0, 60, 120

$$+ 60 \quad + 60$$

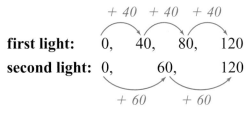

Both lights turn red again in 120 seconds.

When you multiply a number by the whole numbers 0, 1, 2, 3, 4, and so on, you get **multiples** of the number. The **least common multiple (LCM)** of two or more numbers is the least of their common multiples, other than 0. The least common multiple of 40 and 60 is 120.

You can use one of two methods to find a least common multiple.

Method 1   Make a list.
List several multiples of each number. Then identify the common multiples. The least of these is the LCM.

Method 2   Use prime factorization.
Write the prime factorization of each number. Identify all common prime factors. For each prime factor, write it down the greatest number of times it appears in any of the numbers. The product is the LCM.

Examples

1   Use Method 1 to find the LCM of 6 and 9.

multiples of 6: 6, 12, 18, 24, 30, 36, . . .

multiples of 9: 9, 18, 27, 36, 45, 54, 63, . . .

Zero is a multiple of every number, but cannot be the LCM.

The LCM of 6 and 9 is 18.

2   Use Method 2 to find the LCM of 9, 12, and 15.

$9 = 3 \times 3 = 3^2$

$12 = 2 \times 2 \times 3 = 2^2 \times 3$

$15 = 3 \times 5$

The prime factors are 2, 3, and 5. The greatest number of times 2 appears is twice (in 12), so write it down twice. The greatest number of times 3 appears is twice (in 9), so write it down twice. The greatest number of times 5 appears is once (in 15), so write it down once. The LCM of 9, 12, and 15 is $2 \times 2 \times 3 \times 3 \times 5$ or 180.

3 Mentally find the LCM of 4, 5, and 8 by listing multiples.

> *Think:* The positive multiples of 8 are 8, 16, 24, 32, 40, 48, ...
>
> *Ask:* What is the least of these multiples that is divisible by both 4 and 5?
>
> *Reason:* The number has to be even and end in 0. The first such number in the list is 40. And 40 is divisible by 8.

So, the LCM of 4, 5, and 8 is 40.

CONNECTION

4 Earth Science In 1833, the great Leonid meteor shower was seen all over North America. These showers occur every 33 years. The Schwassmann-Wachmann Comet occurs every 15 years. If both these occurrences coincided one year, use your calculator to find in how many years this would happen again.

Explore   You know that the comet occurs every 15 years and the meteor shower occurs every 33 years. Find the least common multiple of 15 and 33.

Plan   Write the multiples of the greater number and find the LCM that is divisible by the lesser number.

Solve   multiples of 33: 33, 66, 99, 132, 165, 198. . .

Which multiple is divisible by 15? Start with the least multiple, 33.

$33 \div 15 = 2.2$ $132 \div 15 = 8.8$

$66 \div 15 = 4.4$ $165 \div 15 = 11$ ✓

$99 \div 15 = 6.6$

The LCM of 15 and 33 is 165. Thus, these solar occurrences would coincide again in 165 years.

Examine   Use prime factorization to find the LCM.

$15 = 3 \cdot 5$

$33 = 3 \cdot 11$

LCM: $3 \cdot 5 \cdot 11 = 165$

The answer is correct.

CHECK FOR UNDERSTANDING

Communicating Mathematics

Read and study the lesson to answer each question.

1. ***Tell*** why the LCM of 18, which is 2×3^2, and 24, which is $2^3 \times 3$, must have factors of 2^3 and 3^2. What is the LCM of 18 and 24?

2. ***Explain*** how the LCM of two numbers can be one of the numbers.

Math Journal

3. ***Write*** the steps you would take to find the LCM of 3, 5, and 9 using pencil and paper. Compare it to solving the problem mentally.

Find the LCM of each set of numbers by listing multiples.

4. 35, 14   **5.** 45, 15   **6.** 2, 3, 10

Find the LCM of each set of numbers by writing prime factorizations.

7. 4, 8, 12   **8.** 18, 8, 36   **9.** 15, 25, 45

10. *Government* Presidential elections are held every four years. Senators are elected every six years. If a senator was elected in the presidential election year 2000, in what year would he or she campaign again during a presidential election year?

EXERCISES

Practice   **Find the LCM of each set of numbers by listing multiples.**

11. 12, 60   **12.** 2, 3, 5   **13.** 30, 15

14. 10, 20, 30   **15.** 18, 300   **16.** 9, 10, 4

17. 24, 12, 6   **18.** 44, 33, 22   **19.** 10, 12, 15

Find the LCM of each set of numbers by writing prime factorizations.

20. 16, 176   **21.** 28, 49, 16   **22.** 6, 12, 18

23. 17, 6, 34   **24.** 33, 44, 55   **25.** 30, 625

26. 12, 15, 28   **27.** 42, 16, 7   **28.** 35, 25, 49

29. Mentally compute the LCM of 4, 5, and 10 by listing multiples.

30. List the multiples that 9 and 15 have in common above 180.

Applications and Problem Solving

31. *Scheduling* Margo has piano lessons every two weeks. Her brother Roberto has a soccer tournament every three weeks. Her sister Rocio has an orthodontist appointment every four weeks. If they all have activities this Friday, how long will it be before all of their activities fall on the same day again?

32. *Number Sense* Explain when the LCM of two numbers is their product. Give two examples to illustrate your rule.

33. *Critical Thinking* Write a set of three numbers whose LCM is the product of the numbers.

Mixed Review

34. **Standardized Test Practice** In a school raffle, one ticket will be drawn out of a total of 500 tickets. If the Stevenson family has 12 tickets, what is the probability that they will win? *(Lesson 4-8)*

   A 1 in 500   **B** 3 in 125   **C** 1 in 12   **D** 1 in 488   **E** 12 in 1,000

35. Find the greatest common factor of 72 and 270. *(Lesson 4-4)*

36. Write 0.4141414141 . . . using bar notation. *(Lesson 2-7)*

37. *Recycling* The math team collected 25 pounds of newspaper during a school newspaper drive. The band collected 19 pounds. If each organization was paid $0.35 per pound of newspaper, how much more money did the math team raise than the band? *(Lesson 2-4)*

For **Extra Practice,** see page 579.

Comparing and Ordering Fractions

What you'll learn

You'll learn to compare and order fractions.

When am I ever going to use this?

Knowing how to order fractions will help you compare quantities.

Word Wise

common denominator
least common denominator (LCD)

Kenya is a country in East Africa. Although the official language is Swahili, only 1% of its inhabitants speak this language. In this country, $\frac{1}{5}$ of the people speak *Kikuyu,* and $\frac{7}{50}$ of the people speak *Luo.* Which of these languages is spoken by more Kenyans? *This problem will be solved in Example 2.*

To compare fractions such as $\frac{1}{5}$ and $\frac{7}{50}$, rewrite each fraction using the same denominator. Then you only need to compare the numerators.

A **common denominator** is a common multiple of the denominators of two or more fractions. The **least common denominator (LCD)** is the least common multiple (LCM) of the denominators of two or more fractions.

Examples

1 Compare $\frac{5}{6}$ and $\frac{7}{9}$.

The LCM of the denominators 6 and 9 is 18.

Rewrite each fraction using the LCD, 18.

$$\frac{5}{6} = \frac{5 \times 3}{6 \times 3} = \frac{15}{18} \qquad \frac{7}{9} = \frac{7 \times 2}{9 \times 2} = \frac{14}{18}$$

Since $\frac{15}{18} > \frac{14}{18}$, then $\frac{5}{6} > \frac{7}{9}$.

CONNECTION

2 **Geography** Refer to the beginning of the lesson. Is Kikuyu or Luo spoken by more Kenyans?

You need to find which fraction is greater, $\frac{1}{5}$ or $\frac{7}{50}$. First, find the LCD by listing the multiples of each denominator.

multiples of 5: 5, 10, 15, 20, 25, 30, 35, 40, 45, 50, 55, . . .

multiples of 50: 50, 100, 150, 200, 250, . . .

The LCD of $\frac{1}{5}$ and $\frac{7}{50}$ is 50, since 50 is the LCM of 5 and 50. So, rewrite $\frac{1}{5}$ using a denominator of 50.

$$\overset{\times\,10}{\underset{\times\,10}{\frac{1}{5} = \frac{10}{50}}}$$

Now, compare $\frac{10}{50}$ and $\frac{7}{50}$. Since 10 > 7, then $\frac{10}{50} > \frac{7}{50}$. More Kenyans speak Kikuyu than Luo.

You can also compare fractions by seeing how they relate to the nearest one half or by expressing them as decimals.

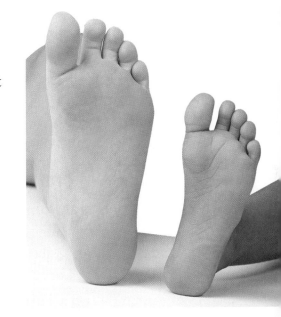

Examples

3 **Faith's foot is $3\frac{3}{16}$" wide. Her friend Raine's foot is $3\frac{7}{8}$" wide. Whose foot is wider?**

Since both widths are at least 3 inches, we will compare only the fractions.

$\frac{3}{16}$ is nearest to 0.

$\frac{7}{8}$ is nearest to 1.

So, $\frac{3}{16} < \frac{7}{8}$ and $3\frac{3}{16} < 3\frac{7}{8}$.

Raine's foot is wider than Faith's.

APPLICATION

4 **Safety** More than a half-million bike riders go to hospital emergency rooms each year. A recent survey showed that 7 out of 10 bike riders use safety equipment. LaShanda found that 9 of 16 friends that she talked to used safety equipment when they rode their bikes. Are LaShanda's friends safer than most?

Since $\frac{9}{16}$ and $\frac{7}{10}$ are difficult to compare, express each fraction as a decimal and then compare.

$9 \div 16 = 0.5625$ $\qquad\qquad\qquad$ $7 \div 10 = 0.7$

Since $0.5625 < 0.7$, then $\frac{9}{16} < \frac{7}{10}$. LaShanda's friends wear safety equipment less than the average bike rider does. So they are not safer than most.

Communicating Mathematics

Read and study the lesson to answer each question.

1. **Tell** what method you would use to compare $\frac{3}{4}$ and $\frac{5}{6}$ and why you chose that method.

2. **You Decide** Darnell thinks that $\frac{34}{70}$ is less than $\frac{31}{60}$. Jennifer disagrees. Who is correct? Explain your reasoning.

3. **Write** the fractions that are modeled at the right. Which fraction is greater? Explain how you know.

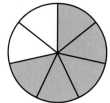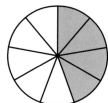

Guided Practice

Find the LCD for each pair of fractions.

4. $\frac{8}{9}, \frac{7}{12}$

5. $\frac{8}{5}, \frac{15}{13}$

6. $\frac{7}{18}, \frac{11}{36}$

7. $\frac{1}{6}, \frac{3}{8}$

inter NET CONNECTION

For the latest statistics on American Indian tribes, visit:

www.glencoe.com/sec/math/mac/mathnet

Replace each ● with <, >, or = to make a true sentence.

8. $\frac{1}{6}$ ● $\frac{1}{3}$

9. $\frac{5}{9}$ ● $\frac{11}{15}$

10. $\frac{3}{4}$ ● $\frac{5}{8}$

11. $\frac{4}{7}$ ● $\frac{5}{9}$

12. **Population** The Choctaw tribe makes up $\frac{13}{300}$ of all American Indian tribes. The Pueblo tribe makes up $\frac{7}{250}$ of all tribes. Which of these two tribes makes up a greater fraction of American Indian tribes?

Practice

Find the LCD for each pair of fractions.

13. $\frac{4}{5}, \frac{8}{9}$

14. $\frac{3}{5}, \frac{3}{6}$

15. $\frac{11}{16}, \frac{3}{4}$

16. $\frac{1}{3}, \frac{3}{10}$

17. $\frac{5}{4}, \frac{9}{8}$

18. $\frac{2}{15}, \frac{1}{6}$

19. $\frac{7}{12}, \frac{13}{36}$

20. $\frac{15}{21}, \frac{6}{7}$

21. $\frac{2}{3}, \frac{17}{24}$

22. $\frac{3}{10}, \frac{2}{12}$

23. $\frac{13}{17}, \frac{3}{4}$

24. $\frac{5}{24}, \frac{3}{8}$

Replace each ● with <, >, or = to make a true sentence.

25. $\frac{4}{7}$ ● $\frac{5}{8}$

26. $\frac{1}{3}$ ● $\frac{3}{15}$

27. $\frac{9}{13}$ ● $\frac{14}{20}$

28. $\frac{9}{17}$ ● $\frac{9}{13}$

29. $\frac{5}{9}$ ● $\frac{8}{15}$

30. $\frac{3}{13}$ ● $\frac{4}{26}$

31. $\frac{16}{20}$ ● $\frac{40}{50}$

32. $\frac{4}{5}$ ● $\frac{6}{7}$

33. $\frac{3}{8}$ ● $\frac{5}{12}$

34. $\frac{45}{90}$ ● $\frac{15}{30}$

35. $\frac{7}{5}$ ● $\frac{14}{11}$

36. $\frac{2}{3}$ ● $\frac{19}{27}$

37. Find the LCD for $\frac{5}{14}$ and $\frac{11}{18}$.

38. Which is greater, $\frac{12}{32}$ or $\frac{16}{40}$?

Applications and Problem Solving

39. **School** Suppose 17 of the 28 students in math class and 15 of 25 students in music class sign up for the school olympics. Which class has a greater portion of students participating?

40. **Number Sense** Is $1\frac{1}{8}$, or $\frac{19}{18}$ nearest to 1? Explain your reasoning.

41. Video Games Recently, $\frac{2}{25}$ of kids surveyed played video games 5–6 hours per week, and $\frac{3}{50}$ played video games 7–8 hours per week. Did more kids surveyed play video games 5–6 hours or 7–8 hours each week?

42. Critical Thinking When is the least common denominator of two fractions equal to one of the denominators? Give two examples.

Mixed Review

43. Find the LCM of 14 and 21. *(Lesson 4-9)*

44. Volunteering The graph shows the percent of people in the Peace Corps. Express each percent as a fraction in simplest form. *(Lesson 4-7)*

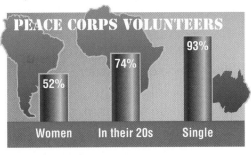

PEACE CORPS VOLUNTEERS

52% Women   74% In their 20s   93% Single

Source: Peace Corps

For **Extra Practice**, see page 580.

45. Statistics Construct a line plot for 75, 65, 82, 93, 75, 72, 81, and 90. *(Lesson 3-3)*

46. Order 3, 0.3, 3.33, 0.33, 3.03 from least to greatest. *(Lesson 2-1)*

47. Standardized Test Practice Josie was born in 1979. Amanda was born 12 years later. In what year was Amanda born? *(Lesson 1-5)*

A 1996   **B** 1991   **C** 1984   **D** 1967

Let the Games Begin

Fractions and Ladders

Math Skill

Comparing and Ordering Fractions

Get Ready This game is for two players.

calculator   posterboard

Get Set Draw a ladder with 12 rungs on the posterboard. Label the bottom rung 0 and the top rung 1.

Go
- Use your calculator to build a ladder of fractions between 0 and 1 so that each fraction on the ladder is greater than the fraction below it. You may use any numbers in the numerators and denominators of your fractions.

- The first player enters a fraction at the bottom of the ladder. The opponent enters a larger fraction on the next rung of the ladder. He or she must show that this fraction is larger by using one of the strategies described in this lesson.

- Alternate turns. A player wins if the opponent is not able to find a larger fraction and he or she can. A player, in his or her turn, can have only one chance to name a fraction for the next rung before their opponent gets to try. If neither player succeeds, they each try again.

inter**NET** CONNECTION Visit www.glencoe.com/sec/math/mac/mathnet for more games.

 **inter NET CONNECTION** Chapter Review **For additional review, visit:** www.glencoe.com/sec/math/mac/mathnet

Vocabulary

After completing this chapter, you should be able to define each term, concept, or phrase and give an example or two of each.

Number and Operations
common denominator (p. 172)
composite number (p. 138)
divisible (p. 133)
factor (p. 133)
factor tree (p. 138)
greatest common factor (GCF) (p. 150)
least common denominator (LCD) (p. 172)

least common multiple (LCM) (p. 169)
multiple (p. 169)
percent (p. 158)
prime factorization (p. 138)
prime number (p. 138)
ratio (p. 154)
simplest form (p. 154)

Problem Solving
make a list (p. 148)

Patterns and Functions
arithmetic sequence (p. 142)
geometric sequence (p. 143)
sequence (p. 142)
term (p. 142)

Probability
event (p. 165)
probability (p. 165)
random (p. 165)

Understanding and Using the Vocabulary

State whether each sentence is *true* or *false*. If false, replace the underlined word or number to make a true sentence.

1. The prime factorization of 24 is $\underline{2^3 \times 3^2}$.

2. The numbers in a sequence are called <u>terms</u>.

3. The greatest common factor of 15 and 18 is <u>3</u>.

4. A <u>ratio</u> is a comparison of two numbers by division.

5. The simplest form of $\frac{16}{24}$ is $\underline{\frac{4}{6}}$.

In Your Own Words

6. *Explain* the difference between a prime number and a composite number.

Objectives & Examples

Upon completing this chapter, you should be able to:

● use divisibility rules *(Lesson 4-1)*
Determine whether 336 is divisible by 2, 4, or 10.
 2: Yes. The ones digit, 6, is even.
 4: Yes. The number formed by the last two digits, 36, is divisible by 4.
10: No. The ones digit is not 0.

Review Exercises

Use these exercises to review and prepare for the chapter test.

Determine whether the first number is divisible by the second number.

 7. 452; 4 **8.** 727; 3 **9.** 123; 6

Determine whether each number is divisible by 2, 3, 4, 5, 6, 9, or 10.

 10. 221 **11.** 1,225 **12.** 630

Objectives & Examples

find the prime factorization of a composite number *(Lesson 4-2)*

Find the prime factorization of 63.

$$\begin{array}{c} 63 \\ 7 \times 9 \\ 7 \times 3 \times 3 \end{array}$$

The prime factorization of 63 is 7×3^2.

recognize and extend a pattern for sequences *(Lesson 4-3)*

Find the next three terms in the sequence 32, 16, 8, 4, 2, ...

This is a geometric sequence created by multiplying the previous term by $\frac{1}{2}$. The next three terms are $1, \frac{1}{2}, \frac{1}{4}$.

find the greatest common factor of two or more numbers *(Lesson 4-4)*

Find the GCF of 45 and 75.

$45 = 3 \times 3 \times 5$

$75 = 3 \times 5 \times 5$

GCF: 3×5 or 15

express fractions and ratios in simplest form *(Lesson 4-5)*

Write $\frac{45}{81}$ in simplest form.

$$\frac{45}{81} = \frac{45 \div 9}{81 \div 9} = \frac{5}{9}$$

illustrate the meaning of percent using models or symbols *(Lesson 4-6)*

Express 18 in 100 as a percent.

18 in $100 = \frac{18}{100} = 18\%$

Review Exercises

Find the prime factorization of each number.

13. 1,000
14. 144
15. 950
16. 77
17. 96
18. 300
19. 2,800
20. 1,450

Describe the pattern in each sequence. Identify the sequence as *arithmetic*, *geometric*, or *neither*. Then find the next three terms.

21. 16, 21, 26, 31, 36, . . .
22. 1, 4, 16, 64, 256, . . .
23. 0, 3, 9, 18, 30, 45, . . .
24. 10, 100, 1,000, 10,000, . . .

Find the GCF of each set of numbers.

25. 36, 81
26. 40, 65
27. 252, 336
28. 57, 240
29. 56, 280, 400
30. 8, 12, 16

Express each fraction or ratio in simplest form.

31. $\frac{56}{70}$
32. 250:750
33. 26:39
34. $\frac{18}{60}$
35. $\frac{77}{121}$
36. 57:95

Express each ratio as a percent.

37. $\frac{56}{100}$
38. 73 out of 100
39. 49 in 100
40. 24:100

Objectives & Examples

Review Exercises

express fractions as percents, and percents and decimals as fractions *(Lesson 4-7)*

Express $\frac{6}{25}$ as a percent.

$$\frac{6}{25} = \frac{24}{100} = 24\%$$

Express 25% as a fraction.

$$25\% = \frac{25}{100} = \frac{1}{4}$$

Express each fraction as a percent.

41. $\frac{14}{25}$ **42.** $\frac{9}{20}$

43. $\frac{2}{5}$ **44.** $\frac{7}{7}$

Express each percent or decimal as a fraction in simplest form.

45. 15%   **46.** 0.27

47. 0.58   **48.** 72%

find the probability of a simple event *(Lesson 4-8)*

Find P(odd) when a number cube is rolled.

$$P(\text{odd}) = \frac{3}{6} \quad \begin{array}{l} \leftarrow \textit{odd numbers} \\ \leftarrow \textit{total number of outcomes} \end{array}$$

$$= \frac{1}{2} \text{ or } 0.5$$

A bag contains 6 red, 3 pink, and 3 white bows. If you draw a bow at random, what is the probability of drawing each of the following? Express each ratio as both a fraction and a percent.

49. red   **50.** pink

51. white   **52.** a red or a white

find the least common multiple of two or more numbers *(Lesson 4-9)*

Find the LCM of 4 and 18.

$4 = 2^2$

$18 = 2 \times 3^2$

LCM: $2^2 \times 3^2$ or 36

Find the LCM of each set of numbers.

53. 6, 15   **54.** 42, 56

55. 16, 40   **56.** 15, 125, 600

57. 21, 81, 147   **58.** 48, 81, 270

compare and order fractions *(Lesson 4-10)*

Compare $\frac{5}{9}$ and $\frac{4}{6}$.

$$\frac{5}{9} = \frac{10}{18} \quad \frac{4}{6} = \frac{12}{18}$$

Since $10 < 12$, then $\frac{5}{9} < \frac{4}{6}$.

Find the LCD for each pair of fractions.

59. $\frac{2}{3}, \frac{3}{4}$ **60.** $\frac{11}{12}, \frac{8}{9}$

Replace each ● with <, >, or = to make a true sentence.

61. $\frac{3}{8}$ ● $\frac{5}{12}$ **62.** $\frac{7}{10}$ ● $\frac{13}{25}$

Applications & Problem Solving

63. Make a List Find the greatest common factor of the twelfth and fifteenth terms of the Fibonacci sequence, 1, 1, 2, 3, 5, 8, 13, 21, ... *(Lesson 4-4A)*

64. School Beth scored 21 out of 25 on her spelling test. Ted scored 37 out of 40 on his spelling test. Who scored higher? *(Lesson 4-10)*

65. Earth Science About 97.3 gallons of every 100 gallons of water on Earth are salt water. Write the percent of the water that is salt water. *(Lesson 4-6)*

66. Activities The graph shows students' favorite activities. Write each percent as a fraction in simplest form. *(Lesson 4-7)*

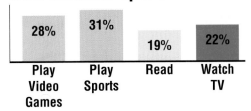

How Students Spend Free Time

Play Video Games	Play Sports	Read	Watch TV
28%	31%	19%	22%

Alternative Assessment

● **Open Ended**

Suppose you are a grocery store manager. You have decided to sell fresh bagels in the store bakery and to charge 45¢ for a bagel. If a customer orders more than one bagel, each additional one costs 40¢. Show how you will find the cost of one dozen bagels.

Make a chart listing the price of 1 through 12 bagels.

● **Completing the CHAPTER Project**

Use the following checklist to make sure your car survey is complete.

☑ table and graph of your car production data

☑ table and graph of your car color survey

☑ the probability that a new student's favorite car color will be white, based on your survey

☑ a paragraph explaining how a car manufacturer would use your survey results

PORTFOLIO Review the items in your portfolio. Make a table of contents of the items, noting why each item was chosen. Replace any items that are no longer appropriate.

A practice test for Chapter 4 is provided on page 610.

Section One: Multiple Choice

There are ten multiple-choice questions in this section. Choose the best answer. If a correct answer is *not here,* choose the letter for Not Here.

1. The length of this paper clip is about —

A 5 millimeters.

B 5 centimeters.

C 5 meters.

D 5 kilometers.

2. Jamal pays $126.82 each month on his student loan. What is this amount rounded to the nearest ten dollars?

F $120.00

G $126.80

H $126.90

J $130.00

3. Lu-Chan performed a probability experiment by spinning a spinner 100 times. The results of his experiment are in the chart.

Color	Number of Spins
Red	40
Blue	20
Green	40

If the spinner is divided into 5 equal sections, how many sections would you expect to be colored green?

A 2

B 3

C 4

D 5

Please note that Questions 4–10 have five answer choices.

4. Ms. Rogers buys 6 pounds of strawberries each day for her daycare center. If there are between 15 and 20 strawberries in one pound, what is a reasonable total for the number of strawberries that she buys each day?

F 75

G 100

H 150

J 175

K 215

5. Rachel earns $5.25 per hour for baby-sitting. Which is the best estimate of her earnings if she baby-sits 13 hours in one week?

A $60

B $70

C $80

D $90

E $100

6. The girls' chorale placed a bank in the cafeteria to collect money for a trip to sing at the statehouse. At the end of the first week, the bank contained 15 quarters, 10 dimes, and 35 nickels. Which number sentence represents the amount of money M in the bank?

F $M = (15 \times 0.25) + (10 \times 0.10) + (35 \times 0.05)$

G $M = (15 + 25) \times (10 + 10) \times (35 + 5)$

H $M = (15 + 0.25) \times (10 + 0.10) \times (35 + 0.05)$

J $M = (15 \times 0.25) \times (10 \times 0.10) \times (35 \times 0.05)$

K $M = (15 \div 25) + (10 \div 0.10) + (35 \div 0.05)$

7. Paul's middle school had a drawing for a family pack of movie tickets. A total of 200 tickets was put in a container, and one ticket was drawn. If Paul had 5 tickets, what was the probability that Paul won?

   A $\frac{1}{200}$ **B** $\frac{1}{40}$

   C $\frac{1}{5}$ **D** $\frac{1}{100}$

   E $\frac{1}{195}$

8. Marco rides his bike to and from work every day. If the round trip is 3.23 miles, how far will Marco ride in 7 days?

   F 45.22 mi

   G 29.07 mi

   H 22.61 mi

   J 19.38 mi

   K Not Here

9. Serina bought 2 packages of hamburger on sale at $1.10 per pound. One package weighed 2.59 pounds, and the other weighed 1.12 pounds. What was the total weight of hamburger that Serina purchased?

   A 2.20 lb

   B 2.71 lb

   C 4.08 lb

   D 7.42 lb

   E Not Here

10. The student yearbook pages are 8 inches wide and 11 inches long. If one page has only one photo that is 3 inches by 5 inches, which sentence can be used to find s, the amount of space left for autographs?

    F $s = (8 + 11) - (3 + 5)$

    G $s = 8 \times 11 \times 3 \times 5$

    H $s = 2(8 \times 11) - 2(3 \times 5)$

    J $s = (8 \times 11) - (3 \times 5)$

    K $s = (8 \times 11) \div (3 \times 5)$

Test-Taking Tip THE PRINCETON REVIEW

Most standardized tests are given with test books that can be written in. Use the test book for scratch work and to mark questions that you will return to if you have time. You will not receive credit for answers that you write in the test book. Remember to transfer answers to the answer sheet as you take the test.

Section Two: Free Response

This section contains five questions for which you will provide short answers. Write your answers on your paper.

11. Identify the sequence 14, 26, 38, 50, . . . as *arithmetic, geometric,* or *neither.* Then find the next three terms.

12. What is the least common multiple of 25 and 45?

13. If $n + 3.9 = 4.2$, what is the value of n?

14. Write a percent to represent the shaded area.

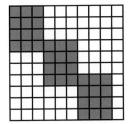

15. At the carnival, 10 prizes were placed inside paper bags. Each bag contained just one prize. If there were 2 one-dollar bills, 3 pencils, 1 ten-dollar bill, and 4 pens, what is the probability that a pen would be chosen?

CHAPTER 5

Algebra: Using Integers

What you'll learn in Chapter 5

- to read, write, and compare integers and find the opposite and absolute value of an integer,
- to graph points on a coordinate plane,
- to add, subtract, multiply, and divide integers,
- to solve problems by looking for a pattern, and
- to graph transformations on a coordinate plane.

CHAPTER Project

LATITUDE VS. TEMPERATURE

In this project, you will make a table comparing latitudes and temperatures of cities. You will graph the data and determine whether there is a relationship between the latitude of cities in the United States and their temperatures. You will then write a paragraph summarizing your findings, describing any relationships or patterns that you noticed.

Getting Started

- Look at the table of cities. Which city had the highest temperature in 1995? Which city had the lowest temperature?
- The higher the latitude number is, the farther the location is from the equator. Use the table to determine which city is closest to the equator. Which city is farthest from the equator?

City	Latitude (nearest degree)	High Temp. (°F, 1995)	Low Temp. (°F, 1995)
New York, NY	41	102	6
Honolulu, HI	21	94	56
Fairbanks, AK	65	88	−48
Seattle, WA	48	96	22
Nashville, TN	36	99	9
San Diego, CA	33	90	43
Bismarck, ND	47	98	−28
Atlanta, GA	34	102	13
Minneapolis, MN	45	101	−11
Denver, CO	40	99	−7

Source: *The World Almanac and Book of Facts*

Technology Tips

- Use a **spreadsheet** to organize your data and to find the difference between the highest and lowest temperatures.
- Use **computer software** to make graphs.
- Use a **word processor** to summarize and analyze your data.

 **Research** For up-to-date information on geography, visit:

www.glencoe.com/sec/math/mac/mathnet

Working on the Project

You can use what you'll learn in Chapter 5 to help you make your graph.

Page	Exercise
190	25
194	33
205	45
221	Alternative Assessment

Integers

After the 1990 census, Texas gained 3 seats in the United States House of Representatives. Michigan lost 2 seats. You can use the **integers** +3 and −2 to describe these situations, respectively.

Integers	An integer is any number from the set $\{\ldots, -4, -3, -2, -1, 0, +1, +2, +3, +4, \ldots\}$.

Integers greater than 0 are **positive integers**. Integers less than 0 are **negative integers**. Zero is neither positive nor negative. Positive integers usually are written without the + sign, so +5 and 5 are the same. You can represent integers using counters or a number line.

Method 1 Counters

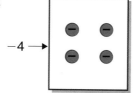

$-4 \rightarrow$

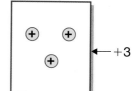 $\leftarrow +3$

Method 2 Number Line

1 **Games** *Mancala* is an African game that is over 3,000 years old. The object is for a player to capture the opponent's stones. Suppose a player removes 12 opponent's stones. Write an integer for this situation.

Since there is a decrease in stones, the negative integer, −12, represents this situation.

Two numbers are **opposites** of one another if they are represented by points that are the same distance from 0, but on opposite sides of 0. The number line below shows that −4 and 4 are opposites.

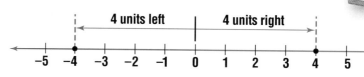

4 units left   4 units right

Absolute value is helpful when adding integers.

Remember, distance is never negative.

Absolute Value	The absolute value of an integer is its distance from 0 on a number line.

The absolute value of n is written as $|n|$. So, $|-4| = 4$ and $|4| = 4$.

Example **2**
a. **Find the opposite of −8.**
b. **Find the absolute value of −8.**

a. On the number line, −8 is at the point 8 units to the left of 0. The opposite of −8 would be at the point 8 units to the right of 0.

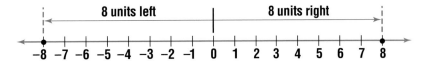

So, the opposite of −8 is 8.

b. The point that represents −8 is 8 units from 0, so the absolute value of −8 is 8. Write $|-8| = 8$.

CHECK FOR UNDERSTANDING

Communicating Mathematics

Read and study the lesson to answer each question.

1. **Tell** why −7.5 is not an integer.

2. **Write** each of the following as an integer over a whole number.
$$7, -3,\ 2\tfrac{1}{3}, -1\tfrac{1}{2},\ 0,\ 0.029, -2.5$$

3. **You Decide** Conner says that the integers represented by A and B are opposites. Keisha thinks they are not. Who is correct? Explain.

Guided Practice

Write an integer for each situation.

4. a loss of 6 yards

5. a profit of $5

6. a deposit of $10

7. 76°F below 0

Write the integer represented by the point for each letter. Then find its opposite and its absolute value.

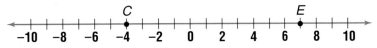

8. C

9. E

10. **Football** On one play, the Baltimore Ravens gained 18 yards. On the next play, they lost 10 yards. Write integers to represent the yards gained and lost.

Practice

Write an integer for each situation.

11. a gain of 9 points

12. a withdrawal of $25

13. 120 feet below sea level

14. 8°C

15. the year A.D. 1600

16. 3 seconds before liftoff

17. down 4 strokes

18. a growth of 2 inches

19. no gain on 1st down

20. the year 130 B.C.

21. 15° F above 0

22. a gain of 11 pounds

Write the integer represented by the point for each letter. Then find its opposite and its absolute value.

23. A

24. B

25. C

26. D

27. E

28. F

Applications and Problem Solving

29. What is the absolute value of 0?

30. Find $|x|$ if $x = -10$.

31. *Geography* Angel Falls in Venezuela is 3,212 feet above sea level. The Caribbean Sea, which borders Venezuela on the north, has an average depth of 8,685 feet below sea level. Use integers to express these elevations.

32. *Recreation* In a travel game, the first person to identify a certain car by color gets 1 point. A certain van is worth 3 points. If you make a mistake, such as getting the color wrong or not seeing the car at all, you lose the points. On a recent trip, Raylene's dad intentionally missed two cars and one van. Use integers to express each of his scores.

33. *Calculators* If you press 7 $\boxed{+\!\bigcirc\!-}$, you get -7.

  a. What is the result when you press 7 $\boxed{+\!\bigcirc\!-}$ $\boxed{+\!\bigcirc\!-}$?

  b. What is the result when you press 7 $\boxed{+\!\bigcirc\!-}$ $\boxed{+\!\bigcirc\!-}$ $\boxed{+\!\bigcirc\!-}$?

  c. What can you conclude about the number of times you press $\boxed{+\!\bigcirc\!-}$ and the result?

34. *Critical Thinking* The distance from San Jose north to San Francisco is about the same as the distance from San Jose south to Santa Cruz. Which distance would have a greater absolute value?

Mixed Review

35. Replace ● with $<$, $>$, or $=$ to make $\frac{15}{24}$ ● $\frac{17}{32}$ a true sentence. *(Lesson 4-10)*

36. **Standardized Test Practice** Find the LCM of 27 and 30. *(Lesson 4-9)*

  A 3 　　　　**B** 810 　　　　**C** 270 　　　　**D** 300

37. *Statistics* Find the mean, median, and mode for the following set of data: 3, 5, 2, 3, 3, 4, 3, 2, 2. *(Lesson 3-4)*

For **Extra Practice**, see page 580.

38. *Write* 3.075×10^4 in standard form. *(Lesson 2-9)*

METEOROLOGY

Edward Lorenz
METEOROLOGIST AND MATHEMATICIAN

Edward Lorenz, working in the field of meteorology, made discoveries that led to a new branch of mathematics called Chaos. One of Lorenz's most famous results is the "Butterfly Effect," which states that a butterfly flapping its wings in the Amazon forest might ultimately lead to a tornado in Texas.

A person who is interested in becoming a meteorologist should take high school courses in mathematics, physics, and chemistry. The ability to recognize patterns and make predictions is based on a good understanding of mathematics. Typically, a bachelor's degree in meteorology is required to obtain a job. An advanced degree is required for work in teaching and research.

For more information:
American Meteorological Society
45 Beacon Street
Boston, MA 02108

inter NET CONNECTION
www.glencoe.com/sec/math/mac/mathnet

The weather amazes me. I wonder how I could learn to predict the weather.

Your Turn
Keep a record, using graphs, tables, and charts, of weather forecasts given on television, radio, or in the newspaper for one week. Then record the actual weather for each day including high and low temperatures and precipitation. Compare the forecasts with the actual conditions. How accurate were the forecasts?

Comparing and Ordering Integers

What you'll learn

You'll learn to compare and order integers.

When am I ever going to use this?

Knowing how to order integers can help you compare historical events.

The table shows the population growth for five cities. A negative number means a decrease in population. If Hartford and Newark started out with the same population at the beginning of that period, which would be greater at the end?

City	Population Growth (%)
Brownsville, TX	14
Chattanooga, TN	0
Durham, NC	5
Hartford, CT	−11
Newark, NJ	−6

Source: *Statistical Abstract of the United States*

You can use a number line.

H means Hartford, N means Newark, and so on.

Since Newark is to the right of Hartford on the number line, its population "increase" (−6) represents a greater amount than Hartford's (−11). In symbols, you can write −6 > −11 or −11 < −6. So, by the end of the period, Newark would have a greater population than Hartford.

Examples

1 **Replace ● with < or > in −7 ● −1 to make a true sentence.**

Draw the graph of each integer on a number line.

Since −7 is to the left of −1 on the number line, −7 < −1.

2 **Order the integers −4, 7, 3, −2, and 1 from least to greatest.**

Draw the graph of each integer on a number line.

Order the integers by reading from left to right.

$$-4, -2, 1, 3, 7$$

INTEGRATION

3 **Statistics** Mars, Venus, and Mercury are the planets most like Earth. Find the median of the temperatures in the chart.

List the temperatures in order from least to greatest.

Planet	Average Temperature (°F)
Earth	50
Mars (day)	72
Mars (night)	−94
Mercury (day)	800
Mercury (night)	−300
Venus	900

−300, −94, 50, 72, 800, 900

There are two middle numbers, 50 and 72. So the median is $\frac{50 + 72}{2}$ or 61. The median temperature is 61°F.

LOOK BACK

You can refer to Lesson 3-4 for information on median.

Communicating Mathematics

Read and study the lesson to answer each question.

1. *Draw* a number line showing $-5 < -2$.

2. *Tell* how to determine when one integer is greater than another integer.

Guided Practice

Replace each ● with < or > to make a true sentence.

3. -3 ● -12
4. -17 ● 7
5. 0 ● -5

6. Order $43, -4, 29, -12, -17,$ and -76 from least to greatest.

7. *Earth Science* The table below gives the record low temperatures in Boise, Idaho, that occurred through 1994. Find the median of the temperatures.

Month	J	F	M	A	M	J	J	A	S	O	N	D
Temperature (°F)	−17	−15	6	19	22	31	35	34	23	11	−3	−25

EXERCISES

Practice

Replace each ● with < or > to make a true sentence.

8. -121 ● -21
9. -23 ● -32
10. -6 ● 0
11. 3 ● -4
12. 1 ● -1
13. 7 ● -89
14. -38 ● -83
15. -4 ● -8
16. -49 ● $|50|$

Order the integers from least to greatest.

17. $5, -3, 0, -59, -43, 11$

18. $-6, 18, -16, 33, 3, -44, 1$

19. Which is greater, -12 or 7?

20. Order $78, -80, 69, -20, 0,$ and 17 from greatest to least.

Applications and Problem Solving

Undersea Craft	Altitude (ft)
Bathyscaph	−35,800
Newt Suit	−1,000
Manned Submersible	−13,000
Submarine	−4,000
Undersea Robot	−20,000

21. *Earth Science* The table shows how deep various undersea craft can go. List the depths in order from deepest to most shallow.

22. *Inventions* Order the dates from earliest to most recent.

1600 B.C. Bronze plows are developed in Vietnam.

A.D. 1992 8-year-old Abbey Fleck invents a gadget for cooking bacon in the microwave oven.

A.D. 1885 Gottlieb Daimler invents the gasoline engine.

3000 B.C. SeLing-She invents silk cloth.

Abbey Fleck

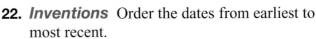

23. *Earth Science* The highest summit in Arizona is Humphreys Peak at 12,633 feet. The table shows how the highest summits in other states compare to it.

Peak	Height Compared to Humphreys Peak (ft)
Granite Peak, MT	+166
Mt. Hood, OR	−1,394
Mt. Elbert, CO	+1,800
Mt. Rogers, VA	−6,904
Spruce Knob, WV	−7,770
Wheeler Peak, NM	+528

Source: *High Points of the United States, A Guide to the 50 State Summits*

  a. Which summit is the highest?

  b. Name the three summits that are closest in height to Humphreys Peak.

24. *Earth Science* The map shows the all-time record low temperatures in ten cities for the month of January.

  a. Order the temperatures from lowest to highest.

  b. Which city had the coldest temperature?

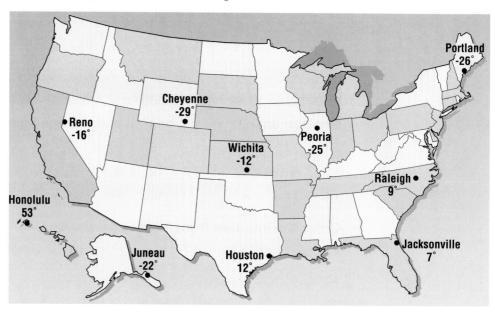

25. *Working on the* CHAPTER Project Refer to the table on page 183.

  a. Make a table listing the cities from least to greatest latitude. Show latitude and low temperatures.

  b. What do you notice about the relationship between latitudes and low temperatures?

26. *Critical Thinking* If 0 is the greatest integer in a set of five, what can you conclude about the other four integers?

Mixed Review

27. **Standardized Test Practice** Marco dove off a 10-foot diving board into 12 feet of water and touched the bottom. Which integer describes the number of feet below the surface of the water that he dove? *(Lesson 5-1)*

  A −12   **B** −10   **C** 10   **D** 12

28. Express $\frac{39}{81}$ in simplest form. *(Lesson 4-5)*

For **Extra Practice,** see page 580.

29. *Physical Science* An experiment requires 3 milligrams of potassium chloride. How many grams of potassium chloride are needed? *(Lesson 2-8)*

5-3

Integration: Geometry
The Coordinate System

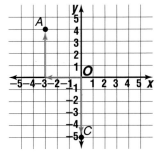

What you'll learn

You'll learn to graph points on a coordinate plane.

When am I ever going to use this?

Knowing how to graph points on a coordinate plane can help you find cities on a map.

Word Wise

coordinate system
origin
x-axis
y-axis
quadrant
ordered pair
x-coordinate
y-coordinate

In-car navigation systems use global positioning satellites (GPS) to pinpoint their positions. Street intersections are often used to locate addresses on a map.

In mathematics, a **coordinate system**, or coordinate plane, is used to graph points in a plane. It is made up of a horizontal number line and a vertical number line that intersect at their zero points. This intersection point is called the **origin**.

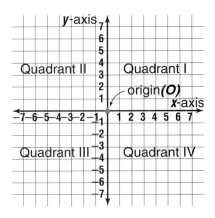

The horizontal line is called the **x-axis**, and the vertical line is called the **y-axis**. Together, they make up a coordinate system that separates the plane into four **quadrants**.

You can identify any point on a coordinate system using an **ordered pair** of numbers. The first number in an ordered pair is the **x-coordinate**, and the second number is the **y-coordinate**.

The ordered pairs (2, 4) and (−4, −3) are graphed at the right.

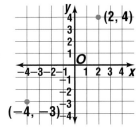

The point at (2, 4) is in quadrant I. The point at (−4, −3) is in quadrant III.

Examples

1 Name the ordered pair for point *A* and identify its quadrant.

- Start at the origin *O*. Locate point *A* by moving left 3 units along the *x*-axis. The *x*-coordinate is −3.

- Now move up 4 units. The *y*-coordinate is +4.

- The ordered pair is (−3, 4). Point *A* is in quadrant II.

2 Name the ordered pair for point *C* and identify its quadrant.

The ordered pair is (0, −5). Point *C* is not in a quadrant because it is on the *y*-axis.

Example **3**

Real World APPLICATION

In this book when no numbers are given on the x- and y-axis, you can assume that each grid square is one unit.

Shopping The graph shows the ratings of different stores' fashions. In which quadrant is Store C? What does this quadrant represent?

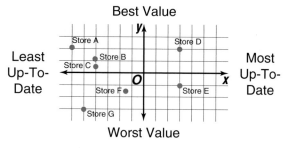

The ordered pair for Store C is $(-4, 0.5)$. This is in quadrant II. Stores in quadrant II have fashions that are less up-to-date, but are a better value.

To graph a point in a coordinate system, draw a dot at the location of its ordered pair.

Example **4**

Graph the point $D(2, 3)$.

First, draw a coordinate system. Start at the origin O. Move 2 units to the right. Then move 3 units up to locate the point. Draw a dot and label it $D(2, 3)$.

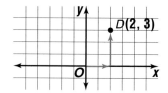

Let the Games Begin

Tic-Tac-Toe

Get Ready This game is for two players.

☐ sheet of paper ▦ grid paper

Get Set Each person should draw a coordinate plane on grid paper.

Go This is like tic-tac-toe, except you must get four Xs or four Os in a row. Players take turns naming two numbers. Each number must be between −10 and 10. The first number is the x-coordinate of an ordered pair, and the second is the y-coordinate. The other player writes these ordered pairs on a sheet of paper and places an X or O in the correct place on the coordinate plane.

● Once you say a number, you may not change it.
● If you say a coordinate that was already used, you lose a turn.

The first player to get four Xs or Os in a row is the winner.

Math Skill

Graphing Points on a Coordinate Plane

Example

	x	y
O	1	3
X	0	0
O	1	−2
X	−1	1

Visit www.glencoe.com/sec/math/mac/mathnet for more games.

Communicating Mathematics

Read and study the lesson to answer each question.

1. **Name** the quadrant in which the x-coordinate is negative and the y-coordinate is positive.

2. **Explain** why the point $H(7, -10)$ is different from the point $K(-10, 7)$.

3. Refer to the beginning of the lesson. **Write** a few sentences describing how a GPS device uses a coordinate system.

Guided Practice

Name the x-coordinate and the y-coordinate for each point labeled at the right. Then tell in which quadrant each point lies.

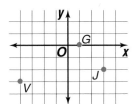

4. V 5. G 6. J

On graph paper, draw a coordinate plane. Then graph and label each point.

7. $P(-3, 5)$ 8. $R(3, -2)$ 9. $S\left(2, \frac{1}{2}\right)$

10. **Geology** Geologists use latitude (horizontal) and longitude (vertical) lines to graph volcanoes on a world map. Cotopaxi, for example, is at 10°S latitude and 70°W longitude, or (10°S, 70°W). Write the location of each volcano as an ordered pair (latitude, longitude).

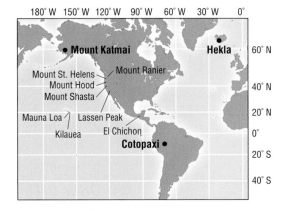

a. Mount Katmai   b. Hekla

Practice

Name the x-coordinate and the y-coordinate for each point labeled at the right. Then tell in which quadrant each point lies.

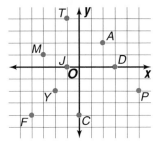

11. A 12. J 13. P

14. F 15. C 16. M

17. T 18. D 19. Y

On graph paper, draw a coordinate plane. Then graph and label each point.

20. $T(0, -3)$ 21. $B(-2, 7)$ 22. $R(-2, -3)$

23. $A(7, 4.5)$ 24. $M(5, -1.5)$ 25. $C\left(1\frac{1}{2}, 4\right)$

26. $H(6, -3)$ 27. $E(-5, -5)$ 28. $W(-4, 0)$

29. In which quadrant is $T(23, -18)$ located?

30. Name the coordinates of the point where the *x*-axis and the *y*-axis intersect. What is the name of this point?

31. *Olympics* The graph shows a map of the Olympic Village during the 1996 Summer Olympics. Name the ordered pair for each point.

    a. Centennial Park

    b. Georgia Dome

    c. Omni

    d. Olympic Center

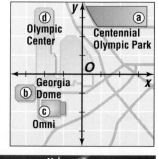

32. *Earth Science* Certain patterns of stars, called *constellations,* resemble animals and objects. The constellations at the right are the hunter Orion and his dog Canis Major.

    a. Name the coordinates of Betelgeuse, the star in the right shoulder of Orion.

    b. Name the coordinates of Sirius, the brightest star in the winter sky.

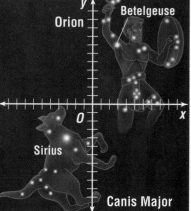

33. *Working on the* **CHAPTER Project**
Refer to Exercise 25 on page 190.

    a. Write the latitude and the low temperature of each city as an ordered pair.

    b. Graph the ordered pairs on a coordinate plane.

34. *Critical Thinking* Suppose for any ordered pair the *x*-coordinate and the *y*-coordinate are always the same integer. Where are the possible locations for these sets of ordered pairs?

Mixed Review

35. Order the integers 5, -1, 3, and -5 from least to greatest. *(Lesson 5-2)*

36. **Standardized Test Practice** Koleka has 5 dimes in her pocket. The dates on the dimes are 1995, 1992, 1987, 1983, and 1978. If she chooses 1 dime from her pocket without looking, what is the probability that it will have a date in the 1990s? *(Lesson 4-8)*

    A $\frac{1}{5}$ **B** $\frac{4}{5}$ **C** $\frac{2}{3}$ **D** $\frac{2}{5}$

37. *Money Matters* A 39-ounce jar of peanut butter costs $1.95. What is the cost per ounce? *(Lesson 2-6)*

38. *Life Science* It is believed that a dog ages 7 human years for every calendar year that it lives. *(Lesson 1-3)*

    a. Write an expression for determining a dog's age in human years. Let *y* represent the number of calendar years the dog has lived.

    b. Find the human age of a dog that has lived for 12 calendar years.

For **Extra Practice,**
see page 581.

TECHNOLOGY LAB

GRAPHING CALCULATORS

5-3B Graphing Points

A Follow-Up of Lesson 5-3

📟 graphing calculator

The graphics screen of a graphing calculator can represent a coordinate plane. You can graph points on a graphing calculator just as you do on a coordinate grid.

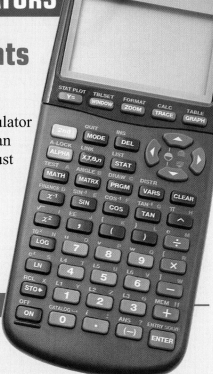

TRY THIS

Work with a partner.

Use the TI-83 graphing calculator to graph points $A(1, -2)$, $B(7, 1)$, $C(-3, -4)$, and $D(5, 5)$ in the standard viewing window $[-10, 10]$ by $[-10, 10]$.

First, press ⎡2nd⎤ [DRAW] 1 ⎡ENTER⎤ to clear the drawing screen. Then graph point A.

Enter: ⎡2nd⎤ [DRAW] ▶ ⎡ENTER⎤ 1 ⎡,⎤ ⎡(−)⎤ 2 ⎡)⎤ ⎡ENTER⎤

Press ⎡2nd⎤ [QUIT]. Then enter the coordinates of points B, C, and D in the same way. Be sure to press ⎡2nd⎤ [QUIT] after graphing each point.

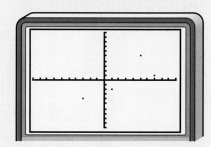

ON YOUR OWN

Use the program to graph each set of points on a graphing calculator. Then sketch the graph on a separate piece of paper.

1. $H(5, -7)$, $I(9, 1)$, $J(-5, -2)$

2. $W(4, 4)$, $X(-2, 1)$, $Y(8, 1)$, $Z(-6, 5)$

3. $L(3, -5)$, $M(-3, -4)$, $N(-9, 2)$, $O(7, -8)$

4. Explain how to change the viewing window so that you could see the graph of the point $P(-12, 15)$.

COOPERATIVE LEARNING

5-4A Adding Integers

A Preview of Lesson 5-4

⚫ counters

☐ integer mat

You can use counters to model the addition of integers.

TRY THIS

Work with a partner.

1 Model $6 + 2$.

Combine a set of 6 positive counters and a set of 2 positive counters on the mat.

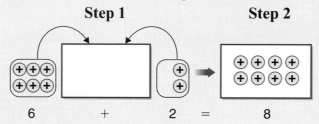

So, $6 + 2 = 8$.

2 Model $5 + (-1)$.

Place 5 positive counters on the mat. Then place 1 negative counter on the mat. When one positive counter is paired with one negative counter, the result is a *zero pair*. Remove all the zero pairs. *You can add or remove zero pairs from a mat because adding or removing zero does not change the value of the counters on the mat.*

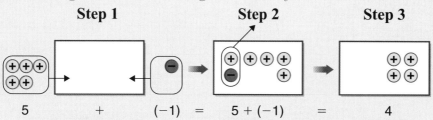

So, $5 + (-1) = 4$.

ON YOUR OWN

1. Write the addition problem that is modeled below.

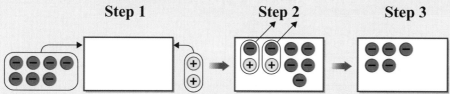

Use counters to find each sum. Use your result to write an addition sentence.

2. $5 + (-3)$ **3.** $-4 + 1$ **4.** $-6 + (-5)$

5. ***Look Ahead*** Tell whether $8 + (-12)$ is *positive, negative,* or *zero.*

5-4

Adding Integers

What you'll learn

You'll learn to add integers.

When am I ever going to use this?

Knowing how to add integers can help you check your savings account balance.

Word Wise

additive inverse

Cultural Kaleidoscope

The papyrus documents were discovered in Egypt in 1899. They had been wrapped around mummified crocodiles.

In 1996, scientists at the University of California discovered how to preserve a 2,300-year-old papyrus document. An *ionizer* was used to add negatively and positively charged atoms to neutralize the charges and eliminate static. This allowed the sheets of the document to be separated.

Like charged ions, integers with opposite signs can be added. What happens when you add integers that are opposites, like 2 and −2? Adding the opposite of an integer "undoes" the integer, and the result is 0. For this reason, two integers that are opposites are called **additive inverses**.

Additive Inverse Property	**Words:** The sum of any number and its additive inverse is 0.
	Symbols: **Arithmetic** $2 + (-2) = 0$ **Algebra** $a + (-a) = 0$

One way to add integers is by using arrows on a number line. Start at 0. Positive integers are represented by arrows pointing *right*. Negative integers are represented by arrows pointing *left*.

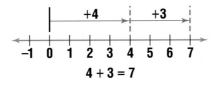

$$4 + 3 = 7$$

$$-4 + (-3) = -7$$

Adding Integers with the Same Sign	The sum of two positive integers is positive. The sum of two negative integers is negative.

Example

1 Use counters to solve $-6 + (-2) = x$.

Combine a set of 6 negative counters and a set of 2 negative counters.

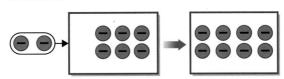

$$-6 + (-2) = x$$
$$-8 = x$$

Let's look at the sum $4 + (-3)$.

Remove all of the zero pairs.

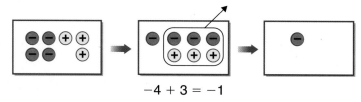

$$4 + (-3) = 1$$

Now, look at the sum $-4 + 3$.

$$-4 + 3 = -1$$

Adding Integers with Different Signs	To add integers with different signs, subtract their absolute values. The sum is • positive if the positive integer has the greater absolute value. • negative if the negative integer has the greater absolute value.

Examples

2 Solve $w = -40 + 15$.

$|-40| > |15|$, so the sum is negative.
The difference of 40 and 15 is 25.
So, $w = -25$.

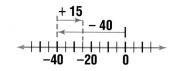

INTEGRATION

3 Algebra Evaluate $-13 + a$ if $a = 28$.

$-13 + a = -13 + 28$ *Replace a with 28.*

$|-13| < |28|$, so the sum is positive.
The difference of 13 and 28 is 15. So, $-13 + a = 15$.

Let the Games Begin

War of Integers

Get Ready This game is for two players.   🂠 playing cards 1–10

Get Set In this game, black cards 1 through 10 are positive integers, and red cards 1 through 10 are negative integers. Shuffle and deal all of the cards.

Go Each player turns over his or her top two cards and finds the sum. The player with the greater total gets all of the cards. Play continues until all cards are used. The winner is the one with the most cards.

> **Math Skill**
> Adding Integers

inter NET CONNECTION Visit www.glencoe.com/sec/math/mac/mathnet for more games.

Example ——④ **Cycling** In the Extreme Games, Joe Guel rode his bicycle 17 feet down one side of a ramp and then rode 31 feet up the other side and into the air to do a back flip. How far above his starting position did Joe perform his trick?

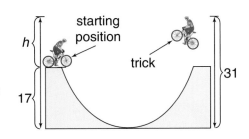

Explore The height of the ramp Joe went down was 17 feet. You can write this drop as -17 feet.

Plan Let h = the height above the ramp. Solve the equation $h = 31 + (-17)$.

Solve $|31| > |-17|$, so the sum is positive. The difference of 31 and 17 is 14. So, $h = 14$. Joe performed his trick 14 feet above his starting position.

Examine To check your solution, use a calculator.
31 [+] 17 [+○−] [=] *14* ✔

CHECK FOR UNDERSTANDING

Communicating Mathematics

Read and study the lesson to answer each question.

1. *Explain* whether or not three integers are all negative if their sum is negative.

2. *Write* the addition sentence shown by each model.

   a.
   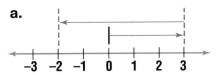

   b.
   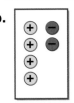

HANDS-ON MATH

3. *Model* the sum of -5 and -2 using counters.

Guided Practice

Tell whether the sum is *positive*, *negative*, or *zero*.

4. $9 + (-11)$ 5. $-8 + 8$ 6. $-5 + 12$

Solve each equation.

7. $y = 4 + (-13)$ 8. $-8 + (-2) = g$ 9. $h = -7 + 14$

Evaluate each expression if $a = 6$, $b = -2$, and $c = 5$.

10. $a + (-4)$ 11. $-9 + c$ 12. $0 + b$

13. *Earth Science* The greatest temperature change recorded in a 24-hour period was a drop of 100°F in 1916 at Browning, Montana. If the highest temperature was 44°F, what was the lowest temperature?

Practice

Tell whether the sum is *positive*, *negative*, or *zero*.

14. $13 + (-8)$ **15.** $-4 + (-4)$ **16.** $-5 + 5$

17. $5 + (-2)$ **18.** $-7 + 15$ **19.** $-11 + (-3)$

20. $-10 + (-8)$ **21.** $9 + (-9)$ **22.** $-15 + 12$

Solve each equation.

23. $-7 + 11 = w$ **24.** $s = -6 + 6$ **25.** $r = 19 + (-13)$

26. $n = 5 + (-18)$ **27.** $g = 3 + 10$ **28.** $m = -11 + 13$

29. $9 + (-8) = p$ **30.** $-30 + 20 = v$ **31.** $-47 + 28 = x$

Evaluate each expression if $a = 10$, $b = -10$, and $c = -5$.

32. $a + (-3)$ **33.** $-7 + b$ **34.** $c + (-2)$

35. $-8 + a$ **36.** $a + (-10)$ **37.** $16 + c$

38. $0 + b$ **39.** $a + b$ **40.** $c + a$

41. Evaluate $d + 7$ if $d = -93$.

42. What is the value of $m + n$ if $m = -3$ and $n = -17$?

Applications and Problem Solving

43. ***Football*** The Panthers lost 7 yards on one play and then completed a pass for 16 yards on the next play. What was the total number of yards gained or lost?

44. ***Personal Finance*** Sean thought he had $27 in his checking account and wrote a check for $12 to buy a new CD. On the way home, Sean remembered that he had withdrawn $20 the day before. What is the minimum amount Sean must quickly deposit in his account to avoid a bank service charge for insufficient funds?

45. ***Critical Thinking*** Marsha's friend Yori asked to borrow $7 so she could go see a movie. When Marsha said she didn't have it, Yori concluded, "O.K., so you owe me $7." What's wrong with Yori's reasoning?

Mixed Review

46. Name the *x*-coordinate and the *y*-coordinate for each point labeled at the right. Then tell in which quadrant each point lies. *(Lesson 5-3)*

     a. C **b.** L

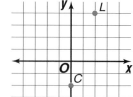

47. Find the opposite and the absolute value of -21. *(Lesson 5-1)*

For **Extra Practice**, see page 581.

48. ***Probability*** If a number cube is rolled, find the probability that the result is an even number. Express the probability as a percent. *(Lesson 4-8)*

49. ***Statistics*** The number of absent students during the past two weeks at Lincoln Junior High were 15, 24, 31, 14, 7, 9, 10, 13, 9, and 3. Make a stem-and-leaf plot for the data. *(Lesson 3-5)*

50. ***Standardized Test Practice*** How many square feet of wallpaper are needed to cover a wall that measures 9 feet by 16 feet? *(Lesson 1-7)*

     A 48 ft^2 **B** 64 ft^2 **C** 96 ft^2 **D** 144 ft^2

5-5A Subtracting Integers

A Preview of Lesson 5-5

You can use counters to model the subtraction of integers.

○ counters

□ integer mat

TRY THIS

Work with a partner.

1 Model $8 - 4$.

Place 8 positive counters on the mat and then remove 4.

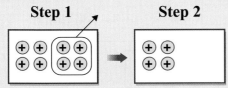

Step 1   **Step 2**

So, $8 - 4 = 4$.

2 Model $7 - (-3)$.

Start with a set of 7 positive counters and try to remove 3 negative counters. Since there are no negative counters, add 3 zero pairs to the set. Now you can remove 3 negative counters.

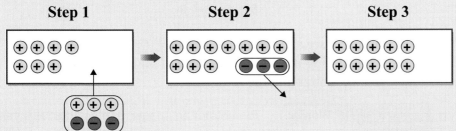

Step 1   **Step 2**   **Step 3**

So, $7 - (-3) = 10$.

ON YOUR OWN

1. Write the subtraction problem that is modeled below.

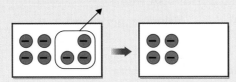

Use counters to find each difference. Use your result to write a subtraction sentence.

2. $6 - 2$ **3.** $-4 - (-1)$ **4.** $5 - (-3)$ **5.** $-6 - 1$

6. Look Ahead Find $t = -8 - 6$ without using models.

Subtracting Integers

What you'll learn

You'll learn to subtract integers.

When am I ever going to use this?

Knowing how to subtract integers can help you find how much temperatures have increased or decreased.

On July 4, 1997, *Sojourner,* the first-ever interplanetary rover, landed on Mars. The temperature on Mars can be −94°F at night and 72°F during the day. What is the range of the temperature *Sojourner* encountered as it explored the surface of Mars? *This problem will be solved in Example 1.*

Cruise stage and entry vehicle

To see how addition and subtraction of integers are related, compare the subtraction 6 − 3 to the addition 6 + (−3).

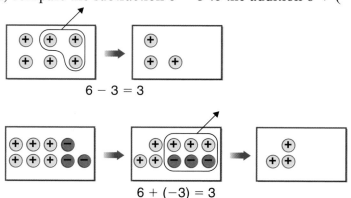

$6 - 3 = 3$

$6 + (-3) = 3$

Did you know The *Sojourner* is named after Sojourner Truth, an African American who promoted freedom for all people during the Civil War.

The diagrams above show that $6 - 3 = 6 + (-3)$. This example suggests that adding the additive inverse of an integer produces the same result as subtracting the integer.

Subtracting Integers	**Words:**	To subtract an integer, add its additive inverse.	
	Symbols:	**Arithmetic**	**Algebra**
		$6 - 3 = 6 + (-3)$	$a - b = a + (-b)$

Example ①

CONNECTION

Earth Science Refer to the beginning of the lesson. What is the range of the temperatures *Sojourner* encountered as it explored the surface of Mars?

Let r = the range of the temperatures.

$r = 72 - (-94)$

$r = 72 + 94$ *To subtract −94, add 94.*

$r = 166$

The range of the temperatures *Sojourner* encountered was 166 degrees.

2 Solve $x = -9 - 5$.

$x = -9 - 5$

$x = -9 + (-5)$ *To subtract 5, add -5.*

$x = -14$

3 Solve $-15 - (-7) = w$.

$-15 - (-7) = w$

$-15 + 7 = w$ *To subtract -7, add 7.*

$-8 = w$

INTEGRATION **4** **Algebra** Evaluate $a - b$ if $a = 5$ and $b = 13$.

$a - b = 5 - 13$ *Replace a with 5 and b with 13.*

$= 5 + (-13)$ *To subtract 13, add -13.*

$= -8$

Interplanetary rover, Sojourner

CHECK FOR UNDERSTANDING

Communicating Mathematics

Read and study the lesson to answer each question.

1. **Write** $a - (-b)$ as an addition expression.

2. **Write** a subtraction sentence shown by the model.

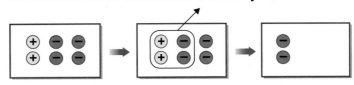

3. **You Decide** Antwan says that to solve $15 - 24 = p$, you can rewrite the equation as $24 - (15) = p$. Ellen thinks this is not right. Who is correct? Explain your reasoning.

Guided Practice

Solve each equation.

4. $m = 5 - (-10)$
5. $-7 - 6 = k$
6. $s = 9 - (-11)$
7. $10 - 16 = j$
8. $n = -22 - 5$
9. $4 - (-19) = p$

Evaluate each expression if $b = -1$, $q = 6$, and $y = 3$.

10. $8 - q$
11. $3 - b$
12. $-y - 10$

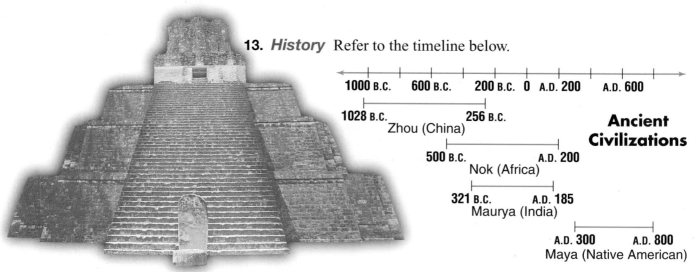

13. *History* Refer to the timeline below.

Ancient Civilizations

Zhou (China): 1028 B.C. — 256 B.C.
Nok (Africa): 500 B.C. — A.D. 200
Maurya (India): 321 B.C. — A.D. 185
Maya (Native American): A.D. 300 — A.D. 800

Mayan Temple of the Giant Jaguar

a. For how many years was the Maurya civilization in existence?

b. How many years were there between the beginning of the Nok civilization and the end of the Maya civilization?

EXERCISES

Practice

Solve each equation.

14. $7 - 13 = t$
15. $27 - (-8) = c$
16. $16 - 7 = q$

17. $m = 2 - (-8)$
18. $x = -12 - 9$
19. $14 - 15 = p$

20. $-5 - (-5) = b$
21. $z = -11 - 42$
22. $-17 - (-9) = r$

23. $-18 - 14 = s$
24. $h = 13 - (-10)$
25. $c = 19 - (-21)$

26. $18 - 100 = y$
27. $b = 24 - (-14)$
28. $-25 - (-10) = d$

29. $z = 5 - (-27)$
30. $17 - (-50) = t$
31. $-42 - (-23) = w$

Evaluate each expression if $m = -3$, $t = 8$, and $a = 4$.

32. $10 - a$
33. $m - 5$
34. $-15 - t$

35. $a - 7$
36. $6 - m$
37. $t - a$

38. $t - m$
39. $m - a$
40. $-a - m$

41. Find the value of $-x - y$ if $x = -3$ and $y = 5$.

42. Find b if $a - (-b) = 10$ and $a = 8$.

Applications and Problem Solving

43. *Environment* The highest temperature recorded in North America is 134°F in Death Valley, California, and the lowest recorded temperature is −87°F in Northice, Greenland. What is the range of temperatures?

44. *Business* Mr. Sanchez is a stockbroker in New York. He uses a table to determine whether it is a good time to contact various brokers' offices around the world.

a. If it is 7 A.M. in his office, what time is it in Honolulu?

b. Do you think it is a good time for Mr. Sanchez to make a call to Honolulu? Why?

When's a Good Time?	
City	Hours
Honolulu	−5
Anchorage	−4
San Francisco	−3
Chicago	−1
New York	0
Paris	6
Tokyo	14

45. Working on the  **CHAPTER Project** Refer to the table on page 183.

    a. Find the difference between the high and low temperatures for each city. Write the latitude and the temperature difference of each city as an ordered pair.

    b. Graph the ordered pairs on a coordinate plane. Let the *x*-axis represent the latitude and the *y*-axis represent the temperature difference.

46. Critical Thinking *True* or *false*? When a is a negative integer, $a - a = 0$.

Mixed Review

47. Standardized Test Practice Leonardo leaves his house and walks east 12 blocks to the store. He then turns around and walks west 17 blocks to his friend's house. Which integer describes Leonardo's final position? *(Lesson 5-4)*

    A -5 **B** 5   **C** 15   **D** 29

48. Find the GCF of 36 and 48. *(Lesson 4-4)*

49. Determine whether 135 is divisible by 2, 3, 4, 5, 6, 9, or 10. *(Lesson 4-1)*

50. Data Analysis The graph shows the number of orders taken each hour one day at a fast food restaurant. What time of the day appears to be the least busy at this restaurant? *(Lesson 3-2)*

For **Extra Practice,** see page 581.

51. Write $\frac{3}{8}$ as a decimal. *(Lesson 2-7)*

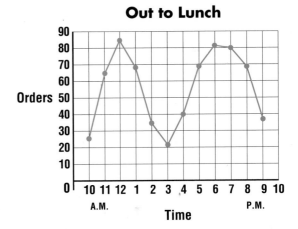

Out to Lunch

Mid-Chapter Self Test

Write an integer for each situation. *(Lesson 5-1)*

  1. a gain of 7 pounds   **2.** 4 points lost

Replace each ● with < or > to make a true sentence. *(Lesson 5-2)*

  3. -16 ● -18 **4.** -12 ● 7 **5.** 4 ● -5

Name the *x*-coordinate and the *y*-coordinate for each point graphed. Then tell in which quadrant each point lies. *(Lesson 5-3)*

  6. *A*   **7.** *B*

Solve each equation. *(Lesson 5-4)*

  8. $-6 + 2 = t$ **9.** $g = 5 + (-14)$

10. Earth Science Mauna Kea in Hawaii is the tallest mountain on Earth. Its height is 33,000 feet, but it's not all visible because 19,204 feet are below the ocean. How much of Mauna Kea is visible above water? *(Lesson 5-5)*

COOPERATIVE LEARNING

5-6A Multiplying Integers

A Preview of Lesson 5-6

⚫ counters

▭ integer mat

You can use counters to model the multiplication of integers.

TRY THIS

Work with a partner.

❶ Model $2 \times (-3)$.

Start with an empty mat. Place 2 sets of 3 negative counters on the mat.

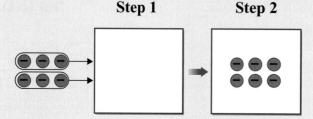

Step 1   **Step 2**

So, $2 \times (-3) = -6$.

❷ Model $-2 \times (-4)$.

Start with an empty mat. Since -2 is the opposite of 2, $-2 \times (-4)$ means to *remove* 2 sets of 4 negative counters. Since there are no negative counters, place 8 zero pairs on the mat. Now you can remove 2 sets of 4 negative counters.

Step 1   **Step 2**   **Step 3**

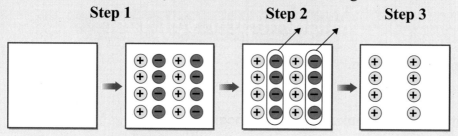

So, $-2 \times (-4) = 8$.

ON YOUR OWN

Use counters to find each product. Use your result to write a multiplication sentence.

1. 2×3

2. $2 \times (-4)$

3. $-2 \times (-3)$

4. $3 \times (-4)$

5. 4×0

6. $-1 \times (-5)$

7. *Look Ahead* Find $9 \times (-3)$ without using models.

Multiplying Integers

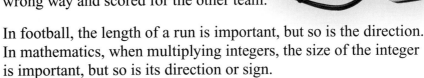

In 1964, Jim Marshall of the Minnesota Vikings picked up a fumble and made a 66-yard run for what he thought was a touchdown. Unfortunately, he ran the wrong way and scored for the other team.

In football, the length of a run is important, but so is the direction. In mathematics, when multiplying integers, the size of the integer is important, but so is its direction or sign.

You can find $3(-2)$ by using counters or by looking for a pattern.

Method 1   Use counters.

3 sets of 2 negative counters is 6 negative counters or -6.

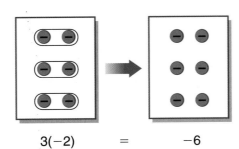

$$3(-2) \quad = \quad -6$$

Method 2   Look for a pattern.

$$3 \cdot 2 = 6$$
$$3 \cdot 1 = 3 \qquad \searrow {-3}$$
$$3 \cdot 0 = 0 \qquad \searrow {-3}$$
$$3 \cdot (-1) = -3 \qquad \searrow {-3}$$
$$3 \cdot (-2) = -6 \qquad \searrow {-3}$$

When two integers have different signs, the following rule applies.

Multiplying Integers with Different Signs	The product of two integers with different signs is negative.

Examples

1 Solve $a = 7(-4)$.

The two integers have different signs. The product will be negative.

$$a = 7(-4)$$
$$a = -28$$

2 Solve $-9(5) = h$.

The two integers have different signs. The product will be negative.

$$-9(5) = z$$
$$-45 = z$$

How would you find $-3(-2)$? You can use the same methods to multiply two integers with the same sign. When using counters, $-3(-2)$ means that you will remove 3 sets of 2 negative counters.

Method 1   Use counters.

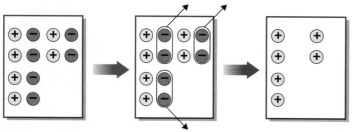

Remove 3 sets of
2 negative counters.

$-3(-2) = 6$

Method 2   Look for a pattern.

$-3 \cdot 2 = -6$
$\quad\quad\quad\quad\quad +3$
$-3 \cdot 1 = -3$
$\quad\quad\quad\quad\quad +3$
$-3 \cdot 0 = 0$
$\quad\quad\quad\quad\quad +3$
$-3 \cdot (-1) = 3$
$\quad\quad\quad\quad\quad +3$
$-3 \cdot (-2) = 6$

So, $-3(-2)$ is 6. This suggests the following rule.

Multiplying Integers with the Same Sign	The product of two integers with the same sign is positive.

Examples

3 **Solve** $a = -5(-6)$.

The two integers have the same signs. The product will be positive.

$a = -5(-6)$
$a = 30$

4 **Solve** $(-7)^2 = z$.

The exponent says there are two factors of -7. The product will be positive.

$(-7)^2 = z$
$(-7)(-7) = z$
$49 = z$

INTEGRATION

5 **Algebra**   Evaluate abc if $a = -3$, $b = 7$, and $c = -2$.

$abc = (-3)(7)(-2)$ *Replace a with -3, b with 7, and c with -2.*
$\quad = [(-3)(7)](-2)$ *First multiply -3 by 7. Then multiply the*
$\quad = (-21)(-2)$ *product by -2.*
$\quad = 42$

CHECK FOR UNDERSTANDING

Communicating Mathematics

Read and study the lesson to answer each question.

1. *Tell* what you can say about two integers if their product is negative.

2. *Show* a pattern to explain why $7(-3)$ must be -21.

HANDS-ON MATH

3. *Model* $-3 \times (-4)$ using counters and write a multiplication sentence.

Guided Practice

Solve each equation.

4. $b = 10(-4)$

5. $(-5)(-7) = x$

6. $y = 8(-4)$

7. $-9(12) = p$

8. $a = 15(-3)$

9. $(-6)^2 = m$

Evaluate each expression if $x = -8$, $y = -3$, $z = 2$, and $w = 4$.

10. $-4z$ **11.** xy **12.** x^2

13. *Write a Problem* in which you need to find the product of 6 and -13. Then find the product.

EXERCISES

Practice — Solve each equation.

14. $y = -7(-13)$ **15.** $(-3)^2 = a$ **16.** $(-13)(-13) = c$

17. $-14(4) = j$ **18.** $b = -16(-5)$ **19.** $10(-2) = s$

20. $x = -11(5)$ **21.** $p = -4(-4)$ **22.** $h = (-9)^2$

23. $v = 21(-3)$ **24.** $-4(-17) = t$ **25.** $r = (-10)(10)$

26. $-8(12) = n$ **27.** $-7(18) = m$ **28.** $-25(4) = q$

Evaluate each expression if $a = -9$, $b = 2$, $c = -5$, and $d = 6$.

29. $-4d$ **30.** $7a$ **31.** $-5ac$ **32.** c^2

33. $-13d$ **34.** $-3b^2$ **35.** $10bc$ **36.** $-2cd$

Applications and Problem Solving

37. *Archaeology* Archaeologists often use delicate instruments as small as toothbrushes to excavate sites of ancient treasures. Suppose an archaeologist removes 2 cubic meters of sand per day.

   a. Write an equation to find how much sand is removed at the end of 14 days.

   b. Solve this equation.

38. *Submarines* A submarine is diving from the surface at a rate of 76 feet per minute. What is the depth of the submarine after 5 minutes?

39. *Critical Thinking* If the product of three integers is negative, what can you conclude about the signs of the integers? Write a rule for determining the sign of the product of three nonzero integers.

Mixed Review

40. *Hiking* A Girl Scout troop is hiking on a trail that is 75 feet above sea level. They hike into a canyon that is 12 feet below sea level. Find the difference in altitudes. *(Lesson 5-5)*

41. Order 6, -3, 0, 4, -8, 1, -4 from least to greatest. *(Lesson 5-2)*

42. **Standardized Test Practice** The advertisement shows athletic shoes on sale for 20% off the regular price. What fraction of the regular price is this? *(Lesson 4-7)*

   A $\frac{1}{20}$ **B** $\frac{1}{4}$ **C** $\frac{1}{5}$ **D** $\frac{1}{2}$

43. Find 2.5×0.3. *(Lesson 2-4)*

44. Evaluate $3 \cdot 0.50 + 4 \cdot 0.75$. *(Lesson 1-2)*

For **Extra Practice**, see page 582.

20% OFF our entire stock
men's
women's
children's
Athletic Shoes

PROBLEM SOLVING

5-7A Look for a Pattern

A Preview of Lesson 5-7

I just got an e-mail message from Angela in Costa Rica!

Good idea! Then they can forward it to other people.

I wonder if Alysia and Joyce have heard from her? Why don't you forward her message to them?

After 10 minutes, Ramon forwarded the e-mail message to 2 of his friends. After 10 more minutes, those 2 friends each forwarded the message to 2 more friends. If the message was forwarded like this every 10 minutes, how many people received Angela's e-mail message after 40 minutes?

Ramon

Melanie

You can solve this problem by looking for a pattern.

1,   2,   4,   ?,   ?
  × 2   × 2

To continue the pattern, multiply 2 by the previous term.

$$4 \times 2 = 8 \qquad 8 \times 2 = 16$$

So, after 40 minutes, $1 + 2 + 4 + 8 + 16$ or 31 people got the message.

Time (min)	People Receiving Message
0	1
10	2
20	4
30	?
40	?

THINK ABOUT IT

Work with a partner.

1. **State** the pattern and find the next three terms in the sequence 6, 18, 54, 162, ...

2. **Determine** how many people would receive Angela's e-mail message at the end of 40 minutes if it is forwarded to 3 people instead of 2 each time.

3. **Apply** the **look for a pattern** strategy to solve the following problem.
   A display of cereal boxes at Quik Mart is stacked in the shape of a pyramid. There are 4 boxes in the top row, 6 boxes in the next row, 8 boxes in the next row, and so on. The display contains 7 rows of boxes. How many boxes are in the display?

For **Extra Practice,** see page 582.

ON YOUR OWN

4. The second step of the 4-step plan for problem solving asks you to *plan* the solution. *Explain* how you can use the look for a pattern strategy to help you plan a solution.

5. *Write a Problem* in which looking for a pattern would help you solve it. Explain your answer.

6. *Look Ahead* Study the pattern in the table.

$6 \times (-1) = -6$	$\rightarrow$	$-6 \div (-1) = 6$
$6 \times (-2) = -12$	$\rightarrow$	$-12 \div (-2) = 6$
$6 \times (-3) = -18$	$\rightarrow$	$-18 \div (-3) = 6$
$6 \times (-4) = -24$	$\rightarrow$	$-24 \div (-4) = 6$
$6 \times (-5) = -30$	$\rightarrow$	$-30 \div (-5) = 6$

Find $-36 \div (-6)$.

MIXED PROBLEM SOLVING

STRATEGIES

Look for a pattern.
Solve a simpler problem.
Act it out.
Guess and check.
Draw a diagram.
Make a chart.
Work backward.

Solve. Use any strategy.

7. Health Minya has decided to start an exercise program. She plans to begin by working out for 5 minutes and then doubling her exercise time each day for 1 week. Write a sequence showing the length of time she exercises each day. Is her plan reasonable? Why or why not?

8. Life Science The table shows about how many times a firefly flashes at different temperatures. Estimate how many times a firefly will flash when the temperature is 36°C.

Outside Temperature (°C)	Flashes Per Minute
20	9
24	11
28	14
32	18

9. Physical Science The 176-pound pendulum at the Center of Science and Industry museum in Columbus, Ohio, swings back and forth, knocking down two pegs every 15 minutes. How many pegs are knocked down after 3 hours?

10. Life Science The graph shows how this summer's rainfall compares to normal.

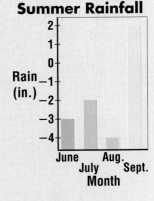

Summer Rainfall

a. What does 0 on this graph represent?

b. Write an integer to represent the rainfall for each month shown.

c. Write a sentence that summarizes the message this graph conveys about this summer's rainfall.

11. Earth Science The giant kelp seaweed is found in the Pacific Ocean. One plant grows 3 feet the first two days. If it continues to grow at the same rate, what would be the length of the seaweed at the end of 80 days?

12. Standardized Test Practice Joan took a bag of cookies to the school musical rehearsal. Half were given to the musicians and five to the director of the play. Joan was left with 15 cookies. How many cookies did she take to rehearsal?

A 30   **B** 35   **C** 40

D 45   **E** Not Here

What you'll learn

You'll learn to divide integers.

When am I ever going to use this?

Knowing how to divide integers can help you split the cost of a gift among any number of people.

Did you know? The Grand Canyon was carved out as a result of 60 million years of erosion by the Colorado River.

The ocean waves cause some coastlines to recede every year. Suppose a beach receded 8 centimeters in 4 years. If the beach receded at a steady rate, what was the average change in the coastline per year?

Let b represent the average change per year. To find b, divide -8 by 4.

$$-8 \div 4 = b$$

Division of integers is related to multiplication. The division sentence $-8 \div 4 = b$ can be written as the multiplication sentence $4 \times b = -8$.
Think: 4 times what number equals -8?

$$4(2) = 8$$
$$4(-2) = -8 \quad \checkmark$$

So, $b = -2$. The beach receded 2 centimeters per year.

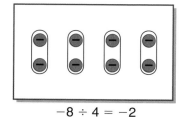

$-8 \div 4 = -2$

Let's see how some other division sentences are related to multiplication sentences.

$$2 \times 3 = 6 \quad \rightarrow \quad 6 \div 2 = 3$$
$$2\,(-3) = -6 \quad \rightarrow \quad -6 \div 2 = -3$$
$$-2 \times 3 = -6 \quad \rightarrow \quad -6 \div (-2) = 3$$
$$-2 \times (-3) = 6 \quad \rightarrow \quad 6 \div (-2) = -3$$

The pattern suggests the following rule to determine the sign of a quotient.

Dividing Integers	The quotient of two integers with the same sign is positive. The quotient of two integers with different signs is negative.

Examples

Solve each equation.

1 $a = -48 \div (-4)$

$a = -48 \div (-4)$ *The signs are the same.*

$\quad = 12$ *The quotient is positive.*

2 $-20 \div 5 = d$

$-20 \div 5 = d$ *The signs are different.*

$\quad -4 = d$ *The quotient is negative.*

You can solve problems in algebra by dividing integers.

Example 3 **Algebra** Evaluate $\frac{h}{jk}$ if $h = 36, j = 2$, and $k = -3$.

INTEGRATION

$\frac{h}{jk} = \frac{36}{2(-3)}$ $h = 36, j = 2, k = -3$

$\quad = \frac{36}{-6}$ $\frac{36}{-6}$ means $36 \div (-6)$.

$\quad = -6$

CHECK FOR UNDERSTANDING

Communicating Mathematics

Read and study the lesson to answer each question.

1. *Write* two division sentences related to the multiplication sentence $4 \times (-3) = -12$.

2. *Explain* why $\frac{45}{-9}$ and $\frac{-45}{9}$ are equal.

Guided Practice

Solve each equation.

3. $-20 \div (-5) = k$ 4. $t = -63 \div 7$ 5. $12 \div (-2) = j$

6. $c = -42 \div (-6)$ 7. $n = 51 \div (-17)$ 8. $-100 \div 20 = b$

Evaluate each expression if $x = -18, y = 6$, and $q = -3$.

9. $\frac{96}{q}$ 10. $-108 \div y$ 11. $\frac{y^2}{x}$

12. **Geometry** The formula for the area A of a parallelogram is $A = bh$, where $b =$ base and $h =$ height. Find the height of the parallelogram if the area is 54 cm² and the base is 9 cm.

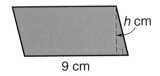

h cm

9 cm

EXERCISES

Solve each equation.

Practice

13. $15 \div (-3) = n$ 14. $p = -44 \div 11$ 15. $a = -14 \div (-1)$

16. $c = 72 \div (-2)$ 17. $-108 \div (-9) = v$ 18. $t = -64 \div 4$

19. $27 \div (-3) = s$ 20. $-90 \div 6 = d$ 21. $-45 \div (-15) = k$

22. $-100 \div (-100) = y$ 23. $q = -56 \div 8$ 24. $300 \div (-25) = h$

25. $r = -220 \div (-1)$ 26. $68 \div (-17) = j$ 27. $-140 \div 7 = m$

Evaluate each expression if $a = -27$, $b = 9$, and $c = -3$.

28. $a \div b$

29. $a \div (-3)$

30. $69 \div c$

31. $a^2 \div b$

32. $\dfrac{126}{b}$

33. $\dfrac{a}{bc}$

34. $\dfrac{a}{c}$

35. $\dfrac{a^2}{bc}$

36. Find the quotient of -36 and -3.

Applications and Problem Solving

37. *Gifts* The six class officers at Fair Middle School have decided to buy a class gift for their faculty advisor. If the gift costs $24, write an equation that can be used to represent how much each student must contribute towards the purchase of the gift.

38. *Sales* The graph shows four magazines that had losses between 1989 and 1999. The negative numbers represent how many fewer magazines were sold in 1999 than in 1989. Find the mean of these four numbers.

39. *Critical Thinking* List all the numbers by which -30 is divisible.

Newsstand Losses
EXTRA
Magazines
A   B   C   D
−1,793,000
−3,821,100   −3,618,200
−4,012,100

Mixed Review

40. **Standardized Test Practice** An oil rig is drilling at a rate of 7 feet per minute. How far has the oil rig dug after 42 minutes? *(Lesson 5-6)*

A 35 ft   **B** −49 ft   **C** −264 ft   **D** −294 ft   **E** Not Here

41. Replace ● in $\dfrac{1}{6}$ ● $\dfrac{2}{15}$ with $<$, $>$, or $=$ to make a true sentence. *(Lesson 4-10)*

42. Describe the pattern in the sequence 4, 12, 36, 108, ... Then find the next three terms. *(Lesson 4-3)*

43. *Nutrition* The table shows the Calorie information for two meals from an Indian restaurant. Find how many of the Calories in each meal come from fat. *(Lesson 2-4)*

a. Meal A

b. Meal B

Meal	Total Calories	Portion of Calories from Fat
A Nan Shrimp biryani Raita Tamata salat	480	0.24
B Samaso Chicken tandoori Peas and rice Sang paneer	752	0.30

Source: *Vitality*

44. *Music* A band is putting together a new CD. It can use up to 60 minutes of music. The band members have selected 6 songs that are each 4 minutes long and 8 songs that are each 5 minutes long. Will all of their selections fit on the CD? Explain. *(Lesson 1-1)*

For **Extra Practice,** see page 582.

5-8

Integration: Geometry
Graphing Transformations

What you'll learn

You'll learn to graph transformations on a coordinate plane.

When am I ever going to use this?

You'll be able to see transformations in art and nature.

Word Wise

transformation
reflection
translation

Many flags have a variety of geometric **transformations**. That is, one geometric shape is used in many different positions to form a pattern or design.

Flag of St. Vincent and the Grenadines

There are many ways to move a geometric shape on a coordinate plane. A figure can be flipped, turned, slid, stretched, or shrunk. When a figure is flipped, it is called a **reflection**. When it is slid, it is called a **translation**. Both of these transformations can be described using ordered pairs and their graphs.

Example

INTEGRATION

1

Geometry Triangle *CDE* has vertices $C(2, 1)$, $D(5, 3)$, and $E(3, 4)$. Graph its reflection over the *y*-axis.

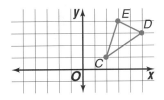

Explore   Graph $\triangle CDE$ by graphing each ordered pair and connecting the points to form $\triangle CDE$. Label each vertex.

Plan   For a figure reflected over the *y*-axis, the *y*-coordinates are exactly the same, and the *x*-coordinates are the opposite of each other. So, multiply the *x*-coordinate of each ordered pair by -1. Write the new ordered pair.

Solve

Vertices of △CDE	Multiply the x-coordinate by −1.	New Ordered Pairs
C(2, 1)	(2 × (−1), 1)	C′(−2, 1)
D(5, 3)	(5 × (−1), 3)	D′(−5, 3)
E(3, 4)	(3 × (−1), 4)	E′(−3, 4)

Graph the new ordered pairs and label each point. Connect these points to form $\triangle C'D'E'$.

Examine   The two triangles have the same shape and size. Triangle $C'D'E'$ is the result of flipping $\triangle CDE$ over the *y*-axis.

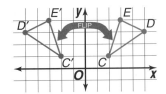

Study Hint

Reading Math C' is read as *C prime*, D' as *D prime*, and E' as *E prime*.

The transformation in Example 1 is a reflection over the *y*-axis. You can also do a reflection over the *x*-axis by multiplying the *y*-coordinates by -1. *You will reflect $\triangle ABC$ over the x-axis in Exercise 1.*

In a translation, a figure slides from one location to the next without changing its orientation.

Example ②

INTEGRATION

Geometry Graph △QRS with vertices Q(6, −2), R(0, 3), and S(−2, −1). Translate △QRS 7 units left and 4 units down.

Graph △QRS. Since each x-coordinate is moved 7 units to the left and each y-coordinate is moved 4 units down, the new ordered pairs can be written as $(x, y) + (−7, −4) = (x − 7, y − 4)$.

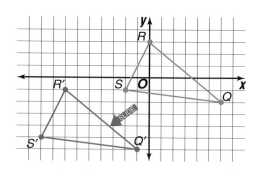

Vertices of △QRS	(x + (−7), y + (−4))	New Ordered Pairs
Q(6, −2)	(6 + (−7), −2 + (−4))	Q'(−1, −6)
R(0, 3)	(0 + (−7), 3 + (−4))	R'(−7, −1)
S(−2, −1)	(−2 + (−7), −1 + (−4))	S'(−9, −5)

CHECK FOR UNDERSTANDING

Communicating Mathematics

Read and study the lesson to answer each question.

1. **Draw** the result of reflecting △CDE from Example 1 over the x-axis.

2. **Tell** what type of transformation is shown by the vertices in the table.

△LMN	△L'M'N'
L(4, −2)	L'(−4, −2)
M(0, 7)	M'(0, 7)
N(−6, 1)	N'(6, 1)

3. **Write** about the reflections and/or translations found in the St. Vincent flag at the beginning of the lesson.

Guided Practice

4. Classify the graph as a *reflection* or a *translation*.

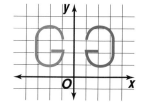

Graph each triangle and its transformation. Write the ordered pairs for the vertices of the new triangle.

5. △ABC with vertices A(1, 2), B(1, −4), and C(5, −1) reflected over the y-axis

6. △HIJ with vertices H(0, −3), I(2, 4), and J(−4, 3) translated 2 units left and 3 units down

7. **Art** Quillwork was a Native American art in which quills of a porcupine were "embroidered" in patterns on moccasins, belts, and bags. Describe the transformations that were used to create the quilled pattern at the right.

Practice

Classify each graph as a *reflection* or a *translation*.

8.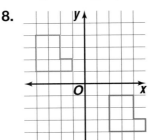

9.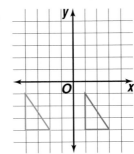

10.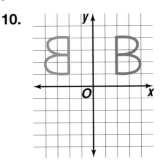

Graph each figure and its transformation. Write the ordered pairs for the vertices of the new figure.

11. $\triangle XYZ$ with vertices $X(1, 3)$, $Y(5, 1)$, and $Z(5, 8)$ reflected over the x-axis

12. $\triangle KLM$ with vertices $K(-3, -1)$, $L(-5, -6)$, and $M(-1, -6)$ translated 4 units right and 2 units down

13. $\triangle WXY$ with vertices $W(2, 1)$, $X(1, 6)$, and $Y(-2, 4)$ translated 5 units right and 3 units up

14. $\triangle ABC$ with vertices $A(5, 2)$, $B(1, 1)$, and $C(2, 4)$ reflected over the x-axis

15. $\triangle JKL$ with vertices $J(-7, 1)$, $K(-2, 5)$, and $L(-5, 7)$ reflected over the y-axis

16. rectangle $MNOP$ with vertices $M(-4, 0)$, $N(-4, -3)$, $O(-2, -3)$, and $P(-2, 0)$ translated 5 units right and 2 units down

Applications and Problem Solving

Real World

17. *Games* When playing chess, you can move game pieces up or down, left or right, or diagonally.
    a. What type of transformation is used in this game?
    b. Write the movement of the game piece at the right as an ordered pair.

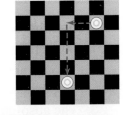

18. *Flags* Describe the transformation in the Philippines flag at the right.

19. *Critical Thinking* Graph $\triangle JKL$ with vertices $J(-7, 4)$, $K(-7, 1)$, and $L(-1, 1)$. Then graph $\triangle J'K'L'$ if both the x- and y-coordinates in $\triangle JKL$ are multiplied by -1. Describe this transformation.

For **Extra Practice**, see page 583.

Mixed Review

20. Solve $z = 360 \div (-6)$. *(Lesson 5-7)*

21. Use a factor tree to find the prime factorization of 630. *(Lesson 4-2)*

22. **Standardized Test Practice** Round 16.2573 to the nearest tenth. *(Lesson 2-2)*

    A 16.2   **B** 20   **C** 16.26   **D** 16.3

 **Chapter Review** For additional lesson-by-lesson review, visit:
www.glencoe.com/sec/math/mac/mathnet

Vocabulary

After completing this chapter, you should be able to define each term, concept, or phrase and give an example or two of each.

Number and Operations
absolute value (p. 185)
additive inverse (p. 197)
integer (p. 184)
negative integer (p. 184)
opposite (p. 184)
positive integer (p. 184)
zero pair (p. 196)

Geometry
reflection (p. 215)
transformation (p. 215)
translation (p. 215)

Algebra
coordinate system (p. 191)
ordered pair (p. 191)
origin (p. 191)
quadrant (p. 191)
x-axis (p. 191)
x-coordinate (p. 191)
y-axis (p. 191)
y-coordinate (p. 191)

Problem Solving
look for a pattern (p. 210)

Understanding and Using the Vocabulary

Choose the correct term or number to complete each sentence.

1. Integers less than 0 are (positive, negative) integers.
2. Two numbers represented by points that are the same distance from 0 are (opposites, integers).
3. The absolute value of 7 is $(7, -7)$.
4. The opposite of $(12, -12)$ is -12.
5. The (coordinate system, origin) is the point where the horizontal and vertical number lines intersect.
6. The horizontal number line is called the (x-axis, y-axis).
7. The x-axis and the y-axis separate the plane into four (quadrants, coordinates).
8. The first number in an ordered pair is the (x-coordinate, y-coordinate).
9. The second number in an ordered pair is the (x-coordinate, y-coordinate).
10. The sum of two (positive, negative) integers is negative.
11. The sum of any number and its additive inverse is (zero, positive).
12. A transformation is a design or pattern where a geometric shape is used in (one, many different) position(s).

In Your Own Words

13. *Explain* the difference between a reflection and a translation.

218 Chapter 5 Algebra: Using Integers

Objectives & Examples

Upon completing this chapter, you should be able to:

● read and write integers and find the opposite and absolute value of an integer *(Lesson 5-1)*

Write 4°F below 0 as an integer.
$$-4°F$$

Find the opposite and the absolute value of -7.

The opposite of -7 would be at the point 7 units to the right of 0. So, the opposite of -7 is 7. The point that represents -7 is 7 units from 0. So, $|-7| = 7$.

● compare and order integers *(Lesson 5-2)*

Replace the ● with $<$ or $>$ to make a true sentence.
$$2 \bullet -6$$

Since 2 is to the right of -6 on the number line, $2 > -6$.

● graph points on a coordinate plane *(Lesson 5-3)*

Name the ordered pair for point A and identify its quadrant.

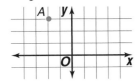

The ordered pair is $(-2, 3)$.
Point A is in quadrant II.

Review Exercises

Use these exercises to review and prepare for the chapter test.

Write an integer for each situation.

14. a withdrawal of $123

15. a gain of 14 yards

16. a deposit of $60

17. a loss of 5 pounds

Write the integer represented by the point for each letter. Then find its opposite and its absolute value.

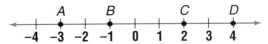

18. A **19.** B

20. C **21.** D

Replace each ● with $<$ or $>$ to make a true sentence.

22. $-18 \bullet -19$ **23.** $12 \bullet -12$

24. $-100 \bullet -10$ **25.** $0 \bullet -8$

Order the integers from least to greatest.

26. $7, -3, 10, 25, -41, 15$

27. $-13, 8, -11, 0, 10, 5$

Name the x-coordinate and the y-coordinate for each point labeled at the right. Then tell in which quadrant each point lies.

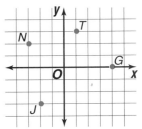

28. N **29.** T

30. G **31.** J

On graph paper, draw a coordinate plane. Then graph and label each point.

32. $L(-2, 0)$ **33.** $Z(3, -4)$

34. $Q(1.5, 3)$ **35.** $B(-4, -2)$

36. $G(0, -3)$ **37.** $R(-1, 2)$

Objectives & Examples

Review Exercises

add integers *(Lesson 5-4)*

Solve $p = 4 + (-3)$.

$|4| > |-3|$, so the sum is positive. The difference of 4 and 3 is 1. So, $p = 1$.

Solve each equation.

38. $c = 6 + (-2)$ **39.** $-10 + 4 = r$

40. $-5 + 12 = m$ **41.** $-7 + (-6) = t$

Evaluate each expression if $a = 6$, $b = 12$, and $c = -6$.

42. $b + (-4)$ **43.** $a + c$

---

subtract integers *(Lesson 5-5)*

Solve $y = -2 - 4$.

$y = -2 - 4$

$y = -2 + (-4)$ *To subtract 4, add −4.*

$y = -6$

Solve each equation.

44. $-13 - 4 = q$ **45.** $12 - (-12) = p$

46. $a = -4 - 6$ **47.** $z = 6 - (-2)$

Evaluate each expression if $q = -4$, $r = -2$, and $s = 9$.

48. $-7 - s$ **49.** $r - q$

---

multiply integers *(Lesson 5-6)*

Solve $h = -2(5)$.

The integers have different signs, so the product will be negative.

$h = -2(5)$

$h = -10$

Solve each equation.

50. $b = -6(-2)$ **51.** $c = (-3)^2$

52. $-8(4) = g$ **53.** $j = -5(5)$

Evaluate each expression if $w = 3$, $x = -6$, and $y = 4$.

54. $-3y$ **55.** $8w$ **56.** $5xy$

---

divide integers *(Lesson 5-7)*

Solve $-15 \div (-3) = a$.

The signs are the same, so the quotient is positive.

$-15 \div (-3) = a$

$5 = a$

Solve each equation.

57. $v = 45 \div (-9)$ **58.** $s = -10 \div 10$

59. $-12 \div (-2) = b$ **60.** $-52 \div (-4) = g$

Evaluate each expression if $j = -16$, $k = 6$, and $h = 2$.

61. $k \div (-3)$ **62.** $j \div h$

---

graph transformations on a coordinate plane *(Lesson 5-8)*

Graph $\triangle QRS$ with $Q(-3, 4)$, $R(-3, 1)$, and $S(0, 1)$. Then translate $\triangle QRS$ 2 units left and 3 units down.

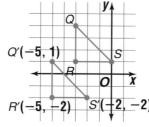

Graph each triangle and its transformation. Write the ordered pairs for the vertices of the new triangle.

63. $\triangle ABC$ with vertices $A(4, -2)$, $B(-2, -3)$, and $C(-1, 6)$ translated 3 units right and 4 units up

64. $\triangle RST$ with vertices $R(-1, 3)$, $S(2, 6)$, and $T(6, 1)$ reflected over the x-axis

Applications & Problem Solving

65. *Weather* During a winter storm, a weather announcer stated that the "windchill factor" was 18°F below 0. Write an integer for this situation. *(Lesson 5-1)*

66. *Earth Science* The table gives weekly water-level readings of Turtle Creek Lake over an 8-week period. The readings indicate the number of feet above or below flood level. Find the median of the readings. *(Lesson 5-2)*

Week	1	2	3	4	5	6	7	8
Water Level	−8	−8	−6	−5	−2	−5	0	3

67. *Football* On the first play of the fourth quarter, the Bearcats lost 12 yards. On the second play, they ran for 8 yards. What was the total number of yards gained or lost? *(Lesson 5-4)*

68. *Look for a Pattern* Find the next three numbers in the sequence 34, 26, 18, 10, ... *(Lesson 5-7A)*

69. *Fund-raising* The Debate Team, the Chess Club, and the Pep Club participated in a joint fund-raiser. If $327 was raised, write an equation that can be used to represent how much each club should receive. *(Lesson 5-7)*

Alternative Assessment

Open Ended

Suppose you are the treasurer of a club. The club has a checking account with a balance of $74. You have written checks in the amounts of $17, $14, and $25 for supplies. You have collected dues of $6 from 7 members. You are to deposit the collected dues into the account. How can you find the new balance?

If the bank charged a $10 service fee, find the current balance.

PORTFOLIO Select one of the assignments from this chapter that you found particularly challenging. Place it in your portfolio.

A practice test for Chapter 5 is provided on page 611.

Completing the CHAPTER Project

Find the latitude, highest temperature, and lowest temperature for your city and four other cities in the United States. (If your city is one of the selected cities, add another city from your state.) Add this information to your table and the graphs you made in the exercises. Explain whether the new data are consistent with your original conclusions. Use the following checklist to make sure your project is complete.

☑ The table of latitudes and highest and lowest temperatures has a total of 15 cities.

☑ You have two graphs.

☑ You have a paragraph describing any relationships or interesting patterns that you found by studying your graphs and table.

Section One: Multiple Choice

There are ten multiple choice questions in this section. Choose the best answer. If a correct answer is *not here,* choose the letter for Not Here.

1. Which is equivalent to 5^4?
   A 20
   B 125
   C $4 \cdot 4 \cdot 4 \cdot 4 \cdot 4$
   D $5 \cdot 5 \cdot 5 \cdot 5$

2. If 4 computers are needed for every 7 students in a grade, how many computers are needed for 280 students?
   F 40
   G 32
   H 160
   J 280

3. At the school raffle, 20 prizes were placed inside a box. There were 12 dinner gift certificates, 6 movie tickets, and 2 baseball tickets. What is the probability of picking a dinner gift certificate?
   A $\frac{15}{20}$
   B $\frac{1}{4}$
   C $\frac{3}{5}$
   D $\frac{3}{20}$

4. Write the addition sentence shown by the number line.

   F $-3 + 7 = 4$
   G $3 + 4 = 7$
   H $-3 + 4 = -1$
   J $-3 + (-7) = -10$

Please note that Questions 5–10 have five answer choices.

Use the chart below to answer Questions 5 and 6.

Several civic groups in Westview collected newspapers to raise money for a spring planting and clean up day. The chart shows the number of pounds collected by each group during a 3-week time period.

Group	Week 1	Week 2	Week 3
Girl Scouts	ЖЖ	Ж III	Ж III
Boy Scouts	Ж II	Ж Ж	Ж I
Keep Our City Clean	III	Ж	Ж II
Garden Club	ЖЖ	Ж II	Ж IIII

5. If the groups were paid $0.08 per pound for the newspaper, how much more did the Garden Club earn than the Keep Our City Clean group?
   A $2.08
   B $1.20
   C $0.88
   D $0.48
   E Not Here

6. How many more pounds of newspaper did the Girl Scouts and Boy Scouts collect together than the Keep Our City Clean group and the Garden Club together?
   F 89 lb
   G 49 lb
   H 41 lb
   J 30 lb
   K Not Here

7. Manuel had a bag of marbles. He gave half of them to Olivia and one-third of the remaining marbles to Angela. He had 18 marbles left. How many marbles did Manuel have in the bag to start with?

A 54

B 48

C 27

D 36

E Not Here

8. At the school cafeteria, Jeremy bought a turkey sandwich for $1.99, potato chips for $0.76, and milk for $0.35. How much did his lunch cost assuming there was no tax?

F $3.20

G $3.10

H $3.00

J $2.90

K Not Here

9. Darlene has $18.48 in her purse. If she gets paid $15 from her paper route, how much money will she have?

A $12.48

B $13.48

C $23.48

D $33.48

E $17.48

10. A submarine is diving from the surface at the rate of 79 feet per minute. What is the depth of the submarine after 8 minutes?

F −71 ft

G −87 ft

H −602 ft

J −632 ft

K Not Here

Test-Taking Tip THE PRINCETON REVIEW

Most standardized tests have a time limit, so you must budget your time carefully. Some questions will be easier than others. If you cannot answer a question within a few minutes, go on to the next one. If there is still time left when you get to the end of the test, go back to the questions that you skipped.

Section Two: Free Response

This section contains five questions for which you will provide short answers. Write your answers on your paper.

11. Marcy has a piece of string that is 85 inches long. What is the greatest number of 12-inch pieces that she can cut from the string?

12. Joshua had $53.33 in his savings account. He took out $15 to spend at the movies. How much money did he have left?

13. $|-86| =$

14. Space shuttles encounter temperatures that range from −250°F while in orbit to 3,000°F during reentry of Earth's atmosphere. Find the range of temperatures that space shuttles encounter.

15. The human body uses 65% of the oxygen it takes in. Express this as a fraction in simplest form.

 **Test Practice** For additional test practice questions, visit:

www.glencoe.com/sec/math/mac/mathnet

CHAPTER 6

Algebra: Exploring Equations and Functions

What you'll learn in Chapter 6

- to solve addition, subtraction, and multiplication equations,
- to write simple algebraic expressions and equations,
- to solve inequalities,
- to graph functions and linear equations by plotting points, and
- to solve problems by working backward.

CHAPTER Project

AMERICA'S SCREAM MACHINES

Cedar Point, a world-famous amusement park on the coast of Lake Erie in Ohio, has earned the nickname "America's Roller Coast." The park built its first roller coaster in 1892 and has continued to add roller coasters ever since.

In this project, you'll use equations to find which of Cedar Point's roller coasters has the fastest average speed. You'll also choose one of the roller coasters and graph the distance it travels and the number of riders it can carry. Finally, you'll do research about your roller coaster and prepare a display of statistics and information.

Getting Started

- Look at the table. Which roller coaster travels the greatest distance? Which takes the least amount of time?
- Which roller coaster can accommodate the most riders per hour?

Roller Coasters at Cedar Point			
Name	Track Length (feet)	Time of Ride	Rider Capacity (per hour)
Blue Streak	2,558	1 min 45 s	1,400
Cedar Creek	2,540	2 min 42 s	2,400
Corkscrew	2,050	2 min	1,800
Gemini	3,935	2 min 20 s	3,300
Iron Dragon	2,800	2 min	2,000
Magnum	5,106	2 min	2,000
Mantis	3,900	2 min 40 s	1,800
Mean Streak	5,427	2 min 45 s	1,600
Raptor	3,790	2 min 16 s	1,800
Wildcat	1,837	1 min 25 s	900

Technology Tips

- Use a **spreadsheet** to find the average speed of each roller coaster.
- Use **computer software** to make graphs.
- Use a **graphing calculator** to make tables and graphs.

 *inter*NET **Data Update** For up-to-date CONNECTION information on Cedar Point, visit:

www.glencoe.com/sec/math/mac/mathnet

Working on the Project

You can use what you'll learn in Chapter 6 to help you find the average speeds and make your graphs.

Page	Exercise
236	32
252	8
257	30
261	Alternative Assessment

COOPERATIVE LEARNING

6-1A Solving Equations Using Models

A Preview of Lesson 6-1

cups and counters

equation mat

If you have ever balanced on a seesaw, you've used a simple machine called a lever. When a seesaw is balanced, the forces on the left side and right side are equal. A seesaw is also a model for an equation.

In this lab, you will use models to solve simple equations.

TRY THIS

Work with a partner.

1 Solve the equation $x + 3 = 5$ using models.

- Use a cup to represent the unknown value, x. Place 1 cup and 3 yellow counters on the left side of an equation mat. Place 5 yellow counters on the right side of the mat.

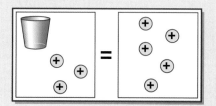

- The goal is to determine what's in the cup. To do so, get the cup by itself on one side of the mat. Then the counters on the other side will be the value of the cup, or x.

- To find the value of x, remove 3 counters from the left side and 3 counters from the right side. *You need to remove the same amount from both sides to maintain the balance.*

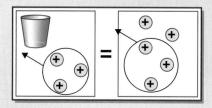

- The cup is by itself on one side of the mat. There are two counters on the other side. Therefore, the solution is 2.

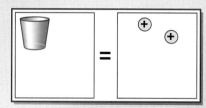

ON YOUR OWN

Solve each equation using models.

1. $x + 1 = 6$ **2.** $x + 2 = 5$ **3.** $x + 5 = 10$

4. $x + 4 = 7$ **5.** $x + 0 = 3$ **6.** $x + 3 = 3$

7. In what way is a balanced seesaw like an equation?

8. Explain how solving equations using cups and counters is similar to keeping a seesaw in balance.

Some equations are solved by using *zero pairs*. A zero pair consists of a positive counter and a negative counter. You may add or subtract a zero pair from either side of an equation without changing its value. Remember, yellow counters represent positive integers, and red counters represent negative integers.

TRY THIS

2 Solve the equation $x + 2 = -1$ using models.

- Place 1 cup and 2 yellow counters on the left side of an equation mat. Place 1 red counter on the right side.

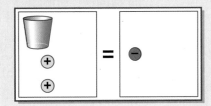

- The goal is to get the cup by itself on one side of the mat. However, it is not possible to remove 2 yellow counters from each side of the mat. Instead, add 2 red counters to each side of the mat, which forms zero pairs.

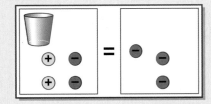

- Now, remove 2 zero pairs from the left side.

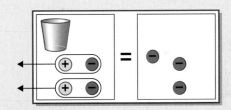

- The cup is by itself on one side of the mat. There are 3 negative counters on the other side. Therefore, the solution is -3.

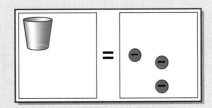

ON YOUR OWN

Solve each equation using models.

9. $x + 3 = -2$ **10.** $x + 4 = 1$ **11.** $-2 = x + 1$

Write each subtraction expression as an addition expression. Then solve the equation. For example, write $x - 3 = -2$ as $x + (-3) = -2$.

12. $x - 3 = -2$ **13.** $x - 1 = -3$ **14.** $4 = x - 2$

15. Explain why you can remove a zero pair from one side of a mat without changing the value of the equation.

Solving Addition and Subtraction Equations

What you'll learn

You'll learn to solve addition and subtraction equations.

When am I ever going to use this?

In geometry, you'll use equations to find missing angle measures.

Word Wise

subtraction property of equality
addition property of equality

On April 13, 1997, golfer Tiger Woods broke the course record for the Masters Championship with a final score of 18 under par, or -18. For each of the first three rounds, his scores were -2, -6, and -7. What was his score for the fourth round? *This problem will be solved in Example 4.*

You have solved equations by using models or by using mental math skills. However, you may not be able to solve all equations using these methods. The following examples describe methods you can use to solve all addition and subtraction equations.

Did you know Tiger Woods is also the youngest golfer ever to win the Masters Championship.

Example ① Solve $x + 4 = 6$.

Method 1 Use symbols.

$$x + 4 = 6$$

$$x + 4 - 4 = 6 - 4$$

> *Subtracting 4 from each side of the equation is like removing 4 yellow counters from each side of the equation mat.*

$$x = 2$$

The solution is 2.

Method 2 Use models.

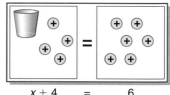

$x + 4 \quad = \quad 6$

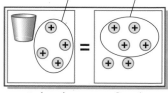

$x + 4 - 4 \quad = \quad 6 - 4$

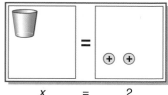

$x \quad = \quad 2$

In Example 1, you used the **subtraction property of equality**.

Subtraction Property of Equality	**Words:** If you subtract the same number from each side of an equation, then the two sides remain equal.
	Symbols: Arithmetic Algebra
	$4 = 4$ $a = b$
	$4 - 3 = 4 - 3$ $a - c = b - c$
	$1 = 1$

There is a similar property when addition is used.

Addition Property of Equality	**Words:** If you add the same number to each side of an equation, then the two sides remain equal.
	Symbols: Arithmetic Algebra
	$5 = 5$ $a = b$
	$5 + 4 = 5 + 4$ $a + c = b + c$
	$9 = 9$

Examples

2 Solve $x - 2 = 3$. Check your solution.

Method 1   Use symbols.

$$x - 2 = 3$$

$$x - 2 + 2 = 3 + 2$$

Adding 2 to each side of the equation is like adding 2 yellow counters to each side of the equation mat.

$$x = 5$$

When you remove zero pairs, you are simplifying the expressions.

Method 2   Use models.

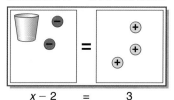

$x - 2$ = 3

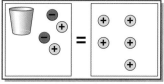

$x - 2 + 2$ = $3 + 2$

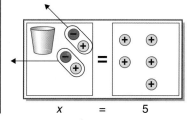

x = 5

> **Study Hint**
>
> **Mental Math** It is always wise to check your solution. You can often use arithmetic facts to check the solutions of simple equations.

To check your solution, replace x with 5 in the original equation.

$x - 2 = 3$

$5 - 2 \stackrel{?}{=} 3$ *Is this sentence true?*

   $3 = 3$ ✓ The solution is 5.

3 Solve $x - 5.7 = 6.3$. Check your solution.

$$x - 5.7 = 6.3$$

$$x - 5.7 + 5.7 = 6.3 + 5.7 \quad \textit{Add 5.7 to each side of the equation.}$$

$$6.3 \boxed{+} 5.7 \boxed{=} \mathit{12}$$

$$x = 12$$

Check: $x - 5.7 = 6.3$

   $12 - 5.7 \stackrel{?}{=} 6.3$ *Replace x with 12.*

      $6.3 = 6.3$ ✓ The solution is 12.

Example **4**

APPLICATION

Golf Refer to the beginning of the lesson. Find Tiger Woods' score for the fourth round.

Explore You know that the scores for the first three rounds were -2, -6, and -7. You know that the final score was -18. You need to find his score for the fourth round.

Plan Let s represent the fourth round score. The final score is the sum of the scores for each round. You can write an equation for this problem.

$$\underbrace{(-2) + (-6) + (-7)}_{} \quad + \quad s \quad = \quad -18$$

$$\qquad\quad \downarrow \qquad\qquad\qquad \downarrow \qquad\quad \downarrow$$

scores for first   *score for*   *final score*
three rounds   *fourth round*

LOOK BACK

You can refer to Lesson 5-4 to review adding integers.

Solve
$$(-2) + (-6) + (-7) + s = -18$$
$$(-15) + s = -18$$
$$(-15) + 15 + s = -18 + 15 \qquad \textit{Add 15 to each}$$
$$s = -3 \qquad\qquad\qquad \textit{side.}$$

Tiger's score for the fourth round was -3.

Examine Check the solution by adding.
$$-2 + (-6) + (-7) + (-3) = -18 \quad \checkmark$$

CHECK FOR UNDERSTANDING

Communicating Mathematics

Read and study the lesson to answer each question.

1. *Draw* a model that shows the equation $x - 4 = -2$.

2. *Tell* how to check your solution to an equation.

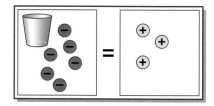

Math Journal

3. *Write* the equation shown by the model. Then explain how to solve the equation by using models and by using the properties of equality.

Guided Practice

Solve each equation. Use models if necessary. Check your solution.

4. $n + 6 = 11$ 5. $21 = r + 18$ 6. $x + 1.2 = 3.5$

7. $z - 5 = -3$ 8. $t - 8 = 4$ 9. $0 = b + 8$

10. The sum of a number and 3 is -2. This means $n + 3 = -2$. Solve the equation to find the number.

11. *Tourist Attractions* The Gateway to the West Arch in St. Louis, Missouri, is 630 feet tall. It is 75 feet higher than the Washington Monument in Washington, D.C. Use the equation $t + 75 = 630$ to find the height of the Washington Monument.

Practice

Solve each equation. Check your solution.

12. $a + 3 = 12$ 13. $m - 8 = 13$ 14. $27 = 18 + g$

15. $32 + c = 24$ 16. $34 = m + 18$ 17. $y + 43 = 68$

18. $k - 7.2 = 4.5$ 19. $-5 + v = 3$ 20. $m - 5 = -9$

21. $-9 = 3 + r$ 22. $-8 = y - 7$ 23. $-34 = t + 9$

24. $-9 + w = -12$ 25. $a + 3.9 = 5.6$ 26. $e + 11.8 = 13.1$

27. $13.2 = p + 4.7$ 28. $s - 5.9 = 4.8$ 29. $y - 16 = -5$

30. If you decrease a number by 4, the result is -5. This means $n - 4 = -5$. Solve the equation to find the number.

31. Negative 10 is the sum of a number and -6. Solve $-10 = n + (-6)$ to find the number.

Applications and Problem Solving

32. *Aviation* Orville and Wilbur Wright flew their airplane called *Flyer I* in Kitty Hawk, North Carolina, on December 17, 1903. Wilbur's flight was 364 feet, which was 120 feet longer than Orville's flight. Solve the equation $f + 120 = 364$ to find the length of his flight.

33. *Geometry* The sum of the measures of the angles of a triangle is 180°. Find the missing measure.

34. *Write an Equation* Write about a real-life situation involving the photo at the left that can be represented by the equation $x - 3 = 7$.

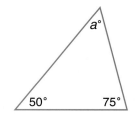

35. *Critical Thinking* Write two different equations that have -2 as a solution.

Mixed Review

36. **Standardized Test Practice** Which graph shows a translation of the letter *Z*? *(Lesson 5-8)*

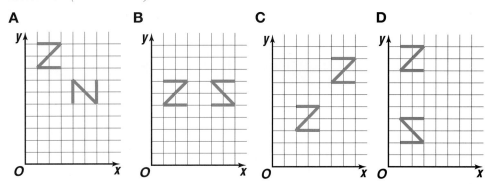

37. Find $12 - (-4)$. *(Lesson 5-5)*

38. On his most recent math test, Ricardo scored 84 out of 100 points. Express his score as a fraction in simplest form. *(Lesson 4-5)*

39. *Statistics* The mean income for a group of accountants was $26,266.67. Their incomes were $17,500, $26,100, $19,800, $23,400, $21,300, and $49,500. In what way is the mean misleading? *(Lesson 3-7)*

For **Extra Practice**, see page 583.

THINKING LAB

PROBLEM SOLVING

6-1B Work Backward

A Follow-Up of Lesson 6-1

Mike and Heather found a puzzle about famous toys in a magazine at the school library. How would you solve it?

- In the early 1900s, a toy company names its stuffed bears Teddy's Bears after President Teddy Roosevelt.
- Forty-nine years later, wire coil toys start rolling down stairs everywhere.
- Four years later, England exports miniature toy cars to the United States.
- Forty years pass until 1996, when small beanbag animals capture everyone's attention.

In what year was the Teddy Bear named?

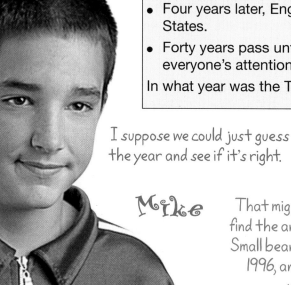

I suppose we could just guess the year and see if it's right.

Mike

That might work. But we can find the answer another way. Small beanbag animals were in 1996, and miniature toy cars were 40 years before. So, 1996 – 40 is 1956.

I get it. We'll keep working backward until we get to the Teddy Bear!

Heather

THINK ABOUT IT

Work with a partner.

1. In what year was the wire coil toy introduced?

2. In what year was the Teddy Bear named?

3. *Explain* why the guess-and-check strategy isn't the best choice for solving this problem.

4. Apply the **work backward** strategy to solve this problem.

Jaime rented 3 times as many videotapes as Phyllis last month. Phyllis rented 4 fewer than Marva, but 4 more than Paloma. Marva rented 10 videotapes. How many videotapes did each person rent?

For **Extra Practice,** see page 583.

ON YOUR OWN

5. Explain how you would solve the equation $x + 13 = 25$ by working backward.

6. Write a Problem that can be solved by working backward.

7. Explain how you can find the answer to Exercise 32 on page 231 by working backward.

MIXED PROBLEM SOLVING

STRATEGIES
Look for a pattern.
Solve a simpler problem.
Act it out.
Guess and check.
Draw a diagram.
Make a chart.
Work backward.

Solve. Use any strategy.

8. Numbers I'm thinking of a number. If I multiply it by 5 and add 17, the result is 67. What is the number?

9. Money Matters Antoine is on vacation and is planning to send postcards and letters to his friends. He has $2.66 to spend on postage. A stamp for a letter costs 33¢, and a stamp for a postcard costs 20¢. If he is going to spend the entire $2.66 on postage, how many postcards and letters can he send?

10. Food Mr. Roberts is delivering cartons of cereal to supermarkets. At the first market, he drops off half of the cartons he has in the truck. At each of the other markets, he drops off half of the cartons he has left. Then, at the eleventh market, he drops off one carton, which is the last one in the truck. How many cartons were originally in the truck?

11. Food When a certain chocolate bar was first introduced, it was not wrapped in paper. Two years later, in 1896, rolled chocolate candy became the first paper-wrapped candy bars. In what year was the chocolate bar introduced?

12. Money Matters Crystal and her sister, Ebony, each own an equal number of shares of stock. Crystal sells one third of her shares for $2,700. What was the total value of Crystal's and Ebony's stock before the sale?

13. Patterns Look at the model of the triangular numbers. How many dots would be in a triangle that has 10 dots on a side?

1   3   6   10

14. Games In a popular board game, players make words from tiles printed with letters of the alphabet. The squares on the board are pink, dark blue, light blue, red, or gray.

- There are 8 less pink than light blue.
- There are twice as many light blue as dark blue.
- There are $\frac{1}{6}$ as many dark blue as red.
- There are 28 fewer red than gray.

There are 100 gray squares on the board. How many are pink?

15. Standardized Test Practice The #6 bus runs every 8 minutes, the #9 bus runs every 10 minutes, and the #10 bus runs every 12 minutes. If all three buses leave the station at 8:00 A.M., when is the next time all three buses leave the station at the same time?

A 9:00 A.M.

B 10:00 A.M.

C 11:00 A.M.

D 12:00 P.M.

6-2 Solving Multiplication Equations

What you'll learn

You'll learn to solve multiplication equations.

When am I ever going to use this?

In physical science, you'll use the formula $d = rt$ to study velocity.

Word Wise

coefficient
division property
 of equality

Equations like $2x = -8$ are called multiplication equations because the expression $2x$ means *2 times the value of x.* You can also use models to solve multiplication equations.

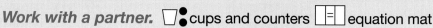

MINI-LAB

Work with a partner. � cups and counters ⬛ equation mat

Solve $2x = -8$.

- Place two cups on the left side of an equation mat. Place 8 red counters on the right side of the mat.

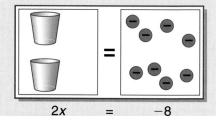

$$2x \qquad = \qquad -8$$

- Each cup must contain the same number of counters. Arrange the counters into two equal groups to correspond to the two cups.

The solution is -4.

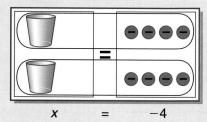

$$x \qquad = \qquad -4$$

Try This

1. Solve each equation using models.
   - **a.** $5x = 20$
   - **b.** $-12 = 4x$
   - **c.** $3x = 3$

Talk About It

2. What operation did you use to find each solution?
3. The **coefficient** of an expression like $3x$ is the numerical part, 3. How can you use the coefficient to solve the equation $3x = 12$?
4. How would you solve $2x = 5$ without using models?

In the Mini-Lab, you placed an equal number of counters in each cup. This suggests the operation of division.

Division Property of Equality	**Words:**	If you divide each side of an equation by the same nonzero number, then the two sides remain equal.	
	Symbols:	**Arithmetic**	**Algebra**
		$8 = 8$	$a = b$
		$\dfrac{8}{2} = \dfrac{8}{2}$	$\dfrac{a}{c} = \dfrac{b}{c}, c \neq 0$
		$4 = 4$	

Notice that the division problem $8 \div 2$ is written as the fraction $\dfrac{8}{2}$.

Solve each equation. Check your solution.

1 $12 = 3x$

Method 1

Use symbols.

Dividing each side of the equation by 3 is like placing an equal number of counters in each cup.

$12 = 3x$

$\dfrac{12}{3} = \dfrac{3x}{3}$ *Divide each side of the equation by 3.*

$4 = x$ $\quad 12 \div 3 = 4$

Check: $\quad 12 = 3x$

$\qquad 12 \stackrel{?}{=} 3 \cdot 4$

$\qquad 12 = 12 \ \checkmark$

The solution is 4.

Method 2

Use models.

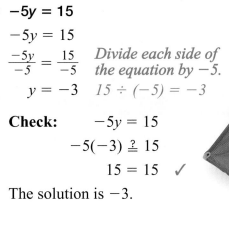

$\quad 4 \quad = \quad x$

2 $-5y = 15$

$-5y = 15$

$\dfrac{-5y}{-5} = \dfrac{15}{-5}$ *Divide each side of the equation by −5.*

$y = -3$ $\quad 15 \div (-5) = -3$

Check: $\qquad -5y = 15$

$\qquad\qquad -5(-3) \stackrel{?}{=} 15$

$\qquad\qquad\qquad 15 = 15 \ \checkmark$

The solution is −3.

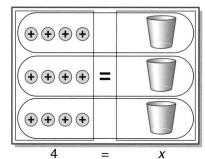

APPLICATION

Real World

3 **Kites** You can make a simple kite with newspaper, bendable sticks, and some string. The longer stick must be 1.5 times the length of the shorter stick. If 36 inches is the length of the longer stick, what should be the length of the shorter stick? Use the equation $1.5s = 36$, where s is the length of the shorter stick.

$1.5s = 36$

$\dfrac{1.5s}{1.5} = \dfrac{36}{1.5}$ *Divide each side of the equation by 1.5.*

$36 \div 1.5 = 24$

$s = 24$

Check: $1.5 \times 24 = 36 \ \checkmark$

The shorter stick should be 24 inches long.

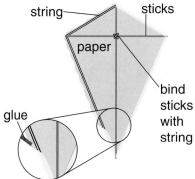

string — sticks

paper

bind sticks with string

glue

Fold paper over string and glue.

Cultural Kaleidoscope

Kites were originally used for military purposes. Around 1200 B.C., the Chinese used kites to send secret codes between army camps. The Chinese were also the first to fly kites for fun.

Communicating Mathematics

Read and study the lesson to answer each question.

1. *Write* the equation shown by the model. Then find the solution.

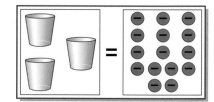

2. *Tell* whether 3 is a solution of $-7b = 21$.

HANDS-ON MATH

3. *Use models* or make a drawing to solve $12 = 2t$.

Guided Practice

Solve each equation. Use models if necessary. Check your solution.

4. $7c = 49$ 5. $9e = -54$ 6. $36 = -4m$

7. $-10x = -120$ 8. $5z = 4.5$ 9. $1.2c = 7.2$

10. *Life Science* An elephant, going at a steady speed of about 5 miles per hour, can walk faster than a human. A herd of elephants can easily cover 50 miles in one day. Find the time it takes for the herd to cover 50 miles. Use the formula $d = rt$, where d represents distance traveled, r represents the rate or average speed, and t represents the time.

EXERCISES

Practice

Solve each equation. Check your solution.

11. $3c = 21$ 12. $12y = 60$ 13. $34 = -2g$

14. $-5m = 35$ 15. $-8n = -16$ 16. $-24 = 6r$

17. $-196 = 4s$ 18. $-168 = 3w$ 19. $6 = 1.2s$

20. $1.3x = 1.56$ 21. $-15s = -225$ 22. $-4d = 64$

23. $3.9y = 18.33$ 24. $4x = 9.2$ 25. $1.8a = 9.72$

26. $2.6b = 2.08$ 27. $5.4 = 0.3p$ 28. $0.792 = 0.6c$

29. When a number is multiplied by -9, the result is 45. This can be represented by the equation $-9n = 45$. Solve to find the number.

30. Four tenths times a number is 16. Find the solution of $0.4n = 16$.

Applications and Problem Solving

31. *Life Science* The cougars that are found in the colder regions of North and South America are about 75 inches long. They are about 1.5 times longer than the cougars that are found in the tropical jungles of Central America. Use the equation $1.5c = 75$ to find c, the length of the tropical cougar.

32. *Working on the* **CHAPTER Project** Refer to the table on page 225.
    a. Use the formula $d = rt$ to find the average speed of each roller coaster in feet per second. In the formula, d represents distance in feet, t represents time in seconds, and r represents the rate or average speed in feet per second. Round each speed to the nearest whole number.
    b. Make a graph that shows the average speed of each roller coaster.

33. *Earth Science* Scientists determine the epicenter of an earthquake by measuring the time it takes for surface waves to travel between two places. How long would it take surface waves to travel from Los Angeles to Phoenix, which is a distance of 600 km, if they travel about 6 km/s through Earth's crust? Use the formula $d = rt$. (See Exercise 32.)

34. *Critical Thinking* Solve $3|x| = 6$.

Mixed Review

35. *Algebra* Solve $t - 3.6 = 4$. *(Lesson 6-1)*

36. **Standardized Test Practice** The owner of a pie shop buys bags of apples every day for apple pie. There are between 15 and 18 apples in a bag, and he always gets 9 bags. Which is a reasonable total for the number of apples that the shop owner buys every day? *(Lesson 5-6)*

   A 100   **B** 120   **C** 150   **D** 170   **E** Not Here

37. Solve $y = 18 + (-17)$. *(Lesson 5-4)*

For **Extra Practice,** see page 584.

38. *Statistics* Find the mean, median, and mode for this set of data. $10, $18, $15, $6, $13, $12, $10 *(Lesson 3-4)*

Math-O

Get Ready This game is for two, three, or four players.   ✂ scissors

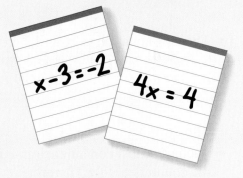

Math Skill
Solving Equations

📇 26 index cards   🖍 4 different-colored markers

Get Set Cut each index card in half, making 52 cards. To make a set of four cards, use the markers to put a different-colored stripe at the top of each card. Then write a different equation on each card. The solution of each equation should be 1. Continue to make sets of four cards having equations with solutions of 2, 3, 4, 5, 6, 0, −1, −2, −3, −4, and −5. Mark the remaining set of four cards "Wild".

Go
- The dealer shuffles the cards and deals five to each person. The remaining cards are placed in a pile facedown in the middle of the table. The dealer turns the top card faceup.

- The player to the left of the dealer plays a card with the same color or solution as the faceup card. Wild cards can be played any time. If the player cannot play a card, he or she takes a card from the pile and plays it, if possible. If it is not possible to play, the player places the card in his or her hand, and it is the next player's turn.

- The winner is the first person to play all cards in his or her hand.

 Visit www.glencoe.com/sec/math/mac/mathnet for more games.

COOPERATIVE LEARNING

6-3A Solving Two-Step Equations

A Preview of Lesson 6-3

□: cups and
counters

□=□ equation mat

In this lab, you will use what you know about solving one-step equations like $x + 3 = -4$ and $2x = 52$ to solve two-step equations like $2x - 3 = 1$.

TRY THIS

Work with a partner.

- First, let's build the equation $2x - 3 = 1$ using models. On the left side of the mat, place 2 cups and 3 negative counters. On the right side, place 1 positive counter.

  Remember, the goal is to get the cups by themselves on one side of the mat.

- Add 3 positive counters to each side of the equation to create zero pairs on the left side. Remove the zero pairs, since their value is 0.

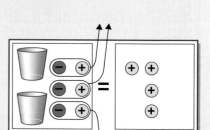

- The new equation is $2x = 4$.

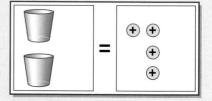

- Each cup must contain the same number of counters. Divide the counters evenly. Therefore, the solution is 2.

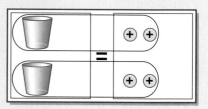

ON YOUR OWN

Solve each equation using models.

1. $3x + 1 = 7$ **2.** $2x - 4 = 2$ **3.** $9 = 4x + 1$

4. $5 = 3x - 4$ **5.** $2x - 3 = -3$ **6.** $2x + 3 = -3$

7. Why is an equation like $2x + 3 = 7$ called a two-step equation?

8. *Look Ahead* Solve $2x + 3 = 7$ without using models.

Solving Two-Step Equations

What you'll learn

You'll learn to solve two-step equations.

When am I ever going to use this?

You can use two-step equations to change temperatures from Celsius to Fahrenheit and vice versa.

Word Wise

term

In a popular movie, a young girl and her father lead a flock of orphaned geese from their home in Canada to the geese's winter home in North Carolina. In order to return to Canada on their own in the spring, the geese had to remember their original flight path and fly it in reverse order.

In algebra, you face a similar task when you solve a two-step equation like $2x + 1 = 17$.
This is a two-step equation because it has two **terms**, $2x$ and 1. It involves two different operations, multiplication and addition. To solve the equation, "undo" the operations in reverse order.

Here's how to solve the equation.

Example ① **Solve $2x + 1 = 17$.**

Method 1

Use symbols.

$$2x + 1 = 17$$
$$2x + 1 - 1 = 17 - 1 \quad \textit{Subtract 1 from each side.}$$

$$2x = 16$$
$$\frac{2x}{2} = \frac{16}{2} \quad \textit{Divide each side by 2.}$$
$$x = 8$$

Check: $2x + 1 = 17$
$$2 \cdot 8 + 1 \stackrel{?}{=} 17 \quad \textit{Replace x with 8.}$$
$$16 + 1 \stackrel{?}{=} 17$$
$$17 = 17 \checkmark$$

The solution is 8.

Method 2

Use models.

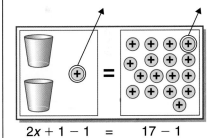

$$2x + 1 - 1 = \quad 17 - 1$$

Notice the cups are by themselves on the left side of the mat.

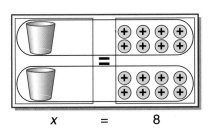

$$x \qquad = \qquad 8$$

2 Solve $-3r + 8 = -4$.

$$-3r + 8 = -4$$

$-3r + 8 - 8 = -4 - 8$ *Subtract 8 from each side.*

$-3r = -12$ $-4 - 8 = -4 + (-8)$ *or* -12

$\dfrac{-3r}{-3} = \dfrac{-12}{-3}$ *Divide each side by -3.*

$r = 4$ *Check the solution.*

The solution is 4.

INTEGRATION   **3** **Measurement** Temperature is usually measured on the Fahrenheit scale (°F) or the Celsius scale (°C). The highest temperature ever recorded in Orlando, Florida, was 102°F in May, 1945. Find this temperature in degrees Celsius by using the formula $F = 1.8C + 32$.

$$F = 1.8C + 32$$

$102 = 1.8C + 32$ *Replace F with 102.*

$102 - 32 = 1.8C + 32 - 32$ *Subtract 32 from each side.*

$70 = 1.8C$

$\dfrac{70}{1.8} = \dfrac{1.8C}{1.8}$ *Divide each side by 1.8.*

$70 \; \boxed{\div} \; 1.8 \; \boxed{=} \; 38.88888889$

$38.9 \approx C$ *Check the solution.*

The temperature was about 39°C.

CHECK FOR UNDERSTANDING

Communicating Mathematics

Read and study the lesson to answer each question.

1. **Show** how to use the model to solve the equation $3x + 2 = -4$.

2. **Explain** how the work backward strategy is used in solving two-step equations.

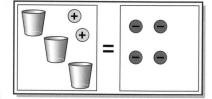

3. **You Decide** Sherita says that the first step in solving $5x - 4 = 16$ with models is to put 5 counters in each cup. Hector says it is to add 4 counters to each side of the mat. Who is correct? Explain your reasoning.

Guided Practice

Solve each equation. Check your solution.

4. $3n - 5 = 16$ 5. $2t + 7 = -1$ 6. $2x - 3 = 7$

7. $-11 = 3m + 1$ 8. $16 = 0.5r - 8$ 9. $5w + 9.2 = 19.7$

10. Three times a number plus 8 is -7. This can be represented by $3n + 8 = -7$. Solve the equation to find the number.

Practice

Solve each equation. Check your solution.

11. $4x + 5 = 13$

12. $3w - 4 = 8$

13. $8m - 12 = -36$

14. $2y + 1 = -3$

15. $13 = 4s + 1$

16. $2x + 5 = -13$

17. $2r - 3.1 = 1.7$

18. $-2y - 7 = 3$

19. $-3n - 8 = 7$

20. $21 = 13w - 5$

21. $16b - 6.5 = 9.5$

22. $85 = 4d + 5$

23. $19 = -4y + 3$

24. $0.2n + 3 = 8.6$

25. $17 = 17 + 8z$

26. $-2x - 7.2 = 18.2$

27. $28 = 7.5s - 2$

28. $1.5x - 16 = 8$

29. Add 3 to the product of a number and 4. The result is 15. Solve $3 + 4n = 15$ to find the number.

30. Multiply a number by -2 and then subtract 5. The result is 9. Find the solution of $-2n - 5 = 9$ to find the number.

Applications and Problem Solving

31. *Money Matters* Benny's Balloons charges $2 for each balloon in an arrangement. There is also a $5 fee for making the arrangement. If you have $15 to spend, how many balloons would you get? Solve the equation $15 = 2b + 5$, where b is the number of balloons.

32. *Measurement* Travelers to the 2000 Olympics in Sydney, Australia, will find that Australians measure temperature in degrees Celsius. Find the high temperature in your city yesterday and convert it to degrees Celsius using the formula $F = 1.8C + 32$. Is your answer warmer or cooler than the average high temperature for this month in Sydney?

Temperatures for Sydney, Australia (degrees Celsius)

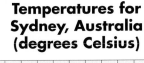
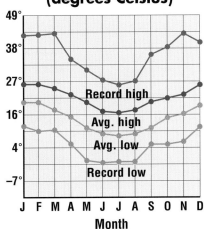

Source: *The Weather Almanac*

33. *Modeling* Show how you could use cups and counters to solve the equation $4x - 3 = x + 6$.

34. *Critical Thinking* Write two different two-step equations that have 1.2 as their solution.

Mixed Review

35. *Standardized Test Practice* Use the formula $A = bh$ to find the height of a parallelogram with a base of 34 mm and an area of 612 mm². *(Lesson 6-2)*

    A 20,800 mm   **B** 646 mm   **C** 578 mm   **D** 18 mm

36. *Algebra* Solve $p - 14 = 27$. *(Lesson 6-1)*

37. Express 9,800 in scientific notation. *(Lesson 2-9)*

38. Divide 0.0081 by 0.09. *(Lesson 2-6)*

For **Extra Practice,** see page 584.

Writing Expressions and Equations

Did you ever feel like these characters from *Peanuts*?

In this lesson, you will learn how to translate verbal phrases such as *three years older than the daughter* into algebraic expressions.

Words and phrases often suggest addition, subtraction, multiplication, and division. Here are some examples.

Addition or Subtraction		Multiplication or Division	
plus	minus	times	divided
sum	difference	product	quotient
more than	less than	multiplied	
increased by	less	of	
total	decreased by	twice	

Example

1 Write each phrase as an algebraic expression.

a. three runs less than the Pirates scored

Let *r* represent the number of runs the Pirates scored. The words *less than* suggest subtraction. So, the expression is $r - 3$.

b. twice as many tomatoes as last year

Let *t* represent the number of tomatoes last year. The word *twice* suggests multiplication by two. So, the expression is $2t$.

Remember, an equation is a sentence in mathematics that contains an equals sign. When you write a sentence as an equation, the word *equals* or *is* can be represented by the equals sign.

Example ➋ Write each sentence as an algebraic equation.

a. Five more than a number is 25.

Five more than a number is 25.
$$5 \qquad + \qquad n \quad = 25$$
The equation is $5 + n = 25$.

b. 17 is equal to four less than three times a number.

17 is equal to three times a number minus four.
$$17 \qquad = \qquad 3 \quad \times \quad n \quad - \quad 4$$
The equation is $17 = 3n - 4$.

One strategy you can use to solve a real-life problem is to solve an equation. First, choose a variable to represent one of the unknowns in a problem. This is called **defining the variable**. Then use the variable to write and solve an equation.

Example ➌

APPLICATION

Money Matters Shopping networks on television are a popular way to shop. In addition to the cost of the items, you usually pay a shipping fee. Koko wants to order several pairs of running shorts that cost $12 each. The total shipping fee is $7. How many shorts can she order with $55?

Explore You know that the shorts cost $12 each. There will be an additional fee of $7. You need to find how many shorts she can order for $55.

Plan Define a variable. Then write and solve an equation.

Solve Let s represent the number of shorts.
Then $12s$ represents the cost of the shorts.

cost of shorts   plus   shipping fee   is   total cost of order
$$12s \qquad + \qquad 7 \qquad = \qquad 55$$

$$12s + 7 = 55$$
$$12s + 7 - 7 = 55 - 7 \qquad \text{\textit{Subtract 7 from each side.}}$$
$$12s = 48$$
$$\frac{12s}{12} = \frac{48}{12} \qquad \text{\textit{Divide each side by 12.}}$$
$$s = 4$$

Examine Koko can order 4 pairs of running shorts.
Examine this solution.

Communicating Mathematics

Read and study the lesson to answer each question.

1. *Choose* the algebraic expression for the phrase *5 less than a number.*

   a. $5 - n$ **b.** $5 + n$ **c.** $n - 5$ **d.** $n \div 5$

2. *Explain* what it means to define the variable.

3. *Write* a sentence explaining the meaning of the expression $a + 3$, if a represents someone's age.

Guided Practice

Write each phrase as an algebraic expression.

4. seven more than t

5. eight less than p

6. 4 times as many bees

7. −9 increased by some number

Write each sentence as an algebraic equation.

8. Ten more than a number is 25.

9. The product of a number and 20 is 120.

10. If you double the number of eggs, the result is 6.

11. *Statistics* The median age of people in the United States was 33.0 years in 1990. This is 10.1 years more than the median age in 1900. Write and solve an equation to find the median age in 1900.

EXERCISES

Practice

Write each phrase as an algebraic expression.

12. the difference of g and 4

13. the product of x and two

14. the quotient of b and 5

15. nine increased by x

16. seventeen less than p

17. your age divided by 3

18. five years older than Paul

19. twice as many apples

20. Jamila's salary plus $1,100

21. Sue's score increased by 10

22. a number decreased by $\frac{1}{2}$

23. the square of some number divided by 10

Write each sentence as an algebraic equation.

24. The sum of a number and three is −8.

25. The product of a number and 8 is −64.

26. Three more than twice a number is 13.

27. Five less than seven times a number is 37.

28. Six less than the number of cookies is 18.

29. The number of students increased by 15 is 220.

30. His allowance less $3 is $12.

31. Eight more than twice her age is 60.

32. Ten inches less than her height is 50 inches.

Applications and Problem Solving

33. *Snacks* Potato chips were the top snack food in 1995, with 1.9 billion pounds consumed. This was 0.2 billion pounds more than the pounds consumed in 1992. Write and solve an equation to find the amount consumed in 1992.

Family Activity

Find yesterday's high and low temperature from a newspaper or television report. Let t represent the low temperature. Write an expression for the high temperature.

34. *Greeting Cards* The graph shows the average number of greeting cards purchased yearly by the average person in the United States. Let c represent the number of cards purchased by Americans ages 35 to 44.

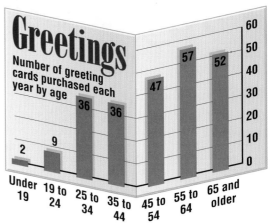

Source: American Greetings Corporation

a. Write an expression using c to represent the number of cards purchased by Americans ages 45 to 54.

b. Which age group would be represented by the expression $\frac{c}{4}$?

35. *Money Matters* A taxi company charges $1.50 per mile, plus a $10 fee. Suppose Eva can afford to spend $19 for a taxi ride from her apartment to the mall. Write and solve an equation to find the distance she can travel, so she will know whether to take this taxi or try to find a cheaper one.

36. *Critical Thinking* If x is an odd number, how would you represent the odd number immediately following it? preceding it?

Mixed Review

37. *Algebra* Solve the equation $2q + 6 = -20$. *(Lesson 6-3)*

38. *Standardized Test Practice* Mark had several baseball cards. He sold 15 of the cards and had 46 cards left. To find the number of cards he started with, Mark wrote the equation $c - 15 = 46$. How many cards did Mark start with? *(Lesson 6-1)*

    A 690   **B** 61   **C** 31   **D** 26

39. Solve $m = 6 - (-12)$. *(Lesson 5-5)*

40. *Geometry* Find the area of a parallelogram having a base of 2.3 centimeters and a height of 1.6 centimeters. *(Lesson 1-7)*

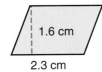

1.6 cm

2.3 cm

For **Extra Practice,** see page 584.

41. *Algebra* Evaluate b^5 if $b = 3$. *(Lesson 1-4)*

CHAPTER 6

Mid-Chapter Self Test

Solve each equation. Check your solution. *(Lessons 6-1, 6-2, and 6-3)*

1. $41 + w = 71$ **2.** $s - 5 = -11$ **3.** $z + 3.5 = 8.7$

4. $11b = 121$ **5.** $1.5y = 18$ **6.** $-24 = -6c$

7. $2m - 7 = -5$ **8.** $5x + 11 = 26$ **9.** $-8 = 6m + 16$

10. *Aviation* An airplane is flying at an altitude of t feet before it increases its altitude by 1,000 feet to avoid a thunderstorm. Write an expression for its new altitude. *(Lesson 6-4)*

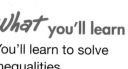

Inequalities

What you'll learn

You'll learn to solve inequalities.

When am I ever going to use this?

You'll use inequalities when you try to determine whether you have enough money to buy an item.

Word Wise

inequality

At age 14, Tara Lipinski became the youngest winner of the Ladies' World Figure Skating Championship. The ages of Tara and two other previous winners are graphed below.

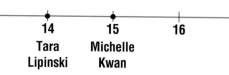

| 14 | 15 | 16 | 17 | 18 | 19 |
| Tara Lipinski | Michelle Kwan | | | | Kristi Yamaguchi |

All of the other winners were older than Tara. Their ages are graphed to the right of Tara's age. If a represents any of those ages, you can use the inequality $a > 14$ to show the relationship between their ages and Tara's age.

An **inequality** is a mathematical sentence that contains the symbols $<$, $>$, $\leq$, or $\geq$.

Words	Symbols
x is greater than 4.	$x > 4$
y is less than 10.	$y < 10$
m is greater than or equal to -5.	$m \geq -5$
r is less than or equal to 8.	$r \leq 8$

Inequalities may have many solutions.

LOOK BACK
You can refer to Lesson 1-5 to review solutions and replacement sets.

HANDS-ON MINI-LAB

Work with a partner.   ☐ sheet of paper   colored pencils

The number line shows whole numbers from 0 through 10. The large dots show whole number solutions of $x > 6$.

0 1 2 3 4 5 6 7 8 9 10

Try This

Draw a number line with the whole numbers 0 to 10. Draw a large dot at each number that is a solution of the inequality.

1. $x \geq 2$ **2.** $x + 2 > 3$ **3.** $2x > 3$ **4.** $x + 5 \leq 6$

Talk About It

5. Number lines include all numbers, not just whole numbers. How could you show all of the solutions of the inequality $x < 5$?

All numbers, not just whole numbers, are included in the graph of an inequality on a number line.

Example

CONNECTION

1 Health According to the *Mayo Clinic Health Letter,* a healthful breakfast cereal contains less than 3 grams of fat, at least 3 grams of fiber, and no more than 5 grams of sugar per serving. Write an inequality for each situation. Then graph the solution on a number line.

a. less than 3 grams of fat

The inequality is $f < 3$, where f is the number of grams of fat.

To show the solution, draw an open circle at 3. Then draw a thick arrow over the numbers to the left.

An open circle shows that this point is not included.

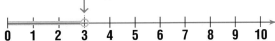

In Example 1, the number lines start at 0 because real-world quantities like the number of grams of fat cannot have a negative value.

b. at least 3 grams of fiber

At least 3 grams means 3 grams or more. Use the symbol $\geq$.

The inequality is $g \geq 3$, where g is the number of grams of fiber.

To show the solution, fill in a circle at 3. Then draw a thick arrow over the numbers to the right.

A filled-in circle shows that this point is included.

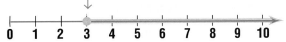

c. no more than 5 grams of sugar

No more than means 5 grams or less. Use the symbol $\leq$.

The inequality is $s \leq 5$, where s is the number of grams of sugar.

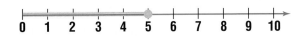

Inequalities and equations are solved in a similar manner.

Example

2 Solve $x - 4 > 3$. Check your solution. Then graph the solution.

$$x - 4 > 3$$
$$x - 4 + 4 > 3 + 4 \quad \text{\textit{Add 4 to each side of the inequality.}}$$
$$x > 7$$

Check: Try 10, a number greater than 7.

$$x - 4 > 3$$
$$10 - 4 > 3 \quad \text{\textit{Replace x with 10.}}$$
$$6 > 3 \quad \checkmark$$

The solution is $x > 7$, all numbers greater than 7.

Communicating Mathematics

Read and study the lesson to answer each question.

1. ***Write*** the inequality graphed on the number line at the right.

$$-3 \quad -2 \quad -1 \quad 0 \quad 1 \quad 2 \quad 3$$

2. ***Explain*** the difference between $x < 10$ and $x \leq 10$.

HANDS-ON MATH

3. ***Draw*** a number line that shows *all numbers greater than or equal to 4*.

Guided Practice

Solve each inequality. Graph the solution on a number line.

4. $5 + y < 13$ 5. $a - 3 \geq -5$ 6. $3t > 12$ 7. $x + 6 \leq 0$

8. ***Food*** One dozen jumbo eggs must weigh at least 30 ounces. Write this sentence as an inequality.

EXERCISES

Practice

Solve each inequality. Graph the solution on a number line.

9. $x + 3 > -4$ 10. $g - 5 > 2$ 11. $6d \geq 24$ 12. $t - 3 < -2$

13. $5 + p > 3$ 14. $2y \leq 15$ 15. $3r < 24$ 16. $b + 4 \leq 3$

17. $y + 1.3 < 4.5$ 18. $3x + 8 < 23$ 19. $2a - 5 > 9$ 20. $9d + 4 \geq 22$

Write an inequality for each sentence. Then solve the inequality.

21. Five times a number is greater than 60.

22. The sum of a number and 5 is less than or equal to -9.

Applications and Problem Solving

23. ***Civics*** The 26th amendment to the United States Constitution guarantees the right to vote to citizens who are eighteen years of age or older. Write an inequality showing the age of all voters.

24. ***Money Matters*** Luther earns $4 an hour doing yard work so he can buy a portable CD player that costs $115. Write an inequality for the least number of hours he needs to work to reach his goal. Then solve the inequality.

25. ***Critical Thinking*** Sandstone is formed by grains of sand that have become cemented together. These grains are between 0.06 millimeter and 2 millimeters in size. Write an inequality for the size of a grain of sand.

For **Extra Practice,** see page 585.

Mixed Review

26. ***Algebra*** Write an algebraic expression for the phrase *6 less than w.* *(Lesson 6-4)*

27. ***Algebra*** Solve the equation $2x - 3 = 19$. *(Lesson 6-3)*

28. ***Standardized Test Practice*** Lou has a section of tubing that is 83 inches long. What is the greatest number of 9-inch pieces that he can cut from the tubing? *(Lesson 5-7)*

    A 3   **B** 5   **C** 7   **D** 8   **E** Not Here

What you'll learn

You'll learn to represent functions as ordered pairs.

When am I ever going to use this?

In social studies, you'll use graphs to study recent trends in population.

Word Wise

function

When Christian Laettner played basketball for Duke University, he was 83 inches, or 211 centimeters, tall. Certainly this is taller than the average 18-year-old. The chart shows the average height for males ages 8 to 18.

This information can be represented by using an ordered pair (age, height). You can then graph the ordered pairs on a coordinate plane. Age is graphed on the horizontal axis, and height is graphed on the vertical axis.

Average Height for Males	
Age (yr)	Height (cm)
8	124
9	130
10	135
11	140
12	145
13	152
14	161
15	167
16	172
17	174
18	178

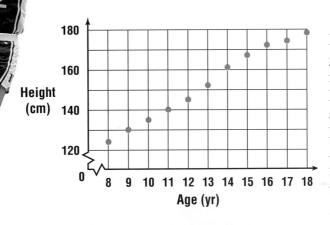

It is clear from the graph that height increases with age. Height is a **function** of age, which means that height *depends* on age. A function describes a relationship between two quantities.

Example

APPLICATION

Law Enforcement The Pennsylvania Bureau of Highway Safety and Traffic Engineering reports these data for cars.

a. Graph the ordered pairs (speed, distance) on a coordinate plane.

b. Describe how the stopping distance is related to the speed.

a.

Dry Pavement Stopping Distance for Cars	
Speed (mph)	Distance (feet)
55	289
60	332
65	378
70	426

b. The stopping distance depends on the speed. So stopping distance is a function of the speed.

2 **Life Science** A scientist was studying how temperature affects viruses. The chart at the right shows the results.

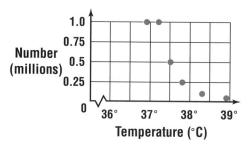

Viruses	
Temperature (°C)	Number (millions)
36.9	1.0
37.2	1.0
37.5	0.5
37.8	0.25
38.3	0.10
38.9	0.05

a. Graph the ordered pairs (temperature, number) on a coordinate plane.

b. What appears to be the relationship between temperature and viruses?

a.

b. The graph shows that the number of viruses decreases as the temperature increases. The number of viruses is a function of the temperature.

How long would it take all of the students in your school to complete one cycle of "the wave"? In the Mini-Lab, you will do an experiment to estimate the time.

HANDS-ON MINI-LAB

Work as a class.   🕙 stop watch

Try This

- Begin with five students, sitting in a row.
- At the timer's signal, the first student stands up, waves his or her arms overhead, and sits down. Each student repeats the wave. When the last student sits down, the timer records the time in seconds.
- Repeat for 10, 15, 20, and 25 students.

Talk About It

1. Graph the ordered pairs (number of students, time) on a coordinate plane.
2. How long would it take 30 students to complete the wave? 50 students? the number of students in your school?
3. Complete the sentence: The time it takes to do the wave is a function of __?__ .

Communicating Mathematics

Read and study the lesson to answer each question.

1. **Describe** a function.

2. **Name** two ways to represent functions.

HANDS-ON MATH

3. Refer to the Mini-Lab. How would your graph change if each student clapped twice and turned around before doing the wave?

Guided Practice

For Exercises 4 and 5, graph the ordered pairs in each table on a coordinate plane. Then write a sentence describing each relationship as a function.

4. **Health** The table shows the average number of heartbeats per minute for an adult who is working to improve aerobic conditioning.

Age	25	30	35	40	45	50	55	60	65
Beats per Minute	147	143	139	135	131	127	124	120	115

Source: *The Heart Rate Monitor Book*

5. **Geography** The population of the state of Arizona over the last several decades is given in the table.

Year	1930	1940	1950	1960	1970	1980	1990
Population (millions)	0.4	0.5	0.7	1.3	1.8	2.7	3.7

Source: U.S. Census

EXERCISES

Applications and Problem Solving

6. **Sports** Starting in 1991, NCAA Division I Men's Basketball instituted three-point shooting. The table shows the average number of attempts per game and the accuracy of the shots.

Three-point Shots		
Year	Attempts	Accuracy
1991	28.0	0.355
1992	29.8	0.354
1993	33.0	0.345
1994	34.3	0.345
1995	34.0	0.341

Source: NCAA

   a. Graph the ordered pairs (year, attempts) on one coordinate plane and (year, accuracy) on another coordinate plane.

   b. Write a statement that describes the trends in attempts and accuracy.

   c. Graph the ordered pairs (attempts, accuracy) on a third coordinate plane. What is the relationship between attempts and accuracy?

7. *Entertainment* The table shows average production costs per film for several years.

Year	1	2	3	4	5
Production Cost (millions of dollars)	42.3	44.0	50.4	54.1	59.7

Source: Motion Picture Association of America

a. Graph the ordered pairs (year, cost).

b. Write a statement that describes the trend in production costs.

8. *Working on the* **CHAPTER Project** Refer to the table on page 225.

a. Select one of the ten roller coasters. Find the average distance traveled by the roller coaster after 1, 2, 3, and up through 10 seconds. Write ordered pairs (time, distance) and graph them on a coordinate plane.

b. Is distance a function of the time? Explain your reasoning.

9. *Health* The graph shows the percent of people who snore or talk in their sleep. Write a statement that tells how the chances of doing either change as you grow older.

Dreamland

Percent who talk in their sleep or snore

17% 30% 42% 47%
29% 23% 15% 9%
18-24 25-34 35-49 50 plus
Age

Talk   zᶻZ Snore

Source: The Better Sleep Council

10. *Current Events* Find data in a recent newspaper or magazine that can be expressed as ordered pairs. Graph the data and write a statement explaining the relationship of the values.

11. *Critical Thinking* Refer to the beginning of the lesson. For which ages is the rate of growth fastest? Explain your reasoning.

Mixed Review

12. **Standardized Test Practice** Choose the graph that is the solution of $x \geq 1$. *(Lesson 6-5)*

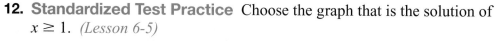

A −3 −2 −1 0 1 2 3

B −3 −2 −1 0 1 2 3

C −3 −2 −1 0 1 2 3

D −3 −2 −1 0 1 2 3

For **Extra Practice,** see page 585.

13. *Algebra* Evaluate $15(xy) - (x + y)$ if $x = 4$ and $y = 1$. *(Lesson 1-3)*

6-7A A Function of Time

A Preview of Lesson 6-7

⏱ stopwatch
or watch with
second hand

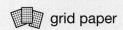 grid paper

On a cold morning, you may "see" your breath as the water vapor condenses when you exhale. But otherwise, you probably don't think much about breathing. Breathing is something we do naturally, without even thinking about it. In this lab, you will learn how breathing is a function of time.

TRY THIS

Work with a partner.

- One person does the breathing, and the other is the timer. When the timer says "Go," the first student begins to count the number of times he or she breathes *out*. After one minute, the timer says "Stop," and the first student records the number of breaths.

- Repeat five more times and record each result.

- Find the mean of the six results. This is the average number of breaths per minute.

- Copy the table and use this average to complete it.

Minutes	Minutes × Average Breaths per Minute	Total Breaths
1	1 × _?_	_?_
2	2 × _?_	_?_
3	3 × _?_	_?_
4	4 × _?_	_?_
5	5 × _?_	_?_
6	6 × _?_	_?_

In this table, the minutes are called *input* values, and the total breaths are called *output* values. The middle column contains the *function rule*. When you input a value into the function rule, you get an output value. On what does the total number of breaths depend? In this case, the total number of breaths depends on the number of minutes. So, the number of breaths is a function of time.

ON YOUR OWN

1. Graph the ordered pairs (minutes, breaths) on a coordinate plane.

2. Describe any patterns in the graph.

3. How can you use the graph to estimate the number of breaths you take in 8 minutes?

4. Let *m* be the number of minutes. Write an expression for the total number of breaths you take in *m* minutes.

5. Repeat this lab using number of heartbeats per minute.

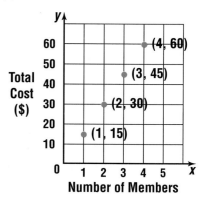

6-7 Functions and Equations

What you'll learn

You'll learn to solve equations with two variables and graph the solution.

When am I ever going to use this?

When you do experiments in science, you will use more than one variable.

Word Wise

linear equation

At the end of each summer, the Westerville Band performs at Cedar Point in Sandusky, Ohio. Each band member pays an admission price of $15, which allows them to ride all of Cedar Point's twelve roller coasters.

The band's total admission cost is a function of the number of members who go to Cedar Point. There are several ways to represent the function.

Method 1 Use a verbal description.
$15 times the number of members equals the total cost.

Method 2 Make a table of values.
The table lists the total cost for various numbers of members.

Number of Members	Multiply by 15	Total Cost
1	15 × 1	15
2	15 × 2	30
3	15 × 3	45
4	15 × 4	60

Method 3 Write an equation.
It's not convenient to list all of the possibilities in a table, so you can use variables. Let x represent the number of members and y represent the total cost.

number of members ⟶ ⟵ *total cost*
$$15x = y$$

Method 4 Draw a graph.
The solution of an equation with two variables consists of two numbers, one for each variable. The solution is usually written as an ordered pair, (x, y).

Based on the table above, four solutions of the equation $15x = y$ are (1, 15), (2, 30), (3, 45), and (4, 60). Graph the ordered pairs on a coordinate plane.

Notice that all four points lie on a line.

Example — 1

Find four solutions of $y = 2x - 1$. Write the solutions as ordered pairs and graph them.

Select any four values for x. We chose 2, 1, 0, and -1. Substitute these values for x to find y and complete the table of values.

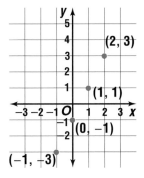

LOOK BACK
You can refer to Lesson 5-3 to review how to graph ordered pairs.

x	2x − 1	y	(x, y)
2	2(2) − 1	3	(2, 3)
1	2(1) − 1	1	(1, 1)
0	2(0) − 1	−1	(0, −1)
−1	2(−1) − 1	−3	(−1, −3)

Four solutions are $(2, 3)$, $(1, 1)$, $(0, -1)$, and $(-1, -3)$.

Notice that all four points in the graph above lie on a line. Draw a line through the points to graph *all* solutions of the equation $y = 2x - 1$.

The graph of $(3, 5)$ is on the line. Let's check whether $(3, 5)$ is also a solution.

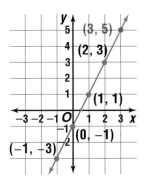

$y = 2x - 1$

$5 = 2(3) - 1$ *Replace x with 3 and y with 5.*

$5 = 5$ ✓

$(3, 5)$ is a solution.

An equation like $y = 2x - 1$ is called a **linear equation** because its graph is a straight line. Only two points are needed to graph the line. However, graph more points to check accuracy.

Example — 2
INTEGRATION

Geometry The equation $A = 5w$, where A is the area and w is the width, can be used to show the area of all rectangles whose length is 5 units. Graph the equation $A = 5w$.

Select any four values for w. Since w represents the width of a rectangle, choose only positive numbers. We chose 1, 2, 3, and 4.

Make a table of values.

w	5w	A	(w, A)
1	5(1)	5	(1, 5)
2	5(2)	10	(2, 10)
3	5(3)	15	(3, 15)
4	5(4)	20	(4, 20)

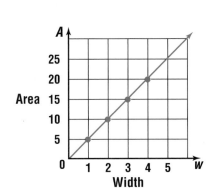

Graph the ordered pairs and draw a line through the points.

Communicating Mathematics

Read and study the lesson to answer each question.

1. **Choose** the equation for the graph.

   a. $y = 2x$ **b.** $y = x$

   c. $y = x - 1$ **d.** $y = 2x - 2$

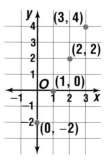

2. **List** four ways to show the relationship between two variables.

3. **You Decide** Beatriz thinks $(1, -1)$ is a solution of $y = 2x - 1$. Grace thinks it is *not* a solution. Who is correct? Explain your reasoning.

Guided Practice

Copy and complete each table. Then graph the ordered pairs.

4. $y = 2x + 1$

x	2x + 1	y
1		
2		
3		
4		

5. $y = 3x$

x	3x	y
-1		
0		
1		
2		

6. $y = -2x + 3$

x	-2x + 3	y
-1		
0		
1		
2		

Graph each equation.

7. $y = 3x - 1$ 8. $y = x - 2$

9. $y = -2x$ 10. $y = 1.5x$

11. **Money Matters** Angel earns $6 per hour at the Ice Cream Shop.
    a. Make a table that shows her total earnings for working 3, 5, 7, and 9 hours.
    b. Write an equation in which x represents the number of hours and y represents Angel's total earnings.
    c. Graph the equation.

Practice

Graph each equation.

12. $y = 2x + 3$ 13. $y = 4x - 1$ 14. $y = 0.25x$

15. $y = 0.5x - 1$ 16. $y = x - 3$ 17. $y = -2x + 2$

18. $y = -3x - 1$ 19. $y = x + 0.5$ 20. $y = 2x - 1.5$

21. $y = 2x - 5$ 22. $y = 6x$ 23. $y = 0.1x$

Make a table of values for each sentence. Then write an equation. Let x represent the first number and y represent the second number.

24. The second number is three more than the first.

25. The second number is twice the first.

26. The second number is the product of -3 and the first number.

27. The sum of the numbers is 10.

28. *Geometry* The formula for the perimeter of a square is $P = 4s$, where P is the perimeter and s is the length of a side. Graph the equation.

29. *Age* The table shows how Jared's age and his sister Emily's age are related.

Jared's age	1	2	3	4	5
Emily's age	7	8	9	10	11

   a. Write a verbal expression to describe how the ages are related.

   b. Write an equation for the verbal expression. Let x represent Jared's age and y represent Emily's age.

   c. Predict how old Emily will be when Jared is 10.

   d. Graph the equation.

30. *Working on the* **CHAPTER Project** Refer to the table on page 225.

   a. Select one of the ten roller coasters. Find the total number of passengers that could have ridden the roller coaster after 1, 2, 3, and up through 12 hours. Write the ordered pairs (time, number of passengers) and graph them on a coordinate plane.

   b. Write an equation for the graph. Let x represent the time in hours and y represent the total number of passengers.

31. *Scuba Diving* When you swim underwater, the pressure you feel in your ears is a function of the depth at which you are swimming. The equation that gives the pressure p, in lb/in², as a function of the depth d, in feet, is $p = 0.43d$.

   a. Find the pressure at a depth of 10 feet.

   b. Graph the equation.

   c. Use your graph to predict the pressure at 30 feet.

32. *Write a Problem* about a real-life situation that can be represented by the equation $y = 3x$.

33. *Critical Thinking* Refer to the beginning of the lesson. Explain whether it makes sense to draw a line joining the points on the graph.

Mixed Review

34. *Recreation* The graphs show recent trends in tennis and softball. Which activity has had the greatest decrease in recent years? *(Lesson 6-6)*

35. *Algebra* Solve $6p = 72$. *(Lesson 6-2)*

36. *Measurement* Complete the sentence 2.33 km = _?_ m. *(Lesson 2-8)*

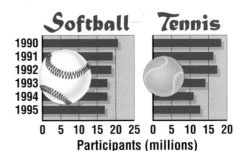

Softball   Tennis

Participants (millions)

Source: National Sporting Goods Association

37. **Standardized Test Practice** Eduardo bought a 2-liter bottle of cola. He drank 0.735 liter for lunch. How much cola was left? *(Lesson 2-3)*

   A 0.652 L   **B** 1.265 L   **C** 1.652 L   **D** 1.865 L   **E** Not Here

inter NET CONNECTION Chapter Review **For additional lesson-by-lesson review, visit:**
www.glencoe.com/sec/math/mac/mathnet

Vocabulary

After completing this chapter, you should be able to define each term, concept, or phrase and give an example or two of each.

Algebra

addition property of equality (p. 229)
coefficient (p. 234)
defining the variable (p. 243)
division property of equality (p. 234)
inequality (p. 246)
linear equation (p. 255)
subtraction property of equality (p. 228)
term (p. 239)
zero pairs (p. 227)

Problem Solving

work backward (p. 232)

Patterns and Functions

function (p. 249)
function rule (p. 253)
input (p. 253)
output (p. 253)

Understanding and Using the Vocabulary

State whether each sentence is *true* or *false*. If *false*, replace the underlined word or number to make a true sentence.

1. Choosing a <u>number</u> to represent an unknown in a problem is called defining the variable.
2. The words "more than" suggest the operation of <u>multiplication</u>.
3. An <u>inequality</u> is a mathematical sentence that contains the symbols $<$, $>$, $\leq$, or $\geq$.
4. When graphing $t < 2$ on a number line, an <u>open circle</u> should be used to show that 2 is not included in the solution.
5. A <u>function</u> describes a relationship between two quantities.
6. An equation is called a linear equation if its graph is a <u>point</u>.
7. The solution of $m + 5 = 12$ is <u>17</u>.
8. The solution of $g - 4 = 18$ is <u>22</u>.
9. The solution of $-8w = 48$ is <u>6</u>.
10. The solution of $2y \leq 18$ is $y \leq$ <u>16</u>.

In Your Own Words

11. ***Explain*** the addition property of equality.

Objectives & Examples

Upon completing this chapter, you should be able to:

● solve addition and subtraction equations *(Lesson 6-1)*

Solve $c - 32 = 112$.

$$c - 32 = 112$$
$$c - 32 + 32 = 112 + 32 \quad \textit{Add 32 to}$$
$$c = 144 \qquad\qquad \textit{each side.}$$

● solve multiplication equations *(Lesson 6-2)*

Solve $33y = 132$.

$$33y = 132$$
$$\frac{33y}{33} = \frac{132}{33} \quad \textit{Divide each side by 33.}$$
$$y = 4$$

● solve two-step equations *(Lesson 6-3)*

Solve $6t - 5 = 19$.

$$6t - 5 = 19$$
$$6t - 5 + 5 = 19 + 5 \quad \textit{Add 5 to each side.}$$
$$6t = 24$$
$$\frac{6t}{6} = \frac{24}{6} \quad \textit{Divide each side by 6.}$$
$$t = 4$$

● write simple algebraic expressions and equations from verbal phrases and sentences *(Lesson 6-4)*

Translate "4 times the price" into an algebraic expression.

Let p represent the price. The algebraic expression is $4p$.

Review Exercises

Use these exercises to review and prepare for the chapter test.

Solve each equation. Check your solution.

12. $x + 15 = 14$ **13.** $w + 13 = -25$

14. $10.9 + r = 11$ **15.** $54 = m - 9$

16. $t - 3.6 = 10.1$ **17.** $s + 3.75 = 5.25$

Solve each equation. Check your solution.

18. $4b = 32$ **19.** $64 = 16a$

20. $-4q = 48$ **21.** $-8w = -72$

22. $5.9r = 0.59$ **23.** $-1.3t = 3.9$

Solve each equation. Check your solution.

24. $3p - 4 = 8$ **25.** $2x + 5 = 3$

26. $8 - 6w = 50$ **27.** $5m + 6 = -4$

28. $6 = 3y - 12$ **29.** $-15 = 5 + 2t$

30. $-4y + 5.1 = 8.3$ **31.** $-1.5b + 1 = 7$

Write each phrase as an algebraic expression.

32. the sum of x and 5

33. 13 less than s

34. the quotient of z and 15

Write each sentence as an algebraic equation.

35. The product of 14 and a number is 56.

36. Four less than 5 times a number is 19.

Objectives & Examples

solve inequalities *(Lesson 6-5)*

Solve $d - 5 \geq 7$. Graph the solution on a number line.

$$d - 5 \geq 7$$
$$d - 5 + 5 \geq 7 + 5 \quad \text{Add 5 to each side.}$$
$$d \geq 12$$

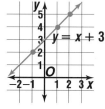

represent functions as ordered pairs *(Lesson 6-6)*

Infant's Age (months)	1	2	3	4
Weight (pounds)	10.5	12.3	14.5	16.2

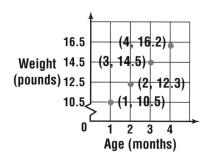

As the age increases, the weight also increases.

solve equations with two variables and graph the solution *(Lesson 6-7)*

Graph $y = x + 3$.

x	$x + 3$	y
-1	$-1 + 3$	2
1	$1 + 3$	4
2	$2 + 3$	5

Three solutions are $(-1, 2)$, $(1, 4)$, and $(2, 5)$.

Review Exercises

Solve each inequality. Graph the solution on a number line.

37. $8 + g \leq 10$

38. $4j < 28$

39. $m - 7 \geq -10$

40. $3h + 2 > 17$

Graph the ordered pairs in each table on a coordinate plane. Then write a sentence describing each relationship as a function.

41.

Length of Side	1	2	3	4	5
Area of Square	1	4	9	16	25

42.

Minutes a Candle is Burned	30	60	90	120	150
Height (inches)	11	10	9	8	7

Graph each equation.

43. $y = 2x$

44. $y = -0.5x$

45. $y = 3x + 2$

46. $y = x + 4$

Applications & Problem Solving

47. Work Backward Four friends collect stamps. They are comparing how many Mexican stamps they each have. Jeff has 3 times as many as Fina. Mario has 4 fewer stamps than Danielle, but 3 more than Fina. Fina has 9 stamps. How many stamps does each friend have? *(Lesson 6-1B)*

48. Money Matters Pedro earned $65 in January shoveling snow. The total was 4 times more than his January earnings last winter. How much did he earn last January? *(Lesson 6-4)*

49. Catering Christina's Catering Service charges $12.50 per person for a sit-down dinner. *(Lesson 6-7)*

   a. Write an equation that represents the cost, *y,* for *x* people.

   b. What is the cost for 40 people?

50. Earth Science The graph shows one factor of the Fujita scale, which rates the intensity of tornados. Graph the ordered pairs (F-scale, maximum wind) on a coordinate plane. Then write a statement that describes the relationship as a function. *(Lesson 6-6)*

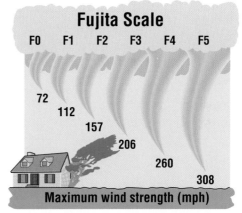

Fujita Scale

F0   F1   F2   F3   F4   F5

72
112
157
206
260
308

Maximum wind strength (mph)

Source: National Audubon Society Field Guide to North American Weather

Alternative Assessment

Open Ended

Suppose your family is moving to a new home. Two moving companies are called so rates can be compared. The first company charges a flat rate of $150 plus $12 per hour for a 2-person crew. The second company charges $28 per hour for a 2-person crew. If it should take 8 hours for the move, how can you determine which company is less expensive?

Suppose the move actually took 12 hours. Which company would have charged the least?

A practice test for Chapter 6 is provided on page 612.

Completing the CHAPTER Project

Use the following checklist to make sure your project is complete.

☑ You have included a graph with the average speed of each roller coaster.

☑ You have graphed the distance it travels and number of riders for one of the roller coasters.

☑ You have included some interesting facts about your roller coaster.

PORTFOLIO   Select an item from this chapter that shows your creativity and place it in your portfolio.

Section One: Multiple Choice

There are eleven multiple-choice questions in this section. Choose the best answer. If a correct answer is *not here* choose the letter for Not Here.

1. Which is equivalent to $3m < 39$?

   A $m > 13$

   B $m > 36$

   C $m = 13$

   D $m < 13$

2. What is the least common multiple of 30 and 45?

   F 15

   G 75

   H 90

   J 1,350

3. Viho performed a probability experiment by rolling a cube that has its faces colored blue, yellow, and red, 120 times. The results of his experiment follow.

Color	Number of Rolls
blue	20
yellow	60
red	40

   How many sides of the cube would you expect to be colored yellow?

   A 2

   B 3

   C 4

   D 5

4. Which decimal is equivalent to $\frac{19}{25}$?

   F 1.31

   G 1.11

   H 0.96

   J 0.76

5. Which is the graph of the equation $y = 3x + 2$?

   A 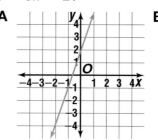 **B**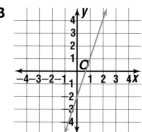

   C 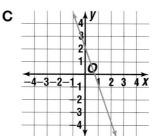 **D**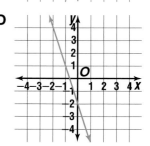

Please note that Questions 6–11 have five answer choices.

6. Tickets for a musical cost $9.50 for adults and $6.75 for children. Which equation could be used to find the total cost in dollars, d, of tickets for any number of adults, a, and children, c?

   F $d = 9.50 + 6.75 + a + c$

   G $d = (9.50 + 6.75) \times (a + c)$

   H $d = (9.50 \times 6.75) + (a \times c)$

   J $d = (9.50 \times c) + (6.75 \times a)$

   K $d = (9.50 \times a) + (6.75 \times c)$

7. The range of points that can be earned on a single lab in physical science is from 5 to 9 points. Lenora has turned in 7 labs. What is a reasonable estimate of the number of points she has earned so far in physical science lab?

   A less than 14

   B between 14 and 35

   C between 35 and 63

   D between 63 and 90

   E greater than 90

8. The stem-and-leaf plot shows the number of points scored by the Bears in each of their basketball games this season.

Stem	Leaf
1	8 9
2	0 2 3 3 6 8 8 9
3	0 1 4 4 5 6 8 9
4	0 1 2 $1|8 = 18$ *points*

In how many games did they score at least 30 points?

F 8   **G** 9

H 11   **J** 20

K Not Here

9. The student council collected a total of $1,100 during the last 4 years for a charity donation. What is the average yearly amount collected over the 4-year period?

A $225

B $275

C $450

D $675

E Not Here

10. Jack packed 396 crayons into 18 boxes. If each box contains the same number of crayons, how many are in each box?

F 20

G 23

H 25

J 52

K Not Here

11. Domingo earned $50 in one week by working 6.25 hours. How much was Domingo paid per hour?

A $6.25

B $6.50

C $8.50

D $10.50

E Not Here

Test-Taking Tip THE PRINCETON REVIEW

As part of your preparation for a standardized test, review basic definitions. For example:
- A number is prime if it has no factors except itself and one.
- 7 is a prime number. 8 is not prime because it has factors other than one and itself.

Section Two: Free Response

This section contains six questions for which you will provide short answers. Write your answers on your paper.

12. Write three numbers less than 10 that are factors of 1,215.

13. Let $y = wx$. If $y = 15$ and $w = 5$, find the value of x.

14. Write an expression to represent the phrase *eight more than a number.*

15. Claire worked 20 hours last week. She earned $6.00 per hour. Write an equation to find her total earnings, E.

16. Kyung is reading a 258-page novel. If he reads 8 pages an hour, about how long will it take him to read the entire book?

17. What number should replace y in the table?

x	$2x - 5$
1	−3
2	y
4	3
8	11
16	27

Test Practice For additional test practice questions, visit:

www.glencoe.com/sec/math/mac/mathnet

Interdisciplinary Investigation

"A" IS FOR APPLE

Have you ever played games where you must build words? Did you groan when you drew a Q or an X? Did you ever say to yourself, "What word can I make with that letter? Now I can't possibly win!" Are some letters used more than others in our language?

What You'll Do

In this investigation, you will collect data using printed materials you read every day. You will determine which letters are most common and design your own letter set.

Materials  magazine calculator

 photocopy machine highlighters

Procedure

1. Work in a group of three or four. Have each member select a page of written text from a different source like newspapers, magazines, novels, or textbooks. Photocopy the page and highlight with a marker about 80 to 100 consecutive words. This is your text sample.

2. Work individually. Tally the number of each letter in your text sample and find the total number of letters in the sample. Write a ratio comparing the number of each letter to the total number of letters. Express each ratio as a percent.

3. Work in your group. Combine the information from all of the text samples into one large sample. Which letters are most common? least common?

4. Work in your group. Some businesses use signs with removable letters to advertise their products. Suppose you are going to make a letter set to sell to these businesses. The set should have at least 100, but no more than 150 letters. As a group, decide which letters and how many of each to include.

Technology Tips

- Use a **spreadsheet** to calculate the percent for each letter.

- Use a **spreadsheet** to help you plan the letter set.

- Surf the **Internet** as a research tool.

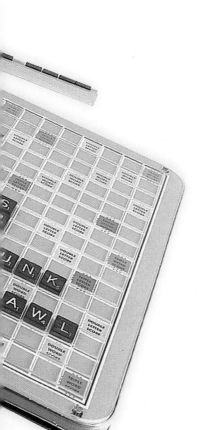

Making the Connection

Use the data collected from your text samples as needed to help in these investigations.

Language Arts

You may be studying a language such as Spanish or French in school or speak a language other than English at home. Repeat steps 1 through 3 of the investigation using a text sample from another language. Which letters are most common?

Music

A musical score is made up of notes, just as words are made up of letters. Count the number of each note in a page of a musical piece. Find the percent each note represents of the total notes in the score.

Social Studies

Research the history of the English alphabet. Investigate *alphabetic writing* and *Phoenician writing*.

Go Further

- Compare the percent of letters in a word game to the percent of letters you found in your text samples.

- Investigate the lengths of words. Use different types of printed material. What word length is most common?

interNET CONNECTION **Research** For current information on frequency distributions of letters in the alphabet, visit the following website.

Data Collection and Comparison To share and compare your data with other students in the U.S., visit:

www.glencoe.com/sec/math/mac/mathnet

PORTFOLIO You may want to place your work on this investigation in your portfolio.

CHAPTER 7

Applying Fractions

What you'll learn in Chapter 7

- to estimate with fractions and mixed numbers,
- to add, subtract, multiply, and divide fractions and mixed numbers,
- to solve problems by eliminating possibilities,
- to change units in the customary system, and
- to find perimeter and circumference.

CHAPTER Project

UPS AND DOWNS

In this project, you will use fractions in a report summarizing the stock prices of four companies. To do this, you will research the stock history of each company and keep track of the value of their stocks.

Getting Started

- Research what it means to own stock in a company.
- Choose the stocks of four companies to track. Include at least one from the table.
- Create a table that you can use to keep track of your stocks for one month on a daily basis.

Company	Close	Change
A	$39\frac{1}{16}$	$-\frac{3}{4}$
B	$58\frac{3}{4}$	$-1\frac{5}{16}$
C	12	$+\frac{1}{8}$
D	$76\frac{5}{8}$	$-2\frac{1}{4}$
E	$43\frac{1}{16}$	$-\frac{9}{16}$
F	$93\frac{13}{16}$	$-\frac{3}{8}$
G	$22\frac{11}{16}$	$+\frac{1}{16}$
H	$33\frac{13}{16}$	$-\frac{7}{8}$
I	50	$-1\frac{17}{16}$
J	$59\frac{1}{2}$	$-2\frac{11}{16}$
K	$35\frac{3}{4}$	$-\frac{1}{2}$
L	$99\frac{1}{16}$	$+\frac{9}{16}$

Technology Tips

- Use an **electronic encyclopedia** to do your research.
- Use a **word processor** to write your report.
- Surf the **Internet** for stock prices.

interNET
CONNECTION
Data Update For up-to-date information on the stock market, visit:

www.glencoe.com/sec/math/mac/mathnet

Working on the Project

You can use what you'll learn in Chapter 7 to help you complete your report.

Page	Exercise
271	51
287	36
307	35
311	Alternative Assessment

Thankfully, Some Things Never Change.

What do you know about your stocks right now?

Estimating with Fractions

What you'll learn

You'll learn to estimate sums, differences, products, and quotients of fractions and mixed numbers.

When am I ever going to use this?

Knowing how to estimate with fractions will help you approximate the amount of ingredients needed for a recipe.

In August, 1996, an article in a kid's magazine described how kids turned their hobbies into dream jobs. Boaz Frankel earns $5 per hour for computer tutoring in Oregon. If he tutors $9\frac{3}{4}$ hours one month, estimate how much money he will make.

To estimate the sum, difference, or product of mixed numbers, round each mixed number to the nearest whole number. Estimate the product of 5 and $9\frac{3}{4}$.

$$5 \times 9\frac{3}{4} \quad \rightarrow \quad 5 \times 10 = 50$$

Boaz will make *about* $50.   *Since $9\frac{3}{4} < 10$, he will make a little less than $50.*

Example ① Estimate the sum of $4\frac{1}{5}$ and $1\frac{1}{2}$.

When a mixed number contains $\frac{1}{2}$, the number is usually rounded up.

$$4\frac{1}{5} + 1\frac{1}{2} \quad \rightarrow \quad 4 + 2 = 6$$

$4\frac{1}{5} + 1\frac{1}{2}$ is *about* 6.   *Is the actual sum more or less than 6?*

To estimate the sum or difference of fractions, round each fraction to 0, $\frac{1}{2}$, or 1, whichever is closest. Sometimes fraction models can help you decide how to round.

Fractions Close to 0	**Fractions Close to $\frac{1}{2}$**	**Fractions Close to 1**

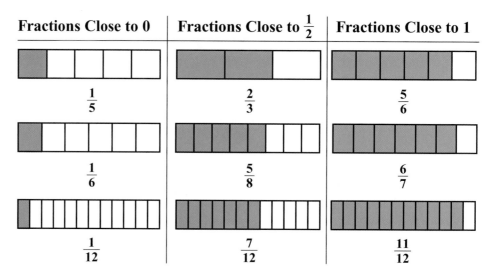

$\frac{1}{5}$	$\frac{2}{3}$	$\frac{5}{6}$
$\frac{1}{6}$	$\frac{5}{8}$	$\frac{6}{7}$
$\frac{1}{12}$	$\frac{7}{12}$	$\frac{11}{12}$

Examples

Estimate.

2 $\frac{1}{7} + \frac{5}{8}$

$\frac{1}{7}$ *is about 0.*

$\frac{5}{8}$ *is about* $\frac{1}{2}$.

$\frac{1}{7} + \frac{5}{8} \rightarrow 0 + \frac{1}{2} = \frac{1}{2}$

$\frac{1}{7} + \frac{5}{8}$ is *about* $\frac{1}{2}$.

3 $\frac{5}{6} - \frac{1}{2}$

$\frac{5}{6} - \frac{1}{2} \rightarrow 1 - \frac{1}{2} = \frac{1}{2}$ $\frac{5}{6}$ *is about 1.*

$\frac{5}{6} - \frac{1}{2}$ is *about* $\frac{1}{2}$.

4 $\frac{1}{8} \times 15$ $\frac{1}{8} \times 15$ *means* $\frac{1}{8}$ *of 15.*

$\frac{1}{8} \times 15 \rightarrow \frac{1}{8}$ of 16 or 2 *Round 15 to 16, since 16 is divisible by 8.*

$\frac{1}{8}$ of 15 is *about 2.*

APPLICATION

5 **Baking** Katrina made $7\frac{1}{4}$ pounds of chocolate fudge that she wants to split into $1\frac{1}{2}$-pound portions to give away as gifts. About how many gifts of fudge can she give away?

Explore You want to estimate how many $1\frac{1}{2}$-pound portions Katrina can get from $7\frac{1}{4}$ pounds of fudge.

Plan To find the number of portions, divide $7\frac{1}{4}$ by $1\frac{1}{2}$. First, round $1\frac{1}{2}$ to 2. Then replace the dividend with a number that is easy to divide mentally.

Solve $7\frac{1}{4} \div 1\frac{1}{2} \rightarrow 7\frac{1}{4} \div 2$ *Round $1\frac{1}{2}$ to 2.*

$\rightarrow$ 8 ÷ 2 or 4 *8 is divisible by 2.*

Katrina can give away about 4 gifts of fudge.

Examine Since four 2-pound gifts would equal 8 pounds of fudge, the estimate is reasonable.

CHECK FOR UNDERSTANDING

Communicating Mathematics

Read and study the lesson to answer each question.

1. *Write* each fraction modeled and tell if it is closest to 0, $\frac{1}{2}$, or 1.

    a.   b.

2. *Draw* an illustration representing $\frac{4}{7}$ and use it to round the fraction to 0, $\frac{1}{2}$, or 1.

Guided Practice

Round each fraction to 0, $\frac{1}{2}$, or 1.

3. $\frac{1}{8}$ 4. $\frac{10}{11}$ 5. $\frac{2}{5}$

Round to the nearest whole number.

6. $3\frac{1}{2}$ 7. $9\frac{1}{3}$ 8. $12\frac{6}{7}$

Estimate.

9. $\frac{1}{2} + \frac{5}{6}$ 10. $\frac{5}{8} - \frac{1}{10}$

11. $\frac{7}{8} \times 11$ 12. $4 + 3\frac{4}{5}$

13. $5\frac{1}{3} \times 2\frac{2}{3}$ 14. $8\frac{1}{2} \div 3\frac{1}{4}$

15. *Horticulture* Carmen uses $\frac{3}{4}$ cup of liquid fertilizer for each shrub in her plant nursery. Estimate how much liquid fertilizer she will need for 24 shrubs.

EXERCISES

Practice

Round each fraction to 0, $\frac{1}{2}$, or 1.

16. $\frac{1}{6}$ 17. $\frac{3}{5}$ 18. $\frac{9}{10}$ 19. $\frac{4}{5}$

20. $\frac{2}{5}$ 21. $\frac{1}{7}$ 22. $\frac{3}{10}$ 23. $\frac{5}{6}$

Round to the nearest whole number.

24. $9\frac{1}{6}$ 25. $2\frac{1}{2}$ 26. $11\frac{2}{3}$ 27. $4\frac{1}{10}$

28. $7\frac{1}{3}$ 29. $5\frac{3}{4}$ 30. $10\frac{2}{9}$ 31. $6\frac{7}{8}$

Estimate.

32. $\frac{1}{2} + \frac{7}{8}$ 33. $\frac{3}{8} - \frac{1}{10}$ 34. $\frac{1}{8} \times \frac{3}{4}$

35. $\frac{4}{5} \div \frac{7}{8}$ 36. $\frac{1}{3} + \frac{1}{8}$ 37. $5\frac{1}{3} - 2\frac{3}{4}$

38. $9\frac{7}{8} + 2\frac{3}{4}$ 39. $\frac{1}{2} \times 17$ 40. $5\frac{5}{7} \times 8\frac{2}{3}$

41. $11\frac{1}{2} - 1\frac{5}{6}$ 42. $2\frac{4}{5} \times \frac{8}{9}$ 43. $21\frac{1}{2} \div 1\frac{3}{4}$

44. Estimate $14\frac{1}{7}$ minus $\frac{5}{6}$.

45. Estimate the product of $\frac{1}{3}$ and $\frac{4}{9}$.

46. Estimate $5\frac{5}{6}$ divided by 3.

47. Estimate the sum of $1\frac{7}{9}$, $\frac{4}{5}$, and $6\frac{1}{8}$.

Applications and
Problem Solving

48. *Construction* *About* how many $3\frac{1}{2}$-foot shelves can a carpenter cut from a 12-foot board for a bookcase?

49. *Life Science* Komodo dragons are the largest lizards ever to have lived. A 250-pound komodo dragon can eat enough in one sitting to increase its weight by $\frac{3}{4}$. Estimate $\frac{3}{4} \times 250$ to find how much weight a komodo dragon would gain after eating.

Komodo dragon

Interview a family member who cooks, sews, or does carpentry work about how they use fractions. Make a list of the situations where they might estimate fractional amounts.

Computer Monitor (as advertised)	Actual Diagonal Measure (in.)
17-inch	$15\frac{4}{5}$
15-inch	$13\frac{1}{2}$
14-inch	$13\frac{1}{10}$

50. *Technology* The advertised diagonal size of a computer monitor is larger than the actual viewing diagonal.

   a. Estimate the difference between the sizes for each computer monitor.

   b. What conjecture could you make, based on your estimates?

51. *Working on the* **CHAPTER Project** Refer to the table on page 267. For each of the companies that you have chosen to track, estimate the cost of buying 32 shares of stock on August 18, 1997.

52. *Critical Thinking* If the number being divided is rounded up and the divisor is rounded down, what is the effect on the answer?

Mixed Review

53. *Algebra* Find four solutions of $y = 3x + 1$. Write the solutions as ordered pairs. *(Lesson 6-7)*

54. *Algebra* Translate the phrase *65 less than w* into an algebraic expression. *(Lesson 6-4)*

55. *Algebra* Solve $18 = m - 5$. *(Lesson 6-1)*

56. Solve $s = -5(12)$. *(Lesson 5-6)*

For **Extra Practice**, see page 586.

57. Replace ● in $\frac{1}{4}$ ● $\frac{2}{7}$ with $<$, $>$, or $=$ to make a true sentence. *(Lesson 4-10)*

58. **Standardized Test Practice** A pair of jeans is on sale for 30% off the regular price. What fraction of the regular price is this? *(Lesson 4-7)*

  A $\frac{3}{100}$ **B** $\frac{3}{10}$ **C** $\frac{3}{5}$ **D** $\frac{1}{3}$

Adding and Subtracting Fractions

What **you'll learn**

You'll learn to add and subtract fractions.

When **am I ever going to use this?**

Knowing how to add and subtract fractions can help you determine how much paint you need to buy in order to paint a room in your home.

The graph shows the different times of the year that people join health clubs. What fraction of the people join during the first six months of the year? *This problem will be solved in Example 1.*

To add or subtract fractions, the denominators must be the same.

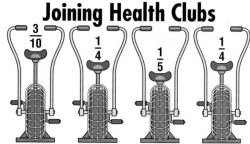

Joining Health Clubs

$\frac{3}{10}$ $\frac{1}{4}$ $\frac{1}{5}$ $\frac{1}{4}$

Jan.-March April-June July-Sept. Oct.-Dec.
New Memberships

Source: National Health Club Association

HANDS-ON MINI-LAB

Work with a partner.   grid paper   markers

Use squares to model each fraction.

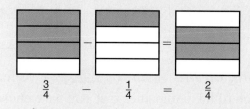

$$\frac{3}{4} - \frac{1}{4} = \frac{2}{4}$$

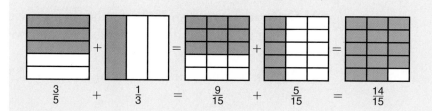

$$\frac{3}{5} + \frac{1}{3} = \frac{9}{15} + \frac{5}{15} = \frac{14}{15}$$

Try This

Model each sum or difference.

1. $\frac{4}{5} - \frac{2}{5}$ **2.** $\frac{1}{3} + \frac{1}{2}$ **3.** $\frac{5}{6} - \frac{3}{4}$

Talk About It

4. Explain how to take two fractions with different denominators and draw them so that they have the same denominator.

5. Explain how to add two fractions that have different denominators.

APPLICATION

1 **Health** Refer to the beginning of the lesson. Find the fraction of the people who join health clubs during the first six months of the year.

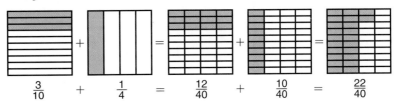

$$\frac{3}{10} \quad + \quad \frac{1}{4} \quad = \quad \frac{12}{40} \quad + \quad \frac{10}{40} \quad = \quad \frac{22}{40}$$

You need to find the sum of $\frac{3}{10}$ and $\frac{1}{4}$.

So, $\frac{22}{40}$ or $\frac{11}{20}$ of the people who join health clubs do so during the first six months of the year.

Study Hint

Technology You can use a calculator to add and subtract fractions. To find $\frac{3}{10} + \frac{1}{4}$ enter:

3 $\boxed{/}$ 10 $\boxed{+}$ 1 $\boxed{/}$ 4

$\boxed{=}$. To get an answer in simplest form, press $\boxed{SIMP}$ $\boxed{=}$ until N/D → n/d no longer appears on the screen.

LOOK BACK

Refer to Lesson 4-10 to review LCD.

Adding and Subtracting Fractions with Unlike Denominators	To add or subtract fractions: 1. Rename the fractions with a common denominator as necessary. 2. Add or subtract the numerators. 3. Simplify.

The least common denominator (LCD) can be used to rename fractions for addition and subtraction. Remember that the LCD is the LCM of the denominators.

Example

2 Find $\frac{5}{6} - \frac{3}{8}$. Write the difference in simplest form.

Estimate: $1 - \frac{1}{2} = \frac{1}{2}$

$$\begin{array}{ccc} \frac{5}{6} & LCD: 24 & \frac{5 \times 4}{6 \times 4} \rightarrow & \frac{20}{24} \\ -\frac{3}{8} & \rightarrow & \frac{3 \times 3}{8 \times 3} \rightarrow & -\frac{9}{24} \\ \hline & & & \frac{11}{24} \end{array}$$

So, $\frac{5}{6} - \frac{3}{8} = \frac{11}{24}$. $\frac{11}{24}$ *is close to the estimate, $\frac{1}{2}$.*

3 Solve $\frac{4}{9} + \frac{11}{12} = a$. Write the solution in simplest form.

Estimate: $\frac{1}{2} + 1 = 1\frac{1}{2}$

$$\frac{4}{9} + \frac{11}{12} = a$$

$$\frac{4 \times 4}{9 \times 4} + \frac{11 \times 3}{12 \times 3} = a \quad \textit{The LCD is 36.}$$

$$\frac{16}{36} + \frac{33}{36} = a$$

$$\frac{49}{36} = a$$

Rename $\frac{49}{36}$ as $1\frac{13}{36}$.

So, $\frac{4}{9} + \frac{11}{12} = 1\frac{13}{36}$.   *Compare to the estimate,* $1\frac{1}{2}$.

CHECK FOR UNDERSTANDING

Communicating Mathematics

Read and study the lesson to answer each question.

1. *Tell* why you must have a common denominator to add or subtract fractions.

2. *Describe* a common unit of measure that can be used to add 4 inches and 1 yard.

HANDS-ON MATH

3. *Model* $\frac{3}{8} + \frac{1}{6}$ using grid paper and find the sum.

Guided Practice

Add or subtract. Write each sum or difference in simplest form.

4. $\frac{1}{7}$
   $+\frac{3}{7}$

5. $\frac{3}{5}$
   $+\frac{1}{15}$

6. $\frac{5}{6}$
   $-\frac{1}{9}$

7. $\frac{3}{8} - \frac{1}{8}$

8. $\frac{7}{10} - \frac{1}{6}$

9. $\frac{5}{9} + \frac{5}{6}$

10. Solve $\frac{1}{2} + \frac{5}{12} = d$. Write the solution in simplest form.

11. *Transportation* When Ke Min started his trip to his sister's house, he had $\frac{1}{2}$ of a tank of gas. When he arrived, he had $\frac{1}{8}$ of a tank. How much gasoline did Ke Min use during his trip?

EXERCISES

Practice

Add or subtract. Write each sum or difference in simplest form.

12. $\frac{4}{9}$
    $+\frac{2}{9}$

13. $\frac{9}{10}$
    $-\frac{1}{6}$

14. $\frac{5}{8}$
    $-\frac{1}{2}$

15. $\frac{3}{7}$
    $+\frac{4}{5}$

16. $\frac{3}{7} + \frac{9}{14}$

17. $\frac{2}{6} - \frac{1}{6}$

18. $\frac{7}{15} + \frac{5}{9}$

19. $\frac{4}{5} - \frac{1}{6}$

20. $\frac{5}{8} - \frac{7}{12}$

21. $\frac{4}{9} + \frac{2}{15}$

22. Find $\frac{5}{8}$ minus $\frac{5}{12}$.

23. Find the sum of $\frac{9}{10}$ and $\frac{4}{15}$.

Solve each equation. Write the solution in simplest form.

24. $\frac{19}{24} - \frac{1}{4} = c$

25. $x = \frac{8}{9} + \frac{7}{15}$

26. $\frac{5}{8} - \frac{5}{36} = k$

27. $t = \frac{3}{8} - \frac{1}{12}$

28. $\frac{3}{4} + \frac{7}{20} = p$

29. $\frac{7}{9} + \frac{5}{6} = w$

30. *Algebra* Evaluate $b + \frac{11}{12}$ if $b = \frac{9}{20}$.

31. *Algebra* Find the value of $\frac{3}{4} - c$ if $c = \frac{3}{11}$.

Applications and Problem Solving

32. *Earth Science* Carbon dioxide gas is said to be responsible for $\frac{1}{2}$ of the greenhouse effect, which is the warming of Earth's surface. Chlorofluorocarbons are said to account for another $\frac{1}{6}$ of it. Together, how much are these two gases responsible for the greenhouse effect?

33. *Entertainment* For a popular movie, shoes were dyed emerald green for the residents of Emerald City. If $\frac{1}{3}$ of the shoes dyed were for the townspeople and $\frac{4}{15}$ were for the shopkeepers, what part of the shoes dyed for the movie were for these two groups?

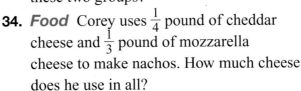

34. *Food* Corey uses $\frac{1}{4}$ pound of cheddar cheese and $\frac{1}{3}$ pound of mozzarella cheese to make nachos. How much cheese does he use in all?

35. *Critical Thinking* Does $\frac{1}{4} + \frac{5}{7} - \frac{3}{8} = \frac{5}{7} + \frac{3}{8} - \frac{1}{4}$? Explain.

Mixed Review

36. Round $\frac{2}{15}$ to 0, $\frac{1}{2}$, or 1. *(Lesson 7-1)*

37. *Algebra* Solve $23 = 14w - 5$. *(Lesson 6-3)*

38. *Scuba Diving* Find the distance between two divers if one diver is 27 feet below sea level and the other diver is 13 feet below sea level. *(Lesson 5-5)*

39. **Standardized Test Practice** Which percent represents the shaded area? *(Lesson 4-6)*

    A 57%   **B** 47%

    C 43%   **D** 38%

For **Extra Practice,** see page 586.

Adding and Subtracting Mixed Numbers

What you'll learn

You'll learn to add and subtract mixed numbers.

When am I ever going to use this?

Knowing how to add and subtract mixed numbers can help you make alterations when sewing.

Leathersmith "Wild" Bill Cleaver of Vashon Island, Washington, makes frontier jeans, cowboy shirts, gloves, and vests using designs from the 1800s. If Bill has a shirt pattern for a back measurement of $18\frac{1}{2}$ inches, but his his client's back measures $16\frac{5}{8}$ inches, how much smaller than the pattern does he need to make the shirt? *This problem will be solved in Example 4.*

Adding and Subtracting Mixed Numbers	To add or subtract mixed numbers: 1. Add or subtract the fractions. If necessary, rename the fractions first. 2. Add or subtract the whole numbers. 3. Simplify.

Examples

Add or subtract. Write each sum or difference in simplest form.

① $9\frac{4}{5} - 2\frac{1}{5}$

Estimate: 10 − 2 = 8

$$\begin{array}{r} 9\frac{4}{5} \\ -2\frac{1}{5} \\ \hline 7\frac{3}{5} \end{array}$$

The difference, $7\frac{3}{5}$, is close to the estimate.

② $14\frac{5}{6} + 17\frac{9}{10}$

Estimate: 15 + 18 = 33

$$\begin{array}{r} 14\frac{5}{6} \\ +17\frac{9}{10} \\ \hline \end{array} \rightarrow \begin{array}{r} 14\frac{25}{30} \\ +17\frac{27}{30} \\ \hline 31\frac{52}{30} \end{array}$$

$$31\frac{52}{30} = 31 + \frac{52}{30}$$
$$= 31 + 1\frac{22}{30}$$
$$= 32\frac{22}{30} \text{ or } 32\frac{11}{15}$$

Study Hint

Technology You can use a calculator to add and subtract mixed numbers. To find $12\frac{1}{5} + 6\frac{3}{4}$, enter:

12 [UNIT] 1 [/] 5 [+]

6 [UNIT] 3 [/] 4 [=]

[Ab/c] *18U19/20*. The answer is $18\frac{19}{20}$.

Sometimes, when you subtract two mixed numbers, the fraction in the first mixed number is less than the fraction in the second. In this case, you need to rename the first mixed number before subtracting.

Examples

③ Find $4\frac{1}{3} - 2\frac{2}{3}$.

Estimate: 4 − 3 = 1

$\frac{1}{3}$ is less than $\frac{2}{3}$, so you need to rename $4\frac{1}{3}$.

Think: $4\frac{1}{3} = 3\frac{\blacksquare}{3}$

Use circle diagrams.

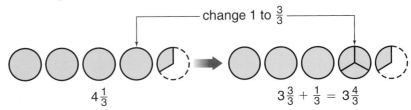

change 1 to $\frac{3}{3}$

$4\frac{1}{3}$

$3\frac{3}{3} + \frac{1}{3} = 3\frac{4}{3}$

Now find the difference.

$$3\frac{4}{3} - 2\frac{2}{3} = 1\frac{2}{3}$$

So, $4\frac{1}{3} - 2\frac{2}{3} = 1\frac{2}{3}$. *Compare to the estimate.*

APPLICATION

④ **Sewing** Refer to the beginning of the lesson. Find how much smaller than the pattern Bill must make the shirt.

Explore   You need to find $18\frac{1}{2} - 16\frac{5}{8}$. *Estimate: $19 - 17 = 2$*

Plan   Rename $18\frac{1}{2}$ as $18\frac{4}{8}$. So, $18\frac{4}{8} - 16\frac{5}{8}$. Since $\frac{4}{8}$ is less than $\frac{5}{8}$, you need to rename $18\frac{4}{8}$.

$$18\frac{4}{8} = 17\frac{8}{8} + \frac{4}{8} = 17\frac{12}{8}$$

Solve $17\frac{12}{8} - 16\frac{5}{8} = 1\frac{7}{8}$

The shirt needs to be $1\frac{7}{8}$ inches smaller than the pattern.

Examine   Compare to the estimate. The answer is reasonable.

CHECK FOR UNDERSTANDING

Communicating Mathematics

Read and study the lesson to answer each question.

1. **Show** that $3\frac{1}{4} = 2\frac{5}{4}$ using circle models.

2. **Explain** whether $5\frac{2}{3} + 1\frac{3}{4}$ is greater than, less than, or equal to $4\frac{1}{3} + \frac{5}{8}$. How do you know?

3. **Write** an explanation comparing renaming of mixed numbers with renaming of whole numbers in subtraction.

Guided Practice

Complete. Use circle diagrams if necessary.

4. $3\frac{6}{4} = 4\frac{\blacksquare}{2}$

5. $5\frac{1}{6} = 4\frac{\blacksquare}{6}$

6. $2\frac{12}{9} = \blacksquare\frac{1}{3}$

Lesson 7-3 Adding and Subtracting Mixed Numbers **277**

Add or subtract. Write each sum or difference in simplest form.

7. $3\frac{1}{6} + 5\frac{1}{6}$

8. $8\frac{7}{9} - 3\frac{1}{9}$

9. $3\frac{1}{2} - 1\frac{3}{4}$

10. Solve $g = 7\frac{3}{8} + 9\frac{1}{6}$. Write the solution in simplest form.

11. *Jewels* In 1908, the world's largest diamond was cut. Two of the gems that came from the cutting were Cullinan I, which weighed $530\frac{1}{5}$ carats, and Cullinan II, which weighed $317\frac{2}{5}$ carats. How much more did Cullinan I weigh than Cullinan II?

EXERCISES

Practice **Complete. Use circle diagrams if necessary.**

12. $3\frac{10}{7} = 4\frac{\blacksquare}{7}$ **13.** $6\frac{1}{2} = 5\frac{\blacksquare}{2}$ **14.** $7\frac{14}{10} = 8\frac{\blacksquare}{10}$ **15.** $8\frac{3}{5} = 7\frac{\blacksquare}{5}$

16. $9\frac{9}{8} = 10\frac{\blacksquare}{8}$ **17.** $4\frac{5}{6} = 3\frac{\blacksquare}{6}$ **18.** $9\frac{2}{3} = \blacksquare\frac{5}{3}$ **19.** $12\frac{9}{5} = \blacksquare\frac{4}{5}$

Add or subtract. Write each sum or difference in simplest form.

20. $9\frac{1}{8} + 2\frac{5}{8}$ **21.** $7\frac{5}{6} - 3\frac{1}{6}$ **22.** $7\frac{5}{6} + 9\frac{3}{8}$

23. $6\frac{5}{6} - 2\frac{1}{3}$ **24.** $9\frac{4}{5} - 2\frac{3}{10}$ **25.** $13\frac{7}{8} + 15\frac{7}{10}$

26. $3\frac{7}{12} + 8\frac{3}{4}$ **27.** $7\frac{1}{3} - 3\frac{5}{9}$ **28.** $8\frac{3}{4} - 1\frac{7}{10}$

29. What is the sum of $2\frac{1}{6}$, $3\frac{1}{2}$, and $5\frac{7}{8}$?

30. Find $13\frac{1}{8}$ minus $1\frac{7}{10}$.

Solve each equation. Write the solution in simplest form.

31. $5\frac{5}{6} - 3\frac{2}{3} = a$ **32.** $6\frac{13}{15} + 2\frac{3}{5} = y$ **33.** $q = 4\frac{3}{10} - 1\frac{3}{4}$

34. *Algebra* Evaluate $k + 8\frac{5}{12}$ if $k = 11\frac{1}{4}$.

35. *Algebra* Find the value of $6\frac{1}{2} - m$ if $m = 3\frac{4}{9}$.

36. **Stock Market** The table shows the 52-week high and low prices of airline stocks as of April, 1997. Find the difference between the high and low price of each stock.

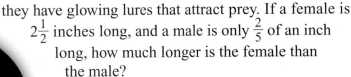

Airline Stocks Soar	52-week
COMPANY	high / low
Airline A	95³/₈ / 66³/₄
Airline B	21³/₈ / 5⁵/₁₆
Airline C	33¹/₄ / 20⁵/₈

  a. Airline A

  b. Airline B

  c. Airline C

37. **Measurement** You can make your own window-washing solution by mixing $1\frac{1}{3}$ cups of ammonia and $1\frac{1}{2}$ cups of vinegar with baking soda and water. Will the solution fit in a $\frac{1}{2}$-quart pan?

38. **Life Science** Female anglerfish are larger than males, and they have glowing lures that attract prey. If a female is $2\frac{1}{2}$ inches long, and a male is only $\frac{2}{5}$ of an inch long, how much longer is the female than the male?

39. **Critical Thinking** A string is cut in half, and one of the halves is used to bundle newspapers. Then one-fifth of the remaining string is cut off and used to tie a balloon. The piece left is 8 feet long. How long was the string originally?

Anglerfish

Mixed Review

40. Add $\frac{5}{6}$ and $\frac{2}{3}$. *(Lesson 7-2)*

41. **Algebra** Find four solutions of $y = -3x + 7$. Write the solutions as ordered pairs. *(Lesson 6-7)*

42. **Standardized Test Practice** Martin's job is to pack calculators into boxes. One day, Martin packed 448 calculators into 16 boxes. If each box contains the same number of calculators, how many calculators did Martin put into each box? *(Lesson 5-7)*

  A 10   **B** 25   **C** 28   **D** 52   **E** 57

43. **Food** The graph shows what percent of Americans chose the given pizza toppings as their favorite. Express the percent of people who chose each topping as a fraction in simplest form. *(Lesson 4-7)*

  a. onions

  b. vegetables

  c. mushrooms

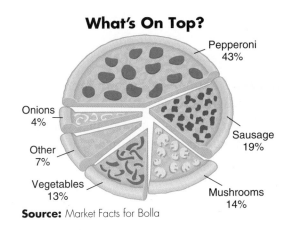

What's On Top?

Pepperoni 43%

Onions 4%

Other 7%

Sausage 19%

Vegetables 13%

Mushrooms 14%

Source: Market Facts for Bolla

For **Extra Practice,**
see page 586.

44. Express $\frac{7}{25}$ as a decimal. *(Lesson 2-7)*

7-3B Eliminate Possibilities

A Follow-Up of Lesson 7-3

Look at this problem. What answer do you get?

Anne has a 5-gallon cooler that she fills with juice and takes to her softball games. If the cooler has $2\frac{1}{3}$ gallons of juice in it, how much juice does she need to add to fill it?

A $7\frac{1}{3}$ gal **B** $2\frac{2}{3}$ gal **C** $3\frac{1}{3}$ gal **D** $\frac{2}{3}$ gal

If it holds 5 gallons and it has $2\frac{1}{3}$ gallons in it, then you need about 3 gallons to fill it. So the answer must be either B or C.

Andy

I got Answer A, but that can't be right because it's too big! So, we can eliminate choice A.

I know what you did to get choice A - you added 5 and $2\frac{1}{3}$ to get $7\frac{1}{3}$. Just because the question says, "how much juice does she need to add," doesn't mean that you add the numbers.

Neshawn

THINK ABOUT IT

Work with a partner.

1. **List** different ways to **eliminate possibilities** in solving problems.

2. **Think** of another answer that you could eliminate in the problem above.

3. **Apply** what you have learned to solve the following problem.

   A fishbowl holds $1\frac{1}{2}$ gallons of water. If there is $\frac{1}{3}$ gallon of water in the bowl, how many more gallons are needed to fill the bowl?

   A $1\frac{5}{6}$ gal   B $\frac{2}{3}$ gal

   C $2\frac{1}{6}$ gal   D $1\frac{1}{6}$ gal

For **Extra Practice,** see page 587.

ON YOUR OWN

4. The third step of the 4-step plan for problem solving asks you to *solve* the problem. **Explain** how you use the strategy of eliminating possibilities to solve a problem.

5. **Write a Problem** in which eliminating the possibilities would help you solve it. Explain your answer.

6. **Reflect Back** Explain how you eliminated possibilities in Exercise 3.

MIXED PROBLEM SOLVING

STRATEGIES

Look for a pattern.
Solve a simpler problem.
Act it out.
Guess and check.
Draw a diagram.
Make a chart.
Work backward.

Solve. Use any strategy.

7. **Food** You need $2\frac{1}{2}$ cups of flour and $1\frac{2}{3}$ cups of sugar in a chocolate chip cookie recipe. How many cups of flour and sugar are called for?

8. **Standardized Test Practice** Migina bought pencils, 2 for $0.49; felt-tipped pens, 3 for $1.39; and an eraser for $0.29. Choose the best estimate for the amount of change she will get from $5.

   A $0.30

   B $1.80

   C $2.80

   D $3.30

9. **Technology** A videotape will record 6 hours of programming. Katherine has recorded $2\frac{5}{6}$ hours of a miniseries. She wants to record $3\frac{1}{2}$ hours more on the same tape. Can she do this? Explain your answer.

10. **Money Matters** In 1965, Congress reduced the silver content of half-dollars from $\frac{9}{10}$ to $\frac{2}{5}$. How much less was the silver content after the reduction?

11. **Technology** The cellular phone industry has grown by leaps and bounds since 1995. The graph shows the number of cellular phone antenna sites in thousands.

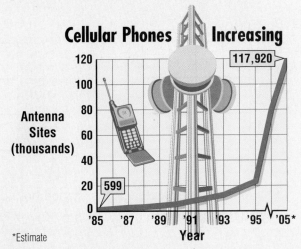

Cellular Phones Increasing

Antenna Sites (thousands)

117,920

599

*Estimate

Source: Cellular Telecommunications Industry Association

Estimate the increase in the number of antenna sites from 1995 to 2005.

12. **Standardized Test Practice** A taxi charges $1.15 for the first 0.2 mile and $0.50 for each additional 0.2 mile. Find the cost of a 4-mile taxi ride.

   A $9.65

   B $10.65

   C $12.65

   D $13.65

   E $13.15

paper

markers

COOPERATIVE LEARNING

7-4A Multiplying Fractions and Mixed Numbers

A Preview of Lesson 7-4

You can use area models to multiply fractions.

TRY THIS

Work with a partner.

1 Model $\frac{1}{4} \times \frac{1}{3}$.

Divide a unit square vertically into fourths and horizontally into thirds.

Color one fourth of the square one color.

Color one third of the square another color. Count the small rectangles that are shaded both colors.

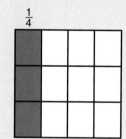

One of the small rectangles is shaded both colors. Since each small rectangle has an area of $\frac{1}{12}$, then $\frac{1}{4} \times \frac{1}{3} = \frac{1}{12}$.

2 Model $2 \times \frac{2}{3}$.

Draw 2 unit squares side by side. Divide them horizontally into thirds.

Color the 2 unit squares one color.

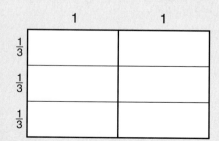

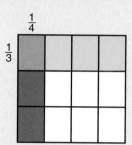

282 **Chapter 7** Applying Fractions

Color two-thirds of the squares another color. Count the small rectangles that are shaded both colors.

Four of the small rectangles are shaded both colors. Add their areas.

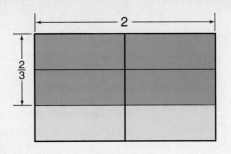

$\frac{1}{3} + \frac{1}{3} + \frac{1}{3} + \frac{1}{3} = \frac{4}{3}$ *Based on 1 unit square, each small rectangle has an area of $\frac{1}{3}$.*

So, $2 \times \frac{2}{3} = \frac{4}{3}$ or $1\frac{1}{3}$.

---

③ Model $1\frac{1}{2} \times \frac{2}{5}$.

Divide 2 unit squares vertically into halves and horizontally into fifths.

Color $1\frac{1}{2}$ of the squares one color.

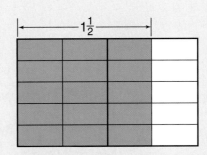

Color two-fifths of the squares another color.

Six of the small rectangles are shaded both colors. Add their areas.

$\frac{1}{10} + \frac{1}{10} + \frac{1}{10} + \frac{1}{10} + \frac{1}{10} + \frac{1}{10} = \frac{6}{10}$ *Based on 1 unit square, each small rectangle has an area of $\frac{1}{10}$.*

So, $1\frac{1}{2} \times \frac{2}{5} = \frac{6}{10}$ or $\frac{3}{5}$.

ON YOUR OWN

Use area models to find each product.

1. $\frac{1}{2} \times \frac{1}{3}$

2. $\frac{2}{3} \times \frac{3}{4}$

3. $3 \times \frac{1}{2}$

4. $2 \times \frac{3}{4}$

5. $1\frac{1}{3} \times \frac{1}{4}$

6. $3\frac{1}{2} \times \frac{1}{2}$

7. **Look Ahead** Find $2\frac{1}{2} \times 1\frac{1}{3}$. Use area models if necessary.

7-4 Multiplying Fractions and Mixed Numbers

What you'll learn

You'll learn to multiply fractions and mixed numbers.

When am I ever going to use this?

Knowing how to multiply fractions and mixed numbers can help you make adjustments in the ingredients in a recipe.

Media ratings measure what part of the U.S. households have their television turned on and also what part of those households were watching a certain program. Suppose $\frac{1}{2}$ of all households have their TV turned on, and $\frac{1}{5}$ of those were watching Program A. You can multiply $\frac{1}{5}$ and $\frac{1}{2}$ to find the part of all households that were watching Program A.

You can multiply fractions like $\frac{1}{2}$ and $\frac{1}{5}$ by using an area model.

$$\frac{1}{2} \times \frac{1}{5} = \frac{1}{10}$$

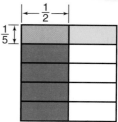

Multiplying Fractions	**Words:** To multiply fractions, multiply the numerators and then multiply the denominators.
	Symbols: **Arithmetic** **Algebra**
	$\frac{1}{2} \times \frac{1}{5} = \frac{1}{10}$ $\frac{a}{b} \times \frac{c}{d} = \frac{ac}{bd}$ $b, d \neq 0$

Example ① **Find** $\frac{1}{3} \times \frac{3}{4}$**.**

$$\begin{array}{c} 1 \times 3 \\ \frac{1}{3} \times \frac{3}{4} = \frac{3}{12} \text{ or } \frac{1}{4} \\ 3 \times 4 \end{array}$$

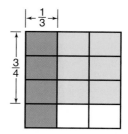

When the numerator and denominator of either fraction have a common factor, you can simplify before you multiply.

Example ② **Multiply** $\frac{1}{6} \times \frac{3}{5}$**.** *Estimate:* $0 \times \frac{1}{2} = 0$

The GCF of 3 and 6 is 3.

$$\frac{1}{6} \times \frac{3}{5} = \frac{1}{\overset{}{6}_2} \times \frac{\overset{1}{3}}{5} \quad \textit{Divide 3 and 6 by 3.}$$

$$= \frac{1}{10}$$

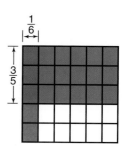

LOOK BACK

Refer to Lesson 4-4 to review GCF.

Example **3** Solve $a = \dfrac{3}{8} \times \dfrac{10}{27}$.

$$a = \dfrac{\overset{1}{\cancel{3}}}{\underset{4}{\cancel{8}}} \times \dfrac{\overset{5}{\cancel{10}}}{\underset{9}{\cancel{27}}}$$ *The GCF of 3 and 27 is 3.*
The GCF of 8 and 10 is 2.

$$a = \dfrac{5}{36}$$

Multiplying Mixed Numbers	To multiply mixed numbers, rename each mixed number as an improper fraction. Then multiply the fractions.

Example **4**
CONNECTION

Life Science Every minute, you inhale about $10\frac{1}{5}$ liters of air. About how much air do you take in each hour?

Estimate: $10 \times 60 = 600$

$$10\frac{1}{5} \times 60 = \frac{51}{5} \times 60 \quad \text{1 hour = 60 minutes}$$

$$= \frac{51}{\underset{1}{\cancel{5}}} \times \frac{\overset{12}{\cancel{60}}}{1} \quad \text{The GCF of 5 and 60 is 5.}$$

$$= 612$$

So, you take in about 612 liters of air each hour.

Let the **Games** Begin

Totally Mental

Get Ready This game is for two players. spinner

Get Set Each player should have a game sheet like the one shown at the right. Use a spinner with the digits 1 through 9.

Go • Each player chooses one of the boxes on his or her sheet and writes the number from the spinner in it. After 4 spins, the player with the greatest product is the winner.

Math Skill
Multiplying Fractions

 Visit www.glencoe.com/sec/math/mac/mathnet for more games.

CHECK FOR UNDERSTANDING

Communicating Mathematics

Read and study the lesson to answer each question.

1. *Explain* the steps you would use to find $16\frac{2}{3} \times 9\frac{5}{9}$.

2. *Write* the multiplication sentence shown by the area model.

HANDS-ON MATH 3. Use area models to show $\frac{4}{5} \times \frac{1}{4}$.

Multiply. Write each product in simplest form.

4. $\frac{1}{5} \times \frac{1}{2}$ **5.** $\frac{3}{7} \times \frac{2}{3}$ **6.** $2 \times \frac{3}{4}$

7. $\frac{2}{3} \times \frac{3}{8}$ **8.** $2\frac{1}{2} \times 2\frac{2}{3}$ **9.** $5\frac{1}{3} \times \frac{4}{5}$

10. Solve $\frac{6}{25} \times \frac{5}{8} = h$. Write the solution in simplest form.

11. *Government* By law, the length of an official United States flag must be $1\frac{9}{10}$ times its width. If the width of a flag is $3\frac{1}{2}$ feet, what is its length?

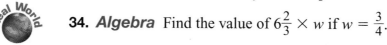

EXERCISES

Practice

Multiply. Write each product in simplest form.

12. $\frac{3}{5} \times \frac{1}{2}$ **13.** $\frac{1}{8} \times \frac{3}{4}$ **14.** $\frac{2}{3} \times \frac{5}{6}$ **15.** $4 \times \frac{2}{5}$

16. $\frac{3}{5} \times \frac{10}{11}$ **17.** $\frac{5}{6} \times \frac{3}{5}$ **18.** $\frac{4}{5} \times \frac{1}{8}$ **19.** $\frac{3}{8} \times \frac{4}{5}$

20. $6 \times \frac{4}{5}$ **21.** $\frac{3}{5} \times \frac{10}{21}$ **22.** $\frac{3}{7} \times \frac{5}{6}$ **23.** $2\frac{1}{2} \times \frac{5}{8}$

24. $\frac{4}{7} \times 4\frac{2}{3}$ **25.** $3\frac{2}{3} \times 9$ **26.** $3\frac{1}{4} \times 2\frac{2}{3}$ **27.** $4\frac{1}{2} \times 1\frac{1}{3}$

28. Find the product of $1\frac{1}{6}$, $\frac{3}{7}$, and $\frac{1}{3}$.

29. What is the product of $4\frac{7}{12}$ and $9\frac{1}{5}$?

Solve each equation. Write the solution in simplest form.

30. $3 \times 2\frac{1}{7} = s$ **31.** $a = 1\frac{4}{7} \times 4\frac{2}{3}$ **32.** $r = 2\frac{1}{4} \times \frac{9}{10}$

33. *Algebra* Evaluate $q \times \frac{2}{9}$ if $q = 1\frac{1}{8}$.

34. *Algebra* Find the value of $6\frac{2}{3} \times w$ if $w = \frac{3}{4}$.

Applications and Problem Solving

35. *History* In 1513, Juan Ponce de León discovered Florida. His route from Puerto Rico to Florida measures about $2\frac{1}{4}$ inches on the map. If 1 inch represents 667 miles, find the approximate length of his route.

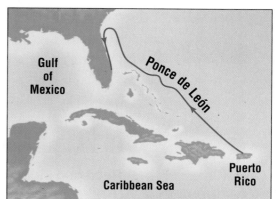

36. Working on the CHAPTER Project Suppose that on your first day of tracking stock prices you had purchased 100 shares of stock in each of your four companies.

   a. For each company, how much would you have paid for 100 shares?

   b. How much money would you have received if you had sold all your shares at yesterday's closing price?

37. Critical Thinking Observe that $3 \times \frac{1}{3} = 1$, $4 \times \frac{1}{4} = 1$, and $8 \times \frac{1}{8} = 1$. What number times $1\frac{1}{2}$ equals 1? What number times $2\frac{1}{2}$ equals 1?

Mixed Review **38.** Find the sum of $6\frac{3}{4}$ and $9\frac{7}{8}$. *(Lesson 7-3)*

39. Algebra Solve the inequality $r + 3 > -5$. *(Lesson 6-5)*

40. Standardized Test Practice Desiree is driving cross-country. If she expects to drive between 350 and 450 miles per day, which number of days is reasonable for her to drive 3,800 miles? *(Lesson 5-7)*

   A fewer than 6 days

   B between 6 and 8 days

   C between 9 and 11 days

   D between 14 and 16 days

   E more than 16 days

41. Earth Science The low temperatures for 10 cities on January 23 are -3, 27, 13, -6, -14, 36, 47, 52, -2, and 0. Order these temperatures from greatest to least. *(Lesson 5-2)*

For **Extra Practice,** see page 587.

42. Music In a survey, 12 out of 78 people preferred classical music to jazz. Write this ratio as a fraction in simplest form. *(Lesson 4-5)*

CHAPTER 7

Mid-Chapter Self Test

Estimate. *(Lesson 7-1)*

1. $\frac{3}{8} + \frac{6}{7}$ **2.** $\frac{4}{5} - \frac{1}{2}$ **3.** $\frac{1}{3} \times 14$

Add or subtract. Write each sum or difference in simplest form. *(Lessons 7-2 and 7-3)*

4. $\frac{5}{8} - \frac{1}{6}$ **5.** $\frac{7}{9} + \frac{5}{12}$ **6.** $\frac{4}{5} - \frac{3}{7}$

7. $8\frac{3}{4} - 2\frac{5}{12}$ **8.** $5\frac{1}{8} + 3\frac{3}{8}$ **9.** $18\frac{3}{10} + 13\frac{5}{6}$

10. History In 1986, *Voyager* became the first plane to fly nonstop around the world without refueling in midair. *Voyager* weighed 2,000 pounds, but at take-off it carried about $3\frac{1}{2}$ times its weight in fuel. How many pounds of fuel did *Voyager* carry at take-off? *(Lesson 7-4)*

COOPERATIVE LEARNING

7-4B Fractal Patterns

dot paper

A Follow-Up of Lesson 7-4

In Lesson 1-6, you learned about fractal patterns. You can use multiplication of fractions and mixed numbers to examine these patterns.

TRY THIS

Work with a partner.

Step 1
Draw a square with sides measuring 9 units.

Step 2
Divide each side into 3 equal lengths to form 9 smaller squares. Shade the middle square.

Step 3
Divide each unshaded square into 9 smaller squares and shade the middle squares.

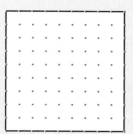

Stage 0

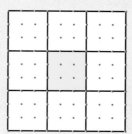

Stage 1

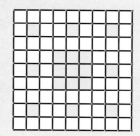

Stage 2

ON YOUR OWN

1. Think of the shaded squares as holes. The area of Stage 0 is 81 square units. Find the areas of Stage 1 and Stage 2. *Do not include the holes.*

2. Write the areas of Stages 0–2. Study the pattern. You can multiply each area by what fraction to get the next area?

3. **Make a conjecture** about the area of each fractal stage.

4. Use your conjecture to find the area of Stage 3.

5. **Reflect Back** Refer to Example 1 in Lesson 1–6. In the Sierpinski triangle, the area of Stage 0 is 1 square unit. Use the equations to find the areas of each stage.

   a. (area of Stage 0) $\times \frac{3}{4}$ = area of Stage 1

   b. (area of Stage 1) $\times \frac{3}{4}$ = area of Stage 2

   c. (area of Stage 2) $\times \frac{3}{4}$ = area of Stage 3

Integration: Measurement
Changing Customary Units

What you'll learn

You'll learn to change units in the customary system.

When am I ever going to use this?

Knowing how to change units will help you compare prices of grocery items.

Word Wise

ounce   pint
pound   quart
ton   gallon
cup

In 1996, the *Olmec Art of Ancient Mexico* exhibit was on display at the National Gallery of Art in Washington, D.C. Included among the 3,000 year-old works was a sculpture of a head that weighed $9\frac{1}{2}$ tons. How many pounds did the sculpture weigh? *This problem will be solved in Example 1.*

Customary units of weight are **ounce**, **pound**, and **ton**. The relationships among these units is shown in the table.

> **1 pound (lb) = 16 ounces (oz)**
> **1 ton (T) = 2,000 pounds**

When you change from a larger unit to a smaller unit, multiply. *There will be more smaller units than larger units.*

Example 1
CONNECTION

Study Hint
Estimation $9\frac{1}{2}$ is about 10. Since there are 2,000 lb in 1 ton, the answer should be close to $10 \times 2,000$ or 20,000 lb.

Art In the beginning of the lesson, how many pounds did the sculpture weigh?

You need to change $9\frac{1}{2}$ tons to pounds.

$9\frac{1}{2}$ T = __?__ lb   *larger unit → smaller unit*

$9\frac{1}{2} \times 2,000 = 19,000$ *Since 2,000 lb = 1 ton, multiply by 2,000.*

The sculpture weighed 19,000 pounds.

Sometimes you will need to convert from a smaller unit to a larger unit. In this case, divide. *There will be fewer larger units than smaller units.*

Example 2

How many pounds is 72 ounces?

72 oz = __?__ lb   *smaller unit → larger unit*

$72 \div 16 = 4.5$ *Since 16 oz = 1 lb, divide by 16.*

72 ounces is equal to 4.5 pounds.

Customary units of liquid capacity are **cup**, **pint**, **quart**, and **gallon**. The relationships among these units is shown in the table.

| 1 cup (c) = 8 fluid ounces (fl oz) |
| 1 pint (pt) = 2 cups |
| 1 quart (qt) = 2 pints |
| 1 gallon (gal) = 4 quarts |

Examples

Real World APPLICATION

3 **Fish** A fish tank contains 2 quarts less water than 9 gallons. How many quarts of water are in the tank?

$9 \cdot 4 = 36$ *Multiply by 4 since there are 4 quarts in a gallon.*

There are 36 quarts in 9 gallons. So, the tank contains $36 - 2$, or 34 quarts of water.

4 **Food** A popular drink contains 64 ounces of soda. How many cups is this?

64 fl oz = _?_ c   *smaller unit → larger unit*
$64 \div 8 = 8$ *Divide by 8 since there are 8 fl oz in a cup.*

There are 8 cups of soda in the drink.

CHECK FOR UNDERSTANDING

Communicating Mathematics

Read and study the lesson to answer each question.

1. *Tell* which operation is needed to convert from cups to pints. Explain how you know.

2. *Write* about a real-life situation in which you would need to change units in the customary system.

Guided Practice

Complete.

3. 5 lb = _?_ oz
4. 12 qt = _?_ gal
5. 4.5 pt = _?_ c
6. 4,000 lb = _?_ T
7. 3 c = _?_ fl oz
8. 1 pt = _?_ qt

9. *Life Science* A newborn hooded seal pup gains about 56 ounces per day for the first four days of its life. How many pounds will the average pup gain in one day?

EXERCISES

Practice

Complete.

10. 2 gal = _?_ qt
11. 5 T = _?_ lb
12. 128 oz = _?_ lb
13. 12 c = _?_ pt
14. 16 qt = _?_ gal
15. 8 pt = _?_ qt
16. 96 oz = _?_ lb
17. 3 lb = _?_ oz
18. 5 pt = _?_ c
19. 2.5 qt = _?_ pt
20. 15 pt = _?_ qt
21. 2.5 lb = _?_ oz
22. 2 T = _?_ lb
23. 4.5 pt = _?_ c
24. 6,000 lb = _?_ T
25. 2 fl oz = _?_ c
26. 5 c = _?_ pt
27. 0.5 gal = _?_ qt

28. How many tons are in 4,200 pounds?

29. If 4 cups = 1 quart, then 9 cups = __?__ quarts?

30. If 36 inches = 1 yard, then 2.3 yards = __?__ inches?

31. **Write an Equation** that you can use to change *c* cups to fluid ounces. Use *f* for fluid ounces. Then use the equation to find the number of fluid ounces in 6.5 cups.

Applications and Problem Solving

32. *Life Science* An adult has about 5 quarts of blood. If a person donates 1 pint of blood, how many pints are left?

33. *Life Science* Mammoths have been discovered deep-frozen in the ice of the Arctic tundra. When these animals lived over 10,000 years ago, they weighed up to 15,500 pounds. How many tons did they weigh?

34. *Safety* A bridge has a $3\frac{1}{2}$-ton weight limit. If a truck and its cargo weigh 6,800 pounds, can the truck cross over the bridge? Explain.

35. *Food* Is a 2-quart pitcher large enough to hold 1 batch of cherry punch?

36. *Critical Thinking* Make a table that shows the number of ounces in 1, 2, 3, and 4 pounds. Graph the ordered pairs (pounds, ounces) on a coordinate plane. Describe the graph.

Cherry Punch
$2\frac{1}{2}$ c cherry juice
2 c orange juice
$1\frac{1}{2}$ c pineapple juice
3 c ginger ale

Mixed Review

37. **Standardized Test Practice** A box of books weighs $8\frac{2}{3}$ pounds. How much do $4\frac{1}{2}$ boxes weigh? *(Lesson 7-4)*

   A 19 pounds   **B** $27\frac{2}{3}$ pounds   **C** $32\frac{1}{2}$ pounds   **D** 39 pounds

38. *Employment* Mallory gets paid a flat rate of $75 plus $4.50 per hour for cleaning. The equation $4.5h + 75 = 165$, where *h* is the number of hours, describes the number of hours she must work to make $165. How many hours must she work to make $165? *(Lesson 6-3)*

39. Find the LCM of 16 and 20. *(Lesson 4-9)*

40. *Statistics*
Rainfall is rarest in the southwest region of the nation. The data show nine southwestern cities and their average number of days with rain per year. Find the mean, mode, and median of these data.
(Lesson 3-4)

interNET
CONNECTION
For the latest statistics on rainfall in the United States, visit: www.glencoe.com/sec/ math/mac/mathnet

It Rarely Rains
Rainfall (days)

17 Yuma, Ariz.
26 Las Vegas, Nev.
29 Bishop, Calif
30 Santa Barbara, Calif.
32 Long Beach, Calif.
35 Los Angeles, Calif.
36 Phoenix, Ariz.
37 Bakersfield, Calif.
42 San Diego, Calif.

Source: *USA TODAY Weather Almanac*

For **Extra Practice,** see page 587.

Integration: Geometry
Perimeter

What you'll learn

You'll learn to find perimeter.

When am I ever going to use this?

Knowing how to find perimeter can help you frame pictures.

Word Wise

perimeter

In Washington, D.C., the Capitol and the White House are $1\frac{1}{2}$ miles apart. This symbolizes the separation of powers between Congress and the President. The Jefferson and Lincoln Memorials are also arranged in a symbolic position. Find the **perimeter** around these four historical structures.

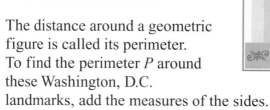

The Federal City

United States Capitol

N

$1\frac{1}{2}$ mi $1\frac{1}{2}$ mi

White House   Washington Monument   Jefferson Memorial

1 mi   1 mi

Lincoln Memorial

Washington, D.C.

The distance around a geometric figure is called its perimeter. To find the perimeter P around these Washington, D.C. landmarks, add the measures of the sides.

$P = 1\frac{1}{2} + 1\frac{1}{2} + 1 + 1$

$P = 5$ The perimeter is 5 miles.

Perimeter of a Rectangle	**Words:**	The perimeter of a rectangle is the sum of the measures of the sides. It can also be expressed as two times the length (ℓ) plus two times the width (w).	
	Symbols: $P = \ell + w + \ell + w$ $P = 2\ell + 2w$	**Model:**	

Model diagram: rectangle with ℓ on top and bottom, w on left and right sides.

Examples

Find the perimeter of each rectangle.

1 length = 9 meters
width = 5 meters

$P = 2\ell + 2w$ *Replace ℓ with 9*
$P = 2(9) + 2(5)$ *and w with 5.*
$P = 18 + 10$ or 28

The perimeter is 28 meters.

2 length = 2 feet
width = 18 inches

$P = 2\ell + 2w$
$P = 2(2) + 2(1.5)$ *18 in. = 1.5 ft*
$P = 4 + 3$ or 7

The perimeter is 7 feet.

3 Find the perimeter of the figure.

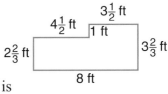

Estimate: 3 + 5 + 1 + 4 + 4 + 8 = 25

$P = 2\frac{2}{3} + 4\frac{1}{2} + 1 + 3\frac{1}{2} + 3\frac{2}{3} + 8$

$P = 23\frac{1}{3}$ The perimeter is $23\frac{1}{3}$ feet. This is close to the estimate of 25 feet.

INTEGRATION **4** **Measurement** Find the perimeter of the rectangle. Measure to the nearest eighth inch.

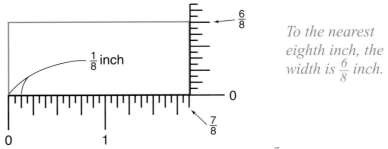

To the nearest eighth inch, the width is $\frac{6}{8}$ inch.

To the nearest eighth inch, the length is $1\frac{7}{8}$ inches.

Estimate: 2 + 1 + 2 + 1 = 6

$P = 2\ell + 2w$

$= \left(2 \times 1\frac{7}{8}\right) + \left(2 \times \frac{6}{8}\right)$ *Replace ℓ with $1\frac{7}{8}$ and w with $\frac{6}{8}$.*

$= \left(2 \times \frac{15}{8}\right) + \left(2 \times \frac{6}{8}\right)$ *Change $1\frac{7}{8}$ to $\frac{15}{8}$.*

$= \frac{30}{8} + \frac{12}{8}$ *Multiply within each set of parentheses.*

$= \frac{42}{8}$ *Add.*

$= 5\frac{2}{8}$ or $5\frac{1}{4}$ The perimeter is $5\frac{1}{4}$ inches.

CHECK FOR UNDERSTANDING

Communicating Mathematics

Read and study the lesson to answer each question.

1. *Show* where $8\frac{3}{8}$ is located on a ruler.

2. *Write,* in your own words, how to find the perimeter of a square figure.

3. *Write* about a situation in which you would need to find the perimeter of an object.

Guided Practice

Find the perimeter of each figure shown or described. Estimate to check your answer.

4.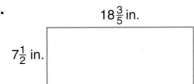

18$\frac{3}{5}$ in.

7$\frac{1}{2}$ in.

5.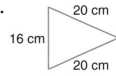

20 cm

16 cm

20 cm

6. rectangle: $\ell = 2\frac{1}{2}$ feet
    $w = 15$ inches

7. rectangle: $\ell = 1.7$ yards
    $w = 3.5$ yards

8. Find the perimeter of the figure. Use a ruler to measure to the nearest fourth inch.

9. *Flag Day* A giant cake decorated as an American flag was displayed in front of Independence Hall in Philadelphia. The rectangular cake, zwhich was created to commemorate Flag Day, measured 60 feet by 90 feet. What was the perimeter of the cake?

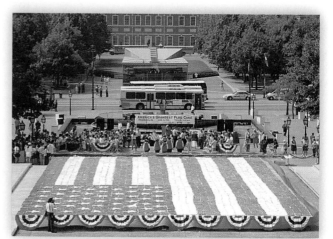

EXERCISES

Practice   **Find the perimeter of each figure shown or described.**

10.
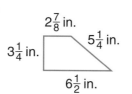
$2\frac{7}{8}$ in.
$3\frac{1}{4}$ in.
$5\frac{1}{4}$ in.
$6\frac{1}{2}$ in.

11.
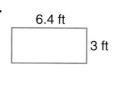
6.4 ft
3 ft

12.

30 mm   30 mm
24 mm

13.

$12\frac{1}{2}$ mi
$12\frac{1}{2}$ mi

14.
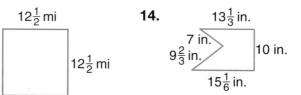
$13\frac{1}{3}$ in.
7 in.
$9\frac{2}{3}$ in.
10 in.
$15\frac{1}{6}$ in.

15.
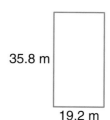
35.8 m
19.2 m

16. rectangle: $\ell = 9\frac{1}{4}$ feet
    $w = 36$ inches

17. rectangle: $\ell = 6\frac{1}{2}$ inches
    $w = 4\frac{3}{8}$ inches

18. rectangle: $\ell = 13\frac{1}{2}$ yards
    $w = 7\frac{3}{4}$ yards

19. rectangle: $\ell = 4.8$ meters
    $w = 7.2$ meters

20. Find the perimeter in feet of a triangle with sides that measure 7 inches, 10 inches, and 1 foot.

21. Find the perimeter of a rectangle with length 17 cm and width 9.1 cm.

22. Find the perimeter of a square with side 21 yards.

23. A rectangle is made with 9 feet of string. One side is $2\frac{5}{16}$ feet long. What is the length of the other side?

Find the perimeter of each figure. Use a ruler to measure to the nearest fourth inch.

24.

25.

26.

27.

Applications and Problem Solving

28. *Pets* Jamal has 38 feet of fencing for a rectangular dog pen. He plans to use 24 feet of the garage wall for one side of the pen.
   a. Draw and label a diagram of the pen.
   b. Find the width of the pen.

29. *Architecture* The drawing is taken from a plan of an octagonal house. If each side measures $16\frac{1}{2}$ feet, find the perimeter of the house.

30. *Landscaping* Ms. Williams is going to plant shrubs across the back and down two sides of her yard. Her yard is 72 feet wide and 120 feet deep.
   a. Draw and label a diagram of Ms. Williams' yard.
   b. How many shrubs will she need to buy if she plants them 4 feet apart?

31. *Critical Thinking* An *irregular* pentagon has five sides that do not all have the same measure. If the perimeter of such a pentagon is $20\frac{1}{3}$ inches and one of its sides measures $4\frac{1}{2}$ inches, what could be the measures of the other sides?

Mixed Review

32. *Measurement* Complete: 40 oz = __?__ lb. *(Lesson 7-5)*

33. *Algebra* Solve $p - 25.5 = 74.4$. *(Lesson 6-1)*

For **Extra Practice,** see page 588.

34. *Probability* A box of pencils contains 7 red, 2 orange, 4 blue, and 3 yellow pencils. If you reach in the box and choose one pencil at random, what is the probability that you will select a blue pencil? Express your answer as both a fraction and a percent. *(Lesson 4-8)*

35. *Standardized Test Practice* If Addie has 1,000 nickels, how much money does she have? *(Lesson 2-4)*
   A $50   **B** $500   **C** $1,000   **D** $20,000

FINANCE

Elouise C. Cobell
BANKER

Elouise Cobell helped found Blackfeet National Bank in Browning, Montana. It is the only national bank located on an Indian reservation and owned by an Indian tribe. She received a "genius grant," which is a fellowship awarded to people who show exceptional creativity and to those who make a significant difference in human thought and action. Today, Ms. Cobell directs a project that is working to improve the management of Indian trust funds.

To work in a bank as a financial manager, you'll need a bachelor's degree in accounting, finance, or business administration. You should have a solid background in mathematics and good computer skills. Financial managers must be able to communicate well, think quickly, and analyze data accurately.

For more information
American Bankers Association
Center for Banking Information
1120 Connecticut Ave. NW.
Washington, DC 20036

interNET
CONNECTION
www.glencoe.com/sec/math/mac/mathnet

I put money in my savings account every week. I'd like to work in a bank someday!

Your Turn
Suppose you started your own bank. Make a poster displaying the various products and services that your bank would provide for its customers.

Integration: Geometry
Circles and Circumference

What you'll learn

You'll learn to find the circumference of circles.

When am I ever going to use this?

You can use circumference to find the distance you travel while riding a bicycle.

Word Wise

circle
center
diameter
radius
circumference

Purdue University's marching band has a bass drum that they claim is the biggest in the world. It takes four band members to pull it and two to pound on it. If the diameter of the drum is 8 feet, what is its circumference? *This problem will be solved in Example 1.*

A **circle** is the set of all points in a plane that are the same distance from a given point called the **center**. The **diameter (d)** is the distance across the circle through its center. The **radius (r)** is the distance from the center to any point on the circle. The **circumference (C)** is the distance around the circle.

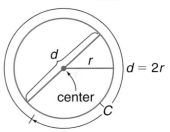

MINI-LAB

Work with a partner.

 ruler string

 circular objects

You can use circular objects to find the relationship between circumference and diameter.

Try This

- Use a ruler to measure the diameter of a circular object. Record your finding.
- Wrap a string around the circular object once. Mark the string where it meets itself.
- Measure the length of the string with your ruler. Record your finding. This is the circumference of the circle.
- Repeat this activity with circular objects of various sizes.

Talk About It

1. For each circular object, graph the ordered pair (diameter, circumference) on a coordinate plane. What do you find?
2. For each object, divide the circumference by the diameter. Compare the results.
3. How is the circumference related to the diameter?

LOOK BACK
Refer to Lesson 6-7 to review graphing on a coordinate plane.

The Greek letter π is used to represent the circumference divided by the diameter $\frac{C}{d}$. Approximations often used for π are 3.14 and $\frac{22}{7}$.

The diameter of a circle is twice the radius.

Circumference of a Circle	**Words:** The circumference of a circle is equal to π times its diameter or π times twice the radius.	
	Symbols: $C = \pi d$ or $C = 2\pi r$	**Model:**

Example 1

Music Refer to the beginning of the lesson. Find the circumference of the drum to the nearest tenth.

Since you know the diameter, use the formula $C = \pi d$.

$C = \pi d$

$C \approx 3.14(8)$ *Use 3.14 for π. Replace d with 8.*

$C \approx 25.12$

The circumference of the drum is about 25.1 feet.

8 ft

One approximate value for π is the fraction $\frac{22}{7}$. Use this value for π when the radius or diameter of the circle is a multiple of 7 or has a multiple of 7 in its numerator.

Example 2

Find the circumference of a circle with a radius of 14 meters.

$C = 2\pi r$

$C \approx 2\left(\frac{22}{7}\right)(14)$

$C \approx 2\left(\frac{22}{\overset{}{7}}\right)\left(\frac{\overset{2}{14}}{1}\right)$

$C \approx 2 \times \frac{22}{1} \times \frac{2}{1}$

$C \approx 88$ The circumference is about 88 meters.

14 m

π usually has its own key on a calculator. You can use this key to find the circumference of a circle. *What is displayed on your calculator when you press* $\boxed{\pi}$?

Example 3

Find the circumference of a circle with a radius of 4.5 inches.

$C = 2\pi r$

$C \approx 2\pi(4.5)$ *Estimate: $2 \times 3 \times 5 = 30$*

2 $\boxed{\times}$ $\boxed{\pi}$ $\boxed{\times}$ 4.5 $\boxed{=}$ 28.27433388

$C \approx 28.3$ *Round to the nearest tenth since the original measurement was only to the tenths place.*

4.5 in.

The circumference is about 28.3 inches.

Communicating Mathematics

Read and study the lesson to answer each question.

1. **Describe** real-life situations in which finding the circumference of a circle would be useful.

2. **Describe** what is meant by an *approximation* when finding the circumference of a circle.

Guided Practice

Find the circumference of each circle to the nearest tenth. Use $\frac{22}{7}$ or 3.14 for π.

3.

9.5 km

4.

$1\frac{3}{4}$ ft

5. $r = 1.5$ yd

6. $d = 17$ ft

7. $d = 10.3$ cm

8. **Cycling** The diameter of a bicycle wheel is 26 inches. How far will you travel after one complete turn of the wheel?

EXERCISES

Practice

Find the circumference of each circle to the nearest tenth. Use $\frac{22}{7}$ or 3.14 for π.

9.

10 ft

10.

$4\frac{1}{2}$ yd

11.

12.2 m

12.

6.8 cm

13.

0.5 mi

14.

$7\frac{1}{4}$ in.

15. $d = 14$ m   16. $r = 10\frac{1}{2}$ in.   17. $r = 6.2$ cm   18. $d = 8\frac{1}{4}$ ft

19. Find the circumference of a circle whose radius is $\frac{2}{5}$ yard.

20. What is the radius of a circle whose diameter is 7 meters?

21. Find the circumference of a circle whose radius is $8\frac{3}{4}$ feet.

22. If the radius of the circle in Exercise 21 is doubled, is the circumference doubled? Why or why not?

23. **Recreation** The pedals on a bicycle from the 1870s were on the front wheel. For speed, the front wheel was large, with a diameter of about 64 inches. The back wheel diameter was about 12 inches.

   a. How far did a cyclist travel each time the pedals made a complete turn?

   b. How many times does the back wheel go around for each complete turn of the front wheel?

24. **Sports** A basketball hoop has a diameter of $18\frac{1}{2}$ inches. Estimate the circumference.

25. **Life Science** The rafflesia plant, found in the rain forests of Malaysia, produces the world's largest flower. It gives off a smell of rotting meat to attract flies and promote pollination. If the center of the flower has a radius of $\frac{3}{4}$ foot, find the circumference.

26. **Life Science** The largest bird nest on record was built by bald eagles near St. Petersburg, Florida. Its estimated weight is more than 2 tons, and it has a diameter of $9\frac{1}{2}$ feet. What is the circumference of the nest?

27. **Critical Thinking** How many segments x will fit on the circumference of the circle?

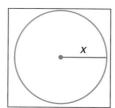

Mixed Review

28. **Physical Fitness** Every evening, Alicia walks around a neighborhood block, which is a rectangle 0.75 mile long and 0.2 mile wide. How far does Alicia walk each evening? *(Lesson 7-6)*

29. **Standardized Test Practice** Belinda had $4\frac{2}{3}$ pounds of chopped walnuts. She used $1\frac{1}{4}$ pounds in a recipe. How much of the chopped walnuts did she have left? *(Lesson 7-3)*

   A $3\frac{5}{12}$ **B** $2\frac{5}{12}$ **C** $3\frac{1}{2}$ **D** $2\frac{1}{3}$

30. **Geometry** In which quadrant do ordered pairs with a positive x-coordinate and a negative y-coordinate lie? *(Lesson 5-3)*

31. **Environment** The graph shows the percent of materials that make up landfills. Express each percent as a fraction in simplest form. *(Lesson 4-7)*

For **Extra Practice**, see page 588.

What's in a Landfill?

Paper, plastic packaging 16%

Non-packaging garbage 35%

Construction/ demolition debris 49%

Source: *ULS Report*

7-8 Properties

What you'll learn

You'll learn to use addition and multiplication properties to solve problems.

When am I ever going to use this?

Knowing how to use properties can help you find how much money you can earn each month.

Word Wise

commutative
associative
identity
multiplicative inverse
reciprocal
distributive

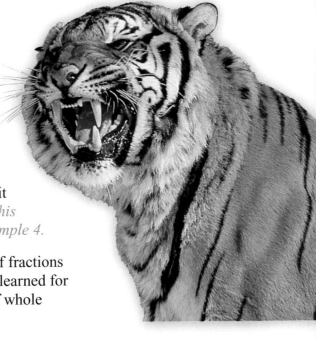

According to Wildlife Fact File, one Siberian tiger traveled 620 miles in search of food. If the tiger traveled an average of $28\frac{1}{5}$ miles each day, how long did it travel before finding food? *This problem will be solved in Example 4.*

Addition and multiplication of fractions have the same properties you learned for addition and multiplication of whole numbers.

Property	Arithmetic	Algebra
Commutative	$\frac{2}{3} + \frac{1}{8} = \frac{1}{8} + \frac{2}{3}$	$a + b = b + a$
	$\frac{2}{3} \times \frac{1}{8} = \frac{1}{8} \times \frac{2}{3}$	$a \times b = b \times a$
Associative	$\left(\frac{1}{5} + \frac{2}{5}\right) + \frac{3}{4} = \frac{1}{5} + \left(\frac{2}{5} + \frac{3}{4}\right)$	$(a + b) + c = a + (b + c)$
	$\left(\frac{1}{5} \times \frac{2}{5}\right) \times \frac{3}{4} = \frac{1}{5} \times \left(\frac{2}{5} \times \frac{3}{4}\right)$	$(a \times b) \times c = a \times (b \times c)$
Identity	$\frac{1}{2} + 0 = \frac{1}{2}$	$a + 0 = a$
	$\frac{1}{2} \times 1 = \frac{1}{2}$	$a \times 1 = a$

Zero has no reciprocal because any number times 0 is 0.

Two numbers whose product is 1 are **multiplicative inverses**, or **reciprocals**.

Multiplicative Inverse Property		
	Words:	The product of a number and its multiplicative inverse is 1.
	Symbols:	**Arithmetic** $\frac{2}{7} \times \frac{7}{2} = 1$
		Algebra For all fractions $\frac{a}{b}$, where $a, b \neq 0$, $\frac{a}{b} \times \frac{b}{a} = 1$.

Example ① Name the multiplicative inverse of $3\frac{1}{4}$.

$3\frac{1}{4} = \frac{13}{4}$ *Rename the mixed number as an improper fraction.*

$\frac{13}{4} \times \blacksquare = 1$ *What number can you multiply by $\frac{13}{4}$ to get 1?*

$\frac{13}{4} \times \frac{4}{13} = 1$

The multiplicative inverse of $3\frac{1}{4}$ is $\frac{4}{13}$.

The **distributive property** ties addition and multiplication together.

Distributive Property	**Words:**	The sum of two addends multiplied by a number is the sum of the product of each addend and the number.
	Symbols:	**Arithmetic** $\frac{1}{5} \times \left(\frac{2}{5} + \frac{3}{4}\right) = \frac{1}{5} \times \frac{2}{5} + \frac{1}{5} \times \frac{3}{4}$
		Algebra $a \times (b + c) = a \times b + a \times c$

You can use the distributive property to solve multiplication problems involving mixed numbers.

Example ② Find $\frac{1}{2} \times 4\frac{2}{5}$.

Estimate: $\frac{1}{2} \times 4 = 2$

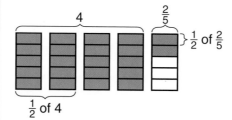

$$\frac{1}{2} \times 4\frac{2}{5} = \frac{1}{2} \times \left(4 + \frac{2}{5}\right) \quad \textit{Distributive property}$$

$$= \left(\frac{1}{2} \times 4\right) + \left(\frac{1}{2} \times \frac{2}{5}\right)$$

$$= 2 + \frac{1}{5} \text{ or } 2\frac{1}{5} \quad 2\frac{1}{5} \textit{ is close to the estimate.}$$

You can "undo" division in an equation by multiplying each side of the equation by the same number.

Multiplication Property of Equality	**Words:**	If you multiply each side of an equation by the same nonzero number, the two sides remain equal.
	Symbols:	**Arithmetic** **Algebra**
		$7 = 7$ $a = b$
		$7 \cdot 2 = 7 \cdot 2$ $ac = bc$
		$14 = 14$

Example ③ **Algebra**   Solve each equation.

INTEGRATION

a. $\frac{a}{3.2} = 5$

$$\frac{a}{3.2} = 5 \qquad \textit{Multiply each side}$$

$$\frac{a}{3.2} \cdot 3.2 = 5 \cdot 3.2 \quad \textit{by 3.2 to undo the}$$

$$a = 16 \qquad \textit{division.}$$

b. $\frac{2}{3}t = 8$

$$\frac{2}{3}t = 8 \qquad \textit{Multiply each side}$$

$$\frac{2}{3}t\left(\frac{3}{2}\right) = 8\left(\frac{3}{2}\right) \quad \textit{by the reciprocal}$$

$$t = 12 \qquad \textit{of } \frac{2}{3}.$$

CONNECTION

4 Life Science Refer to the beginning of the lesson. Use $28\frac{1}{5}d = 620$ to find d, the number of days that the tiger traveled before finding food.

$$28\frac{1}{5}d = 620$$

$$\frac{141}{5}d = 620 \qquad \text{\textit{Rename } } 28\frac{1}{5} \text{ \textit{as} } \frac{141}{5}.$$

$$\left(\frac{5}{141}\right)\frac{141}{5}d = \left(\frac{5}{141}\right)620 \qquad \begin{array}{l}\text{\textit{Multiply each side by } } \frac{5}{141} \text{ \textit{to}} \\ \text{\textit{undo the multiplication by } } \frac{141}{5}.\end{array}$$

$$d = \frac{3,100}{141}$$

$3100 \boxed{\div} 141 \boxed{=} \; 21.9858156 \qquad \text{\textit{Use a calculator.}}$

The tiger traveled about 22 days before finding food.

CHECK FOR UNDERSTANDING

Communicating Mathematics

Read and study the lesson to answer each question.

1. **Tell** whether 5 is a solution of $\frac{t}{4} = 20$.

2. **You Decide** *True* or *false*? Explain. $(12 + 27) \times 3 = 12 + 27 \times 3$

3. **Simplify** each of the following. Then tell the properties you used.

   a. $80 + 276 + 20$ **b.** $-26 + 54 + 26$

   c. $\left(-\frac{5}{9}\right)\left(\frac{11}{11}\right)$ **d.** $(-7)(3)(-15)(-9 + 9)$

Guided Practice

Name the property shown by each statement.

4. $\frac{5}{9} \times \frac{1}{3} = \frac{1}{3} \times \frac{5}{9}$ 5. $\frac{3}{5} + 0 = \frac{3}{5}$

Name the multiplicative inverse of each number.

6. $\frac{8}{11}$ 7. $4\frac{2}{5}$ 8. 14

Solve each equation. Write the solution in simplest form.

9. $\frac{3}{4}a = 24$ 10. $\frac{c}{16} = 2$ 11. $r = 2 \times 5\frac{1}{2}$

12. **Earth Science** In 1996, astronomers discovered a huge planet where they once thought it was impossible for any planet to form. Earth's diameter is only $\frac{5}{86}$ the size of this unnamed planet's diameter. Use the equation $\frac{5}{86}d = 7,970$ to find d, the diameter of the new planet in miles. *7,970 miles is the diameter of Earth.*

Practice

Name the property shown by each statement.

13. $\frac{1}{5} + \frac{3}{5} = \frac{3}{5} + \frac{1}{5}$

14. $\frac{2}{9} + \left(\frac{1}{9} + \frac{6}{4}\right) = \left(\frac{2}{9} + \frac{1}{9}\right) + \frac{6}{4}$

15. $\frac{8}{9} \times 1 = \frac{8}{9}$

16. $\frac{1}{2} \times \left(\frac{1}{5} + \frac{1}{4}\right) = \frac{1}{2} \times \frac{1}{5} + \frac{1}{2} \times \frac{1}{4}$

17. $\frac{1}{2}b = 4$, so $b = 8$

18. $\frac{3}{5} \times 1\frac{2}{3} = 1$

Name the multiplicative inverse of each number.

19. $\frac{2}{3}$

20. $\frac{1}{9}$

21. 12

22. $6\frac{2}{3}$

Solve each equation. Write the solution in simplest form.

23. $\frac{a}{12} = 3$

24. $\frac{1}{2}t = \frac{2}{5}$

25. $\frac{7}{8}x = 21$

26. $2 \times 1\frac{1}{6} = b$

27. $28 = \frac{g}{4}$

28. $\frac{4}{9}r = 84$

29. $4 \times 6\frac{1}{8} = q$

30. $\frac{s}{3.6} = 0.8$

31. What is the reciprocal of $1\frac{1}{4}$?

32. Solve $\frac{5}{6}m = 35.1$.

Applications and Problem Solving

33. **Food** Nick is serving a 12-pound turkey at a dinner party. As a rule, you should allow about $\frac{3}{4}$ of a pound of meat per person. If he invited 15 people, will he have enough turkey? Use the equation $\frac{3}{4}p = 12$, where p is the number of people.

34. **Engineering** In designing gasoline storage tanks, engineers multiply the government-required minimum thickness by a factor of $\frac{5}{2}$ for added safety. Fill in the blank with the appropriate fraction: The minimum thickness of gasoline storage tanks is ___?___ as thick as the engineers use.

35. **Critical Thinking** Use the properties of addition and multiplication to compute $60\left(4\frac{7}{8}\right) + 60\left(2\frac{1}{8}\right)$ mentally.

Mixed Review

36. **Geometry** Find the circumference to the nearest tenth of a circle whose radius is 21 meters. *(Lesson 7-7)*

37. **Standardized Test Practice** A jug contains 3 quarts less water than 4 gallons. How many quarts is this? *(Lesson 7-5)*

   A 12 qt   **B** 16 qt   **C** 1 qt   **D** 13 qt

38. Write an integer to represent decreasing the length by 11 inches. *(Lesson 5-1)*

For **Extra Practice,**
see page 588.

39. **Statistics** Find the range for 45, 29, 31, 38, and 25. Choose an appropriate scale and an interval. *(Lesson 3-1)*

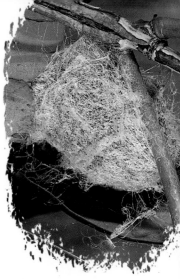

7-9

Dividing Fractions and Mixed Numbers

What you'll learn

You'll learn to divide fractions and mixed numbers.

When am I ever going to use this?

Knowing how to divide fractions and mixed numbers can help you find equal food portions.

A silkworm's cocoon is made up of a single thread about $\frac{1}{2}$ mile long. A thread that is strong enough for spinning contains 3 miles of silkworm threads twisted together. How many cocoons does it take to make a 3-mile long thread for spinning?

To solve this problem, we need to find how many $\frac{1}{2}$ miles are in 3 miles.

Method 1 Use models.

Divide 3 by $\frac{1}{2}$.

So, $3 \div \frac{1}{2} = 6$.

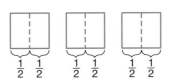

Method 2 Use the multiplicative inverse.

We can also divide by a fraction or mixed number. To do this, multiply by its multiplicative inverse.

$$3 \div \frac{1}{2} = \frac{3}{1} \div \frac{1}{2} \quad \textit{Rename 3 as } \frac{3}{1}.$$

$$= \frac{3}{1} \times \frac{2}{1} \quad \textit{Dividing by } \frac{1}{2} \textit{ is the same as multiplying by } \frac{2}{1}.$$

$$= \frac{6}{1} \textit{ or } 6$$

So, it takes 6 cocoons to make a thread 3 miles long for spinning.

Division of Fractions and Mixed Numbers	**Words:**	To divide by a fraction, multiply by its multiplicative inverse.
	Symbols:	**Arithmetic** $\frac{5}{6} \div \frac{3}{4} = \frac{5}{6} \cdot \frac{4}{3}$
		Algebra $\frac{a}{b} \div \frac{c}{d} = \frac{a}{b} \cdot \frac{d}{c}$, where b, c, and $d \neq 0$.

Example

CONNECTION

1 **Life Science** During the first year, a baby whale gains about $27\frac{3}{5}$ tons. What is the average weight gain per month?

Estimate: $28 \div 14 = 2$

$$27\frac{3}{5} \div 12 = \frac{138}{5} \div \frac{12}{1} \quad \textit{Rename } 27\frac{3}{5} \textit{ as } \frac{138}{5} \textit{ and 12 as } \frac{12}{1}.$$

$$= \frac{\overset{23}{\cancel{138}}}{5} \times \frac{1}{\underset{2}{\cancel{12}}} \quad \textit{Dividing by } \frac{12}{1} \textit{ is the same as multiplying by } \frac{1}{12}.$$

$$= \frac{23}{10} \textit{ or } 2\frac{3}{10}$$

So, a whale gains an average of $2\frac{3}{10}$ tons per month during the first year. *This is close to the estimate of 2 tons.*

2 Find $\frac{1}{2} \div 2\frac{3}{4}$. Write in simplest form.

Estimate: $\frac{1}{2} \div 3 = \frac{1}{6}$

$\frac{1}{2} \div 2\frac{3}{4} = \frac{1}{2} \div \frac{11}{4}$ *Rename $2\frac{3}{4}$ as $\frac{11}{4}$.*

$\qquad\qquad = \frac{1}{2} \times \frac{4}{11}$ *Dividing by $\frac{11}{4}$ is the same as multiplying by $\frac{4}{11}$.*

$\qquad\qquad = \frac{2}{11}$ *Compare to the estimate.*

3 Solve $4\frac{1}{3} \div 1\frac{1}{2} = d$. Write in simplest form.

$4\frac{1}{3} \div 1\frac{1}{2} = d$ *Estimate: $4 \div 2 = 2$*

$\frac{13}{3} \div \frac{3}{2} = d$ *Rename $4\frac{1}{3}$ as $\frac{13}{3}$ and $1\frac{1}{2}$ as $\frac{3}{2}$.*

$\frac{13}{3} \times \frac{2}{3} = d$ *Dividing by $\frac{3}{2}$ is the same as multiplying by $\frac{2}{3}$.*

$\frac{26}{9} = d$ *Rename as a mixed number.*

The solution is $\frac{26}{9}$ or $2\frac{8}{9}$.   *How does this compare to the estimate?*

CHECK FOR UNDERSTANDING

Communicating Mathematics

Read and study the lesson to answer each question.

1. **Write** the steps you would use to find $1\frac{1}{9} \div \frac{8}{9}$.

2. **Draw** a figure or fold a piece of paper to show $4 \div 1\frac{1}{3}$. Explain how to use the figure or paper to find the answer.

3. **You Decide** For a camping trip, Joshua has $2\frac{1}{4}$ pounds of trail mix to split among three friends and himself. He figures each person will get about 1.8 pounds. Omar says that each person will get less than a pound. Who is right? Explain.

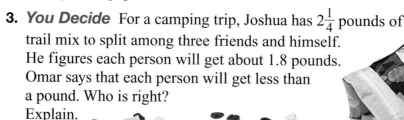

Guided Practice

Divide. Write each quotient in simplest form.

4. $\frac{2}{3} \div \frac{1}{2}$ 5. $6 \div \frac{1}{2}$ 6. $2\frac{2}{3} \div 4$ 7. $4\frac{2}{3} \div \frac{7}{8}$

Solve each equation.

8. $1\frac{2}{3} \div 3 = a$ 9. $k = \frac{3}{5} \div 2\frac{1}{4}$

10. **Consumerism** A box of laundry detergent contains 35 cups. If you use $1\frac{1}{4}$ cups per load of laundry, how many loads do you get from 1 box?

Practice

Divide. Write each quotient in simplest form.

11. $\frac{3}{8} \div \frac{6}{7}$ **12.** $\frac{3}{4} \div \frac{1}{2}$ **13.** $\frac{3}{5} \div \frac{1}{4}$ **14.** $\frac{4}{9} \div 2$

15. $\frac{5}{6} \div \frac{2}{3}$ **16.** $\frac{5}{9} \div \frac{5}{6}$ **17.** $\frac{3}{4} \div \frac{3}{8}$ **18.** $3 \div \frac{6}{7}$

19. $\frac{1}{8} \div \frac{1}{3}$ **20.** $\frac{3}{4} \div 1\frac{1}{2}$ **21.** $\frac{2}{3} \div 2\frac{1}{2}$ **22.** $5 \div 1\frac{1}{3}$

23. What is $1\frac{1}{9}$ divided by $\frac{2}{3}$?   **24.** Divide $\frac{5}{6}$ by $\frac{8}{9}$.

Solve each equation.

25. $b = 5\frac{1}{4} \div 3$ **26.** $\frac{9}{10} \div 2\frac{1}{4} = h$ **27.** $2\frac{1}{4} \div \frac{2}{3} = r$

28. $f = 4\frac{1}{2} \div 6\frac{3}{4}$ **29.** $p = 1\frac{1}{4} \div 3\frac{1}{2}$ **30.** $6\frac{2}{7} \div 3\frac{1}{7} = s$

31. *Algebra* Evaluate $k \div 1\frac{1}{8}$ if $k = \frac{6}{7}$.

32. Solve $m = \left(\frac{1}{4} + \frac{1}{10} \right) \div 2\frac{1}{3}$.

Applications and Problem Solving

33. *Food* Nancy has 5 pounds of mints from which she wants to make $\frac{1}{4}$-pound packages for party favors. How many packages can she make?

34. *Housing*
A contractor has
15 acres of land that she is going to sell as $\frac{3}{4}$-acre lots.
How many lots can she sell?

35. *Working on the* **CHAPTER Project** Refer to the table on page 267.
   a. For each of the companies that you have chosen to track, find how many shares you could have purchased for $1,000 on August 18, 1997.
   b. Using yesterday's closing price for the same companies, how much money would the shares that you purchased in 1997 be worth now?

36. *Critical Thinking* Will the quotient $6\frac{1}{5} \div 2\frac{3}{4}$ be a proper fraction or a mixed number? Why?

Mixed Review

37. Name the multiplicative inverse of $3\frac{5}{8}$. *(Lesson 7-8)*

38. *Geometry* Find the perimeter of a rectangle if the length is 6.4 inches and the width is 5 inches. *(Lesson 7-6)*

39. Express 18 out of 100 as a percent. *(Lesson 4-6)*

40. Tell whether 240 is divisible by 2, 3, 4, 5, 6, 9, or 10. *(Lesson 4-1)*

41. **Standardized Test Practice** Estimate $29.78 - 7.21$ by rounding. *(Lesson 2-3)*

   A 20   **B** 22   **C** 23   **D** 24

For **Extra Practice**, see page 589.

interNET
CONNECTION Chapter Review **For additional lesson-by-lesson review, visit:**
www.glencoe.com/sec/math/mac/mathnet

Vocabulary

After completing this chapter, you should be able to define each term, concept, or phrase and give an example or two of each.

Number and Operations
associative (p. 301)
commutative (p. 301)
distributive property (p. 302)
identity (p. 301)
multiplicative inverse (p. 301)
multiplication property of equality (p. 302)
reciprocal (p. 301)

Geometry
center (p. 297)
circle (p. 297)
circumference (p. 297)
diameter (p. 297)

perimeter (p. 292)
radius (p. 297)

Measurement
cup (p. 290)
gallon (p. 290)
ounce (p. 289)
pint (p. 290)
pound (p. 289)
quart (p. 290)
ton (p. 289)

Problem Solving
eliminate possibilities (p. 280)

Understanding and Using the Vocabulary

Choose the letter of the term that best matches each phrase.

1. the property that states $a \times b = b \times a$
2. the multiplicative inverse of a number
3. the property that states that $a + 0 = a$
4. the distance around a rectangle
5. the distance across a circle through the center
6. the property that states that $(a + b) + c = a + (b + c)$
7. the distance around a circle
8. what to do to convert from a smaller unit of measure to a larger unit
9. what to do to convert from a larger unit of measure to a smaller unit

a. reciprocal
b. perimeter
c. identity
d. circumference
e. commutative
f. inverse
g. multiply
h. radius
i. divide
j. diameter
k. associative
l. distributive

In Your Own Words

10. *Explain* how the distributive property can be used to compute the product of a mixed number and a whole number.

Objectives & Examples

Upon completing this chapter, you should be able to:

● estimate sums, differences, products, and quotients of fractions and mixed numbers *(Lesson 7-1)*

Estimate $15\frac{7}{9} + 3\frac{1}{6}$.

$16 + 3 = 19$

● add and subtract fractions *(Lesson 7-2)*

$$\frac{11}{12} - \frac{1}{6} = \frac{11}{12} - \frac{2}{12}$$
$$= \frac{9}{12} \text{ or } \frac{3}{4}$$

● add and subtract mixed numbers *(Lesson 7-3)*

$$5\frac{3}{4} + 10\frac{1}{3} = 5\frac{9}{12} + 10\frac{4}{12}$$
$$= 15\frac{13}{12} \text{ or } 16\frac{1}{12}$$

● multiply fractions and mixed numbers *(Lesson 7-4)*

$$\frac{3}{5} \times \frac{1}{2} = \frac{3}{10} \qquad 4\frac{3}{4} \times 2\frac{2}{3} = \frac{19}{14} \times \frac{\overset{2}{8}}{3}$$
$$= \frac{38}{3} \text{ or } 12\frac{2}{3}$$

● change units in the customary system *(Lesson 7-5)*

Complete: 14 lb = ___?___ oz

Since 1 lb = 16 oz, multiply by 16.

$14 \cdot 16 = 224$

14 lb = 224 oz

Review Exercises

Use these exercises to review and prepare for the chapter test.

Estimate.

11. $\frac{3}{4} + \frac{8}{9}$ 12. $5\frac{4}{5} + 12\frac{1}{6}$

13. $12 \div \frac{2}{3}$ 14. $2\frac{9}{10} \div 1\frac{1}{8}$

15. $\frac{13}{15} - \frac{1}{4}$ 16. $89\frac{9}{10} - 7\frac{4}{5}$

17. $1\frac{12}{13} \times 16\frac{1}{12}$ 18. $6\frac{1}{9} \times 2\frac{5}{6}$

Add or subtract. Write each sum or difference in simplest form.

19. $\begin{array}{r} \frac{3}{5} \\ +\frac{3}{7} \\ \hline \end{array}$ 20. $\begin{array}{r} \frac{5}{6} \\ -\frac{1}{2} \\ \hline \end{array}$

21. $\frac{3}{4} - \frac{3}{20}$ 22. $\frac{3}{8} + \frac{1}{3}$

Add or subtract. Write each sum or difference in simplest form.

23. $8 + 5\frac{5}{9}$ 24. $5\frac{1}{12} + 2\frac{5}{6}$

25. $3\frac{7}{8} - \frac{5}{6}$ 26. $5\frac{2}{7} - 4\frac{1}{2}$

Multiply. Write each product in simplest form.

27. $\frac{5}{9} \times \frac{3}{5}$ 28. $4\frac{1}{5} \times 3\frac{1}{3}$

29. $5\frac{1}{2} \times 6\frac{2}{7}$ 30. $6 \times 2\frac{3}{8}$

Complete.

31. 4 qt = ___?___ pt
32. 6 gal = ___?___ qt
33. 8,000 lb = ___?___ T
34. 48 oz = ___?___ lb
35. 48 fl oz = ___?___ c
36. 9 c = ___?___ pt

Objectives & Examples

Review Exercises

find perimeter *(Lesson 7-6)*

Find the perimeter of a rectangle $4\frac{3}{4}$ feet long and $3\frac{1}{3}$ feet wide.

$P = 4\frac{3}{4} + 4\frac{3}{4} + 3\frac{1}{3} + 3\frac{1}{3}$

$P = 16\frac{2}{12}$ or $16\frac{1}{6}$ feet

The perimeter is $16\frac{1}{6}$ feet.

Find the perimeter of each figure shown or described.

37.

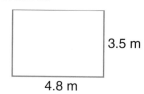

3.5 m

4.8 m

38. rectangle: $\ell = 25\frac{1}{6}$ ft

$w = 10\frac{2}{3}$ ft

find the circumference of circles *(Lesson 7-7)*

If the diameter of a circle is $2\frac{1}{3}$ feet, then find its circumference.

$C = \pi d$

$C \approx \frac{22}{7} \times 2\frac{1}{3}$ *Use $\frac{22}{7}$ because*

$C \approx \frac{22}{7} \times \frac{7}{3}$ *the diameter has a numerator of 7.*

$C \approx 7\frac{1}{3}$

The circumference is about $7\frac{1}{3}$ or 7.3 feet.

Find the circumference of each circle to the nearest tenth. Use $\frac{22}{7}$ or 3.14 for π.

39.

$1\frac{1}{2}$ yd

40.

10.4 m

41. $d = 14$ in.

42. $r = 1.4$ cm

43. $r = \frac{7}{11}$ ft

44. $d = 6\frac{2}{5}$ ft

use addition and multiplication properties to solve problems *(Lesson 7-8)*

$\frac{1}{4} \times 8\frac{2}{3} = \frac{1}{4}\left(8 + \frac{2}{3}\right)$

$= 2 + \frac{2}{12}$ or $2\frac{1}{6}$

Solve each equation.

45. $6 \times 3\frac{1}{8} = a$

46. $\frac{1}{3}t = \frac{1}{5}$

47. $\frac{5}{6}r = 5$

48. $c = 5\frac{5}{6} \times 2$

divide fractions and mixed numbers *(Lesson 7-9)*

$2\frac{4}{5} \div \frac{7}{10} = \frac{\overset{2}{\cancel{14}}}{\underset{1}{\cancel{5}}} \times \frac{\overset{2}{\cancel{10}}}{\underset{1}{\cancel{7}}} = \frac{4}{1}$ or 4

Divide. Write each quotient in simplest form.

49. $\frac{3}{5} \div \frac{1}{10}$

50. $\frac{5}{8} \div \frac{3}{4}$

51. $3\frac{1}{7} \div \frac{2}{5}$

52. $1\frac{5}{6} \div 3\frac{2}{9}$

Applications & Problem Solving

53. *Cooking* A recipe calls for $2\frac{1}{2}$ cups of flour, $1\frac{3}{4}$ cups of sugar, and $1\frac{2}{3}$ cups of brown sugar. How many cups are in the mixture? *(Lesson 7-3)*

54. *Eliminate Possibilities* On a map, 1 inch represents 200 miles. A trip you are planning is $4\frac{1}{2}$ inches on the map. How long is your trip? *(Lesson 7-3B)*

  A 900 miles

  B 450 miles

  C $4\frac{1}{2}$ miles

  D 90 miles

  E 800 miles

55. *Driving* A bridge has a weight limit of 5 tons. If a truck weighs 4,125 pounds, is it safe for the truck to cross the bridge if no other vehicles are on the bridge? *(Lesson 7-5)*

56. *Gardening* Celia wants to fence in her garden whose dimensions are shown in the diagram below. How much fencing will she need? *(Lesson 7-6)*

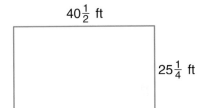

$40\frac{1}{2}$ ft

$25\frac{1}{4}$ ft

Alternative Assessment

Open Ended

Suppose you want to build a deck around a circular swimming pool. The swimming pool has a radius of 66 inches and you want the deck to be 4 feet wide. If you put a fence around the outside of the deck, about how much fencing will you need to the nearest foot? Show your work.

Suppose you decide to put a fence around just the pool instead of the deck. If you save $375 by doing this instead of putting a fence around the outside of the deck, about how much does one foot of fencing cost? Explain.

PORTFOLIO   Select one of the problems you solved in this chapter and place the problem and its solution in your portfolio. Attach a note explaining why you selected it.

Completing the CHAPTER Project

Suppose you had purchased 100 shares of stock on the first day of tracking stock prices. In your report, explain which stock would have made you the most money if you had sold all your shares on the last day. Also, pick the day on which each stock should have been sold to make the most money during the month.

Use the following checklist to make sure your project is complete.

- ☑ You have a report that includes your table of stock prices.
- ☑ You have a graph of the price for each stock.
- ☑ You have a brief explanation of the price trend for each stock over the month.

A practice test for Chapter 7 is provided on page 613.

Section One: Multiple Choice

There are eleven multiple-choice questions in this section. Choose the best answer. If a correct answer is *not here,* choose the letter for Not Here.

1. Which is the equation for the line graphed?

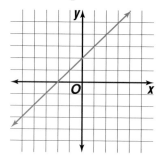

A $y = x + 2$

B $y = x - 2$

C $y = 2x$

D $y = 2x + 2$

2. Change $3\frac{5}{8}$ to an improper fraction.

F $\frac{35}{8}$

G $\frac{20}{8}$

H $\frac{15}{8}$

J $\frac{29}{8}$

3. Which expression is equivalent to $3.2 \times (2.4 \times 5.8)$?

A $(3.2 \times 2.4) + (3.2 \times 5.8)$

B $(3.2 \times 2.4) \times 5.8$

C $3.2 \times (2.4 + 5.8)$

D $2.4 + (3.2 \times 5.8)$

4. A doorway is 5 inches less than 8 feet tall. How many *inches* is this?

F 101 inches

G 96 inches

H 91 inches

J 40 inches

Please note that Questions 5–11 have five answer choices.

5. Margaret can earn tips ranging from $3–$5 per table at a steak restaurant. If she has 9 tables, which is a reasonable estimate of the tips she will earn?

A less than $27

B between $27 and $45

C between $45 and $70

D between $70 and $90

E more than $90

6. Which expression represents *three less than a number*?

F $n - 3$

G $3 - n$

H n

J $\frac{n}{3}$

K $3n - 3$

7. Forest Park Middle School's 52 band students and 7 adult sponsors are planning a trip to an amusement park. Each school bus will carry at most 31 people. Each ticket to the park costs $19.75, but schools get a $3.00 discount per ticket. Which piece of information is *not* needed for the school principal to determine the amount of money required for the amusement park tickets?

A There are 52 students in the band.

B There are 7 adult sponsors.

C Each school bus will carry at most 31 people.

D The price of each ticket is $19.75.

E The discount per ticket is $3.00.

8. Nate had $\frac{7}{8}$ of a tank of gas in the lawn mower. After mowing the grass, he had $\frac{1}{4}$ of a tank. How much gas did Nate use mowing the lawn?

F $\frac{5}{8}$ of a tank

G $\frac{3}{4}$ of a tank

H $\frac{3}{8}$ of a tank

J $\frac{1}{8}$ of a tank

K Not Here

9. A recycling group at Thompson Middle School collected cans for a service project. They collected 122.4 pounds, 88.9 pounds, and 117.02 pounds in the last three weeks. What was the total amount of cans collected for that period?

A 248.15 pounds   **B** 211.13 pounds

C 328.32 pounds   **D** 328.5 pounds

E Not Here

10. The regular price of 3 packages of batteries is $8.85 without tax. Before tax is added, how much can be saved by buying 3 packages on sale for $6.49?

F $3.36   **G** $2.36

H $2.26   **J** $1.06

K Not Here

11. In 1996, Donovan Bailey set a world record of 9.84 seconds for the 100-meter dash. A honeybee can fly the same distance in 20.706 seconds. How many times faster than a honeybee is Donovan Bailey?

A 10.2 times

B 10.9 times

C 4.8 times

D 2.1 times

E Not Here

Test-Taking Tip THE PRINCETON REVIEW

When taking a standardized test, you may be able to eliminate answer choices through estimating. Also, look to see which answers are not reasonable for the information given in the problem.

Section Two: Free Response

This section contains five questions for which you will provide short answers. Write your answers on your paper.

12. $\frac{3}{4} + \frac{1}{2} =$

13. $\frac{1}{4} \times 2\frac{1}{2} =$

14. Solve $3x + 10 < 28$.

15. Name the ordered pairs for points Q, R, and S. Then tell in which quadrant each point lies.

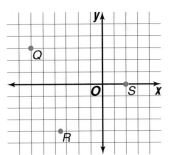

16. The McCanns make cookies for their carpool. If they buy a 12-pound bag of flour and they use $1\frac{1}{3}$ cups of flour per batch of cookies, how many batches of cookies do they get from one bag of flour?

 **inter**NET **Test Practice** For additional
CONNECTION   test practice questions, visit:

www.glencoe.com/sec/math/mac/mathnet

CHAPTER 8

Using Proportional Reasoning

What you'll learn in Chapter 8

- to express ratios as fractions,
- to solve proportions,
- to solve problems by drawing a diagram,
- to solve problems involving scale drawings,
- to express percents as fractions and decimals, and
- to find the percent of a number.

RECYCLE!

GLASS
TO REPORT THEFT OF RECYCLABLES OR CONTAINERS CALL 503-8561

ALUMINUM & TIN CANS / PLASTIC SODA BOTTLES

NEWSPAPER
NO MAGAZINES OR JUNK MAIL PLEASE

CHAPTER Project

WASTE NOT, WANT NOT

In this project, you will use percents to make a poster about trash. You will determine the steps you can take to reduce the amount of trash by increasing the amount of material that is recycled.

Getting Started

- Keep track of the trash your family throws away for one week. Make a chart showing the type and amount of trash you throw away each day. Use fractions to estimate what part of the day's trash each type represents. For example, aluminum cans may be about $\frac{1}{4}$ of your trash on a certain day.
- Find recent statistics about what materials are recycled and what the recycled material becomes.
- Research the recycling programs in your area.

Technology Tips

- Use a **spreadsheet** to keep track of the data you collect.
- Use **graphing software** to make statistical graphs to represent the data.
- Use a **word processor** to make the text for your poster.

 **inter NET** **CONNECTION** **Research** For up-to-date information on recycling, visit:

www.glencoe.com/sec/math/mac/mathnet

Working on the Project

You can use what you learn in Chapter 8 to study trash and recycling.

Page	Exercise
320	42
345	50
355	Alternative Assessment

COOPERATIVE LEARNING

8-1A Equal Ratios

A Preview of Lesson 8-1

◐ counters

A *ratio* is the comparison of two numbers. Ratios are often used to show how large one quantity is compared to another. For example, for every four counters in pile A there is one counter in pile B. So we can say that piles A and B have the ratio 4 to 1.

A　　　　　　**B**

TRY THIS

Work in pairs.

Step 1　Fold a sheet of paper into eighths. Then unfold it and label and place the counters in the sections as shown.

Step 2　Place the counters in sections V, W, and Z so that the ratio of the counters in each column is equal to the ratio of the counters in C and D.

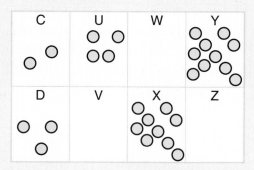

ON YOUR OWN

1. Describe how you decided how many counters to place in sections V, W, and Z.

2. If there were 84 counters in section X, how many counters would you have placed in section W?

3. Fold another piece of paper into twelve sections. Then label and place the counters in the sections as shown. Place the counters in the empty sections so that the ratios in each column are equal to the ratio of E to F to G.

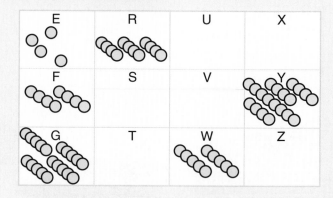

316　**Chapter 8** Using Proportional Reasoning

8-1

Ratios

What you'll learn

You'll learn to express ratios as fractions and determine whether two ratios are equivalent.

When am I ever going to use this?

Ratios are used to report foreign currency exchange rates in the newspaper.

Word Wise

ratio
equivalent ratios

Are potato chips on your list of favorite snacks? Americans eat about 606 million pounds of chips each summer! It takes four pounds of potatoes to make one pound of potato chips. The **ratio** of pounds of potatoes to pounds of potato chips is 4 to 1.

Ratio	**Words:**	A ratio is a comparison of two numbers by division.
	Symbols:	**Arithmetic** 4 to 1 4:1 $\frac{4}{1}$
		Algebra a to b $a{:}b$ $\frac{a}{b}$

A ratio can be written as a fraction. Often, ratios are written as fractions in simplest form.

Example APPLICATION

① **Recycling** For every 207 pounds of waste that is generated in the United States, 45 pounds are recycled into new products. Write the ratio of the pounds of recycled material to the total pounds of waste in simplest form.

pounds recycled $\rightarrow$ $\dfrac{45}{207} = \dfrac{45 \div 9}{207 \div 9}$ *The GCF of 45 and 207 is 9.*
pounds of waste $\rightarrow$

$\qquad\qquad\qquad = \dfrac{5}{23}$

The ratio in simplest form is $\frac{5}{23}$, 5 to 23, or 5:23.

Study Hint

Reading Math
The ratio of 5 to 23 means that for every 5 pounds of recycled material, there was 23 pounds of waste.

You can also write a ratio as a decimal. The ratio in Example 1 can be expressed as a decimal in the following way.

5 ÷ 23 = *0.217391304*

When you simplify a ratio that compares measurements, be sure that the measurements have the same unit of measure.

Example

CONNECTION

LOOK BACK

Refer to Lesson 7-5 for information on changing measurements.

2 **Life Science** The shortest member of the deer family is the southern puda, which is about 14 inches tall. The Alaskan moose is the tallest deer. The tallest Alaskan moose recorded was 7 feet 8 inches tall. Write the ratio of the height of the shortest deer to the height of the tallest deer in simplest form.

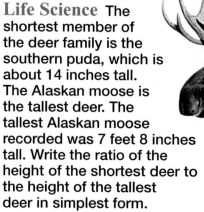

Write both measurements using inches.

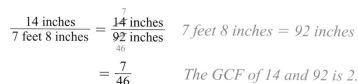

$$\frac{14 \text{ inches}}{7 \text{ feet } 8 \text{ inches}} = \frac{\overset{7}{\cancel{14}} \text{ inches}}{\underset{46}{\cancel{92}} \text{ inches}}$$ *7 feet 8 inches = 92 inches*

$$= \frac{7}{46}$$ *The GCF of 14 and 92 is 2.*

The ratio in simplest form is $\frac{7}{46}$, or 7:46.

Two ratios that have the same value are **equivalent ratios**.

Example

3 Tell whether 4:9 and 36:81 are equivalent ratios.

Express each ratio as a fraction in simplest form.

The GCF of 4 and 9 is 1. $\dfrac{36}{81} = \dfrac{36 \div 9}{81 \div 9}$ *The GCF of 36 and 81 is 9.*

So $\dfrac{4}{9}$ is in simplest form. $= \dfrac{4}{9}$

The ratios in simplest form are equal. So 4:9 and 36:81 are equivalent ratios.

CHECK FOR UNDERSTANDING

Communicating Mathematics

Read and study the lesson to answer each question.

1. *Express* the ratio *15 out of 20 cars* in three different ways.

2. *Demonstrate* how to simplify the ratio 18:64.

3. *Write* how you can determine whether two ratios are equivalent.

Guided Practice

Express each ratio as a fraction in simplest form.

4. $\dfrac{5}{35}$ 5. 16 to 42   6. 36:9

7. 8 hours in a week   8. 3 pounds: 12 ounces

Tell whether the ratios are equivalent. Show your answer by simplifying.

9. $\frac{12}{16}$ and $\frac{21}{28}$

10. $\frac{5}{8}$ and $\frac{65}{100}$

11. $15 for 6 pounds and $90 for 36 pounds

12. *Sports* In baseball, a player's batting average is the ratio of his hits to times at bat. The table shows the hits and times at bat for various professional players in 1999. Find each player's batting average as a fraction in simplest form and as a decimal to the nearest thousandth.

Player	Hits	Times At Bat
Juan Gonzalez, Texas Rangers	183	562
Ken Griffey, Jr., Seattle Mariners	173	606
Kenny Lofton, Cleveland Indians	140	465
Jim Thome, Cleveland Indians	137	494

Source: *Major League Baseball*

EXERCISES

Practice

Express each ratio as a fraction in simplest form.

13. $\frac{21}{45}$

14. 9 to 12

15. 49:14

16. 24:4

17. 125 to 25

18. 27:36

19. 35 to 36

20. 27:15

21. 155 to 220

22. 11 teachers to 275 students

23. 64 inches to $1\frac{1}{2}$ feet

24. 45 minutes to 8 hours

25. 14 ounces to 5 pounds

26. 65¢ to $4.50

27. 18 weeks to one year

Tell whether the ratios are equivalent. Show your answer by simplifying.

28. $\frac{12}{9}$ and $\frac{15}{12}$

29. 150:15 and 3:1

30. $\frac{6}{39}$ and $\frac{2}{13}$

31. $\frac{65}{5}$ and $\frac{1}{13}$

32. $\frac{6}{12}$ and $\frac{0.5}{1}$

33. 14:42 and 58:1,218

34. 6 pounds:72 ounces and 2 pounds:24 ounces

35. 3 days to 4 hours and 9 days to 12 hours

36. 5 miles to 100 feet and 15 miles to 100 yards

37. Write 180 days of school in one year as a fraction in simplest form.

38. Are the ratios 14 inches to 9 feet and 16 feet to 12 yards equivalent?

Applications and Problem Solving

39. *Collectibles* Do you have any small beanbag animals? In 1997, you could buy one of the retired styles, Peanut the Elephant, for about $1,500 at trade shows. The retail price of a small beanbag animal then was $5. Write the ratio of the retail price to Peanut's price as a fraction in simplest form.

40. *Entertainment* In the 1995-1996 television season, the highest-rated show had 21,098,000 of the 95,900,000 television-owning households tuning in. The highest-rated show in the 1985-1986 season had 29,040,000 of the 85,900,000 television-owning households tuned in. Are the ratios of viewers to total TV households for these shows equivalent? Explain.

41. *Geography* The Tobu World Square in Nikko, Japan, has 102 of the world's most recognized landmarks recreated at a fraction of their actual size. The table shows the height of the model and the actual height of a few of the buildings.

Landmark	Model Height	Actual Height
Arch de Triomphe, Paris, France	1.98 m	49.5 m
Great Sphinx, Giza, Egypt	2.64 ft	66 ft
Parthenon, Athens, Greece	72 cm	18 m
White House, Washington, D.C.	40.8 in.	85 ft

a. Find the ratio of the model height to the actual height for each building.

b. Are the models all in the same ratio? Explain.

42. *Working on the* CHAPTER Project Refer to the data you have gathered about your family's trash. Write the ratio of the amount of trash in each category to the total amount of trash.

43. *Critical Thinking* Find the next number in the pattern 5,400, 2,700, 900, 225, ____. Explain your reasoning. (*Hint:* Look at the ratios of consecutive numbers.)

Mixed Review

44. Find $\frac{4}{5} \div 1\frac{1}{2}$. Write the quotient in simplest form. *(Lesson 7-9)*

45. *Algebra* Solve $18x = 54$. Check your solution. *(Lesson 6-2)*

46. Standardized Test Practice Melissa is writing a report on these events.

Year	Event
1200 B.C.	A calendar of 18 months of 20 days developed by the Olemecs in Mesomenia.
46 B.C.	Julius Caesar reforms the Roman calendar to use 365 days per year, plus one extra day every four years.
A.D. 1948	The atomic clock is invented.
8000 B.C.	The Egyptians use a calendar with 12 months of 30 days each.

If she wants to list the events in order from least recent to most recent, which order should she choose? *(Lesson 5-2)*

A Egyptian calendar, atomic clock, Olemec calendar, Roman calendar

B Egyptian calendar, Olemec calendar, Roman calendar, atomic clock

C Egyptian calendar, Roman calendar, Olemec calendar, atomic clock

D Roman calendar, Olemec calendar, atomic clock, Egyptian calendar

47. *Patterns* Determine whether 3,498 is divisible by 2, 3, 4, 5, 6, 9, or 10. *(Lesson 4-1)*

For **Extra Practice**, see page 589.

48. Find 0.65×2.4. *(Lesson 2-4)*

Rates

The first fountains were built in ancient Greece above natural springs. In Renaissance Europe, elaborate pump systems made complicated fountains possible. Many modern fountains use computers to control the water flow and lights. The Prometheus Fountain in Rockefeller Center in New York City is one of the most photographed fountains in the world. It pumps one million gallons of water every 250 minutes.

You often need to compare two quantities with different units. For example, the ratio *one million gallons of water every 250 minutes* compares a number of gallons to a number of minutes. A ratio of this type is called a **rate**.

Rate	A rate is a ratio of two measurements with different units.

How many gallons of water does the Prometheus Fountain pump every minute? To answer this question, you need to find a **unit rate**.

Unit Rate	A unit rate is a rate in which the denominator is 1 unit.

Example **APPLICATION**

Architecture Refer to the beginning of the lesson. How many gallons of water does the Prometheus Fountain pump every minute?

Write the rate as a fraction. Then find an equivalent rate with a denominator of 1.

$$\begin{array}{l}\textit{gallons} \rightarrow \\ \textit{minute} \rightarrow \end{array} \frac{1{,}000{,}000}{250} = \frac{1{,}000{,}000 \div 250}{250 \div 250} = \frac{4{,}000}{1}$$

The Prometheus Fountain pumps 4,000 gallons each minute, or 4,000 gallons *per minute*.

In Example 1, the unit rate is found by writing the rate with a denominator of 1. You can simplify this process by dividing. For example, to find miles per gallon, divide the number of miles driven by the number of gallons of gas used.

Example **2** **Express selling 150 tickets in 5 days as a unit rate.**

Divide the number of tickets by the number of days to find tickets per day.

$150 \div 5 = 30$

The unit rate is 30 tickets per day.

The table shows some of the common unit rates you know.

The U.S. Bureau of the Census studies the changes in the population. One rate they find is the **population density**, which is the population per square mile.

Unit Rate	Abbreviation
miles per gallon	mi/gal (or mpg)
miles per hour	mi/h (or mph)
price per pound	dollars/lb
meters per second	m/s

Example **3**

INTEGRATION

Statistics The U.S. Bureau of the Census recently estimated the population of Virginia to be 6,791,345. If the land area of Virginia is 39,598 square miles, find the population density.

Explore You know the population and the area of Virginia. You need to find the population density.

Plan Divide the population by the land area to find the population per square mile.

Solve $\dfrac{6{,}791{,}345 \text{ people}}{39{,}598 \text{ square miles}}$

6791345  39598 [=] *171.5072731*

Virginia had a population density of about 172 people per square mile.

Examine Check the answer by estimating. The population is about 6,800,000, and the area is about 40,000. So the population density should be about 170 people per square mile. The answer is reasonable.

CHECK FOR UNDERSTANDING

Communicating Mathematics

Read and study the lesson to answer each question.

1. *State* whether each rate is a unit rate. Explain why or why not.
   a. 35 miles in 1.5 gallons
   b. $1.99 per pound
   c. 155 people per square mile

2. *Explain* the difference between a ratio and a rate.

3. *You Decide* Marta says that $3.99 for a 16-ounce bag of candy is a better buy than $2.99 for a 12-ounce bag. April disagrees. Who is correct and why?

Guided Practice

Express each rate as a unit rate.

4. $6.99 for 4 pounds
5. 360 miles in 6 hours
6. 410 miles in 16 gallons
7. 45 people in 3 vans
8. $42 for 8 hours
9. 99¢ for 16 ounces

10. *History* The *Mayflower* set sail from Plymouth, England, on September 16, 1620 bound for America. The tiny ship completed its 3,000-nautical mile journey 67 days later. On the average, what was the *Mayflower's* unit rate of travel per day?

EXERCISES

Practice

Express each rate as a unit rate.

11. 200 miles in 5 hours
12. 2 cups for 36 cookies
13. $350 for 5 days
14. 90,000 seats in 36 sections
15. 18 pounds in 6 weeks
16. 14,960 visitors in 55 days
17. 200 meters in 40 seconds
18. 245,000 people in 1,750 sq mi
19. $1.89 for 6.5 ounces
20. $960 for 16 days
21. $6.20 for 5 pounds
22. 480 miles in 6 days
23. 24 people in 8 cars
24. 1,500 words in 25 minutes
25. 228 feet in 24 seconds
26. 450 Calories in 3 servings
27. 1,080 rotations in 12 minutes
28. 205 students to 8 teachers

29. Find the unit price for a 16-ounce box of cereal that costs $3.92.

30. What is the population density of a city with 425,000 people in 1,872 square miles?

Applications and Problem Solving

31. *Manufacturing* In its first 25 years of business, a company made 54,000,000 of one of the most popular children's toys. On average, how many toys were made each of those years?

32. *Money Matters* Aspirin is sold in boxes of 24 for $3.69, 50 for $5.49, 100 for $8.29, and 150 for $11.99. Which box has the best unit price?

33. *Life Science* The average heart rate of an adult human is 72 beats per minute. For an adult elephant, the average heart rate is 35 beats per minute.
    a. Whose heart beats more times in one hour?
    b. Whose heart makes 1,000,000 beats in less time?

34. Statistics Find the population density for each country.

Country	Population	Area (square miles)
Australia	18,260,863	2,966,200
Belize	219,296	8,867
Japan	125,449,703	145,850
Zaire	46,498,539	905,354

Source: *World Almanac, 1997*

Let a faucet slowly drip for 15 minutes into a measuring cup. Then calculate how much water would be wasted per hour and per day for a similar drip.

35. Critical Thinking A train goes 65 miles per hour and travels 320 miles. How many hours will it take for the train to reach its destination?

Mixed Review

36. Write the ratio 45:81 as a fraction in simplest form. *(Lesson 8-1)*

37. Civics The United States Constitution states that no person who is not at least 25 years old and a citizen of the United States for 7 years may serve as a U.S. Representative. Write an inequality showing the age of a person who may be a U.S. Representative. *(Lesson 6-5)*

38. Statistics Find the mean, mode(s), and median for the data: 2.1, 2.2, 2.2, 2.4, 2.4, 2.4, 2.7, 2.8, 3.0, 3.0, 3.4. *(Lesson 3-4)*

39. Standardized Test Practice Which is equivalent to 6×8^4? *(Lesson 1-4)*

A 6×64

B $48 \times 48 \times 48 \times 48$

C $6 \times 8 \times 6 \times 8 \times 6 \times 8 \times 6 \times 8$

D $6 \times 8 \times 8 \times 8 \times 8$

For **Extra Practice,** see page 589.

MATH IN THE MEDIA

1. Is the rate stated a unit rate? Explain.

2. In 20 feet of freshwater, a scuba diver whose body has an area of 2,880 square inches experiences about 67,104 pounds of pressure. Find the unit rate.

3. Research what pressure has to do with your ears popping when you drive up or down a big hill.

8-3 Solving Proportions

What you'll learn

You'll learn to solve proportions.

When am I ever going to use this?

You can use proportions to change the number of servings made using a recipe.

Word Wise

proportion
cross products

How much lemonade can you drink when you're really thirsty? How about 2,250 liters! In 1994, a company made a giant bottle of lemonade to mark its 200th year. It held 2,250 liters or 9,000 servings of lemonade! A standard bottle holds 2 liters or 8 servings of lemonade.

If you find the ratios of the number of liters to the number of servings of lemonade in each bottle, how do they compare?

Giant bottle: $\dfrac{liters}{servings} \rightarrow \dfrac{2,250}{9,000} = \dfrac{2,250 \div 2,250}{9,000 \div 2,250} = \dfrac{1}{4}$

Standard bottle: $\dfrac{liters}{servings} \rightarrow \dfrac{2}{8} = \dfrac{2 \div 2}{8 \div 2} = \dfrac{1}{4}$

The ratios $\dfrac{2,250}{9,000}$ and $\dfrac{2}{8}$ are equivalent. So, we can write $\dfrac{2,250}{9,000} = \dfrac{2}{8}$. This equation is an example of a **proportion**.

Proportion	**Words:**	A proportion is an equation that shows that two ratios are equivalent.
	Symbols:	**Arithmetic** $\quad$ **Algebra**
		$\dfrac{2}{3} = \dfrac{4}{6} \qquad \dfrac{a}{b} = \dfrac{c}{d}, b \neq 0, d \neq 0$

Study Hint

Reading Math
$\frac{2}{3} = \frac{8}{12}$ is sometimes read as 2 is to 3 as 8 is to 12.

When two ratios form a proportion, the **cross products** are equal. The cross products in the proportion below are 2×12 and 3×8.

$$\frac{2}{3} = \frac{8}{12}$$

 $\quad 2 \times 12 = 24$
$\quad 3 \times 8 = 24$

Property of Proportions	**Words:**	The cross products of a proportion are equal.
	Symbols:	If $\dfrac{a}{b} = \dfrac{c}{d}$, then $ad = bc$. ($b \neq 0, d \neq 0$)

You can use cross products to find a missing term in a proportion. This is known as *solving the proportion*. Solving a proportion is similar to solving an equation.

INTEGRATION

① Algebra Solve $\frac{n}{7} = \frac{18}{42}$.

$$\frac{n}{7} = \frac{18}{42}$$

$n \times 42 = 7 \times 18$ *Find the cross products.*

$$42n = 126$$

$$\frac{42n}{42} = \frac{126}{42}$$ *Divide each side by 42.*

$$n = 3$$

The solution is 3.

APPLICATION

② Models   A toy car is 7.6 centimeters long and 3.2 centimeters wide. A real car is 4.788 meters long. How wide is the real car?

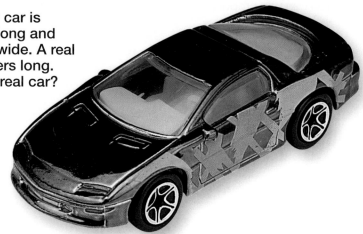

Explore   The length and width of the model car are 7.6 centimeters and 3.2 centimeters. The length of the real car is 4.788 meters or 478.8 centimeters.

Plan   Use w to represent the width of the real car. Then you can write a proportion.

$$\begin{array}{ccc} & \textbf{Model} & \textbf{Real Car} \\ \textit{length} \rightarrow & \frac{7.6}{3.2} = & \frac{478.8}{w} \leftarrow \textit{length} \\ \textit{width} \rightarrow & & \leftarrow \textit{width} \end{array}$$

Find the cross products. Then solve the proportion.

Solve $$\frac{7.6}{3.2} = \frac{478.8}{w}$$

$7.6w = 3.2 \times 478.8$ *Find the cross products.*

$$w = \frac{3.2 \times 478.8}{7.6}$$

3.2 ⊠ 478.8 ⊡ 7.6 ⊟ *201.6*

The width of the real car is 201.6 centimeters or 2.016 meters.

Examine   Express each ratio as a decimal and compare.

$\frac{7.6}{3.2} \rightarrow$ 7.6 ⊡ 3.2 ⊟ *2.375*

$\frac{478.8}{201.6} \rightarrow$ 478.8 ⊡ 201.6 ⊟ *2.375*

Because the decimals are equal, the ratios are equivalent. So, 201.6 centimeters or 2.016 meters is correct.

Study Hint

Problem Solving More than one proportion can be used to solve a problem. Here's another proportion for Example 2.

$$\begin{array}{cc} & \textit{length} \quad \textit{width} \\ \textit{model} \rightarrow & \frac{7.6}{478.8} = \frac{3.2}{w} \begin{array}{l} \leftarrow\textit{model} \\ \leftarrow\textit{real car} \end{array} \\ \textit{real car} \rightarrow & \end{array}$$

Communicating Mathematics

Read and study the lesson to answer each question.

1. *Explain* how you can determine whether two ratios are equivalent.

2. *Demonstrate* how to solve $\frac{8}{12} = \frac{a}{3}$.

Guided Practice

Solve each proportion.

3. $\frac{6}{9} = \frac{4}{x}$

4. $\frac{2}{6} = \frac{5}{n}$

5. $\frac{a}{8} = \frac{3}{4}$

6. $\frac{10}{y} = \frac{2.5}{4}$

7. **a.** If there are 12 wheels, how many bicycles are there?

   b. If there are 35 toes, how many feet are there?

EXERCISES

Practice

Solve each proportion.

8. $\frac{2}{3} = \frac{16}{n}$

9. $\frac{a}{36} = \frac{15}{24}$

10. $\frac{15}{9} = \frac{10}{z}$

11. $\frac{18}{27} = \frac{t}{3}$

12. $\frac{3}{d} = \frac{9}{5}$

13. $\frac{8}{20} = \frac{30}{x}$

14. $\frac{n}{12} = \frac{12}{4}$

15. $\frac{5}{g} = \frac{6}{3}$

16. $\frac{2.6}{13} = \frac{8}{m}$

17. $\frac{21}{b} = \frac{10}{20}$

18. $\frac{6}{1} = \frac{1,200}{t}$

19. $\frac{0.2}{y} = \frac{3}{5}$

20. In the sixth grade, the ratio of boys to girls is 5:6. If there are 88 sixth-grade students, how many are boys?

21. Nikki can word process 7 words in 6 seconds. At this rate, how many words can she word process in 3 minutes?

Applications and Problem Solving

22. *Models* The first jumbo jet was the Boeing 747, which is 70.5 meters long. The wingspan of a 747 is 60 meters. A model 747 has a wingspan of 80 centimeters. What should the length of the model be?

23. *Cooking* Every Saturday, Amber Coffman and her friends make 600 sack lunches for the homeless in Glen Burnie, Maryland.

   a. Use proportions to find the amount of each ingredient Amber would need to make 600 tuna pitas.

   b. *Write a Problem* that can be solved using a proportion and the information in the recipe.

> **Tuna Pitas**
>
> | $6\frac{1}{8}$ oz tuna | 2 pita breads |
> | $\frac{1}{4}$ tsp dill | $\frac{1}{3}$ c plain yogurt |
> | $\frac{1}{8}$ c diced celery | $1\frac{1}{2}$ tsp mustard |
> | 4 lettuce leaves | 1 tomato, sliced |
>
> Combine tuna, yogurt, celery, mustard and dill. Line pitas with lettuce leaves and tomatoes. Divide tuna mixture among pitas. (Serves 4.)

Amber Coffman

24. *Money* Suppose that one French franc is worth 20¢. In Paris, France, a quart of milk costs 9 francs. In Los Angeles, a quart of milk cost $1.65. In which city is the quart of milk more expensive?

25. *Critical Thinking* In some proportions, such as $\frac{3}{6} = \frac{6}{12}$, the same number appears in two of the diagonal positions. In that case, the repeated number is called the *geometric mean* of the other two. Find a pair of numbers other than 3 and 12 for which 6 is the geometric mean.

Mixed Review

26. *Travel* On her summer vacation, Luanda drove 250 miles in 5 hours on the first day. She continued driving at the same rate the second day and drove for 8 hours. How many miles did Luanda drive the second day of her vacation? *(Lesson 8-2)*

27. Multiply 4 and $4\frac{3}{8}$. *(Lesson 7-4)*

28. Standardized Test Practice There are 293 Calories in 1 serving of cookies. If a serving consists of 5 cookies, which is the best estimate of the number of Calories in each cookie? *(Lesson 1-1)*

A 20   **B** 40   **C** 60   **D** 80   **E** 100

For Extra Practice, see page 590.

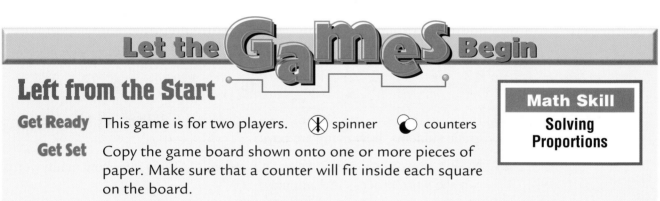

Let the Games Begin

Left from the Start

Math Skill
Solving Proportions

Get Ready   This game is for two players.   ⊗ spinner   ◕ counters

Get Set   Copy the game board shown onto one or more pieces of paper. Make sure that a counter will fit inside each square on the board.

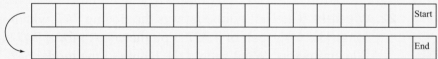

Label equal sections of a spinner with 10, 12, 15, 20, 24, 30, 40, and 60.

Go
- Each player places a counter on the *Start* square.

- One player spins the spinner and substitutes the number on the spinner for *x* in the proportion $\frac{y}{15} = \frac{8}{x}$. The player solves the proportion and moves his or her counter *y* spaces. Then the other player spins the spinner and substitutes the number on the spinner for *y* in the given proportion. He or she solves the proportion and moves his or her counter *x* spaces.

- Trade roles and continue play. The first person to reach the *End* square wins the round. It is not necessary to land on *End* with an exact roll.

 Visit www.glencoe.com/sec/math/mac/mathnet for more games.

328   **Chapter 8** Using Proportional Reasoning

HANDS-ON LAB

COOPERATIVE LEARNING

8-3B Wildlife Sampling

A Follow-Up of Lesson 8-3

In planning for the care of a park or wildlife preserve, it is often important for naturalists to know the size of an animal population. One method used to estimate this is the *capture-recapture* technique. You will model this technique to see how naturalists make reasonable estimates. Dried beans represent deer in a forest and a bowl will represent the forest.

- small bowl
- dried beans
- markers
- paper cup

TRY THIS

Work in groups.

Step 1 Fill a small bowl with dried beans.

Step 2 Use the paper cup to scoop some of the beans. Count the beans selected and record in a table like the one at the right. These represent the captured deer. Mark each bean with an X on both sides.

Step 3 Return the beans to the bowl and mix well with the rest.

Step 4 Scoop another cupful of beans from the bowl. Count the beans selected. This is the first *sample* which represents the deer that are recaptured. Count the beans marked with an X. This is the tagged deer recaptured. Record both numbers.

Step 5 Use the proportion below to estimate the total number of beans in the bowl. Then record the value of *P*.

$$\frac{original\ number\ captured}{total\ population} = \frac{tagged\ in\ sample}{recaptured}$$

Step 6 Return the beans to the bowl.

Step 7 Repeat Steps 4–6 nine times.

Original Number Captured			
Sample	Recaptured	Tagged	P
A			
B			
C			
J			
Total			

ON YOUR OWN

1. Find the average of the estimates. Do you think this is a good estimate of the number of beans in the bowl? Explain your reasoning.

2. Count the number of beans in the bowl. How does the actual count compare to the estimates?

3. Why is it important to return the beans to the bowl and mix well after taking each sample?

4. What would happen to the estimates if one or more of the marks wore off during sampling? What does this represent with deer in the field?

8-4A Draw a Diagram

A Preview of Lesson 8-4

18 U.S.C. 707

Charmaine and Lisa are the president and vice-president of the Grove City 4-H Club. They are discussing the club's service project to shovel snow for elderly people in their community. When it snows, they need to organize quickly. Let's listen in!

> We have 40 members in the club. If you and I are the only ones calling members, it will take forever to get our act together. I think we should call a few and then have them call some to get everyone together faster.

> Good idea. We won't have to say much on each call, so let's guess that it will take about 1 minute each. How about if I call you, the secretary and the treasurer? Then each of you can call three people, and so on.

Lisa

> That makes sense. If I draw a diagram we can make sure everyone gets called and figure out how long the calling will take.

> That looks good! And there are 40 people on it. At a minute a call, each person will be on the phone about 3 minutes. So each level of the tree is 3 minutes.

Charmaine

3 minutes

3 minutes

3 minutes

So, it's 3 × 3 or 9 minutes to activate the troops!

Not bad! Operation "Snow Shovel" is ready for duty!

THINK ABOUT IT

Work with a partner.

1. **Tell** how many people are contacted in the last 3 minutes of the telephone tree.

2. **Write** one or two sentences explaining why **drawing a diagram** can be a useful problem-solving strategy.

3. **Draw** a diagram to find the number of people that could be contacted if the telephone tree was extended to 12 minutes.

For **Extra Practice,** see page 590.

ON YOUR OWN

4. The last step of the 4-step plan for problem solving is to *examine* the solution. **Describe** one or two things that Lisa and Charmaine assumed when they made the telephone tree that could change the solution if the assumptions were wrong.

5. A map is a diagram of a location. **Draw** a map of your route to school.

MIXED PROBLEM SOLVING

STRATEGIES

Look for a pattern.
Solve a simpler problem.
Act it out.
Guess and check.
Draw a diagram.
Make a chart.
Work backward.

Solve. Use any strategy.

6. Money Matters Jill bought some folders for 79¢ each and some spiral-bound notebooks for $1.19 each. If she spent $5.54, how many of each item did Jill buy?

7. Education The scores on a Social Studies test are found by adding 8 points for each correct answer, subtracting 4 points for each incorrect answer, and subtracting 2 points for each unanswered question. There were 15 questions on the test. If Kenji's score was 86, how many of his answers were correct, wrong, and blank?

8. Standardized Test Practice If 4 computers are needed for every 6 students in a class, how many computers are needed for a class of 54 students?

A 18   **B** 27
C 36   **D** 5

9. Civics Chief Justice Melville W. Fuller began the tradition of the "conference handshake" in the U.S. Supreme Court in the late 1800s. Before they take their seats, each justice shakes hands with the others to show that they have a common purpose. If there are nine justices, how many handshakes take place?

10. Sports Sixteen softball teams are participating in a single-elimination tournament; that means that if a team loses one game it is eliminated. How many games will the winning team have played?

11. Physics A ball is dropped from 10 feet above the ground. It hits the ground and bounces up half as high as it fell.
  a. What height does the ball reach on the fourth bounce?
  b. Find the total distance up and down that the ball has traveled when it hits the ground the fifth time.

12. Puzzles Copy the grid below. Start at the 0 and draw a line 1 square long to a 1. Then continue the line 2 squares to a 2, three squares to a 3, and so on. You can move horizontally and vertically, but not diagonally. Find a path to the 8 without revisiting any square.

0	1	3	2	5	5	4	6
1	3	2	4	3	5	6	4
3	3	4	4	5	6	4	5
5	3	5	4	5	7	5	6
5	4	5	6	4	5	6	7
7	7	4	5	7	5	6	7
6	7	4	6	5	4	6	5
6	7	6	7	5	7	6	8

8-4

Scale Drawings

What you'll learn

You'll learn to solve problems involving scale drawings.

When am I ever going to use this?

You can use a scale to find distances between cities on a map.

Word Wise

scale drawing
scale

At the recommendation of Thomas Jefferson, Benjamin Banneker was appointed a part of the team to design Washington, D.C. Mr. Banneker excelled as an astronomer, farmer, mathematician, and surveyor. Some historians believe that when the team leader resigned and left with the plans for the city, he was able to reproduce them from memory.

The city plans that Benjamin Banneker and the design team developed are an example of a **scale drawing**. A scale drawing is used to represent something that is too large or too small for an actual-size drawing to be useful.

You have probably used a map before. A map is a scale drawing. The **scale** on a map is the ratio of the distance on the map to the actual distance. When you know the scale of a map, you can find actual distances by writing and solving proportions.

Example

Real World APPLICATION

Travel Set-Su and Amy are helping plan a field trip to the Johnson Space Center in Houston from their school in Austin. They measured the distance their map shows from Austin to Houston as about 3.25 inches. The scale on the map says that 1 inch represents 55 miles. What is the actual distance between the cities?

Let n represent the actual distance between the cities.
Write and solve a proportion.

$$\begin{array}{cc} & \textbf{key} \qquad \textbf{Austin to Houston} \\ \textit{map distance} \rightarrow & \dfrac{1 \text{ inch}}{55 \text{ miles}} = \dfrac{3.25 \text{ inches}}{n \text{ miles}} \leftarrow \textit{map distance} \\ \textit{actual distance} \rightarrow & \qquad\qquad\qquad\qquad \leftarrow \textit{actual distance} \end{array}$$

$$1 \times n = 55 \times 3.25 \quad \textit{Estimate:}$$
$$\qquad\qquad\qquad\qquad\quad \textit{60} \times \textit{3} = \textit{180}$$
$$n = 178.75$$

The actual distance between Austin and Houston is about 179 miles.

Architecture Architects draw buildings to scale in blueprints. A room in a new home will be 4.6 meters long and 3.3 meters wide. If the scale of the blueprint is 1 centimeter = 2.5 meters, what should the dimensions be in the scale drawing?

Use ℓ for the length and w for the width.

Length

$$\begin{array}{l} scale \rightarrow \\ actual \rightarrow \end{array} \quad \frac{1 \text{ cm}}{2.5 \text{ m}} = \frac{\ell \text{ cm}}{4.6 \text{ m}} \quad \begin{array}{l} \leftarrow scale \\ \leftarrow actual \end{array}$$

$$1 \times 4.6 = 2.5 \times \ell \quad \textit{Find the cross products.}$$

$$\frac{1 \times 4.6}{2.5} = \ell$$

$$4.6 \div 2.5 = 1.84$$

Width

$$\begin{array}{l} scale \rightarrow \\ actual \rightarrow \end{array} \quad \frac{1 \text{ cm}}{2.5 \text{ m}} = \frac{w \text{ cm}}{3.3 \text{ m}} \quad \begin{array}{l} \leftarrow scale \\ \leftarrow actual \end{array}$$

$$1 \times 3.3 = 2.5 \times w \quad \textit{Find the cross products.}$$

$$\frac{1 \times 3.3}{2.5} = w$$

$$3.3 \div 2.5 = 1.32$$

The scale drawing should be 1.84 cm long and 1.32 cm wide.

You can use proportions to make your own scale drawing.

Did you know?

Archaeologists have found an ancient blueprint for the Pantheon in Rome, Italy, in which the scale is 1:1.

HANDS-ON

MINI-LAB

Work with a partner. measuring tape ▱ ruler

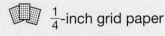 $\frac{1}{4}$-inch grid paper

Try This

- Measure the length of each wall, door, window, and chalkboard in your classroom.
- Round each length to the nearest inch. Record the lengths.
- Make a scale drawing like the one at the right on a piece of grid paper. Use the scale $\frac{1}{4}$-inch:12 inches.

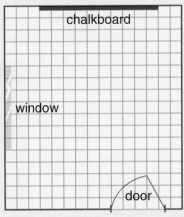

Talk About It

1. Write the proportion you used to find the length of the chalkboard for your scale drawing.
2. How would a scale drawing that used a scale of $\frac{1}{4}$-inch:24 inches compare to your scale drawing?

Communicating
Mathematics

Read and study the lesson to answer each question.

1. *Tell* what important information must be given on a scale drawing in order to use it.

2. *Explain* how to find an actual distance using a scale drawing.

HANDS-ON
MATH

3. *Draw* a scale drawing of a room in your home. Use $\frac{1}{4}$-inch grid paper and a scale of 1 inch = 3 feet.

Guided Practice

Find the distance between each pair of cities, given the map distance and the scale.

4. Charlotte, North Carolina and Knoxville, Tennessee; $12\frac{4}{5}$ inches; 1 inch:18 miles

5. Cape Town, South Africa and London, England; 19.3 cm; 1 cm:500 km

Find the length of each object on a scale drawing with the given scale.

6. a patio door 72 inches wide; 1 inch:2 feet

7. a room 50 feet long; $\frac{1}{2}$ inch:6 inches

Lucio Costa

8. *City Planning* When the Brazilian government moved its capital from Rio de Janeiro, they held a contest to design the city of Brasilia. The winner was Lucio Costa, who designed the city to resemble the curved wings of a jet airplane. A map of Brasilia has a scale of 1 inch to 5 miles. If the city is $2\frac{7}{16}$ inches across on the map, how far is it across the actual city? Use estimation to check your answer.

EXERCISES

Practice

Find the distance between each pair of cities, given the map distance and the scale.

	Cities	Map Distance	Scale
9.	Spokane and Richland, Washington	$6\frac{1}{10}$ inches	1 inch: 22 miles
10.	Baltimore, Maryland and Washington, D.C.	$2\frac{7}{8}$ inches	$\frac{1}{2}$ inch: 6 miles
11.	Dallas, Texas and Montgomery, Alabama	$4\frac{5}{8}$ inches	$\frac{1}{4}$ inch: 35 miles
12.	Chicago, Illinois and Mexico City, Mexico	10.9 cm	1 cm: 250 km
13.	Singapore and New Delhi, India	9.9 cm	3 cm: 1,250 km
14.	Paris, France and Montreal, Canada	16.2 cm	2.2 cm: 750 km

Find the length of each object on a scale drawing with the given scale.

15. a garage door 16 feet wide; 2 inches:1 foot

16. a bridge 26 meters wide; 1 centimeter:5.5 meters

17. a elevator shaft 3 meters wide; 0.5 centimeter:2.5 meters

18. a surgical instrument $5\frac{7}{8}$ inches long; 1 inch:$\frac{1}{2}$ inch

19. a computer circuit board 4 centimeters wide; 1 centimeter: 0.2 centimeter

20. a gear $3\frac{5}{8}$ inches across; 4 inches:$\frac{1}{2}$ inch

Applications and Problem Solving

21. *Technology* Have you ever wished your television showed three-dimensional images? An invention by NASA engineer Valerie Thomas may make that possible. A drawing for her patent of the illusion transmitter is 8.4 centimeters tall. If the scale is 1 cm:2.2 cm, what will the actual height of the transmitter be?

Valerie Thomas

22. *Civil Engineering* The Natchez Trace Bridge in Franklin, Tennessee, is the first and longest bridge of its kind in the United States. A scale drawing of the bridge has a scale of 1 inch:25 feet. How long is the drawing of the bridge if the actual bridge spans 1,500 feet?

23. *Critical Thinking* The distance between Huntington, West Virginia, and Cincinnati, Ohio, is 148 miles. If a map shows the distance as about $9\frac{1}{4}$ inches, what is the scale of the map?

Mixed Review

24. **Standardized Test Practice** Inali earned $157.50 one week by working 30 hours. If he works 35 hours the next week, how much will he earn? *(Lesson 8-3)*

    A $135.00   **B** $160.50   **C** $183.75   **D** $210.25   **E** Not Here

For **Extra Practice**, see page 590.

25. *Geometry* On graph paper, draw coordinate axes. Then graph the point $B(-4, 4)$. *(Lesson 5-3)*

26. Find the GCF of 345, 253, and 115. *(Lesson 4-4)*

CHAPTER 8

Mid-Chapter Self Test

Express each ratio as a fraction in simplest form. *(Lesson 8-1)*

1. 4 to 36

2. $\frac{60}{12}$

3. 5 pounds to 10 ounces

Express each rate as a unit rate. *(Lesson 8-2)*

4. 640 miles in 5 hours

5. $11.90 for 10 disks

6. 5 million people in 410 square miles

Solve each proportion. *(Lesson 8-3)*

7. $\frac{3}{4} = \frac{9}{n}$

8. $\frac{9}{36} = \frac{x}{48}$

9. $\frac{5}{y} = \frac{1}{0.5}$

10. On a scale drawing, 1 centimeter represents 2 meters. What length on the drawing should be used to represent 3.2 meters? *(Lesson 8-4)*

Percents and Fractions

You'll learn to express fractions as percents, and vice versa.

When am I ever going to use this?

You can express your test and homework scores using percents.

Scuba divers wear belts with lead weights so that they will be able to spend time beneath the surface of the water more easily. An average diver's belt should weigh about $\frac{1}{10}$ of his or her weight. What is $\frac{1}{10}$ written as a percent?

Remember that percent means per hundred. Any fraction can also be written as a percent. One way to express a fraction as a percent is to find an equivalent fraction with a denominator of 100.

$$\frac{1}{10} = \frac{1}{10} \times \frac{10}{10}$$
$$= \frac{10}{100} \text{ or } 10\%$$

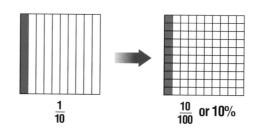

$\frac{1}{10}$ \qquad $\frac{10}{100}$ **or 10%**

So, a diver's belt should be 10% of his or her weight.

You can also use a proportion to express a fraction as a percent.

Examples

Express each fraction as a percent.

1 $\frac{7}{8}$

$\frac{7}{8} = \frac{n}{100}$ *Find the cross products.*

$700 = 8n$

$\frac{700}{8} = \frac{8n}{8}$ *Divide each side by 8.*

$87.5 = n$

So, $\frac{7}{8} = 87.5\%$.

2 $\frac{20}{48}$

$\frac{20}{48} = \frac{n}{100}$ *Find the cross products.*

$2{,}000 = 48n$

$2000 \boxed{÷} 48 \boxed{=} 41.6666667$

So, $\frac{20}{48}$ is about 41.7%.

Study Hint

Estimation Notice that $\frac{7}{8}$ is close to 1. So the percent should be close to 100%. $\frac{20}{48}$ is close to $\frac{1}{2}$. So the percent should be close to 50%.

When you want to write a percent as a fraction, begin with a fraction that has a denominator of 100. Then write the fraction in simplest form.

Examples

Express each percent as a fraction in simplest form.

③ 45%

Estimate: 45% is about 50%, which is $\frac{1}{2}$.

$45\% = \frac{45}{100}$

$= \frac{45 \div 5}{100 \div 5}$ *The GCF is 5.*

$= \frac{9}{20}$

So, $45\% = \frac{9}{20}$.   *Compare to the estimate.*

④ $83\frac{1}{3}\%$

$83\frac{1}{3}\% = \frac{83\frac{1}{3}}{100}$

$= 83\frac{1}{3} \div 100$

$= \frac{250}{3} \div 100$

$= \frac{250}{3} \times \frac{1}{100}$ *To divide by 100, multiply by $\frac{1}{100}$.*

$= \frac{250}{300}$ or $\frac{5}{6}$

So, $83\frac{1}{3}\% = \frac{5}{6}$.

Some percents are used often in everyday situations. It is helpful to memorize these percents and their equivalent fractions.

$20\% = \frac{1}{5}$	$25\% = \frac{1}{4}$	$12\frac{1}{2}\% = \frac{1}{8}$	$16\frac{2}{3}\% = \frac{1}{6}$
$40\% = \frac{2}{5}$	$50\% = \frac{1}{2}$	$37\frac{1}{2}\% = \frac{3}{8}$	$33\frac{1}{3}\% = \frac{1}{3}$
$60\% = \frac{3}{5}$	$75\% = \frac{3}{4}$	$62\frac{1}{2}\% = \frac{5}{8}$	$66\frac{2}{3}\% = \frac{2}{3}$
$80\% = \frac{4}{5}$		$87\frac{1}{2}\% = \frac{7}{8}$	$83\frac{1}{3}\% = \frac{5}{6}$
$100\% = 1$			

Example ⑤
Real World APPLICATION

Economics In 1996, $5\frac{2}{5}\%$ of the American workforce was unemployed. Write the percent as a fraction.

$5\frac{2}{5}\% = \frac{5\frac{2}{5}}{100}$

$= 5\frac{2}{5} \div 100$ *Write the fraction as a division problem.*

$= \frac{27}{5} \times \frac{1}{100}$ *Rewrite $5\frac{2}{5}$ as an improper fraction.*

$= \frac{27}{500}$

In 1996, $\frac{27}{500}$ of the workforce was unemployed.

CHECK FOR UNDERSTANDING

Communicating Mathematics

Read and study the lesson to answer each question.

1. **Demonstrate** how to write $\frac{3}{16}$ as a percent.

2. **Write** a fraction and a percent to represent the shaded portion of the model at the right.

Express each fraction as a percent.

3. $\frac{3}{10}$ **4.** $\frac{7}{20}$ **5.** $\frac{5}{16}$

Express each percent as a fraction in simplest form.

6. 85% **7.** 72% **8.** $17\frac{1}{2}\%$

9. *Life Science* A banana is 75% water. Write 75% as a fraction in simplest form.

EXERCISES

Practice

Express each fraction as a percent.

10. $\frac{9}{20}$ **11.** $\frac{7}{16}$ **12.** $\frac{18}{25}$ **13.** $\frac{23}{200}$ **14.** $\frac{5}{5}$

15. $\frac{5}{12}$ **16.** $\frac{14}{20}$ **17.** $\frac{4}{11}$ **18.** $\frac{1}{16}$ **19.** $\frac{1}{30}$

Express each percent as a fraction in simplest form.

20. 55% **21.** 40% **22.** 18% **23.** 34.5% **24.** 6.2%

25. $62\frac{1}{2}\%$ **26.** $35\frac{1}{4}\%$ **27.** 45.05% **28.** $43\frac{3}{4}\%$ **29.** 100%

30. Write $33\frac{1}{3}\%$ as a fraction.

31. What percent is equivalent to $\frac{40}{125}$?

Applications and Problem Solving

32. *Careers* Only 3% of the people who work in the space industry work for NASA. Write 3% as a fraction in simplest form.

33. *Coins* A quarter is made of one-twelfth nickel, and the rest is copper. Write the portion of a quarter that is nickel as a percent.

34. *Critical Thinking* For what value of x is $\frac{1}{x} = x\%$?

Mixed Review

35. **Standardized Test Practice** Choose the distance between Chattanooga and Memphis if they are 1.75 inches apart on the map and the scale is 1 inch:152 miles. *(Lesson 8-4)*

A 266 miles **B** 86.8 miles **C** 188 miles **D** 322 miles

36. *Measurement* Change 56 ounces to pounds. *(Lesson 7-5)*

37. Solve $-9 - (-6) = t$. *(Lesson 5-5)*

38. Find the least common multiple of 28 and 74. *(Lesson 4-9)*

For **Extra Practice,** see page 591.

8-6 Percents and Decimals

What you'll learn

You'll learn to express decimals as percents, and vice versa.

When am I ever going to use this?

You can write your batting average as a decimal and a percent.

Did you know that jellyfish aren't really fish? They are invertebrates, which means they have no backbone. In fact, jellyfish have no bones at all. A jellyfish's body is made up of 95% water. What is 95% written as a decimal? *This question will be answered in Example 3.*

When you studied fractions in Chapter 4, you learned that any decimal can also be written as a fraction. You can use that fact to express any decimal as a percent.

Examples

Express each decimal as a percent.

① 0.12

$0.12 = \dfrac{12}{100}$ *Write as a fraction.*

$= 12\%$

So, $0.12 = 12\%$.

② 0.119

$0.119 = \dfrac{119}{1,000}$ *Write as a fraction.*

$= \dfrac{119 \div 10}{1,000 \div 10}$ *Divide to make the denominator 100.*

$= \dfrac{11.9}{100}$ or 11.9%

So, $0.119 = 11.9\%$.

Study the pattern in the decimals and the equivalent percents in Examples 1 and 2. Notice that you can write the percent by multiplying the decimal number by 100 and adding the percent symbol.

Writing a Decimal as a Percent	**Words:**	To write a decimal as a percent, multiply the decimal by 100 and add the percent symbol.
	Symbols:	$0.56 = 0.56 = 56\%$

When you want to write a percent as a decimal, reverse the process.

Writing a Percent as a Decimal	**Words:**	To write a percent as a decimal, divide the percent by 100 and remove the percent symbol.
	Symbols:	$68\% = 68 = 0.68$

3 **Life Science** Refer to the beginning of the lesson. Write 95% as a decimal.

$95\% = 95 = 0.95$

A jellyfish's body is 0.95 water.

Express each percent as a decimal.

4 44.9%

$44.9\% = 44.9 = 0.449$

So, $44.9\% = 0.449$.

5 $23\frac{1}{4}\%$

$23\frac{1}{4}\% = 23.25\%$ *Write $\frac{1}{4}$ as 0.25.*

$= 23.25 = 0.2325$

So, $23\frac{1}{4}\% = 0.2325$.

Jellyfish

CHECK FOR UNDERSTANDING

Communicating Mathematics

Read and study the lesson to answer each question.

1. **Choose** the percent that is equivalent to 0.67.

   a. 6.7%   **b.** 67%   **c.** 0.67%

2. **Tell** the steps used to write a percent as a decimal.

3. **Write** an explanation of how the diagram illustrates that any number can be expressed in three ways.

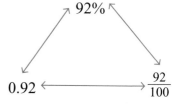

Guided Practice

Express each decimal as a percent.

4. 0.23   **5.** 0.06   **6.** 0.785

Express each percent as a decimal.

7. 65%   **8.** 42%   **9.** $18\frac{1}{2}\%$

10. **Food** What is your favorite vegetable? In a survey, 19.7% of adults said broccoli was their favorite. Write 19.7% as a decimal.

EXERCISES

Practice **Express each decimal as a percent.**

11. 0.17	**12.** 0.08	**13.** 0.85	**14.** 0.675	**15.** 0.099
16. 0.0444	**17.** 0.01	**18.** 0.025	**19.** 0.009	**20.** 1.0

Express each percent as a decimal.

21. 45%   **22.** 70%   **23.** 16%   **24.** 64.5%   **25.** 8.1%

26. $78\frac{1}{2}$%   **27.** $14\frac{1}{4}$%   **28.** 12.35%   **29.** $94\frac{3}{4}$%   **30.** 100%

31. Write the percent that is equivalent to 0.848.

32. What decimal is equivalent to $25\frac{1}{4}$%?

Applications and Problem Solving

33. *Sports* In 1971, 0.04 of high school girls played sports. In 1994, the number had risen to 0.33. Write each decimal as a percent.

34. *Technology* The graph shows the Internet users who find websites from different sources. Write each percent as a decimal.

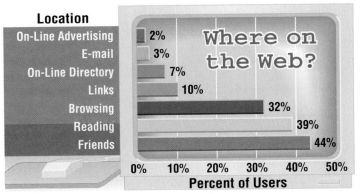

Source: *Electronic Access*

35. *Critical Thinking* Choose the greater number of each pair.

    a. 35%, 3.5   **b.** $1\frac{3}{4}$%, 0.175   **c.** $\frac{23}{40}$, 0.60

Mixed Review

36. *Standardized Test Practice* Antonia and Luis bought a new stereo that was on sale for 38% off. What is this percent written as a fraction? *(Lesson 8-5)*

    A $\frac{1}{38}$ **B** $\frac{9}{25}$ **C** $\frac{19}{50}$ **D** $\frac{38}{75}$

37. *Algebra* Translate *t divided by 8* into an algebraic expression. *(Lesson 6-4)*

38. Express $\frac{28}{98}$ in simplest form. *(Lesson 4-5)*

39. *Statistics* Use the graph. In what age category are most officers of the U.S. military? *(Lesson 3-2)*

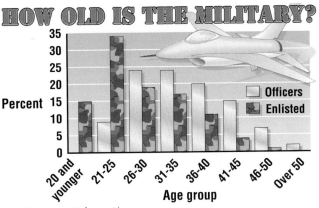

Source: *Defense Almanac*

For **Extra Practice**, see page 591.

Percents Greater Than 100% and Percents Less Than 1%

The North African ostrich is the largest bird living on Earth today. An ostrich can grow as tall as 9 feet! This is 150% the height of an average man. How many times taller than a man is an ostrich? *This question will be answered in Example 1.*

A percent greater than 100% represents a number greater than 1. Percents less than 1% represent numbers less than 0.01 or $\frac{1}{100}$.

HANDS-ON MINI-LAB

Work with a partner. grid paper colored pencils

Try This

- Draw three 10×10 squares on a piece of grid paper. Each large square represents 100% and each small square represents 1%.

- Use two of the models to shade 120 small squares.

- On the third model, shade half of one small square.

Talk About It

1. Which model represents a percent greater than 100%? What is the percent?

2. Which model represents a percent less than 1%? What is the percent?

Example ——1 **Life Science** Refer to the beginning of the lesson. Express 150% as a decimal and as a fraction to find how many times taller the ostrich is than an average man.

CONNECTION

Make a model.

$150\% = 1.50$ or 1.5

$\qquad = 1\frac{1}{2}$

So, the ostrich is 1.5 or $1\frac{1}{2}$ times as tall as a man.

You can use the same method for writing a decimal as a percent to write decimals greater than 1 and decimals less than 0.01 as percents.

Examples

Express each number as a percent.

2 1.45

1.45 = 145 = 145%

So, 1.45 = 145%.

3 0.0016

0.0016 = 00016 = 0.16%

So, 0.0016 = 0.16%.

An alternative way to write a mixed number or a fraction as a percent is to first write the number as a decimal. Then write the decimal as a percent.

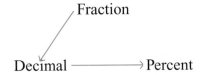
Fraction

Decimal ⟶ Percent

Examples

Express each number as a percent.

4 $7\frac{1}{2}$

$7\frac{1}{2}$ = 7.5

7.5 = 750 = 750%

So, $7\frac{1}{2}$ = 750%.

5 $\frac{2}{500}$

Use a calculator.

2 [÷] 500 [=] *0.004*

0.004 = 0004 = 0.4%

So, $\frac{2}{500}$ = 0.4%.

APPLICATION

Real World

6 **Recreation** The National Sporting Goods Association estimates that in a recent year, the total sales of equipment for water skiing was $51 million. That same year, all of the sporting goods sold in the United States totaled $48,732 million! What percent of the sporting goods sales was related to water skiing? Round to the nearest hundredth of a percent.

Use a calculator.

51 [÷] 48732 [=] *0.00104654*

0.00104654 = 000104654

= 0.104654%

So, about 0.10% of the sporting goods were related to water skiing.

Communicating Mathematics

Read and study the lesson to answer each question.

1. **Explain** why 150% of the height of a person is taller than the person.

2. **Describe** a fraction that is equivalent to a percent less than 1%.

HANDS-ON MATH

3. **Draw** a model to represent each percent.
   a. 175%   **b.** 0.75%

Guided Practice

Express each percent as a decimal.

4. 800%   5. 0.55%   6. 240%   7. $\frac{1}{5}$%

Express each number as a percent.

8. 4.3   9. $15\frac{1}{2}$ 10. 0.005   11. $\frac{1}{350}$

Norma Mankiller

Tell whether each of the following is reasonable. Explain why or why not.

12. Norma Mankiller received 125% of the vote when she was elected chief of the Cherokee Nation.

13. The demand for Egyptian cotton rose in 1996. A popular newspaper reported that farmers there could expect prices to be 120% of the minimum guaranteed by the state.
    Source: *USA TODAY*

14. **Education** Recently, 8,753 of the 1,136,553 bachelor's degrees earned were in architecture. Use a calculator to find the ratio of the degrees in architecture to the degrees earned as a percent. Round your answer to the nearest hundredth of a percent.

Practice

Express each percent as a decimal.

15. 500%   16. 310%   17. 115%   18. 270%

19. 100.5%   20. 1,000%   21. 0.068%   22. 0.012%

23. 0.0025%   24. 0.75%   25. 0.032%   26. $\frac{1}{8}$%

Express each number as a percent.

27. 9   28. $3\frac{1}{2}$ 29. $7\frac{3}{4}$ 30. 6.25

31. 34   32. 2.9   33. 0.009   34. 0.0018

35. $\frac{4}{1,000}$ 36. $\frac{8}{900}$ 37. 0.0001   38. $\frac{12}{5,000}$

39. Express 1.8 as a percent.

40. Write 925% as a decimal.

Tell whether each of the following is reasonable. Explain why or why not.

41. Your height now is 0.1% of your height at 1 year old.

42. Central High School's 1998 enrollment is 105% of its 1997 enrollment.

43. Teenagers are 120% of the attendance at an amusement park.

44. Josefina gave away 130% of her stamp collection.

45. The U.S. population in 1990 is 0.5% of its population in 1790.

46. In 1994, 0.3% of the population of Pittsburgh moved out of the area.

Applications and Problem Solving

47. Economics In 1994, the price of consumer goods in Brazil increased to 2,669% of their 1993 prices. How many times the 1993 prices were the 1994 prices?

48. Earth Science The diameter of the Sun is 865,500 miles. The diameter of Earth is about 0.9% of the Sun's diameter. Write 0.9% as a fraction in simplest form.

49. Life Science The blue whale is the largest mammal that has ever lived. It has been known to reach a length of 110 feet! The smallest mammal is the pygmy shrew, with a total length of just 2.9 inches, including its tail. What percent of a whale's size is a shrew?

50. Working on the **CHAPTER** **Project** Use the information that you gathered about material that is recycled in your community. Write a paragraph and make a graph showing the amounts of different available materials that are recycled.

51. Critical Thinking In 1995, there were 69,036,000 people under the age of 18 in the United States. The U.S. Bureau of the Census predicts that in 2010, the number of people under 18 will be 106.6% of that number. What is the estimated number of people under 18 for 2010?

Mixed Review

52. Express 43% as a decimal. *(Lesson 8-6)*

53. Tell whether the ratios 13 to 39 and 26 to 78 are equivalent. Show your answer by simplifying. *(Lesson 8-1)*

54. Standardized Test Practice The maximum capacity of an elevator is 3,000 pounds. Which number line shows this capacity? *(Lesson 6-5)*

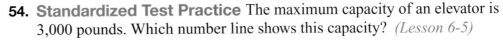
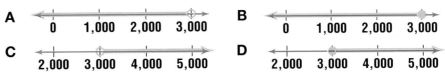

55. Geometry Graph $\triangle XYZ$ with vertices $X(2, 4)$, $Y(1, -1)$, and $Z(3, 5)$ and its reflection over the *x*-axis. Write the ordered pairs for the vertices of the new figure. *(Lesson 5-8)*

56. Patterns Find the next three terms in the sequence 2, 5, 12.5, 31.25, ... *(Lesson 4-3)*

For **Extra Practice,** see page 591.

57. Statistics Construct a stem-and-leaf plot for the data: 95, 83, 66, 81, 92, 85, 62, 90. *(Lesson 3-5)*

Percent of a Number

What you'll learn

You'll learn to find the percent of a number.

When am I ever going to use this?

You'll use percents to find the sale price of an item.

You can learn a lot by reading the back of a packet of seeds. A new gardener can learn how deep to plant the seeds, how far apart to plant the seeds, and how many are expected to germinate or sprout. For example, one packet of beans guarantees that 95% of its 200 seeds will germinate. How many beans are expected to germinate?

In this problem, 95% means that 95 out of every 100 seeds are expected to germinate. There are several ways to solve this problem.

Method 1 Use a model.

The entire square represents 200 seeds.

95% is represented by shading 95 small squares.

Each small square represents 200 ÷ 100 or 2 seeds.

Study Hint

Technology You can find a percent of a number with a calculator. To find 95% of 200, enter

95 [2nd] [%] [×] 200 [=] *190*

95% of 200 is represented by 95 × 2 or 190.
So, 190 seeds are expected to germinate.

Method 2 Use a proportion.

Let *n* represent the number of seeds that are expected to germinate.

$$\begin{array}{c} germinating\ seeds \\ total\ seeds\ in\ packet \end{array} \rightarrow \frac{n}{200} = \frac{95}{100} \leftarrow \begin{array}{c} percent\ germinating \\ total\ percent\ in\ packet \end{array}$$

$$n \cdot 100 = 200 \cdot 95 \quad \text{\textit{Find the cross products.}}$$

$$100n = 19{,}000$$

$$\frac{100n}{100} = \frac{19{,}000}{100} \quad \text{\textit{Divide each side by 100.}}$$

$$n = 190$$

Method 3 Use multiplication.

First express the percent as a decimal. Then multiply.

95% of 200 = 0.95 × 200   *95% = 0.95*

          = 190

All three methods give the same result, 190 seeds.

Examples

① **Find 18.5% of 500 by using a proportion.**

Let n represent the number.

$$\frac{n}{500} = \frac{18.5}{100}$$

$n \cdot 100 = 500 \cdot 18.5$ *Find the cross products.*

$$\frac{n \cdot 100}{100} = \frac{500 \cdot 18.5}{100}$$ *Divide each side by 100.*

$$n = 92.5$$

18.5% of 500 is 92.5.

Study Hint

Mental Math You can find 25% of 120 mentally by using the fraction $\frac{1}{4}$.
$\frac{1}{4}$ of 120 is 30.

② **What number is 25% of 120? Find the number by multiplying.**

25% of $120 = 0.25 \times 120$ *$25\% = 0.25$*

$\qquad\qquad\qquad = 30$

25% of 120 is 30.

INTEGRATION

③ **Statistics** A magazine recently surveyed 603 students to find what fast-food restaurants they preferred. The results are shown in the circle graph. Of the 603 students surveyed, how many preferred restaurant E?

Find 3% of 603.

3% of $603 = 0.03 \times 603$

$\qquad\qquad\qquad = 18.09$

About 18 of the students surveyed prefer restaurant E.

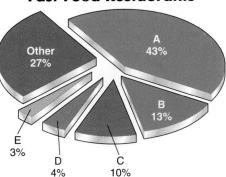

Favorite Fast-Food Restaurants

Other 27%
A 43%
B 13%
C 10%
D 4%
E 3%

CHECK FOR UNDERSTANDING

Communicating Mathematics

Read and study the lesson to answer each question.

1. *Explain* how the model at the right can be used to find 60% of 300.

2. *Write* a proportion that can be used to find 42% of 386.

Guided Practice

Find each number. Round to the nearest tenth if necessary.

3. Find 93% of 215.

4. Find 64% of 88.

5. What number is 140% of 220?

6. 0.5% of 350 is what number?

7. *Money Matters* The purchase price of a bicycle is $140. The state tax rate is 6.5% of the purchase price. How much state tax is charged?

Practice

Find each number. Round to the nearest tenth if necessary.

8. What number is 25% of 560?

9. Find 37.5% of 64.

10. 50% of 128 is what number?

11. What number is 25% of 36?

12. Find 80% of 90.5.

13. What number is 75% of 92?

14. 12% of 16.5 is what number?

15. 130% of 96 is what number?

16. 0.25% of 400 is what number?

17. What number is 20% of twenty?

18. What number is $33\frac{1}{3}$% of 18?

19. Find $16\frac{2}{3}$% of 60.

Applications and Problem Solving

20. *Shopping* A company surveyed 263,000 consumers ages 14 years and older to find how often they buy products from TV shopping shows. How many consumers fall into each category of the circle graph?

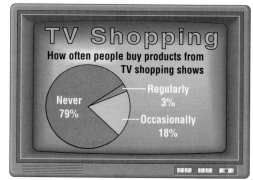

TV Shopping

How often people buy products from TV shopping shows

Never 79%

Regularly 3%

Occasionally 18%

Source: Impact Resources

21. *Money Matters* The standard rate for tipping in a restaurant is 15% of your total bill. A family of six has a bill of $74.80 at a restaurant. What should their tip be?

22. *Civics* On an average day, 1,648 persons immigrate to the United States. Nearly 41% of them will become naturalized American citizens. How many of the daily immigrants to the United States are likely to become citizens?

23. *Business* The leading brand of cosmetics in the United States had about 20% of the total sales in the industry. The total sales were $2.6 billion. Find the leading brand's sales.

24. *Critical Thinking* Suppose you add 10% of a number to the number. Then you subtract 10% of the total. How does the result compare to your original number? Explain your reasoning.

Mixed Review

25. Express 0.6% as a decimal. *(Lesson 8-7)*

26. **Standardized Test Practice** Which of the following has the best unit price? *(Lesson 8-2)*

A 18 ounces for $5.40

B 16 ounces for $4.64

C 12 ounces for $3.72

D 10 ounces for $3.30

For **Extra Practice**, see page 592.

27. *Measurement* How many $\frac{1}{2}$-cup servings of ice cream are there in a gallon of chocolate ice cream? *(Lesson 7-5)*

The Percent Proportion

What you'll learn

You'll learn to solve problems using the percent proportion.

When am I ever going to use this?

You'll use the percent proportion to make and analyze circle graphs.

Word Wise

percentage
base
rate
percent proportion

Are you afraid of spiders? Do insects give you the creeps? It's no wonder, because there are about 854,000 different species of spiders, insects, crustaceans, millipedes, and centipedes on Earth. The graph shows that 88% of the total number of species of arthropods are insects. How many species are insects?

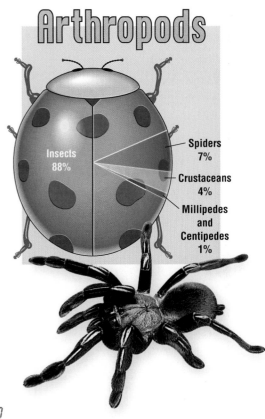

Arthropods

Insects 88%

Spiders 7%

Crustaceans 4%

Millipedes and Centipedes 1%

Let n represent the number of species of insects.

$$\frac{n}{854,000} = \frac{88}{100}$$

$$n \cdot 100 = 854,000 \cdot 88$$

$$\frac{n \cdot 100}{100} = \frac{854,000 \cdot 88}{100}$$

$$n = \frac{854,000 \cdot 88}{100}$$

854000 $\boxed{\times}$ 88 $\boxed{\div}$ 100 $\boxed{=}$ *751520*

There are about 752,000 species of insects on Earth.

In the problem above, 752,000 is the **percentage (P)**. The number 854,000 is the **base (B)**. The ratio $\frac{88}{100}$ is the **rate (r)**. The percentage, base, and rate are related in the **percent proportion**.

Percent Proportion	The percent proportion is $\frac{P}{B} = \frac{r}{100}$, where P represents the percentage, B represents the base, and r represents the number per hundred.

There are three basic kinds of percent problems that can be solved using the percent proportion. Using the proportion $\frac{1}{2} = \frac{50}{100}$, you can see that the types are related as shown below.

Find the percentage.	What number is 50% of 2? *Percentage*	$\frac{\blacksquare}{2} = \frac{50}{100}$
Find the rate.	1 is what percent of 2? *Rate*	$\frac{1}{2} = \frac{\blacksquare}{100}$
Find the base.	1 is 50% of what number? *Base*	$\frac{1}{\blacksquare} = \frac{50}{100}$

Study Hint

Reading Math In a percent problem, the base usually follows the word *of*.

1 Twenty-five is 20% of what number?

$$\frac{P}{B} = \frac{r}{100}$$ *Write the percent proportion.*

$$\frac{25}{B} = \frac{20}{100}$$ *Replace P with 25 and r with 20.*

$$25 \cdot 100 = B \cdot 20$$ *Find the cross products.*

$$2{,}500 = 20B$$

$$\frac{2{,}500}{20} = \frac{20B}{20}$$ *Divide each side by 20.*

$$125 = B$$ 25 is 20% of 125.

APPLICATION

Real World

2 **Sports** Refer to the graph at the right. What percent of the athletes trying out for the U.S. swimming team earned positions on the team?

Find what percent 44 is of 780.

$$\frac{P}{B} = \frac{r}{100}$$

$$\frac{44}{780} = \frac{r}{100}$$

$$44 \cdot 100 = 780r$$

$$\frac{44 \cdot 100}{780} = r$$ $\;44\;$ $\boxed{\times}$ $\;100\;$ $\boxed{\div}$ $\;780\;$ $\boxed{=}$ $\;5.641025641$

$$5.6 \approx r$$

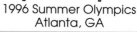

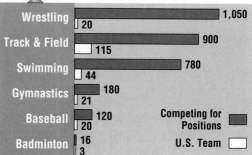

Source: SportsTicker Enterprises

About 5.6% of the athletes who tried out made the team.

CHECK FOR UNDERSTANDING

Communicating Mathematics

Read and study the lesson to answer each question.

1. **Tell** what P, B, and r represent in the percent proportion.

2. **Write** a proportion that can be used to find the rate if the percentage is 18 and the base is 54.

3. **You Decide** Jackson is trying to find what percent 30 is of 20. He uses the proportion $\frac{20}{30} = \frac{r}{100}$. Meredith uses $\frac{30}{20} = \frac{r}{100}$. Who is correct?

Guided Practice

Find each number. Round to the nearest tenth if necessary.

4. What number is 45% of 60?
5. 3 is what percent of 40?
6. 80 is 75% of what number?
7. What percent of 24 is 12?
8. Find 42.5% of 48.
9. 20% of what number is 25?

10. **School** A class picture included 95% of the students. Seven students were missing. How many students were in the class?

EXERCISES

Practice

Find each number. Round to the nearest tenth if necessary.

11. 8 is what percent of 16?

12. 20% of what number is 18?

13. 58.2% of 50 is what number?

14. What number is 105% of 36?

15. What number is 38% of 70?

16. 14 is what percent of 49?

17. 61 is 35% of what number?

18. 7.5% of 48 is what number?

19. What percent of 180 is 30?

20. 12.5% of what number is 24?

21. 63 is what percent of 42?

22. 50% of what number is 15.8?

23. $6\frac{1}{4}$% of 235 is what number?

24. 5% of what number is $6\frac{1}{2}$?

25. What percent of 250 is 25?

26. What number is 40% of 86?

27. 20% of what number is 12?

28. Find 125% of 48.

*inter***NET**
C O N N E C T I O N

For the latest
recommended daily
allowances, visit:
www.glencoe.com/sec/
math/mac/mathnet

**Applications and
Problem Solving**

29. *Health* A nutritional label from a bag of pretzels is shown at the right. One serving of pretzels contains 1.5 grams of fat, which is 3% of the daily amount recommended for a 2,000-Calorie diet. How many grams of fat are recommended?

Nutrition Facts		
Serving Size 1 package (46.8g)		
Servings per container 1		
Amount per serving		
Calories 190 Calories from Fat 15		
		% Daily Value
Total Fat 1.5g		3%
Saturated Fat 0g		0%
Cholesterol 0mg		0%
Sodium 760mg		32%
Total Carbohydrate 37g		12%
Dietary Fiber less than 1g		2%
Sugars 2g		
Protein 5g		
Vitamin A 0%	•	Vitamin C 0%
Calcium 0%	•	Iron 3%
*Percent Daily Values are based on a 2,000 calorie diet. Your daily values may be higher or lower depending on your calorie needs.		
Calories per gram:		
Fat 9 • Carbohydrates 4 • Protein 4		

30. *Civics* Recently, there were 6,954 cases filed in the U.S. Supreme Court. Of these cases, only 82 of them were decided in the Supreme Court. What percent of the cases that were filed were actually decided by the Supreme Court?

31. *Write a Problem* that can be solved by finding 75% of $245.59.

32. *Critical Thinking* Refer to Example 2. The wrestling, baseball, and gymnastics teams had almost the same number of competitors. Order the sports from greatest to least percent of hopefuls who made the cut. Explain how you can determine the order without determining the actual percents.

Mixed Review

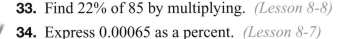

33. Find 22% of 85 by multiplying. *(Lesson 8-8)*

34. Express 0.00065 as a percent. *(Lesson 8-7)*

35. *Standardized Test Practice* A mechanic charges a $35 initial fee and $32.50 for each hour he works. Which equation could be used to find the cost, *c*, of a repair job that lasts *h* hours? *(Lesson 6-4)*

  A $c = 32.5 + 35h$ **B** $c = 35 + 32.5h$ **C** $c = 32.5 - 35h$
  D $c = 35 - 32.5h$ **E** $c = 32.5(35 - h)$

36. *Algebra* Find the solution of $g - 6.9 = 13.3$. *(Lesson 6-1)*

For **Extra Practice,**
see page 592.

CHAPTER 8

Study Guide and Assessment

 inter NET
CONNECTION Chapter Review For additional lesson-by-lesson review, visit:
www.glencoe.com/sec/math/mac/mathnet

Vocabulary

After completing this chapter, you should be able to define each
term, concept, or phrase and give an example or two of each.

Number and Operations
base (p. 349)
cross products (p. 325)
equivalent ratios (p. 318)
percent proportion (p. 349)
percentage (p. 349)
population density (p. 322)
property of proportions (p. 325)
proportion (p. 325)
rate (pp. 321, 349)
ratio (p. 317)
unit rate (p. 321)

Geometry
scale (p. 332)
scale drawing (p. 332)

Statistics and Probability
capture-recapture technique (p. 329)
sample (p. 329)

Problem Solving
draw a diagram (p. 330)

Understanding and Using the Vocabulary

Choose the letter of the term that best matches each phrase.

1. a comparison of two numbers by division

2. two ratios that have the same value

3. a ratio of two measurements with different units

4. an equation that shows that two ratios are equivalent

5. used to present something that is too large or too small for actual-size drawing

6. the ratio of the distance on a map to the actual distance

7. a ratio that compares a number to 100

8. the name for 60 in the percent proportion $\frac{15}{25} = \frac{60}{100}$

9. the name for 25 in the percent proportion $\frac{15}{25} = \frac{60}{100}$

a. rate
b. base
c. scale drawing
d. percent
e. ratio
f. scale
g. unit rate
h. proportion
i. equivalent ratios
j. cross products
k. percentage

In Your Own Words

10. ***Explain*** how to solve $\frac{6}{12} = \frac{n}{4}$.

Objectives & Examples

Upon completing this chapter, you should be able to:

● express ratios as fractions and determine whether two ratios are equivalent *(Lesson 8-1)*

Express 6:18 as a fraction in simplest form.

$$\frac{6}{18} = \frac{6 \div 6}{18 \div 6} = \frac{1}{3}$$

The simplest form is $\frac{1}{3}$.

● to determine unit rates *(Lesson 8-2)*

Find the unit price for a 16-ounce box of pasta on sale for 96 cents.

$$\frac{cents}{ounces} \rightarrow \frac{96}{16} = \frac{96 \div 16}{16 \div 16} = \frac{6}{1}$$

The unit price is 6 cents per ounce.

● to solve proportions *(Lesson 8-3)*

Solve $\frac{6}{9} = \frac{x}{12}$

$$\frac{6}{9} = \frac{x}{12}$$
$$6 \times 12 = 9 \times x$$
$$72 = 9x$$
$$8 = x \qquad \text{The solution is 8.}$$

● to solve problems involving scale drawings *(Lesson 8-4)*

A map scale is 1 inch:80 miles. Find the actual distance for a map distance of $3\frac{1}{4}$ inches.

$$\begin{array}{l} map \rightarrow \\ actual \rightarrow \end{array} \frac{1 \text{ inch}}{80 \text{ miles}} = \frac{3\frac{1}{4} \text{ inches}}{n \text{ miles}} \begin{array}{l} \leftarrow map \\ \leftarrow actual \end{array}$$

$$1 \times n = 80 \times 3\frac{1}{4}$$
$$n = 260$$

The actual distance is 260 miles.

Review Exercises

Use these exercises to review and prepare for the chapter test.

Express each ratio as a fraction in simplest form.

11. 25 to 10   **12.** 14:70   **13.** 11:66

14. 12 to 64   **15.** 90 to 33   **16.** 50:100

Tell whether the ratios are equivalent. Show your answer by simplifying.

17. $\frac{63}{9}$ and $\frac{21}{14}$ **18.** $\frac{5}{10}$ and $\frac{12}{24}$

Express each rate as a unit rate.

19. 16 cups for 4 people

20. 150 people for 5 classes

21. $23.75 for 5 pounds

22. 810 miles in 9 days

23. $38 in 4 hours

24. 24 gerbils in 3 cages

Solve each proportion.

25. $\frac{13}{25} = \frac{39}{m}$ **26.** $\frac{w}{6} = \frac{12}{8}$

27. $\frac{350}{p} = \frac{2}{10}$ **28.** $\frac{45}{5} = \frac{x}{7}$

29. *Algebra* Find the value of x that makes $\frac{5}{x} = \frac{6}{3}$ a proportion.

30. Find the distance between Atlanta and Savannah, Georgia if the map distance is $3\frac{1}{2}$ inches and the scale is 1 inch:70 miles.

31. Find the length of a building 20 yards wide on a scale drawing with a scale of $\frac{1}{2}$ in. = 1 yd.

to express fractions as percents, and vice versa *(Lesson 8-5)*

Express $\frac{18}{20}$ as a percent.

$\frac{18}{20} = \frac{n}{100}$ *Find cross products.*

$1,800 = 20n$ *Divide each side by 20.*

$90 = n$

So, $\frac{18}{20} = 90\%$.

Express each fraction as a percent.

32. $\frac{3}{5}$ **33.** $\frac{5}{8}$

Express each percent as a fraction in simplest form.

34. 65%   **35.** $13\frac{1}{2}\%$

36. $78\frac{3}{4}\%$ **37.** 43.5%

---

to express decimals as percents, and vice versa *(Lesson 8-6)*

Express 48% as a decimal.

$48\% = 048 = 0.48$

Express each decimal as a percent.

38. 0.87   **39.** 0.325

Express each percent as a decimal.

40. 42%   **41.** $15\frac{3}{4}\%$

---

to express percents greater than 100% and percents less than 1% as fractions and as decimals, and vice versa *(Lesson 8-7)*

Write 2.35 as a percent.

$2.35 = 2\frac{35}{100}$

$\quad\quad = \frac{235}{100}$ or 235%

Express each percent as a decimal.

42. 125%   **43.** 0.25%

44. 0.05%   **45.** 563%

Express each number as a percent.

46. 0.002   **47.** 4.75

48. $7\frac{1}{2}$ **49.** 0.0095

---

find the percent of a number *(Lesson 8-8)*

Find 45% of 400.

$\frac{n}{400} = \frac{45}{100}$

$100n = 18,000$

$n = 180$

Find each number. Round to the nearest tenth if necessary.

50. What number is 68% of 320?

51. Find 18% of 90.

52. 0.75% of 80 is what number?

53. What number is 280% of 18?

---

solve problems using the percent proportion *(Lesson 8-9)*

What percent of 90 is 18?

$\frac{P}{B} = \frac{r}{100}$

$\frac{18}{90} = \frac{r}{100}$

$1,800 = 90r$

$20 = r$

Find each number. Round to the nearest tenth if necessary.

54. 6 is what percent of 120?

55. Find 0.8% of 35.

56. What percent of 375 is 40?

57. 15% of what number is 900?

Applications & Problem Solving

58. Geography Brookville has a population of 4,312 people and an area of 127 square miles. To the nearest whole number, how many people per square mile are there in Brookville? *(Lesson 8-2)*

59. Draw a Diagram The advisor for the Spanish Club tells three students about a club meeting. It takes 1 minute for her to tell three students and 1 minute for each of those three students to tell three other students, and so on. How many students will know about the meeting in three minutes? *(Lesson 8-4A)*

60. Statistics Of the 100 teens who answered a survey, 21 owned a cat. What percent of those who answered the survey own a cat? *(Lesson 8-5)*

61. Life Science Did you know that humans are outnumbered on Earth? There are so many insects that the weight of all humans together is just $\frac{1}{3}\%$ of the weight of all the insects! What is $\frac{1}{3}\%$ written as a decimal? *(Lesson 8-7)*

Alternative Assessment

● **Open Ended**

A manager of a telephone sales force found that last month her team sold 436 magazine subscriptions. The team made 15,265 calls that month. How many calls should the manager have her team make if the goal for this month is to sell 550 subscriptions?

One salesperson can make a call every five minutes. A full-time salesperson works forty hours in a week, and a part-time salesperson works 20 hours in a week. How many full-time and part-time salespeople should the manager have on the team to meet the 550 subscription goal?

● **Completing the** 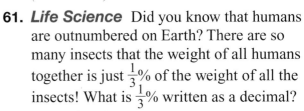 **CHAPTER Project**

Use the following checklist to make sure that your poster is complete.

☑ The chart for the type and amount of trash thrown away each day is clear and easy to read.

☑ The recent statistics about the items you can recycle and local recycling programs are complete and easy to understand.

☑ The percents describing the amount of trash that can be recycled and the decrease in the total amount of trash are included.

● PORTFOLIO Select an item from this chapter that you found challenging. Place it in your portfolio.

A practice test for Chapter 8 is provided on page 614.

Section One: Multiple Choice

There are ten multiple-choice questions in this section. Choose the best answer. If a correct answer is *not here*, choose the letter for Not Here.

1. In a stem-and-leaf plot of the data below, what numbers would be used for the stems?

   Height (Inches)

52	75	49	51	57	66	69
58	59	62	61	61	73	68
76	74	78	49			

   A 0–9

   B 1–9

   C 4–7

   D 4–9

2. There are twenty pieces of candy in a jar on the counter at the bank. Four are orange, seven are root beer, three are lemon, and six are strawberry. If a customer chooses one at random, what is the probability that it will *not* be root beer?

   E $\frac{1}{4}$

   F $\frac{1}{20}$

   G $\frac{7}{20}$

   H $\frac{13}{20}$

3. If 10 overhead projectors are needed for every 14 teachers, how many overhead projectors are needed for 112 teachers?

   A 140

   B 80

   C 60

   D 25

Please note that Questions 4–10 have five answer choices.

4. People were asked to choose a favorite newscast. The graph shows the results of the survey.

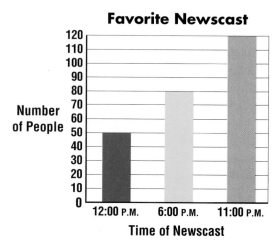

Favorite Newscast

What were the total number of people who were surveyed about the newscasts?

   F 180

   G 210

   H 220

   J 250

   K 280

5. A symphony orchestra held a series of special concerts to raise money for the restoration of the theater where they perform. Sixty percent of the price of each ticket was donated. Each ticket sold for $35. What else do you need to know to find how much money was donated to the theater restoration?

   A the amount of profit on each ticket

   B the number of seats in the theater

   C the number of performances in the series

   D the total number of tickets sold

   E the operating cost of the orchestra

6. The coach paid $476.60 for thirty soccer balls without tax. How much could be saved by buying the thirty soccer balls on sale for $381.36?
   F $95.24
   G $85.24
   H $125.54
   J $114.34
   K $75.54

7. Jay bought a bag of flour for making bread in his bread machine. Each loaf of bread uses $2\frac{1}{2}$ cups of flour. If the bag contains 20 cups of flour, how many loaves will Jay get from the bag?
   A 5   B 6
   C 7   D 8
   E Not Here

8. What is the value of $8 + x^2$ if $x = 12$?
   F 32   G 152
   H 400   J 420
   K Not Here

9. There was $244.87 in the cash register at a video store. The manager removed $150 to deposit in the bank. How much money was left in the cash register?
   A $94.87   B $88.87
   C $84.87   D $74.87
   E Not Here

10. Golf balls are on sale for $9.98 per box of 3. What is the cost of 7 boxes of golf balls before sales tax is added?
    F $29.64   G $29.94
    H $45.86   J $69.86
    K $71.86

Test-Taking Tip
THE PRINCETON REVIEW

When taking a long test, work carefully but quickly through the problems. Skip those problems that take more than the average amount of time allotted for doing a problem. Mark these in the test booklet and come back to them after you have completed the problems that you could easily do.

Section Two: Free Response

This section contains five questions for which you will provide short answers. Write your answers on your paper.

11. What property allows you to say that
    $$\frac{1}{3} + \left(\frac{2}{3} + \frac{4}{5}\right) = \left(\frac{1}{3} + \frac{2}{3}\right) + \frac{4}{5}?$$

12. Carol pays $185.55 each month for her car payment. What is this amount rounded to the nearest ten dollars?

13. Astronomers use angles to describe the positions of objects in the sky. On April 25, 1997, the angle of Comet Hale-Bopp was 30°. The angle of the comet was −10° on July 24, 1997. Find the difference between these angle measures.

14. Find the next three terms in the sequence 6, −12, 18, −24, 30, −36, ... Is the sequence arithmetic, geometric, or neither?

15. Soft drinks are sold in 2-liter bottles. How many milliliters are in a 2-liter bottle?

 **Test Practice** For additional test practice questions, visit:
www.glencoe.com/sec/math/mac/mathnet

Geometry: Investigating Patterns

What you'll learn in Chapter 9

- to classify angles and polygons,
- to use proportions to express the relationship between corresponding parts of similar figures,
- to use logical reasoning to solve problems,
- to classify triangles and quadrilaterals, and
- to create patterns using tessellations, translations, and reflections.

CHAPTER Project

GEOMETRIC ART

In this project, you will use geometric concepts to design
a pattern for a wall-hanging, poster, tie, or T-shirt. You will
create three designs: one with a tessellation, another with
a translation, and a third with a reflection. To complete your
project, you will choose one of your designs to present to
the class.

Getting Started

- Look for examples of geometric patterns
  in fabrics you find in your home.
  Describe the patterns.

- Do research about geometric patterns
  that may be common in your ethnic
  heritage. The patterns can be modern
  designs or may date back many
  centuries.

Technology Tips

- Use an **electronic encyclopedia** to
  do your research.

- Use *Geometer's Sketchpad* or other
  drawing software to complete your
  basic design.

 **interNET** **Research** For up-to-date
CONNECTION information on tessellations, visit:

www.glencoe.com/sec/math/mac/mathnet

Working on the Project

You can use
what you'll
learn in
Chapter 9 to
help you create
your geometric
design.

Page	Exercise
391	14
394	12
397	11
401	Alternative Assessment

9-1A Measuring Angles

A Preview of Lesson 9-1

protractor

dot paper

colored pencils

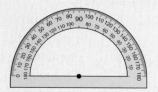

When you want to measure a line, you can use a ruler. When you want to measure an angle, you can use a *protractor*. Angles are measures of rotation and are measured in units called *degrees*.

Notice that the protractor has two scales. The outer scale goes from 0 to 180 from left to right. The inner scale goes from 0 to 180 from right to left.

TRY THIS

1 Follow these steps to measure an angle.

- Place the protractor on the angle so that the center is on the vertex of the angle and one side goes through 0° on the protractor.

- In this case, 0° is on the inner scale. Follow the inner scale to the point where the other side of the angle meets the protractor. Find the inner number. This is the angle's measure. The measure of this angle is 110°.

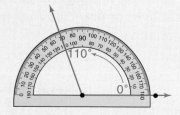

ON YOUR OWN

Copy each angle onto dot paper. Then use a protractor to find its measure. You may need to extend the sides of the angle in order to measure it.

1.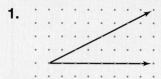

2.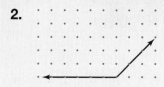

3.

4. How do you know which scale on the protractor to read?

5. Why do you think there are two scales on the protractor?

2 The figure at the right shows parallel lines. The line that intersects the parallel lines is called a *transversal*.

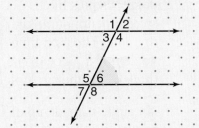

- Copy the figure onto dot paper and label the angles as shown.

- Measure each angle with your protractor and record each measure.

- Using colored pencils, shade any angles with the same measure one color. For example, angle 1 and angle 4 have the same measure, and they are shaded yellow. If there are other angles with a different measure, shade them a second color. Angle 6 is shaded blue because it has a different measure. Continue shading using this pattern.

In the figure above, certain pairs of angles have special names. The angles that make up each pair have the same measure.

alternate interior angles	∠3 and ∠6, ∠4 and ∠5
alternate exterior angles	∠1 and ∠8, ∠2 and ∠7
corresponding angles	∠1 and ∠5, ∠2 and ∠6, ∠3 and ∠7, ∠4 and ∠8
vertical angles	∠1 and ∠4, ∠2 and ∠3, ∠5 and ∠8, ∠6 and ∠7

ON YOUR OWN

Copy each figure onto dot paper. Then use a protractor to find the measure of each angle. Shade angles with equal measures as you did above.

6.

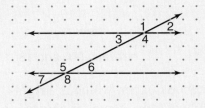

7.

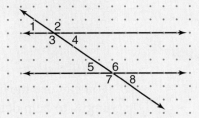

8. Look for a pattern in the shading of your parallel lines. Make a conjecture about the measure of the angles formed when a transversal intersects a pair of parallel lines.

9. In the figure at the right, the measure of angle 2 is 45°. Predict the measure of the other angles in the figure. Then copy the figure onto dot paper and check by using a protractor.

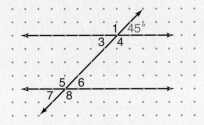

Angles

Many families celebrate Thanksgiving with delicious pies like sweet potato, pumpkin, or apple. The next time you enjoy a piece of pie, think of geometry. A piece of pie is a model of an **angle**.

An angle is made up of two rays with a common endpoint. The rays are called the *sides* of the angle, and the point where they meet is called the **vertex**.

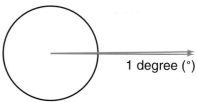

An angle is not measured by the length of its sides, but in units called **degrees**. Imagine that you cut a whole pie into 360 equal-sized pieces. The angle of each piece measures 1 degree.

Angles are classified according to their measure.

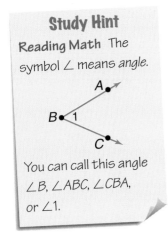
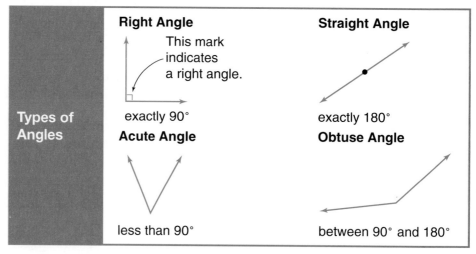

You can use the corner of a sheet of notebook paper, which is a right angle, to help determine whether an angle is right, acute, or obtuse. The edge of the paper can be used to determine straight angles.

Example

1 Classify each angle as right, acute, obtuse, or straight.

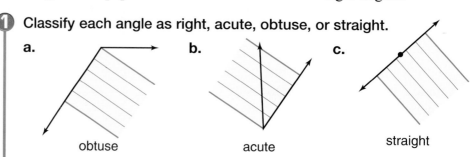

a. obtuse

b. acute

c. straight

Some pairs of angles are **supplementary** or **complementary**.

MINI-LAB

Work with a partner. protractor dot paper

Try This

Copy each pair of angles onto dot paper and measure each angle with a protractor. Look for a pattern.

1.
supplementary

2.
not supplementary

3.
supplementary

4.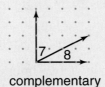
complementary

5.
complementary

6.
not complementary

Talk About It

Tell whether each pair of angles is supplementary.

7.

8.

9.

10. What is true about supplementary angles?

Tell whether each pair of angles is complementary.

11.

12.

13.

14. What is true about complementary angles?

If the sum of the measures of two angles is 180°, the angles are supplementary. If the sum of the measures of two angles is 90°, the angles are complementary.

Supplementary Angles
$m\angle 1 + m\angle 2 = 180°$

Complementary Angles
$m\angle 3 + m\angle 4 = 90°$

Lesson 9-1 Angles   **363**

Example **2** **Algebra** Angles A and B are supplementary.
If $m\angle A = 70°$, find $m\angle B$.

INTEGRATION

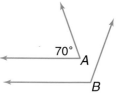

$$m\angle A + m\angle B = 180$$
$$70 + m\angle B = 180 \qquad \textit{Replace } m\angle A \textit{ with 70.}$$
$$70 - 70 + m\angle B = 180 - 70 \qquad \textit{Subtract 70 from each side.}$$
$$m\angle B = 110$$

The measure of $\angle B$ is 110°.

CHECK FOR UNDERSTANDING

**Communicating
Mathematics**

Read and study the lesson to answer each question.

1. *Show* how you can classify an angle using the corner of your paper.

2. *Draw* an example of supplementary angles. Explain how you know they are supplementary.

**HANDS-ON
MATH**

3. The figure at the right shows two intersecting lines. They form two pairs of "opposite" angles called *vertical angles*. Angles 1 and 3 are vertical angles, and angles 2 and 4 are vertical angles. Copy and measure each angle. Make a conjecture about the measures of vertical angles. Test your conjecture with other vertical angles.

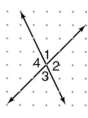

Guided Practice

Classify each angle as *acute, obtuse, right,* or *straight.*

4.   5.

6. 165°   7. 180°

8. Classify the angles at the right as *complementary, supplementary,* or *neither.*

9. *Algebra* Angles X and Y are complementary. If $m\angle X = 15°$, find $m\angle Y$.

EXERCISES

Practice

Classify each angle as *acute, obtuse, right,* or *straight.*

10.   11.   12.

13. 36°   **14.** 170°   **15.** 60°

16. 1°   **17.** 91.5°   **18.** 179.5°

Classify each pair of angles as *supplementary*, *complementary*, or *neither*.

19.    **20.**   **21.**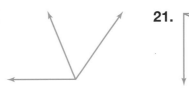

22. Line *l* is parallel to line *m*. Find the missing angles.

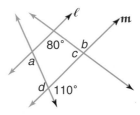

23. *Algebra* Angles *A* and *B* are supplementary. If $m\angle A = 25°$, find $m\angle B$.

Applications and Problem Solving

24. *Geography* Earth rotates 360° degrees in one day. Through how many degrees does it rotate in one hour?

25. *Earth Science* The Big Dipper is the best known group of stars in the sky. The figures below show how the Big Dipper probably looked 100,000 years ago, how it looks today, and how it will look 100,000 years from now.

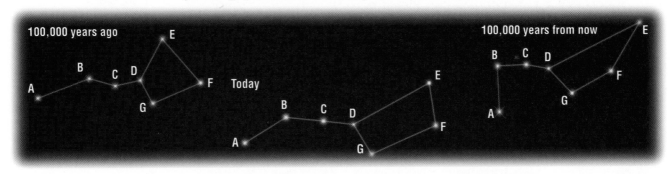

a. Which angle was obtuse 100,000 years ago and will be acute 100,000 years from now?

b. Identify an angle that appears to be a right angle.

26. *Critical Thinking* Suppose two angles are supplementary. If each angle below represents one angle of the pair, what kind of angle is its supplement?

a. acute   b. right   c. obtuse   d. straight

Mixed Review

27. **Standardized Test Practice** 95 of the 273 seventh-graders have turned in permission slips for the field trip. About what percent of the seventh-graders have *not* turned in a permission slip? *(Lesson 8-9)*

A 65%   B 55%   C 45%   D 35%

For **Extra Practice**, see page 592.

28. Find 45% of 1,600. *(Lesson 8-8)*

29. Subtract $1\frac{1}{3}$ from $4\frac{3}{4}$. *(Lesson 7-3)*

COOPERATIVE LEARNING

compass

straightedge

9-1B Perpendicular and Parallel Lines

A Follow-Up of Lesson 9-1

Perpendicular lines are lines that form right angles when they intersect. In the figure, line ℓ is perpendicular to line m. This can also be written as $\ell \perp m$.

You can use a compass and straightedge to construct perpendicular lines.

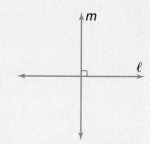

TRY THIS

1 Construct a line perpendicular to line m through point P.

- Draw a line and label it m. Draw a dot on the line and label it point P.

- Place the compass point on P and draw equal arcs to intersect line m twice. Label these points Q and R. *An arc is part of a circle.*

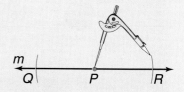

- Open your compass wider. Put the compass point at Q and draw an arc above line m.

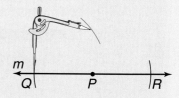

- With the same setting, put the compass point at R and draw an arc to intersect the one you just drew. Label this intersection point S.

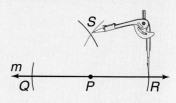

- Use a straightedge to draw a line through S and P. Line PS is perpendicular to line m.

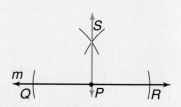

ON YOUR OWN

1. Construct a rectangle by constructing perpendiculars.

In Lesson 9-1A, you learned that some of the angle pairs formed when a transversal intersects parallel lines have the same measure. For example, in the figure at the right, line ℓ is parallel to line m. Angle 1 and angle 2 have the same measure.

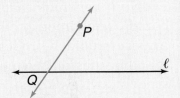

You can use a compass and a straightedge to construct a line parallel to a given line.

TRY THIS

2 Construct a line parallel to line ℓ through point P.
- Draw a line and label it ℓ. Choose any point P, not on the line.
- Draw a line through P that intersects ℓ. Label the point of intersection Q.

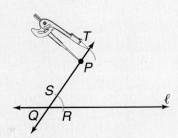

- Place your compass point at Q and draw an arc. Label the points R and S.
- With the same setting, place the compass point point at P and draw an arc. Label the point T.

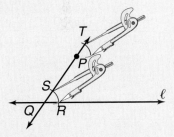

- Use your compass to measure the distance from R to S.
- With the same setting, place your compass at T and draw an arc to intersect the one already drawn. Label this point U.

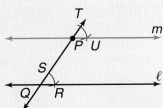

- Draw a line through P and U. Label it m. Line m is parallel to line ℓ.

ON YOUR OWN

2. Name angles in the construction that have the same measure.
3. A parallelogram is a figure with opposite sides parallel. Use a compass and straightedge to construct a parallelogram.

FASHION

Jhane Barnes
FASHION DESIGNER

Did you ever think that you could use geometry to become a successful fashion designer? Jhane Barnes did! While still a student at New York's Fashion Institute of Technology, she started her own business. More than twenty years later, her company remains at the leading edge of fashion design. Along the way, Ms. Barnes has won numerous awards, including a Coty American Fashion Critics' Menswear Award in 1980. At that time, she was the youngest person, and the only female, to win this award.

Ms. Barnes designs textiles that start with a basic geometric element and grow into unique designs through translations, reflections, and rotations. She also uses fractals in her designs. To be a fashion designer, you'll need to take courses in mathematics, business, design, and art. You should be artistically creative, imaginative, and like challenges.

For more information:
International Association
  of Clothing Designers
240 Madison Avenue
12th Floor
New York, NY 10016

inter NET CONNECTION
www.glencoe.com/sec/
math/mac/mathnet

Someday, I'd like to be a famous fashion designer like Jhane Barnes.

Your Turn
Design a brochure that describes how geometry is used in fashion design. Include sketches of your designs.

HANDS-ON LAB

COOPERATIVE LEARNING

9-2A Angles of a Polygon

A Preview of Lesson 9-2

🔺 protractor

✂ scissors

In this lab, you will look for a pattern among the angles of a triangle and use the pattern to find the sum of the angle measures of any figure with more than three sides.

TRY THIS

Work with a partner.

1 • Draw three triangles. One triangle should have three acute angles, one should have an obtuse angle, and the last should have a right angle. Cut out the triangles.

• For each triangle, tear off the angles and arrange as shown.

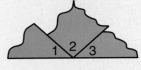

2 • Draw any four-sided figure and cut it out.

• Pick one vertex and draw the diagonal to the opposite vertex. Cut along the diagonal.

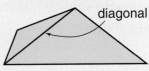
diagonal

ON YOUR OWN

1. For each triangle in Activity 1, what kind of angle is formed where the vertices meet?

2. What is the measure of this type of angle?

3. Complete this statement: The sum of the measures of the angles of a triangle is ___?___ .

4. How many triangles were formed when you cut along the diagonal of the four-sided figure in Activity 2?

5. Predict the sum of the measures of the angles in your four-sided figure. Explain your reasoning.

6. Devise a way to check your prediction.

7. Find the sum of the measures of the angles of a figure with the following number of sides.

   a. 5   **b.** 6   **c.** 8   **d.** 10

8. *Algebra* If *n* is the number of sides of a figure, write an algebraic expression that tells the sum of the measures of the angles of the figure.

Polygons

What you'll learn

You'll learn to identify polygons and regular polygons.

When am I ever going to use this?

You can identify many types of traffic signs by their shape.

Word Wise

polygon
triangle
quadrilateral
pentagon
hexagon
heptagon
octagon
nonagon
decagon
congruent
regular polygon

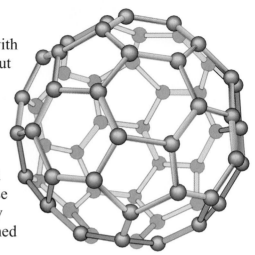

If you love soccer, you are familiar with the unique shapes on a soccer ball. But did you know there is a molecule of carbon with the same shapes? In the early 1990s, scientists discovered that when they fired lasers at carbon molecules, the molecules sometimes formed tiny structures that resembled hollow soccer balls. They named these structures "Buckyballs," because they resembled the geodesic domes designed by Buckminster Fuller.

The five- and six-sided shapes on a soccer ball or Buckyball are examples of **polygons**. A polygon is a *simple closed* figure formed by three or more line segments. *Simple* means that the line segments don't cross each other, and *closed* means that, when you draw the polygon, your pencil ends up where it started. Since the segments meet to form angles and angles have vertices, polygons also have vertices.

Polygons	*Not* Polygons

A polygon is named by the number of its sides or by the number of its angles. The word **triangle** means *three angles*. The word **quadrilateral** means *four sides*. Other common polygons are shown below. Each of these figures is only one example of that type of figure.

pentagon
5 sides

hexagon
6 sides

heptagon
7 sides

octagon
8 sides

nonagon
9 sides

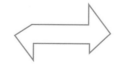

decagon
10 sides

The red slash marks indicate congruent sides, and the red arcs indicate congruent angles.

When two angles or two line segments have the same measure, they are **congruent**. The hexagon at the right has all sides congruent and all angles congruent. This figure is an example of a **regular polygon**.

Examples

Determine which figures are polygons. If the figure is a polygon, name it and tell whether it is a regular polygon. If the figure is *not* a polygon, explain why.

1

The figure is a polygon. It is a regular pentagon.

2

The figure is *not* a polygon because one of its sides is *not* a line segment.

Study Hint

Reading Math The word polygon comes from the prefix *poly-* meaning *many* and the suffix *–gon* meaning *angle*. So, a polygon is a many-angled figure.

3

The figure is a polygon. It is a hexagon. It is *not* regular because its angles are not congruent.

4

The figure is a polygon. It is a decagon. It is *not* regular because its sides and angles are *not* congruent.

In Lesson 9-2A, you found the sum of the measures of the angles of a polygon by drawing all of the diagonals from one vertex and counting the triangles. Then you multiplied that number by 180°. In a hexagon, there are 4 triangles, so the sum of the measures of the angles of a hexagon is 4 · 180° or 720°.

Example

INTEGRATION

5 **Algebra** Use an equation to find the measure of each angle of a regular hexagon.

In a regular hexagon, all of the angles have the same measure. Let a represent the measure of one angle.

The sum of the measures of the six angles is 720°. So, solve the equation $6a = 720$.

$6a = 720$

$\dfrac{6a}{6} = \dfrac{720}{6}$ *Divide each side of the equation by 6.*

$a = 120$ Each angle of a regular hexagon measures 120°.

Communicating Mathematics

Read and study the lesson to answer each question.

1. *Tell* whether the figure is a regular polygon. Explain your reasoning.

2. *Draw* a figure that is not a polygon and explain why it is not.

3. *Draw and label* a pentagon, hexagon, heptagon, octagon, nonagon, and decagon so that three of the figures are regular and three are *not* regular.

Guided Practice

Determine which figures are polygons. If the figure is a polygon, name it and tell whether it is a regular polygon. If the figure is *not* a polygon, explain why.

4.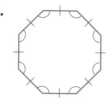

5.

Family Activity

Draw the shape of some other signs or objects you see everyday. Name each shape and identify which type of polygon it is, if any.

6. *Traffic Safety* Part of the exam you'll take to earn a driver's license asks you to identify road signs. Name the shape of each sign below and tell whether the sign represents a regular polygon.

a. b. c. d.

Practice

Determine which figures are polygons. If the figure is a polygon, name it and tell whether it is a regular polygon. If the figure is *not* a polygon, explain why.

7.

8.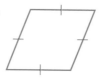

9.

10.

11.

12.

13. A *dodecagon* is a polygon with 12 angles. Draw a dodecagon.

Applications and Problem Solving

14. *Algebra* The sum of the measures of the angles of a regular decagon is 1,440°. Write and solve an equation to find the measure of one of its angles.

15. *Clothing* If you've bought a new T-shirt or pair of pants recently, you may notice something different on the label. Words are no longer used to describe how to care for the garment. Instead, they have been replaced by symbols. Which of the symbols are polygons?

wash

bleach

dry

iron

dry-clean

16. *Earth Science* In about 135 B.C., Chinese author Han Ying recognized that snowflakes are always six-pointed. Name the figure formed by connecting the points of a snowflake.

17. *Critical Thinking* Line *l* is perpendicular to line *m*. Line *p* is parallel to line *n*. Find each of the following angles.

    a. complementary
    b. supplementary
    c. vertical
    d. alternate interior
    e. corresponding
    f. acute
    g. right

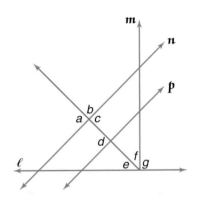

Mixed Review

18. Classify the angles as *supplementary*, *complementary*, or *neither*. *(Lesson 9-1)*

19. *Food* The United States produced about 11 billion pounds of apples in a recent year. Use the information in the graph to find how many pounds of apples were used to make juice. *(Lesson 8-8)*

20. **Standardized Test Practice** Alberta is 3 inches less than 6 feet tall. How tall is she in inches? *(Lesson 7-5)*

    A 67 in.   **B** 69 in.
    C 72 in.   **D** 75 in.

An Apple A Day
Uses of apples in the United States

Juice 41%
Fresh 41%
Other 9%
Apple sauce 9%

Source: International Apple Institute

For **Extra Practice**, see page 593.

COOPERATIVE LEARNING

9-2B Inscribed Polygons

A Follow-Up of Lesson 9-2

 compass

 straightedge

 scissors

 ruler

 protractor

A polygon is an *inscribed* polygon if each of its vertices lies on a circle. You can inscribe a square in a circle by using some simple paper folding or by using a compass and straightedge.

TRY THIS

① To use paper folding, follow these steps.

- Use your compass to draw a circle. Then cut out the circle.

- Fold the circle in half and in half again.

- Open the circle. Use your straightedge to draw line segments that connect the points where the paper folds meet the circle. Connect the points in order.

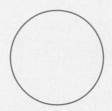

② To use a compass and straightedge, follow these steps.

- Use your compass to draw a circle.

- Draw a diameter through the center of the circle. Then, draw a line perpendicular to the diameter through the center of the circle.

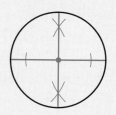

- Use your straightedge to connect the points where the diameter and its perpendicular meet the circle. Connect the points in order.

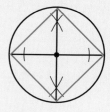

ON YOUR OWN

1. Use your protractor to measure the angles of the quadrilaterals. You may need to extend the sides. Use a ruler to measure the sides. What do you find?

2. Explain how the paper folding method is similar to using a compass and straightedge.

3. Use paper folding to inscribe an octagon in a circle. Explain the steps you used.

You can also inscribe other regular polygons in a circle using a compass and straightedge.

TRY THIS

3 To inscribe a regular hexagon in a circle, follow these steps.

- Use your compass to draw a circle. Leave the compass at the same setting.

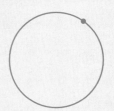

- Put a point on the circle and place your compass on that point. Draw a small arc that intersects the circle.

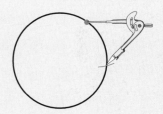

- Place the compass on the point where the arc intersects the circle. Draw another small arc to intersect the circle.

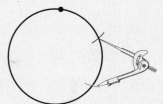

- Continue the process until you come back to the first point. Use the ruler to connect the intersection points in order as shown.

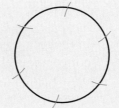 

ON YOUR OWN

4. Use your protractor to measure the angles of your hexagon. You may need to extend the sides. Use a ruler to measure the sides. What do you find? Is the hexagon regular?

5. Inscribe an equilateral triangle in a circle. Explain the steps you used. (*Hint*: Modify the steps you used to inscribe a hexagon in a circle.)

Integration: Algebra
Similar Polygons

What you'll learn

You'll learn to determine whether polygons are similar and find a missing length in a pair of similar polygons.

When am I ever going to use this?

You'll use similar polygons when you enlarge or reduce drawings.

Word Wise

similar
indirect measurement

Have you ever seen a photograph of the United States taken from space? The photo is an example of a **similar** figure. It has the same shape as the United States but is a different size.

HANDS-ON
MINI-LAB

Work with a partner. dot paper   protractor

Try This

• Copy each pair of polygons onto dot paper.

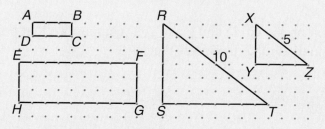

• Measure each angle in degrees and each side in units.

Talk About It

1. Express each ratio in simplest form. *Letters such as AB refer to the measure of the segment with those endpoints.*

    a. $\dfrac{AB}{EF}$, $\dfrac{BC}{FG}$, $\dfrac{DC}{HG}$, $\dfrac{AD}{EH}$ b. $\dfrac{RS}{XY}$, $\dfrac{ST}{YZ}$, $\dfrac{RT}{XZ}$

2. The rectangles are similar and the triangles are similar. What do you notice about the ratios of their corresponding sides?

3. In the triangles above, find corresponding angles. What do you notice about the measure of corresponding angles?

4. Write a definition of similar polygons.

Study Hint

Reading Math Parts of similar figures that "match" are called corresponding parts. In the triangles above, ∠R and ∠X are corresponding angles.

The pattern in the Mini-Lab suggests the following definition.

Similar Polygons	Words:	Two polygons are similar if their corresponding angles are congruent and their corresponding sides are in proportion.
	Symbols:	$\triangle ABC \sim \triangle XYZ$ *The symbol ~ means is similar to.*
	Model:	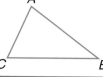

Proportions are useful in finding the missing length of a side in any pair of similar polygons.

Example ① If $\triangle ABC \sim \triangle DEF$, find the length of $\overline{EF}$.

The symbol $\overline{EF}$ means line segment EF.

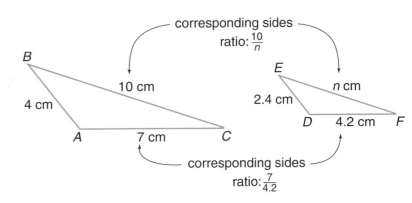

corresponding sides
ratio: $\frac{10}{n}$

corresponding sides
ratio: $\frac{7}{4.2}$

In similar polygons, these two ratios are equal. So, you can write and solve a proportion to find the missing measure.

Let n represent the missing measure.

$$\frac{7}{4.2} = \frac{10}{n}$$

$7n = 4.2(10)$ *Find the cross products.*

$7n = 42$

$n = 6$ *Divide each side by 7.*

The length of $\overline{EF}$ is 6 centimeters.

LOOK BACK

You can refer to Lesson 8-3 to review proportions.

You can use similar triangles to find the height of objects like flagpoles that are too difficult to measure directly. This kind of measurement is called **indirect measurement**.

Example

INTEGRATION

② **Measurement** The height of a flagpole and the length of its shadow are proportional to the height of another object and the length of its shadow. The objects and their shadows form two sides of similar triangles. Find the height of the flagpole shown below.

Let x represent the height of the flagpole.

height of tree → $\dfrac{6}{4} = \dfrac{x}{18}$ ← height of flagpole
length of shadow → ⠀⠀ ← length of shadow

$$6(18) = 4x$$
$$108 = 4x$$
$$27 = x$$

The height of the flagpole is 27 feet.

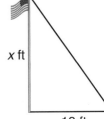

6 ft

4 ft x ft   18 ft

Study Hint

Estimation After setting up the proportion, you can estimate to help you solve the problem.

$5 \times 20 = 4x$

$100 = 4x$

$25 = x$

Communicating Mathematics

Read and study the lesson to answer each question.

1. **Write** another proportion that could have been used to find the value of *n* in Example 1 on page 377.

2. **Draw** two similar rectangles whose corresponding sides are in the ratio of 3:5. Label the sides.

3. **You Decide** Lorena thinks the rectangles at the right are similar. Nikki thinks they are *not* similar. Who is correct? Explain your reasoning.

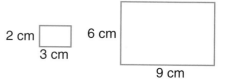

2 in.   3 in.
3 in.   4 in.

Guided Practice

Tell whether each pair of polygons is similar. Justify your answer.

4.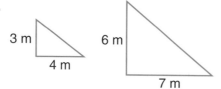

3 m   6 m
4 m   7 m

5.
2 cm   6 cm
3 cm   9 cm

Find the value of *x* in each pair of similar polygons.

6.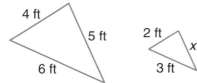

4 ft   5 ft
6 ft

2 ft   *x* ft
3 ft

7.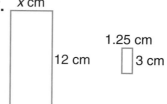

x cm
12 cm

1.25 cm
3 cm

8. Rectangles *F* and *G* are similar. The ratio of rectangle *F*'s width to rectangle *G*'s width is 2:3. The length of rectangle *F* is 15 inches, and its width is 10 inches. Find the perimeter of rectangle *G*.

Practice

Tell whether each pair of polygons is similar. Justify your answer.

9.
1 yd
2 yd

3 yd
6 yd

10.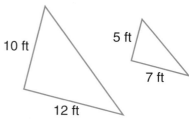

10 ft   5 ft
12 ft   7 ft

Find the value of *x* in each pair of similar polygons.

11.

6 m
3 m
4 m

12 m
6 m
x m

12.
21 in.
x in.

7 in.
1 in.

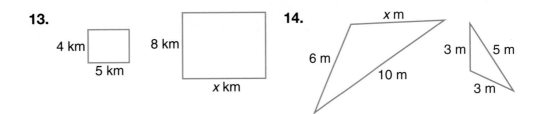

13.
4 km
5 km
8 km
x km

14.
x m
6 m
10 m
3 m
5 m
3 m

15. Triangles *A* and *B* are similar. The ratio of a side of triangle *B* to a corresponding side of triangle *A* is 5:3. The lengths of the sides of triangle *A* are 18 feet, 27 feet, and 30 feet. Find the perimeter of triangle *B*.

16. *Photography* Jamie sizes and positions photos for the Heritage Middle School Yearbook. She needs to reduce a photo that is 3 inches wide and 5 inches long so that it fits into a space that is 2 inches wide. What is the length of the reduced photo?

17. *Write a Problem* about the photo at the left that uses indirect measurement.

18. *Critical Thinking* Two rectangles are similar. The ratio of their corresponding sides is 1:2.
  a. Find the ratio of their perimeters.
  b. Find the ratio of their areas.

Mixed Review

19. **Standardized Test Practice** Which is *not* a polygon? *(Lesson 9-2)*

A    B   C   D

For **Extra Practice,** see page 593.

20. Evaluate 5^3. *(Lesson 1-4)*

CHAPTER 9

Mid-Chapter Self Test

Classify each angle as *acute*, *obtuse*, *right*, or *straight*. *(Lesson 9-1)*

1.

2.

Determine which figures are polygons. If the figure is a polygon, name it and tell whether it is a regular polygon. If the figure is *not* a polygon, explain why. *(Lesson 9-2)*

3.

4.

5. Find the value of *x* in the similar triangles at the right. *(Lesson 9-3)*

4 cm
7 cm
3 cm
x cm

COOPERATIVE LEARNING

9-3B Dilations

A Follow-Up of Lesson 9-3

Some copy machines can reduce and enlarge images. In mathematics, the process of reducing and enlarging a figure is a transformation called a *dilation*.

TRY THIS

- Place a piece of grid paper over a cartoon or picture you want to enlarge. Trace the picture. It may help to put the paper against a window to see the image more clearly.

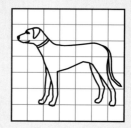

- On another piece of grid paper, use a colored pencil to draw horizontal lines every two squares. Then draw vertical lines every two squares.

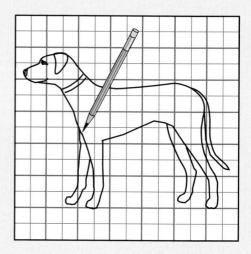

- Now sketch the parts of the figure contained in each small square of your original picture onto each large square of the grid you created.

- Place a piece of white paper over your new drawing. Trace the picture and then color it.

ON YOUR OWN

1. Has the figure been enlarged or reduced? By how much has the height changed?
2. Is the enlargement similar to the original drawing? Explain your reasoning.
3. What type of grid would you use to *reduce* a picture?

9-4A Investigating Triangles and Quadrilaterals

A Preview of Lesson 9-4

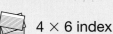 4 × 6 index cards

scissors

brass fasteners

 protractor

In this lab, you will investigate the characteristics of special triangles and quadrilaterals. Before you begin the lab, cut $\frac{1}{2}$-inch wide strips from the index cards. You will need four 3-inch long strips, four 5-inch long strips, and four 6-inch long strips.

TRY THIS

Work with a partner.

1 • Use your paper strips and fasteners to make three triangles *A*, *B*, and *C*.

  Triangle *A* has three congruent sides.

  Triangle *B* has exactly two congruent sides.

  Triangle *C* has no congruent sides.

  • Measure the angles of your triangles.

  • Compare your triangles with those of other groups.

  • What conclusions can you draw about the angles of a triangle with three congruent sides? two congruent sides? no congruent sides?

2 • Choose four congruent paper strips. Join them with brass fasteners to form a quadrilateral with four right angles. This is quadrilateral *D*.

  • Shift your figure into a different-shaped quadrilateral. Call it quadrilateral *E*. Measure its angles. What pattern do you notice?

  • Make quadrilateral *F* from two congruent paper strips and two other congruent paper strips. Form a quadrilateral with four right angles.

  • Shift your figure into a different-shaped quadrilateral. This is quadrilateral *G*. Measure its angles. What pattern do you notice?

  • Compare your findings with those of other groups. What conclusions can you draw about the angles of quadrilaterals like quadrilateral *E*? like quadrilateral *G*?

ON YOUR OWN

1. You were able to shift your quadrilaterals so that the angles changed. Are you able to shift a triangle into a different triangle? Try it by making several different triangles.

2. Suppose you were building a bookcase. The outside is in the shape of a rectangle.

  a. Would you expect the rectangle to be rigid or would you expect it to shift?

  b. How could you make the bookcase more rigid?

3. *Make a conjecture* about why triangles are often used in the construction of buildings.

Triangles and Quadrilaterals

What you'll learn

You'll learn to classify triangles and quadrilaterals.

When am I ever going to use this?

You'll use the properties of triangles and quadrilaterals when you do sewing or carpentry projects.

Word Wise

acute triangle
right triangle
obtuse triangle
scalene triangle
isosceles triangle
equilateral triangle
rhombus
trapezoid

The drawing shows how scientists classify living things. An organism is a single individual. A population is all of the individuals of one species with the same characteristics. A community is made up of populations of different species.

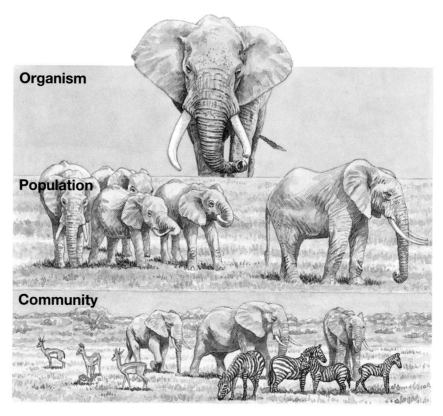

In geometry, you can classify polygons in a similar manner. For example, triangles can be classified by their angle measures. Every triangle has two acute angles. You can classify a triangle using the third angle.

acute
all angles acute

right
1 right angle

obtuse
1 obtuse angle

Triangles can also be classified by the number of congruent sides they have.

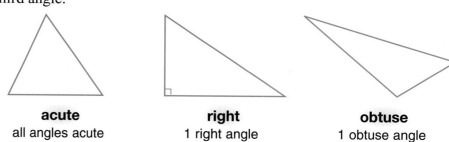

scalene
no congruent sides

isosceles
at least 2 congruent sides

equilateral
3 congruent sides

Example

Real World APPLICATION

① **Architecture** Triangles are used in the design of many office buildings to secure them in case of high winds or earthquakes. Classify each triangle by its angles and by its sides.

a. △ABC

△ABC is acute and isosceles.

b. △BCD

△BCD is obtuse and scalene.

c. △ACE

△ACE is right and scalene.

You are already familiar with three types of quadrilaterals. They are squares, rectangles, and parallelograms. Two other types of quadrilaterals are the **rhombus** and the **trapezoid**. The chart below shows one way to classify quadrilaterals.

The best description of a quadrilateral is the one that is the most specific.

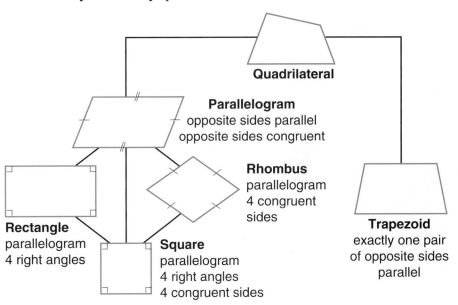

Quadrilateral

Parallelogram
opposite sides parallel
opposite sides congruent

Rhombus
parallelogram
4 congruent sides

Trapezoid
exactly one pair
of opposite sides
parallel

Rectangle
parallelogram
4 right angles

Square
parallelogram
4 right angles
4 congruent sides

Examples

Name every quadrilateral that describes each figure. Then state which name best describes the figure.

②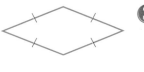

quadrilateral
parallelogram
rhombus

Rhombus best describes this figure.

③

quadrilateral
trapezoid

Trapezoid best describes this figure.

④

quadrilateral
parallelogram
rectangle

Rectangle best describes this figure.

Communicating Mathematics

Read and study the lesson to answer each question.

1. **Draw** a sketch of an isosceles right triangle.

2. **Tell** why all squares are rectangles, but not all rectangles are squares.

3. **Write** a short paragraph that tells how a rhombus and a square are alike and how they are different.

Guided Practice

Classify each triangle by its angles and by its sides.

4.

5.

Name every quadrilateral that describes each figure. Then underline the name that best describes the figure.

6.

7.

8. **Architecture** The figure at the right shows the basic design for a geodesic dome. What kinds of triangles are found in the figure?

Practice

Classify each triangle by its angles and by its sides.

9.

10.

11.

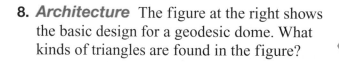

Name every quadrilateral that describes each figure. Then underline the name that best describes the figure.

12.

13.

14.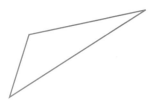

15. Draw a quadrilateral that is a rhombus, but not a rectangle.

16. Which quadrilaterals have four right angles?

17. Three angles of a triangle measure 30°, 60°, and 90°. Classify the triangle by its angles.

18. Three sides of a triangle measure 5 inches, 8 inches, and 8 inches. Classify the triangle by its sides.

19. *Algebra* The sum of the measures of the angles of a triangle is 180°. Find the missing measure(s) in each figure.

a.

b.

c.

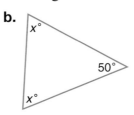

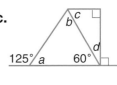

20. *Sewing* Quilting has become a popular art form, but it began for practical purposes. In the American colonies, new fabric was scarce. So, quilt makers stitched together scraps of worn-out clothing to make patterned quilt tops.

   a. Classify the triangles and quadrilaterals you see in the photo at the right.

   b. Explain how knowing the characteristics of different shapes can help in making a quilt.

21. *Critical Thinking* Determine whether each statement is *always*, *sometimes*, or *never* true.

   a. A trapezoid is a quadrilateral.

   b. A quadrilateral is a trapezoid.

   c. A rectangle is a parallelogram.

   d. A rectangle is a square.

   e. A square is an equilateral triangle.

Mixed Review

22. **Standardized Test Practice** Which best represents a pair of similar figures? *(Lesson 9-3)*

A

B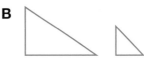

C

D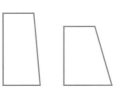

23. Estimate $1\frac{7}{12} \times 12\frac{1}{5}$. *(Lesson 7-1)*

24. *Algebra* The temperature in Buffalo, N.Y. dropped to −5°F. Find this temperature in degrees Celsius by using the formula $F = \frac{9}{5}C + 32$. *(Lesson 6-3)*

For **Extra Practice**, see page 593.

9-4B Using Logical Reasoning

A Follow-Up of Lesson 9-4

It's Friday morning before school, and Vince and Carlos are both studying for quizzes. How do they know they'll be having quizzes? Let's listen in!

Carlos

My science teacher has given us a quiz every Friday for the last five Fridays. I'm sure she'll be giving us a quiz today!

Vince

On the first day of school, my social studies teacher told us that he would give us a quiz every Friday. I know I'll have a quiz today!

THINK ABOUT IT

Work with a partner.

1. **Compare and contrast** Vince's and Carlos' reasoning.

2. When you use *inductive reasoning*, you make a rule after seeing several examples. Who was using inductive reasoning?

3. When you use *deductive reasoning*, you use a rule to make a decision. Who was using deductive reasoning?

4. **Use logical reasoning** to decide whether each situation is an example of inductive or deductive reasoning.

   a. *A number is divisible by 4 if its last two digits make a number that is divisible by 4. So, 15,624 is divisible by 4.*

   b. *For many years, the swallows of San Juan Capistrano have returned to the mission on March 19. So, every year, a celebration is planned for March 19.*

For **Extra Practice,** see page 594.

ON YOUR OWN

5. *Draw* several rectangles and measure their diagonals.

   a. What can you conclude about the diagonals of rectangles?

   b. Did you use deductive or inductive reasoning?

6. *Explain* how the problem-solving strategy *look for a pattern* is like inductive reasoning.

7. *Write* about a situation in which you use either inductive or deductive reasoning.

MIXED PROBLEM SOLVING

Solve. Use any strategy.

STRATEGIES

Look for a pattern.
Solve a simpler problem.
Act it out.
Guess and Check.
Draw a diagram.
Make a chart.
Work backward.

8. *Puzzles* Brett, Conrad, and Alton play safety, running back, and quarterback on a football team, but not necessarily in that order. Brett and the quarterback drove Alton to practice on Saturday. Brett does not play safety. Who is the safety?

9. *Number Theory* The figures represent triangular numbers. Use the pattern to draw figures for the next two triangular numbers.

10. *Geometry* Trisha reads in a geometry book that the diagonals of a square are perpendicular to each other.

   a. If Trisha draws a square and its diagonals, what should she expect to be true about the diagonals?

   b. Is this an example of inductive or deductive reasoning?

   c. Suppose Trisha had drawn several squares before reading her geometry book and found that their diagonals were perpendicular. Was she using inductive or deductive reasoning?

11. Is the average of 96.8, 68.3, 79.3, 125.2, 138.7, and 101.5 about 101.7 or 10.17? Explain your reasoning.

12. *Physical Science* Luisa was trying to find the relationship between the time it took a pendulum to swing back and forth and its length. The data are listed in the table.

Time (seconds)	Length (units)
1	1
2	4
3	9
4	16

Apply inductive reasoning to predict the length of a pendulum if its time is 5 seconds.

13. Standardized Test Practice Which construction is shown in the drawing?

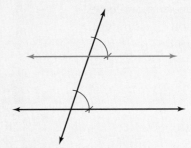

A bisect an angle

B construct a perpendicular line through a point on a given line segment

C construct a line parallel to a given line through a point not on the line

D construct a segment congruent to a given line segment

Tessellations

Checkmate! In 1997, a supercomputer nicknamed "Deep Blue" stunned the world of chess by beating Garry Kasparov. This was the first time a computer was able to defeat the best human player in a chess match.

Although computer chess is a recent development, historians believe that the game of chess was invented in India almost 1,500 years ago. Even though the rules of chess are very complicated, the game is played on a simple game board made of squares. In a chessboard, the squares cover the entire surface without any overlaps or gaps. In mathematics, this is called a **tesselation**. A tessellation can be made of one kind of polygon or several kinds of polygons.

In the Mini-Lab, you'll investigate which regular polygons can be used to make a tessellation.

HANDS-ON MINI-LAB

Work with a partner.   pattern blocks

Try This

1. Arrange the triangles so that they make a tessellation.
2. Make a sketch of the pattern.
3. Find the measure of each angle of the triangle.
4. Find a point in your pattern where the vertices of several triangles meet. How many triangles are around this point?
5. Do equilateral triangles make a tessellation?
6. Repeat Steps 1–5 using squares and then hexagons.
7. Organize your data in a table.

Polygon	Measure of Angle	How Many Polygons Around a Point?	Make a Tessellation?
Triangle Square Hexagon			

Talk About It

8. Find the sum of the measures of the angles of the polygons whose vertices meet at a point.
9. Predict whether a regular pentagon will make a tessellation. Explain your reasoning.

In the Mini-Lab, you found that the sum of the measures of the angles where the vertices meet is 360°.

6 × 60° = 360°

4 × 90° = 360°

3 × 120° = 360°

Example

INTEGRATION

1 **Algebra** The sum of the measures of the angles of a pentagon is 540°. Can a regular pentagon make a tessellation?

Explore   You know that the sum of the measures of the angles of a pentagon is 540°. You need to find if regular pentagons can make a tessellation.

Plan   Each angle of a regular pentagon has a measure of 540 ÷ 5 or 108°.

To find out if a pentagon tessellates, solve $108n = 360$, where n is the number of angles at a vertex.

Solve $108n = 360$ 360 ÷ 108 = *3.333333333*

$n \approx 3.33$

The solution is not a whole number. So, a regular pentagon will not make a tessellation.

Examine   Check your answer by making a drawing.

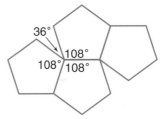

It is often possible to make a tessellation with a combination of polygons.

Example

2 **Make a tessellation with hexagons and triangles.**

Each angle of a regular hexagon measures 120°. Use 1 hexagon.

total at vertex	−	*one angle of hexagon*		
360°	−	120°	=	240°

Now determine how many triangles will fit in 240°. Each angle of an equilateral triangle measures 60°.

240° =   ?   triangles

240° = 4 triangles   *60° × 4 = 240°*

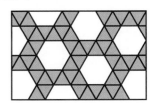

In the tessellation, there are four triangles touching each vertex of each hexagon.

Communicating Mathematics

Read and study the lesson to answer the following.

1. **Tell** how you know when a regular polygon can be used by itself to make a tessellation.

2. **Explain** how you know that the rhombus at the right can be used to make a tessellation.

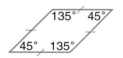

HANDS-ON MATH

3. In Example 2 on page 389, you made a tessellation with 1 hexagon and 4 triangles.
   a. Use pattern blocks to find a different tessellation made with hexagons and triangles.
   b. Write an addition problem showing that the sum of the measures of the angles where the vertices meet is 360°.

Guided Practice

4. The sum of the measures of the angles of a regular octagon is 1,080°.
   a. Determine whether an octagon can be used by itself to make a tessellation.
   b. Verify your results by finding the number of angles at a vertex.

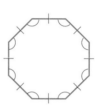

5. **Design** Describe a tessellation that you have seen.

EXERCISES

Practice

Determine whether each polygon can be used by itself to make a tessellation. Verify your results by finding the measures of the angles at a vertex. The sum of the measures of the angles of each polygon is given.

6. heptagon; 900°   7. nonagon; 1,260°   8. decagon; 1,440°

9. Sketch a tessellation made with right triangles.

The following regular polygons tessellate. Determine how many of each polygon you need at each vertex. Then sketch the tessellation.

10. triangles and squares

11. squares and octagons

Applications and Problem Solving

12. **Art** Ancient Greeks used marble, alabaster, and granite for the designs in their mosaics. Identify the shapes in the mosaic shown at the right.

13. *Life Science* One of the most famous tessellations found in nature is a bee's honeycomb. Explain one advantage of using hexagons in a honeycomb.

14. *Working on the* 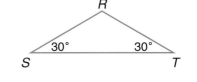 Create a tessellation on isometric dot paper. Use various colors to make a geometric design.

15. *Critical Thinking* You can make a tessellation with equilateral triangles. Can you make a tessellation with any isosceles or scalene triangle? If so, explain your reasoning and make a drawing of your tessellation.

Mixed Review

16. **Standardized Test Practice** Which procedure could best be used to find the measure of angle R? *(Lesson 9-4)*

    A Add 30° to 180°.

    B Subtract 60° from 180°.

    C Subtract 30° from 90°.

    D Add 30° to 90°.

    E Subtract 180° from 60°.

17. *Earth Science* The temperature of the surrounding air cools by 3.5°F for each 1,000 feet that a hot air balloon rises. If the ground temperature is 65°F, the equation $y = -3.5x + 65$ gives the air temperature. In the equation, x is the height of the balloon in thousands of feet and y is the air temperature. Make a table of values for heights of 1,000, 2,000, and 3,000 feet. *(Lesson 6-7)*

For **Extra Practice,** see page 594.

Let the Games Begin

Tic-Tac Squares

Math Skill	
Tessellations	

Get Ready This game is for two players.

    8 red counters   8 yellow counters   dot paper

Get Set Copy the game board onto dot paper.

Go
- The first player covers any black dot with a counter. Then, players alternate turns.

- The object of the game is to be the first player to cover the four vertices of a square. Note that the sides of a square don't have to be vertical or horizontal. The square can be "tilted" to one side.

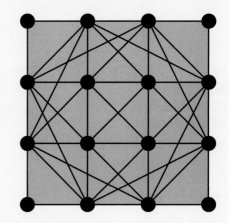

 Visit www.glencoe.com/sec/math/mac/mathnet for more games.

Translations

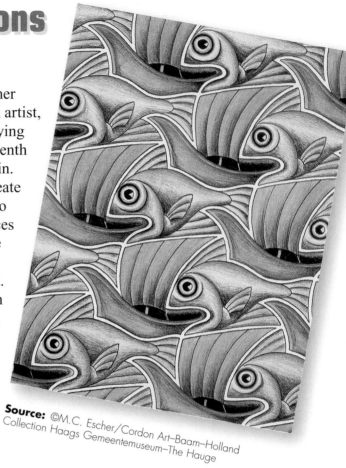

What you'll learn

You'll learn to create Escher-like drawings by using translations.

When am I ever going to use this?

You'll use translations when you make designs.

Word Wise

translation

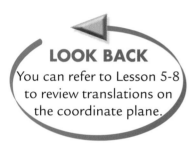

LOOK BACK

You can refer to Lesson 5-8 to review translations on the coordinate plane.

Maurits Cornelis Escher (1898–1972), a Dutch artist, spent many days studying the Alhambra, a thirteenth century palace in Spain. He was inspired to create recognizable figures to fill space like the pieces of stone that filled the surface of the mosaic walls in the Alhambra. His figures were often in the shapes of birds, fish, or reptiles.

Many of Escher's sketches began as tessellations of polygons. You can make Escher-like drawings by making changes in the polygons of the tessellation. One way to do this is by using a **translation**.

A translation is sliding part of a drawing to another place without turning it. The square below has the left side changed. To make sure the pieces, or pattern units, will tessellate, slide or translate that change to the opposite side and copy it.

inter NET
CONNECTION
For more information about Escher designs, visit:
www.glencoe.com/sec/math/mac/mathnet

Now change all of the squares in a tessellation the same way. The tessellation takes on Escher-like qualities when you use different colors or designs.

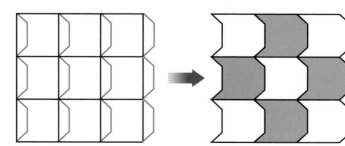

Example ❶

CONNECTION

A cardboard pattern unit can help in creating the tessellation.

Art Draw a tessellation using the change shown at the right.

First, complete the pattern unit.

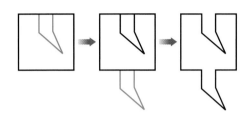

Translate the change to all squares in the tessellation. Use color to complete the effect.

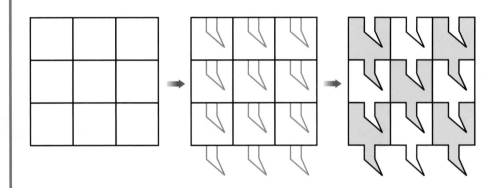

You can make more complex tessellations by doing two translations.

Example ❷

CONNECTION

Art Draw a tessellation using both changes shown at the right.

First, complete the pattern unit.

Study Hint

Reading Math When a figure is translated, the original figure and the translated figure have the same size and shape. They are called congruent figures.

Then complete the tessellation.

Communicating Mathematics

Read and study the lesson to answer each question.

1. *Write* a definition of a translation.

2. *Tell* how a translation can be used to form an Escher-like drawing.

3. *You Decide* Omar thinks that the puzzle piece will make a tessellation. Talutah disagrees. Who is correct? Explain your reasoning.

Guided Practice

Complete the pattern unit for each translation. Then draw the tessellation.

4.

5.

EXERCISES

Practice

Complete the pattern unit for each translation. Then draw the tessellation.

6.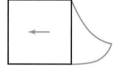

7.

8.

9. Draw a pattern unit that has a parallelogram as the basic tessellation.

Applications and Problem Solving

10. *Animation* In the first full-length film made completely with computer graphics, animators first draw the characters to look like wire skeletons and then change the figures with computers. Explain how translations can be used in computer graphics.

11. *Art* Animals frequently appear in Escher designs. Research Mr. Escher and identify one animal that was used in his designs. Tell what polygon might have been used as the basis for the design.

12. *Working on the* **CHAPTER Project** Use translations to design a pattern for an Escher-like drawing. Then draw the tessellation. Use different colors or textures to create an interesting design.

For **Extra Practice,** see page 594.

13. *Critical Thinking* Is it possible to make a tessellation with translations by using equilateral triangles? Explain your reasoning.

Mixed Review

14. *Art* Marlene wishes to construct a tessellation for a wall-hanging made only from regular decagons. Is this possible? *(Lesson 9-5)*

15. **Standardized Test Practice** Jarina had 4.65 pounds of sugar. She used 0.86 pound in a recipe. How much sugar did she have left? *(Lesson 1-1)*

    A 3.79 lb   **B** 4.35 lb   **C** 4.62 lb   **D** 5.51 lb   **E** Not Here

9-7 Reflections

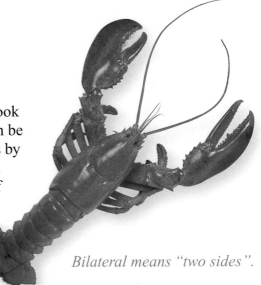

What you'll learn

You'll learn to create Escher-like drawings by using reflections.

When am I ever going to use this?

You'll use symmetry when you classify animals in science.

Word Wise

line symmetry
line of symmetry
reflection

Most animals have bodies that look the same on both sides. They can be divided into right and left halves by drawing an imaginary line down the length of the body. Each half is a mirror image of the other half. Scientists call this bilateral symmetry.

Bilateral means "two sides".

In mathematics, figures that match exactly when folded in half have **line symmetry**. The figures below have line symmetry. Some figures can be folded in more than one way to show symmetry. Each fold line is called a **line of symmetry**.

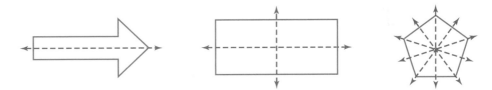

Examples

Determine which figures have line symmetry. Draw all of the lines of symmetry.

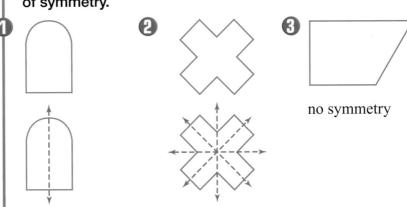

① ② ③ no symmetry

You can create figures that have line symmetry by using a **reflection**. A reflection is a mirror image of a figure across a line of symmetry.

Mr. Escher also used reflections in some of his works. You can create different types of drawings using reflections. However, in these tessellations, two pattern units are used.

Example 4

CONNECTION

Art Complete an Escher-like drawing using the change shown at the right.

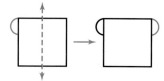

Complete the first pattern unit by drawing the reflection of the design on another side of the square.

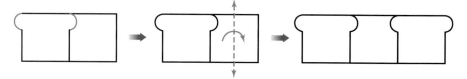

Study Hint

Problem Solving Look for a pattern to determine the two different units of the tessellation.

Now add another square. Reflect the new pattern in the second square.

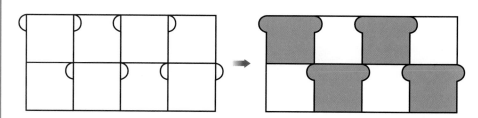

Continue this process to complete the tessellation.

CHECK FOR UNDERSTANDING

Communicating Mathematics

Read and study the lesson to answer each question.

1. **Tell** how a line of symmetry is related to a reflection.

2. **Explain** how a reflection is different from a translation.

Guided Practice

3. The figure below is a regular hexagon. Copy the figure and draw all lines of symmetry.

4. **Art** Complete both pattern units for the reflection shown below. Then draw the tessellation.

Practice **Copy each figure. Draw all lines of symmetry.**

5.  6.   7.

8. How many lines of symmetry does an equilateral triangle have?

Applications and Problem Solving

9. *Art* Complete the tessellation described by the pattern shown at the right.

10. *Crossword Puzzles* Some crossword puzzles are designed so that the pattern of black and white squares looks the same upside down as right-side up. These puzzles have *half-turn symmetry*. In addition, some puzzles also have line symmetry. How many lines of symmetry are there in the puzzle at the right?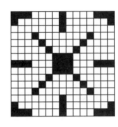

11. *Working on the* **CHAPTER Project** Use reflections to design a pattern for an Escher-like drawing. Then draw the tessellation. Use colors and/or textures to create an interesting pattern.

12. *Critical Thinking* A starfish has *radial symmetry* and certain kinds of sponges are *asymmetrical*. Study the photos and write definitions of radial symmetry and asymmetrical.

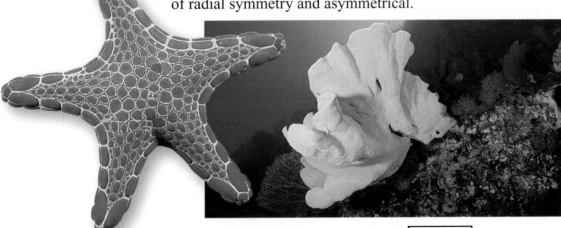

Mixed Review

13. *Art* Complete the pattern unit for the translation at the right. Then draw the tessellation. *(Lesson 9-6)*

14. **Standardized Test Practice** The Wee Folk Furniture Company produces furniture for children that is a reduced version of adult furniture. The top of a full-sized desk measures 54 inches long by 36 inches wide. If the top of the child's desk is 24 inches wide, what is the length? *(Lesson 9-3)*

   A 16 in.   **B** 36 in.   **C** 48 in.   **D** 54 in.   **E** 60 in.

For **Extra Practice,** see page 595.

CHAPTER 9

Study Guide and Assessment

 **interNET**
CONNECTION Chapter Review **For additional lesson-by-lesson review, visit:**
www.glencoe.com/sec/math/mac/mathnet

Vocabulary

After completing this chapter, you should be able to define each
term, concept, or phrase and give an example or two of each.

Geometry
acute angle (p. 362)
acute triangle (p. 382)
angle (p. 362)
complementary (p. 363)
congruent (p. 371)
decagon (p. 370)
degree (p. 362)
dilation (p. 380)
equilateral (p. 382)
heptagon (p. 370)
hexagon (p. 370)
indirect
    measurement (p. 377)
inscribed (p. 374)

isosceles (p. 382)
line of symmetry (p. 395)
line symmetry (p. 395)
nonagon (p. 370)
obtuse angle (p. 362)
obtuse triangle (p. 382)
octagon (p. 370)
pentagon (p. 370)
polygon (p. 370)
protractor (p. 360)
quadrilateral (p. 370)
reflection (p. 395)
regular polygon (p. 371)
rhombus (p. 383)
right angle (p. 362)

right triangle (p. 382)
scalene (p. 382)
similar (p. 376)
straight angle (p. 362)
supplementary (p. 363)
tessellation (p. 388)
translation (p. 392)
transversal (p. 361)
trapezoid (p. 383)
triangle (p. 370)
vertex (p. 362)

Problem Solving
use logical reasoning
    (p. 386)

Understanding and Using the Vocabulary

Choose the correct term or number to complete each sentence.
1. The point where the sides of an angle meet is called the (ray, vertex).
2. An (acute, obtuse) angle has a measure less than 90°.
3. If the sum of the measures of two angles is (90°, 180°), the angles are supplementary
4. A polygon with six sides is called a (heptagon, hexagon).
5. A polygon with congruent sides and congruent angles is called a (regular, similar) polygon.
6. A scalene triangle has (2, 0) congruent sides.
7. An (isosceles, equilateral) triangle has three congruent sides.
8. The sum of the angle measures at the vertex of any tessellation is (180°, 360°).
9. Figures that match exactly when folded in half have (translation, line symmetry).
10. A (reflection, tessellation) is a mirror image of a figure across a line of symmetry.

In Your Own Words
11. *Explain* how a translation can be used in constructing a tessellation.

Objectives & Examples

Upon completing this chapter, you should be able to:

● classify angles *(Lesson 9-1)*

Classify the angle.

The angle above is an acute angle because its measure is less than 90°.

● identify polygons and regular polygons *(Lesson 9-2)*

Name the figure and tell whether it is regular.

The figure above is a six-sided polygon. It is a hexagon. It is not regular because its sides are not congruent.

● determine whether polygons are similar and find a missing length in a pair of similar polygons *(Lesson 9-3)*

Tell whether the pair of polygons is similar.

6 cm

3 cm   2 cm

1 cm

The polygons above are similar because corresponding angles are congruent and corresponding lengths are in proportion.

$$\frac{1}{2} = \frac{3}{6}$$

Review Exercises

Use these exercises to review and prepare for the chapter test.

Classify each angle as *acute*, *obtuse*, *right*, or *straight*.

12. 49°   **13.** 90°

14. 180°   **15.** 113°

16.   **17.**

Determine which figures are polygons. If the figure is a polygon, name it and tell whether it is regular. If the figure is *not* a polygon, explain why.

18.   **19.**

20.   **21.**

Tell whether each pair of polygons is similar. Justify your answer.

22.   **23.**
10 cm 16 cm   3 ft   2 ft
5 cm 8 cm   3 ft   2 ft

Find the value of *x* in each pair of similar polygons.

24. 4 m   *x* m   **25.**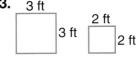
7 m   6 m   4 ft   *x* ft
                    3 ft
              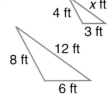
              8 ft   12 ft
                    6 ft

Objectives & Examples

Review Exercises

● classify triangles and quadrilaterals *(Lesson 9-4)*

Classify the triangle.

The figure above has 3 congruent sides, and all of its angles are acute. It is an acute equilateral triangle.

Classify each triangle by its angles and by its sides.

26. **27.**

Name every quadrilateral that describes each figure. Then underline the name that best describes the figure.

28. **29.**

● determine which regular figures can be used to form a tessellation *(Lesson 9-5)*

Determine whether regular pentagons (540°) can be used to make a tessellation.

$$540 \div 5 = 108$$

No, because each angle measures 108° and 108 is not a factor of 360.

Determine whether each polygon can be used by itself to make a tessellation. Verify your results by finding the measures of the angles at a vertex.

30. hexagon; 720° **31.** octagon; 1,080°

32. Equilateral triangles and squares tessellate. Determine how many of each polygon are needed at each vertex. Then sketch the tessellation.

● create Escher-like drawings by using translations *(Lesson 9-6)*

Complete the pattern and then the tessellation.

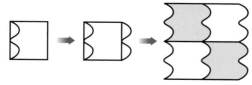

Complete the pattern unit for each translation. Then draw the tessellation.

33. **34.**

● create Escher-like drawings by using reflections *(Lesson 9-7)*

Draw all lines of symmetry in the figure.

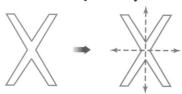

Copy each figure. Draw all lines of symmetry.

35. **36.**

Applications & Problem Solving

37. Food Angelo's Pizza Parlor shapes its pizzas as squares. After baking, the pizzas are cut along one diagonal into two triangles. Describe completely the triangles that result. *(Lesson 9-4)*

38. Logical Reasoning Use the pattern below to draw the next two figures in the sequence. *(Lesson 9-4B)*

39. Crafts Edwyna is piecing together a quilt from fabric pieces in the shapes of hexagons and equilateral triangles. How many of each of the shapes will she need at each vertex in the tessellation created by the fabric pieces? Sketch the tessellation. *(Lesson 9-5)*

40. Sports A tennis court and a badminton court are shaped like rectangles. Their dimensions are shown in the graph. Determine whether the rectangles are similar. *(Lesson 9-3)*

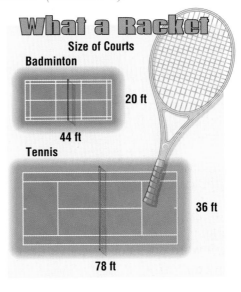

What a Racket

Size of Courts
Badminton
20 ft
44 ft
Tennis
36 ft
78 ft

Alternative Assessment

Open Ended

Suppose you want to find the height of a tall tree. It is not possible to climb the tree to measure it, but you notice the tree is casting a shadow. You also notice that there is a 4-foot fence post next to it casting a shadow. How can you determine the height of the tree? What other information do you need?

If the shadow of the tree is 36 feet long and the shadow of the fence post is 3 feet long, how tall is the tree?

A practice test for Chapter 9 is provided on page 615.

Completing the CHAPTER Project

Use the following checklist to make sure your project is complete.

☑ You have created three designs.

☑ Your presentation includes an explanation of the mathematics that is used in your design.

Add any finishing touches you would like to make your design more attractive.

PORTFOLIO Select some of your work from this chapter that shows your creativity. Place it in your portfolio.

Section One: Multiple Choice

There are ten multiple choice questions in this section. Choose the best answer. If a correct answer is *not here,* choose the letter for Not Here.

1. Which is the equation for the line graphed?

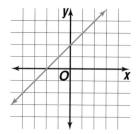

A $y = x + 2$
B $y = x - 2$
C $y = 2x$
D $y = 2x + 2$

2. Which word does not have a vertical line of symmetry?

F H
    O
    W

H M
    A
    T
    T

G B
    O
    Y

J W
    H
    O

3. A doorway is 5 inches less than 8 feet tall. How many inches is this?

A 101 inches

B 96 inches

C 91 inches

D 40 inches

4. Which expression represents *three less than a number?*

F $n - 3$

G $3 - n$

H $3n$

J $\dfrac{n}{3}$

5. The triangles below are congruent. The measures of some of the angles and some of the sides are shown.

What is the value of *y*?

A 55
B 67
C 48
D 65

Please note that Questions 6–10 have five answer choices.

6. The stem-and-leaf plot shows the prices for tickets to the summer concerts at a local amphitheater.

Stem	Leaf
1	5 9
2	2 4 7 8 9
3	0 2 2
4	1
5	
6	2 5 *1|5 = $15*

Into which price range do most of the tickets fall?

F $10–$19

G $20–$29

H $30–$39

J $40–$49

K $50–$59

7. Which expression is equivalent to $3.2 \times (2.4 \times 5.8)$?

A $(3.2 \times 2.4) + (3.2 \times 5.8)$

B $(3.2 \times 2.4) \times 5.8$

C $3.2 \times (2.4 + 5.8)$

D $2.4 + (3.2 \times 5.8)$

E Not Here

8. Colin had $\frac{7}{8}$ of a tank of gas in the lawn mower. After mowing the lawn, he had $\frac{1}{4}$ of a tank. How much gas did Colin use mowing the lawn?

  F $\frac{6}{4}$ of a tank

  G $\frac{3}{4}$ of a tank

  H $\frac{5}{8}$ of a tank

  J $\frac{3}{8}$ of a tank

  K $\frac{1}{8}$ of a tank

9. A recycling group at the middle school collected cans for a service project. They collected 122.4 pounds, 88.9 pounds, and 117.02 pounds in the last three weeks. What was the total amount collected for that period?

  A 248.15 lb

  B 211.13 lb

  C 328.32 lb

  D 328.5 lb

  E Not Here

10. On Monday, Fiona worked $5\frac{7}{12}$ hours. On Wednesday, she worked $6\frac{3}{4}$ hours. How many hours did Fiona work altogether?

  F $13\frac{5}{8}$ h

  G $12\frac{1}{3}$ h

  H $11\frac{5}{8}$ h

  J $11\frac{1}{3}$ h

  K Not Here

Test Practice For additional test practice questions, visit:

www.glencoe.com/sec/math/mac/mathnet

Test-Taking Tip THE PRINCETON REVIEW

When taking a standardized test, you may be able to eliminate answer choices through estimating. Also, look to see which answers are *not* reasonable for the information given in the problem.

Section Two: Free Response

This section contains four questions for which you will provide short answers. Write your answers on your paper.

11. The table shows the values of p and q, where the values of p and q form a proportion.

p	5	10	Z
q	9	Y	63

  What are the values of Y and Z?

12. In the figure, line a is parallel to line b. Name two supplementary angles.

  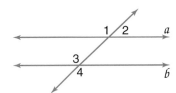

13. Triangles XYZ and RST are similar. Find the value of x.

  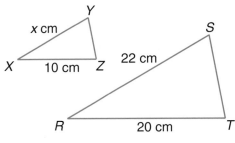

14. Find $\frac{3}{4} + \frac{1}{2}$.

Pi FOR POLYGONS

You have already learned that circles have a special ratio called *pi*. Pi is the ratio of the circumference of a circle to twice its radius. Could there be a special ratio for polygons that compares the perimeter of a polygon to its "radius"?

What You'll Do

In this investigation, you will construct and measure regular polygons to decide whether regular polygons have a special ratio like pi.

Materials protractor ruler compass

 construction paper calculator

Procedure

1. Work in groups of four. Assign each member of your group one pair of these numbers: 3 and 10, 4 and 9, 5 and 8, or 6 and 7. These numbers will represent the number of sides of regular polygons.

2. Work alone. Construct large regular polygons for your pair of numbers. Find the center of each polygon. This is the point in the interior that is equal in distance from each vertex. Measure the distance from the center to a vertex. This length will be called the *radius* of the polygon.

radius

3. Work in your group. Prepare a table similar to the one below for all of the polygons from your group. Use a calculator to find the values in the last two columns.

4. Compare the tables from all of the groups. How do the values in the last column compare?

5. Write expressions for the perimeter of each regular polygon using P for the perimeter and $2r$.

Polygon	Number of Sides	Perimeter (P)	Radius (r)	2r	$\frac{P}{2r}$

ⓜaking the Connection

Use the information from your table as needed to help in these investigations.

Language Arts

Design a poster to display your polygons, table, and formulas. Explain how you found the formulas for the regular polygons.

Social Studies

Research Carl Friedrich Gauss and his theory about constructing regular polygons. How did a heptadecagon affect his life?

Science

Honeycombs contain regular hexagons. Investigate other occurrences of regular polygons in nature.

ⓖo Further

- Using geometry software, construct regular polygons with more than ten sides. Make a conjecture about the ratio of the perimeter of a polygon to twice its radius.

- Investigate the measure of one interior angle of a regular polygon as the number of sides of the polygon increases. Make a conjecture about the angle measure.

Technology Tips

- Use a **spreadsheet** to calculate the values for the last two columns of the table.

- Use **geometry software** to construct and measure regular polygons.

🖥 *inter*NET
CONNECTION For more information on π, visit:
www.glencoe.com/sec/math/mac/mathnet

PORTFOLIO
You may want to place your work on this investigation in your portfolio.

Geometry: Exploring Area

What you'll
learn in Chapter 10

- to solve problems by using guess and check,

- to find the squares and square roots of numbers,

- to find length using the Pythagorean Theorem,

- to find the areas of irregular figures, triangles, trapezoids, and circles, and

- to find probability using area models.

CHAPTER Project

IT'S A SMALL WORLD

In this project, you will make a simple map of Earth and use geometry to approximate the area of landmasses and bodies of water. You will determine what part of Earth is covered in water. You can use colored pencils to draw your map on poster board or a large sheet of paper.

Getting Started

- Find a world map. Using the map as a guideline, sketch the outlines of the landmasses on your poster board or paper. Keep in mind that the existing map may be a different size than the one you are drawing. Be sure to make any necessary adjustments.

- Label the landmasses and the bodies of water with their names.

Technology Tips

- Use an **electronic encyclopedia** to find a map.

- Use a **calculator** to help you approximate the areas of the landmasses and the bodies of water.

- Use a **word processor** to write information about the map.

 **inter**NET
CONNECTION Research For information on world maps, visit:

www.glencoe.com/sec/math/mac/mathnet

Working on the Project

You can use what you'll learn in Chapter 10 to help you make your map.

Page	Exercise
426	16
431	24
441	18
445	Alternative Assessment

10-1A Guess and Check

A Preview of Lesson 10-1

The Los Angeles Kodiaks played the Massachusetts Chariots in the Second National Junior Wheelchair Basketball Tournament. The leading scorer was Willie Arriago with 30 points. He missed 20 of his 36 attempts. All of his shots were worth either one or two points. Quesone and Charo wonder how many one-point shots and how many two-point shots Willie made. Let's listen in!

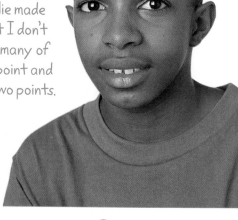

I know that Willie made 36 − 20 or 16 shots, but I don't know how to determine how many of the shots were worth one point and how many were worth two points.

Sometimes when I don't know how to do a problem, I make a guess until I find one that works.

Charo

Quesone

What do you mean?

Let's guess that Willie made 6 one-point shots and 10 two-point shots.

THINK ABOUT IT

Work with a partner.

1. **Analyze** Charo's thinking. How do you think she came up with her two numbers? Does this guess result in 30 points? If not, how should Charo adjust her guess?

2. **Make** a table to show different guesses to solve this problem. How many one-point shots and two-point shots did Willie make?

3. **Apply** the **guess and check** strategy to solve the following problem.

   The Pike's Peak souvenir shop sells standard-sized postcards in packages of 5 and large-sized postcards in packages of 3. If Iku bought 16 postcards, how many packages of each did she buy?

For **Extra Practice,** see page 595.

ON YOUR OWN

4. The third step of the 4-step plan for problem solving tells you to *solve* the problem. *Tell* why, during this step in the guess and check strategy, you must keep a careful record of each of your guesses and their results.

5. *Write a Problem* which you could solve using the guess and check strategy.

6. *Look Ahead* Explain how the guess and check strategy could be used to find a number that when multiplied by itself produces a product of 256.

MIXED PROBLEM SOLVING

STRATEGIES

Look for a pattern.
Solve a simpler problem.
Act it out.
Guess and check.
Draw a diagram.
Make a chart.
Work backward.

Solve. Use any strategy.

7. *Number Theory* Three consecutive integers have a sum of 12 and a product of 60. What are the integers?

8. *Football* In football, if a team makes a 6-point touchdown, it has a chance to try for either 1 or 2 extra points. If the team cannot make a touchdown, it can attempt to kick a 3-point field goal. Suppose 2-point safeties are excluded. List the number of points less than 20 that a football team can accumulate.

9. *Transportation* Mr. Cardona has agreed to drive 4 students to the concert. If he can put one student in the front seat and three students in the back, how many ways can the 4 students be arranged in the car?

10. *Ticket Sales* Kelsey sold tickets to the school musical. She had 12 bills worth $175 for the tickets she sold. If all the money was in $5 bills, $10 bills, and $20 bills, how many of each bill did she have?

11. *Geometry* The large square below has been divided into 9 squares. The lengths of the sides of two squares are given. Find the area of the large square.

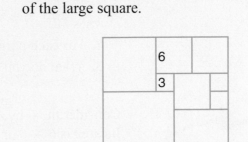

12. *Standardized Test Practice* Derrick used his calculator to divide 5,762,664 by 2,113.6. Which number is a good estimate for the quotient?

A 30

B 300

C 3,000

D 30,000

E 300,000

What you'll learn

You'll learn to find squares of numbers and square roots of perfect squares.

When am I ever going to use this?

Knowing how to find squares of numbers can help you find the amount of floor covering needed for a square room.

Word Wise

square
perfect square
square root
radical sign

There are many rectangles that have a perimeter of 12 units. Which of these rectangles has the greatest area? One way to solve this problem is to use the guess-and-check strategy.

HANDS-ON MINI-LAB

Work with a partner. 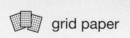 grid paper

The following rectangle has a perimeter of 12 units and an area of 5 square units.

```
┌─────────────────┐
│                 │ 1 unit
└─────────────────┘
   5 units
```

Try This

1. On grid paper, draw other rectangles that have a perimeter of 12 units.

2. Complete the following chart so that it shows all of your rectangles.

Drawing	Perimeter	Area
▭▭▭▭▭	12 units	5 square units

3. Repeat Steps 1–2 for a rectangle with a perimeter of 16 units.

Talk About It

4. For each perimeter, which rectangle has the greatest area?
5. What do you notice about the rectangles with the greatest areas?

> **LOOK BACK**
> You can refer to Lesson 1-4 for information on exponents.

Consider the 6-by-6 square at the right. Its area is 6×6 or 36 square units.

Remember that an *exponent* tells how many times a number, called the *base,* is used as a *factor.* In the expression 6×6, 6 is used as a factor twice. When you compute 6×6 or 6^2, you are finding the **square** of 6.

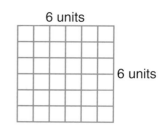

6 units
6 units

$$6^2 = 6 \times 6$$
$$= 36 \quad \textit{The square of 6 is 36.}$$

Examples

1 Evaluate 7^2.

$7 \times 7 = 49$

2 Evaluate 23^2.

23 $\boxed{x^2}$ *529*

Numbers such as 36, 49, and 529 are called **perfect squares** because they are squares of whole numbers.

The 6-by-6 square on page 410 has an area of 36 square units. One way to find the length of a side of a square with an area of 36 square units is to count the number of units on one side of the square. Another way is to find the **square root** of 36.

Square Root	If $a^2 = b$, then a is a square root of b.

Since $6^2 = 36$, one square root of 36 is 6. It is also true that $(-6)^2 = 36$, so another square root of 36 is -6. *Since the length of a side of a square with an area of 36 square units must be positive, each side of the square is 6 units long.*

The symbol $\sqrt{}$, called a **radical sign**, is used to represent a nonnegative square root.

$$\sqrt{36} = 6$$

Examples

3 Find $\sqrt{64}$.

Since $8^2 = 64$, $\sqrt{64} = 8$.

4 Find $\sqrt{961}$.

961 [2nd] [√] *31*

APPLICATION

5 **Ballooning** On a clear day, the distance you can see from a location above Earth can be estimated using the formula $V = 1.22 \times \sqrt{A}$, where V is the distance in miles and A is the altitude in feet. Catalina is riding in a hot air balloon about 196 feet above the ground. She looks off to the horizon and can just see her school in the distance. About how far is she from the school?

$V = 1.22 \times \sqrt{A}$

$V = 1.22 \times \sqrt{196}$ *Replace A with 196.*

$V = 1.22 \times 14$ *196* [2nd] [√] *14*

$V = 17.08$

Catalina is about 17 miles from her school.

CHECK FOR UNDERSTANDING

Communicating Mathematics

Read and study the lesson to answer each question.

1. ***Tell*** why the model shows $\sqrt{16} = 4$.

2. ***Explain*** how finding the square of a number is like finding the area of a square.

3. *Make a model* to show each of the following.
   a. $2^2 = 4$ **b.** $\sqrt{81} = 9$

Guided Practice

Find the square of each number.

  4. 10   **5.** 15   **6.** 33

Find each square root.

  7. $\sqrt{144}$ **8.** $\sqrt{1,225}$ **9.** $\sqrt{324}$

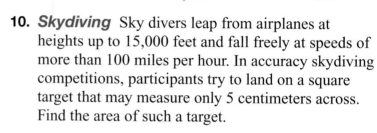

10. *Skydiving* Sky divers leap from airplanes at heights up to 15,000 feet and fall freely at speeds of more than 100 miles per hour. In accuracy skydiving competitions, participants try to land on a square target that may measure only 5 centimeters across. Find the area of such a target.

EXERCISES

Practice

Find the square of each number.

  11. 1   **12.** 13   **13.** 32   **14.** 16
  15. 40   **16.** 55   **17.** 27   **18.** 200

Find each square root.

  19. $\sqrt{121}$ **20.** $\sqrt{625}$ **21.** $\sqrt{289}$ **22.** $\sqrt{900}$
  23. $\sqrt{1,089}$ **24.** $\sqrt{2,601}$ **25.** $\sqrt{576}$ **26.** $\sqrt{90,000}$

  27. What is the square of 38?

  28. Find the positive square root of 361.

  29. Find the length of a side of a square whose area is 484 square inches.

  30. The area of a square is 1,024 square centimeters. Find its perimeter.

  31. Find the greatest possible area for a rectangle whose perimeter is 96 feet.

  32. *Algebra* Use graphs to study area and perimeter.

    a. Let the *x*-axis of a coordinate plane represent the length of a side of a square and the *y*-axis represent the area of the square. Graph the points that represent squares with sides 0, 1, 2, 3, 4, and 5 units long. Draw a line or curve that goes through each point.

    b. On the same coordinate plane, let the *x*-axis represent the length of a side of a square and the *y*-axis represent the perimeter of the square. Graph the points that represent squares with sides 0, 1, 2, 3, 4, and 5 units long. Draw a line or curve that goes through each point.

    c. Compare and contrast the two graphs.

    d. When is the value of the perimeter greater than the value of the area? When are the values equal?

    e. Why do these graphs only make sense in the first quadrant?

Applications and Problem Solving

33. Architecture Because a square house has the least outside wall space for the area, it is the most energy-efficient type of house to build. A house in the shape of an H is the least energy efficient.

  a. Describe the most energy-efficient one-story house you could build with an area of 1,600 square feet.

  b. A two-story home is more energy-efficient than a one-story home, because there is less roof space. Describe the most energy-efficient two-story house you could build with a total area of 2,450 square feet.

34. Sports A boxing ring is actually a square with an area of 400 square feet. What are the dimensions of the ring?

35. Landscaping Gro-Fast fertilizer comes in bags that cover 2,500 square feet.

  a. What are the dimensions of the largest square that could be covered by one bag of fertilizer?

  b. How many bags of fertilizer would you need to buy to cover a square whose side is 75 feet?

36. Critical Thinking The states of Wisconsin, Michigan, Illinois, Indiana, and Ohio made up the Northwest Territory. The Land Ordinance of 1785 divided this territory into townships. Each township was a square 6 miles on a side and was divided into 36 sections. One of these sections, the 16th, was reserved to support public schools. If a square piece of land contained 2,304 square miles, how many sections were reserved for public schools?

Mixed Review

37. Standardized Test Practice Which word, written as shown, does *not* have a vertical line of symmetry? *(Lesson 9-7)*

  A H   **B** M   **C** B   **D** W
     I   O   A   H
     T   T   T   O
                    H

38. Algebra Solve $\frac{3}{x} = \frac{25}{15}$. *(Lesson 8-3)*

39. Algebra Solve the equation $d = -3 + (-2)$. *(Lesson 5-4)*

40. Statistics Refer to the graph of shoe sales. Which color of shoe would you consider to be the least popular? *(Lesson 3-2)*

For **Extra Practice**, see page 595.

Shoe Sales

Tic Tac Root

Math Skill
Finding Square Roots

Get Ready This game is for two to four players.

 index cards

Get Set Use 20 index cards. On each card, write one of the following square roots.

$\sqrt{1}$	$\sqrt{4}$	$\sqrt{9}$	$\sqrt{16}$
$\sqrt{25}$	$\sqrt{36}$	$\sqrt{49}$	$\sqrt{64}$
$\sqrt{81}$	$\sqrt{100}$	$\sqrt{121}$	$\sqrt{144}$
$\sqrt{169}$	$\sqrt{196}$	$\sqrt{225}$	$\sqrt{256}$
$\sqrt{289}$	$\sqrt{324}$	$\sqrt{361}$	$\sqrt{400}$

Each player should draw a tic-tac-toe board on a piece of paper. In each square, place a number from 1 to 20, but do not use any number more than once. See the sample board at the right.

6	13	18
8	12	5
3	17	9

Go

- The dealer shuffles the index cards and places them facedown on the table.

- The player to the left of the dealer chooses the top index card and places it faceup. Any player with the matching square root on his or her board places an X on the appropriate square.

- The next player chooses the top index card and places it faceup on the last card chosen. Players mark their boards accordingly.

- The game continues until a player has three Xs in a row. The other players use the cards in the chosen pile to check the tic-tac-toe.

- The first player to get a correct tic-tac-toe wins the game.

10-2 Estimating Square Roots

What you'll learn

You'll learn to estimate square roots.

When am I ever going to use this?

Knowing how to estimate square roots can help you to solve problems involving car accident investigations.

Does your family have a vegetable garden? The graph shows how the average size of family vegetable gardens decreased in one decade.

Suppose a vegetable garden is planted on a square plot of ground. What is the approximate length of the side of a typical garden for 1981 and for 1991? *This problem will be solved in Exercise 29.*

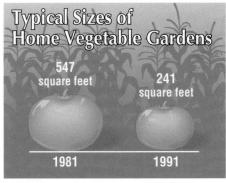

Typical Sizes of Home Vegetable Gardens

547 square feet — 1981

241 square feet — 1991

Source: National Garden Association/Gallup Inc.

You know that the square root of a perfect square, such as 25, is a whole number. You can estimate the square root when the number is not a perfect square.

HANDS-ON MINI-LAB

Work with a partner.

 grid paper

Try This

- On grid paper, draw the largest possible square using no more than 22 small squares.
- On grid paper, draw the smallest possible square using at least 22 small squares.

Talk About It

1. The value of $\sqrt{22}$ is between two consecutive whole numbers. What are these numbers?
2. Which whole number would be the better estimate for $\sqrt{22}$? Why?
3. Explain how you could use grid paper to estimate $\sqrt{39}$.

How does this method of estimating square roots compare to the one in the Mini-Lab?

Since 22 is not a perfect square, estimate $\sqrt{22}$ by finding the two perfect squares closest to 22. List some perfect squares.

$$1, 4, 9, \underbrace{16, 25}, 36, 49, \ldots$$

22 is between 16 and 25.

$$16 < 22 < 25$$
$$\sqrt{16} < \sqrt{22} < \sqrt{25} \quad \textit{Find the square root of each number.}$$
$$4 < \sqrt{22} < 5 \quad \textit{This means that } \sqrt{22} \textit{ is between 4 and 5.}$$

So $\sqrt{22}$, is between 4 and 5. Since 22 is closer to 25 than 16, $\sqrt{22}$ is closer to 5 than to 4. The best whole number estimate for $\sqrt{22}$ is 5.

Example 1

Estimate $\sqrt{85}$ **to the nearest whole number.**

List some perfect squares.

$$1, 4, 9, 16, 25, 36, 49, 64, \underbrace{81, 100}, ...$$

85 is between 81 and 100.

$$81 < 85 < 100$$
$$\sqrt{81} < \sqrt{85} < \sqrt{100} \quad \textit{Find the square root of each number.}$$
$$9 < \sqrt{85} < 10 \quad \textit{This means that } \sqrt{85} \textit{ is between 9 and 10.}$$

Since 85 is closer to 81 than 100, the best whole number estimate is 9.

In real-life situations, calculators are used to find square roots.

Example 2

APPLICATION

Accident Investigations After an accident, police officers can determine the speed of a car before it skidded to a stop. They use the formula $s = \sqrt{30df}$, where s is the speed in miles per hour, d is the length of the skid marks in feet, and f is a value that depends on weather conditions. What was the approximate speed of a car if the skid marks are 90 feet long and f is 1.2?

$$s = \sqrt{30df}$$
$$s = \sqrt{30 \times 90 \times 1.2} \quad \textit{Replace d with 90 and f with 1.2.}$$
$$s = \sqrt{3,240} \quad \textit{3240 \boxed{2nd} } [\sqrt{\ }] \textit{ 56.92099788}$$
$$s \approx 56.9 \quad \approx \textit{means "is approximately equal to."}$$

The car was traveling about 57 miles per hour.

CHECK FOR UNDERSTANDING

Communicating Mathematics

Read and study the lesson to answer each question.

1. ***Explain***, in your own words, why 6 is the best whole number estimate for $\sqrt{34}$.

2. ***Name*** three numbers that have square roots between 3 and 4.

HANDS-ON MATH

3. ***Draw*** the largest possible square using no more than 18 small squares on a piece of grid paper. Then draw the smallest possible square using at least 18 small squares. Which whole number would be the better estimate for $\sqrt{18}$?

Guided Practice

Estimate each square root to the nearest whole number.

4. $\sqrt{10}$ 5. $\sqrt{47}$ 6. $\sqrt{194}$

Use a calculator to find each square root to the nearest tenth.

7. $\sqrt{15}$ 8. $\sqrt{89}$ 9. $\sqrt{230}$

10. *Accident Investigations* Refer to Example 2. Estimate the speed of a car if the skid marks are 60 feet long and *f* is 1.4.

EXERCISES

Practice

Estimate each square root to the nearest whole number.

11. $\sqrt{8}$ 12. $\sqrt{50}$ 13. $\sqrt{65}$ 14. $\sqrt{79}$

15. $\sqrt{140}$ 16. $\sqrt{230}$ 17. $\sqrt{115}$ 18. $\sqrt{580}$

Use a calculator to find each square root to the nearest tenth.

19. $\sqrt{20}$ 20. $\sqrt{55}$ 21. $\sqrt{72}$ 22. $\sqrt{88}$

23. $\sqrt{125}$ 24. $\sqrt{99}$ 25. $\sqrt{645}$ 26. $\sqrt{1,380}$

27. Which is closer to 6, $\sqrt{34}$ or $\sqrt{44}$?

28. Which is closer to $\sqrt{55}$, 7 or 8?

Applications and Problem Solving

29. *Gardening* Refer to the beginning of the lesson. Assume the gardens are square.
   a. Find the approximate length of a side of a typical vegetable garden in 1981.
   b. Find the approximate length of a side of a typical vegetable garden in 1991.

30. *Geometry* You can use Heron's formula to find the area of a triangle if you know the measures of its sides. If the measures of the sides are *a, b,* and *c*, the area *A* equals $\sqrt{s(s-a)(s-b)(s-c)}$, where *s* is half of the perimeter. Suppose a triangle has sides 15 meters, 20 meters, and 27 meters long.

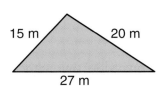

   a. What is the value of *s*?
   b. Find the area of the triangle.

31. *Critical Thinking* The equation $R = \frac{s^2}{A}$ is used to design hang gliders. In this equation, *R* represents the aspect ratio, *s* the wingspan, and *A* the area of the wing. If *R* is 2.5 and *A* is 100 square feet, find the wingspan.

Mixed Review

32. Find $\sqrt{484}$. *(Lesson 10-1)*

33. Express 35% as a fraction in simplest form. *(Lesson 8-5)*

34. **Standardized Test Practice** The wheel on Namid's bicycle has a diameter of 11 inches. About how far does Namid travel in 150 revolutions of the wheel? Use 3.14 for π. *(Lesson 7-7)*
   A 34.54 in.   B 286 in.   C 2,590.5 in.
   D 5,181 in.   E Not Here

For **Extra Practice,** see page 596.

35. Evaluate $2(3 + 5) \div 4 - 2$. *(Lesson 1-2)*

COOPERATIVE LEARNING

10-3A The Pythagorean Theorem

A Preview of Lesson 10-3

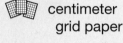
centimeter grid paper

ruler

scissors

The ancient Egyptians used mathematics to lay out their fields with square corners. About 2000 B.C., they discovered a 3-4-5 right triangle. They took a piece of rope and knotted it into 12 equal spaces. Taking three stakes, they stretched the rope around the stakes to form a right triangle. The sides of the triangle had lengths of 3, 4, and 5 units.

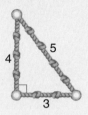

In this lab, you will investigate the relationship that exists among the sides of a right triangle.

TRY THIS

Work with a partner.

Step 1 On grid paper, draw a segment that is 3 centimeters long. At one end of this segment, draw a perpendicular segment that is 4 centimeters long. Draw a third segment to form a triangle. Cut out the triangle.

Step 2 Measure the length of the longest side in centimeters.

Step 3 Cut out three squares: one with 3 centimeters on a side, one with 4 centimeters on a side, and one with 5 centimeters on a side.

Step 4 Place the edges of the squares against the corresponding sides of the right triangle.

Step 5 Find the area of each square.

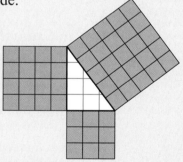

ON YOUR OWN

1. What relationship exists among the areas of the three squares?
2. Do you think the relationship you described in Exercise 1 is true for any right triangle? Repeat the activity for each right triangle whose perpendicular sides have the following measures.
   a. 6 centimeters, 8 centimeters   **b.** 5 centimeters, 12 centimeters
3. *Write* one or more sentences that summarize your findings.
4. *Look Ahead* Suppose the perpendicular sides of a right triangle are 9 inches and 12 inches long. Find the length of the other side.

10-3 The Pythagorean Theorem

What you'll learn

You'll learn to find length using the Pythagorean Theorem.

When am I ever going to use this?

Knowing how to use the Pythagorean Theorem can help you find lengths of sides of right triangles.

Word Wise

hypotenuse
leg
Pythagorean Theorem

Pythagoras

The National Safety Council recommends placing the base of a ladder one foot from the wall for every three feet of the ladder's length. How high can a 15-foot ladder safely reach? *This problem will be solved in Example 1.*

Look at the diagram of the ladder. A right triangle is formed. The longest side of a right triangle is called the **hypotenuse**. It is opposite the right angle. The other two sides, called **legs**, form the right angle.

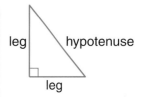

About 2,500 years ago, a Greek mathematician, Pythagoras, formalized a relationship among the sides of any right triangle. It has become known as the **Pythagorean Theorem**.

Pythagorean Theorem	**Words:** In a right triangle, the sum of the squares of the lengths of the legs (*a* and *b*) is equal to the square of the length of the hypotenuse (*c*).
	Symbols:
	Arithmetic $3^2 + 4^2 = 5^2$ **Algebra** $a^2 + b^2 = c^2$

Model:

Example 1

APPLICATION

Safety Refer to the beginning of the lesson. How high can a 15-foot ladder safely reach?

Explore You know that the National Safety Council recommends the base of a ladder be placed one foot from the wall for every three feet of the ladder's length. You also know the ladder is 15 feet long. You need to know how high the ladder can safely reach.

Plan The base of the ladder should be placed 15 ÷ 3 or 5 feet from the wall. Use the Pythagorean Theorem to find how high the ladder can reach.

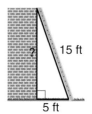

15 ft

5 ft

(continued on the next page)

Solve

$a^2 + b^2 = c^2$ *Pythagorean Theorem*

$5^2 + b^2 = 15^2$ *Replace a with 5 and c with 15.*

$25 + b^2 = 225$

$25 - 25 + b^2 = 225 - 25$ *Subtract 25 from each side.*

$b^2 = 200$

$b = \sqrt{200}$ *Definition of square root*

$b \approx 14.14$ *200* [2nd] [√] *14.14213562*

The ladder can safely reach 14.14 feet above the ground.

Examine Use the Pythagorean Theorem to check the answer.

$a^2 + b^2 = c^2$ *a = 5, b ≈ 14.14, c = 15*

$5^2 + 14.14^2 \stackrel{?}{=} 15^2$

$25 + 199.9396 \stackrel{?}{=} 225$

$224.9396 \approx 225$ ✓

You can also use the Pythagorean Theorem to determine whether a triangle is a right triangle.

Examples

Remember the hypotenuse is always the longest side.

Given the lengths of the sides of a triangle, determine whether each triangle is a right triangle.

2 0.7 meters, 2.4 meters, 2.5 meters

$a^2 + b^2 = c^2$

$0.7^2 + 2.4^2 \stackrel{?}{=} 2.5^2$

$0.49 + 5.76 \stackrel{?}{=} 6.25$

$6.25 = 6.25$ ✓

It is a right triangle.

3 9 inches, 15 inches, 18 inches

$a^2 + b^2 = c^2$

$9^2 + 15^2 \stackrel{?}{=} 18^2$

$81 + 225 \stackrel{?}{=} 324$

$306 \neq 324$

It is *not* a right triangle.

CHECK FOR UNDERSTANDING

Communicating Mathematics

Read and study the lesson to answer each question.

1. *Write* an equation that describes the relationship among the three large squares in the figure.

2. *Describe* how you would use the Pythagorean Theorem to find the length of the hypotenuse of a right triangle when you are given the lengths of its two legs.

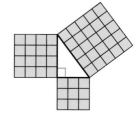

3. **You Decide** Ms. Egan asked two students from her class to come to the board and write an equation that can be used to find x in the triangle. Who wrote the correct answer? Explain your reasoning.

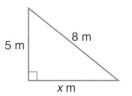

Brian	**Ricardo**
$5^2 + 8^2 = x^2$	$5^2 + x^2 = 8^2$

Guided Practice

Find the missing measure for each right triangle. Round to the nearest tenth.

4. a: 2 ft; b: 5 ft

5. a: 13 cm; c: 27 cm

Write an equation to solve for x. Then solve. Round to the nearest tenth.

6.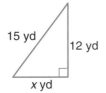

7.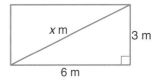

Given the lengths of the sides of a triangle, determine whether each triangle is a right triangle. Write yes or no.

8. 2 m, 3 m, 4 m

9. 7 ft, 24 ft, 25 ft

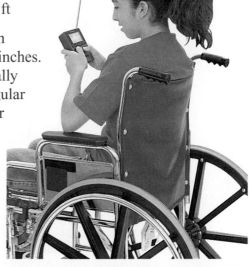

10. **Technology** The sizes of television and computer monitors are given in inches. However, these dimensions are actually the diagonal measures of the rectangular screens. Suppose a 14-inch computer monitor has an actual screen length of 11 inches. What is the height of the screen?

EXERCISES

Practice

Find the missing measure for each right triangle. Round to the nearest tenth.

11. a: 13 mm; b: 9 mm

12. a: 14 ft; b: 8 ft

13. a: 8.2 m; b: 15.6 m

14. a: 5 cm; c: 13 cm

15. b: 24 m; c: 25 m

16. a: 8 in.; c: 14 in.

Write an equation to solve for x. Then solve. Round to the nearest tenth.

17.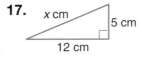

18.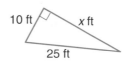

19.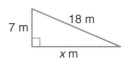

Write an equation to solve for x. Then solve. Round to the nearest tenth.

20.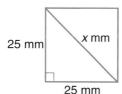
25 mm

x mm

25 mm

21.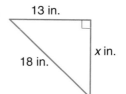
13 in.

18 in.

x in.

22.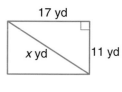
17 yd

x yd

11 yd

Given the lengths of the sides of a triangle, determine whether each triangle is a right triangle. Write *yes* or *no*.

23. 8 in., 11 in., 19 in.

24. 9 cm, 40 cm, 41 cm

25. 30 yd, 40 yd, 50 yd

26. 5 ft, 7 ft, 9 ft

27. The lengths of the sides of a triangle are 9 meters, 12 meters, and 15 meters. Is the triangle a right triangle?

28. A rectangle is 13 centimeters by 8 centimeters. Find the length of one of its diagonals to the nearest tenth of a centimeter.

Applications and Problem Solving

29. *Forestry* A tree was hit by lightning during a storm. The part of the tree still standing is 3 meters tall. The top of the tree is now resting 8 meters from the base of the tree. Assume the ground is level. How tall was the tree before it was hit by lightning?

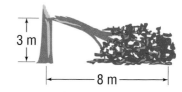
3 m

8 m

30. *Sports* A diamond used for baseball or softball is actually a square. The distance between bases on a major league field is 90 feet. If the catcher has to throw the ball to second base in an attempt to throw out a runner trying to steal the base, how long is the throw?

31. *Critical Thinking* Find the length of the diagonal of the cube.

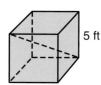
5 ft

Mixed Review

32. Which is closer to $\sqrt{45}$, 6, or 7? *(Lesson 10-2)*

33. **Standardized Test Practice** Is an angle that measures 87° *acute, obtuse, right,* or *straight*? *(Lesson 9-1)*

A acute

B obtuse

C right

D straight

34. *Carpentry* The deck on a house is $25\frac{3}{4}$ feet long and $12\frac{1}{2}$ feet wide. One length of the deck is against the house. How many feet of wood does Jack need to buy to build a railing around the deck? *(Lesson 7-6)*

For **Extra Practice,** see page 596.

35. Write the prime factorization for 36. *(Lesson 4-2)*

Area of Irregular Figures

What you'll learn

You'll learn to estimate the area of irregular figures.

When am I ever going to use this?

Knowing how to estimate the area of irregular figures can help you estimate the area of states.

Word Wise

irregular figure
inner measure
outer measure

You can find the area of a rectangle by multiplying its length by its width. Finding the area of an **irregular figure** is more difficult. Irregular figures do not necessarily have straight sides and square corners.

HANDS-ON
MINI-LAB

Work with a partner.   centimeter grid paper

Try This

Draw an outline of your foot on a piece of grid paper. If necessary, tape two pieces of grid paper together.

Talk About It

1. Find the number of whole squares within the outline of your foot.
2. Find the number of whole squares within and containing part of the outline of your foot.
3. Estimate the area of your foot by finding the mean of the two numbers.
4. Describe another way you can estimate the area of the outline of your foot.

In the Mini-Lab, you estimated the area of an irregular figure by finding the mean of the **inner measure** and the **outer measure**. Inner measure is the number of whole squares within the figure. Outer measure is the number of squares within and containing part of the figure.

Example
Real World APPLICATION

① Skateboarding
Estimate the area of the skateboard.

inner measure: 171 in^2

outer measure: 195 in^2

mean: $\frac{171 + 195}{2} = 183$

An estimate of the area of the skateboard is 183 square inches.

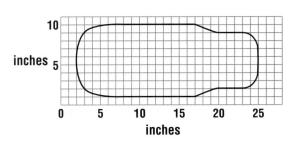

You may be able to divide some irregular figures into shapes that look like squares and rectangles. Then you can add the areas of those figures to estimate the area of the irregular figure.

Example ② Find the area of the shaded region in the figure.

Method 1   Inner/Outer Measures

   inner measure: 100 units^2
   outer measure: 112 units^2
   mean: $\dfrac{100 + 112}{2} = 106$

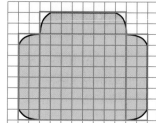

Method 2   Add the areas of the rectangles.

The irregular figure is made up of a square and three figures that look like rectangles.

   square: $8 \times 8 = 64$
   3 rectangles: $3(2 \times 8) = 48$

The area of the figure outlined in red is $64 + 48$ or 112 square units. Since the shaded region is less than the figure outlined in red, the area of the shaded region is a little less than 112 square units.

The area of the shaded area is about 106 square units.

CHECK FOR UNDERSTANDING

Communicating Mathematics

Read and study the lesson to answer each question.

1. **Tell** why the area of some figures cannot be determined by using a formula.

2. **Compare and contrast** two methods to estimate the area of an irregular figure.

HANDS-ON MATH

3. **Draw** an outline of a glove or mitten on a piece of centimeter grid paper. Estimate its area.

Guided Practice

Estimate the area of each figure.

4.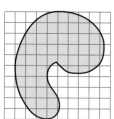

5.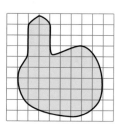

6. **Baking** Mr. Kim bakes cakes for special events. He is making a cake in the shape of a bunny for a child's birthday party. To determine how much icing he needs to make, he estimates the area of the top of the cake. What is a good estimate for this area?

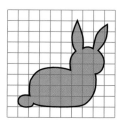

Practice

Estimate the area of each figure.

7.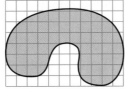

8.

9.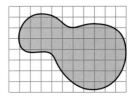

10.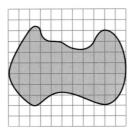

11.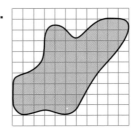

12.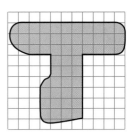

13. Draw a spoon on a piece of centimeter grid paper. Estimate the area of the figure.

Applications and Problem Solving

14. *Geography* Refer to the maps of the states shown below.
   a. Which states most closely resemble a rectangle?
   b. Which states most closely resemble a triangle?
   c. Use estimation to order the areas of the states below from greatest to least.

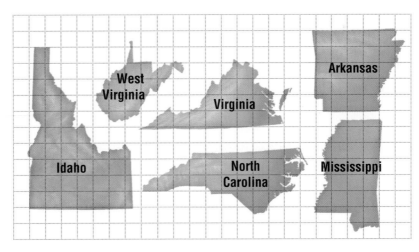

15. *Heating and Air Conditioning (HVAC)*
   Ann Merriman is an HVAC engineer who is designing a new duct system. Part of the template for her design is in the diagram. She needs to estimate the area to be sure it doesn't exceed 120 square inches. If each small square represents one square inch, will her design conform to the specifications? Explain.

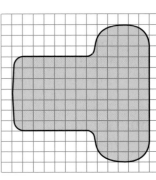

16. **Working on the**  Refer to the map you drew on page 407.

    a. Estimate the area of each landmass and body of water.

    b. Find the actual area of each landmass and body of water. Compare these numbers with your estimates. Make any necessary changes to your map.

17. **Critical Thinking** Suppose a figure has an inner measure of 10 square units and an outer measure of 20 square units.

    a. Draw the figure so that the area is closer to 10 square units than 20 square units.

    b. Draw the figure so that the area is closer to 20 square units than 10 square units.

    c. Draw the figure so that the area is exactly 15 square units.

Mixed Review

18. **Geometry** Each side of a square is 7 inches long. Find the length of one of its diagonals to the nearest tenth of an inch. *(Lesson 10-3)*

19. **Standardized Test Practice** Translate the phrase *12 more than d* into an algebraic expression. *(Lesson 6-4)*

    A $12d$ **B** $d - 12$ **C** $d + 12$ **D** $12 - d$ **E** $12 \div d$

For **Extra Practice,**
see page 596.

CHAPTER 10

Mid-Chapter Self Test

Find the square of each number. *(Lesson 10-1)*

  1. 18   **2.** 22

Find the square root of each number. *(Lesson 10-1)*

  3. $\sqrt{256}$ **4.** $\sqrt{784}$

Estimate each square root to the nearest whole number. *(Lesson 10-2)*

  5. $\sqrt{24}$ **6.** $\sqrt{140}$

Find the missing measure for each right triangle. Round to the nearest tenth. *(Lesson 10-3)*

  7. a: 10 inches; b: 24 inches   **8.** a: 7 meters; c: 9 meters

  9. **Football** The mascot of Smith Middle School carried the school banner from one corner of the football field to the opposite corner of the field. A football field including the end zones is 360 feet long and 160 feet wide. How far did the mascot carry the banner? *(Lesson 10-3)*

10. **Geometry** Estimate the area of the figure. *(Lesson 10-4)*

HANDS-ON LAB

COOPERATIVE LEARNING

 grid paper

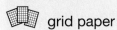 scissors

10-5A Finding the Area of a Triangle

A Preview of Lesson 10-5

You have learned that a triangle is a polygon with three sides. In this lab, you will find the area of a triangle.

TRY THIS

Work with a partner.

Step 1 Draw a parallelogram on a piece of grid paper. Your parallelogram can be of any size or shape.

Step 2 Cut out your parallelogram.

Step 3 Draw a diagonal of the parallelogram.

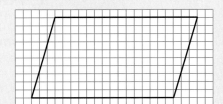

Step 4 Cut along the diagonal.

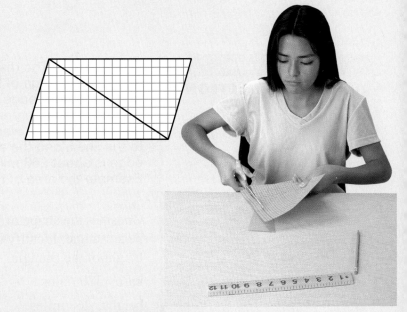

ON YOUR OWN

1. What two shapes are formed?
2. How do the two shapes compare?
3. What is the area of the original parallelogram?
4. What is the area of each triangle?
5. What conclusions can you make about the areas of all triangles?
6. ***Look Ahead*** The formula for the area of a parallelogram is $A = bh$. Using this formula, write a formula for a triangle with the same base and height.

What you'll learn

You'll learn to find the areas of triangles and trapezoids.

When am I ever going to use this?

Knowing how to find the area of triangles and trapezoids can help you find the area of decks and rooms.

Word Wise

triangle
trapezoid

Nicaragua is a country in Central America. Its shape resembles a **triangle**. You can estimate the area of the country by using the formula for the area of a triangle.

The results of the Hands-On Lab on page 427 suggest the formula for the area of a triangle.

Area of a Triangle	**Words:** The area (A) of a triangle is equal to half the product of its base (b) and height (h).
	Symbols: $A = \frac{1}{2}bh$ **Model:**

Example **CONNECTION** 1

Geography The distance from the northern edge of Nicaragua to the southern edge is about 370 miles. The perpendicular distance from the western edge to the line along the eastern edge is about 268 miles. Estimate the area of Nicaragua.

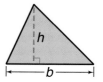

Consider the shape of Nicaragua to be a triangle. Identify the base and height of the triangle.

base: 370 miles

height: 268 miles

$A = \frac{1}{2}bh$ *Formula for the area of a triangle*

$A = \frac{1}{2} \times 370 \times 268$ *Replace b with 370 and h with 268.*

$A = 185 \times 268$

$A = 49{,}580$

The area of Nicaragua is about 49,580 square miles.

Use reference materials to check this estimate with the actual area.

A **trapezoid** is a quadrilateral with exactly one pair of parallel sides.

HANDS-ON MINI-LAB

Work with a partner.

 grid paper scissors

Try This

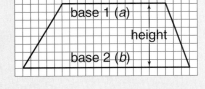 tape

- Draw a trapezoid of any shape or size on a piece of grid paper.
- Cut out your trapezoid. Label the bases and height as shown.
- Measure the lengths of base 1 and base 2. Measure the height. Record the measurements.
- Fold base 1 onto base 2. Unfold.
- Cut the trapezoid on the fold line. Then form a parallelogram.

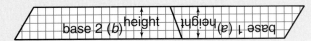

Talk About It

1. Find the length of the base of your parallelogram. How does it compare with the bases of your trapezoid?
2. Measure the height of your parallelogram. How does this height compare with the height of your trapezoid?
3. Find the area of your parallelogram.
4. What is the area of your trapezoid?
5. What conclusions can you make about the areas of all trapezoids?
6. Suppose you know the lengths of base 1 and base 2 and the height of a trapezoid. Write a formula for the area of the trapezoid.

The results of the Mini-Lab suggest the formula for the area of a trapezoid.

| Area of a Trapezoid | **Words:** | The area (*A*) of a trapezoid is equal to half the product of the height (*h*) and the sum of the bases (*a* + *b*). |
| | **Symbols:** $A = \frac{1}{2}h(a + b)$ **Model:** | |

2 **Find the area of the trapezoid.**

First, identify the bases and height.
bases: 12 meters, 18 meters
height: 9 meters

$A = \frac{1}{2} h(a + b)$ *Formula for the area of a trapezoid*

$A = \frac{1}{2} (9)(12 + 18)$ *Replace h with 9, a with 12, and b with 18.*

$A = \frac{1}{\overset{1}{2}} (9)(\overset{15}{30})$ or 135

The area of the trapezoid is 135 square meters.

CHECK FOR UNDERSTANDING

Communicating Mathematics

Read and study the lesson to answer each question.

1. *Describe* the relationship between the area of a parallelogram and the area of a triangle with the same height and base. Explain.

2. *Draw* a trapezoid and label the two bases and the height. In your own words, explain how to find the area of the trapezoid.

HANDS-ON MATH

3. Use grid paper to *draw* a trapezoid with bases of 6 units and 12 units and a height of 4 units. Draw another trapezoid with bases of 6 units and 12 units and a height of 8 units.

   a. Find the area of each trapezoid.

   b. Write a ratio that compares the heights of the trapezoids.

   c. Write a ratio that compares the areas of the trapezoids.

Guided Practice

4. Find the area of the triangle.
   base: 7 ft
   height: 12 ft

5. Find the area of the trapezoid.
   bases: 5 cm, 14 cm
   height: 6 cm

Find the area of each figure to the nearest tenth.

6.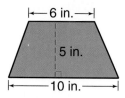

7.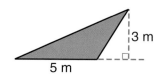

8. *Construction* The Deck and Porch Company has several designs for decks. One of them is shown at the right. Find the area of the deck.

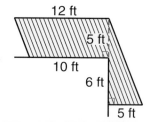

EXERCISES

Practice

Find the area of each triangle to the nearest tenth.

9. base: 16 yd
   height: 12 yd

10. base: 1.4 m
    height: 1.1 m

11. base: 5 ft
    height: 12 ft

Find the area of each trapezoid to the nearest tenth.

12. bases: 4 m, 12 m
height: 10 m

13. bases: 4 in., 8 in.
height: 6 in.

14. bases: 7.3 cm, 9.5 cm
height: 8.8 cm

Find the area of each figure to the nearest tenth.

15.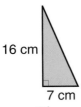
16 cm
7 cm

16.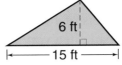
6 ft
15 ft

17.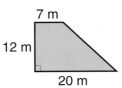
7 m
12 m
20 m

18.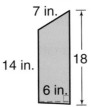
7 in.
14 in.
6 in.
18

19.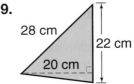
28 cm
22 cm
20 cm

20.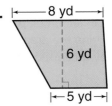
8 yd
6 yd
5 yd

21. Find the area of a right triangle if one leg is 5 inches long and the hypotenuse is 13 inches long.

Applications and Problem Solving

22. *Geography* Delaware is nicknamed the Diamond State, but its shape looks more like a triangle. Nevada has a shape that looks like a trapezoid. Estimate the area of each state.

a. Delaware

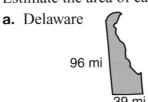
96 mi
39 mi

b. Nevada

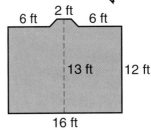
318 mi
206 mi
478 mi

23. *Interior Design* The living room drawn at the right has a bay window. An interior designer is planning to have the hardwood floors in the room refinished. What is the total area that needs to be refinished?

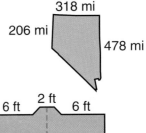
6 ft 2 ft 6 ft
13 ft 12 ft
16 ft

24. *Working on the* CHAPTER Project Refer to the map you drew on page 407. Are any of the landmasses or bodies of water shaped like a triangle or a trapezoid? If so, use the formulas to estimate the areas. How do these estimates compare with the actual areas?

25. *Critical Thinking* A triangle has height *h*. Its base is 4. Find the area of the triangle. (Express your answer in terms of *h*.)

Mixed Review

26. **Standardized Test Practice** What is the best estimate of the area of the figure? *(Lesson 10-4)*
A 10 square units **B** 20 square units
C 30 square units **D** 40 square units

27. *Geometry* What is the name of a polygon with 6 sides? *(Lesson 9-2)*

28. *Measurement* Round 26.394 kilometers to the nearest kilometer. *(Lesson 2-2)*

For **Extra Practice,** see page 597.

10-6 Area of Circles

What you'll learn

You'll learn to find the area of circles.

When am I ever going to use this?

Knowing how to find the area of a circle can help you determine the area affected by an earthquake.

Model Ana Luque paraded through the streets of Torremolinos, Spain, in a very unusual dress. The dress was 28 feet in diameter and had a train 330 feet long. Assume the waist was 2 feet in diameter. What was the area of the skirt of the dress? *This problem will be solved in Example 1.*

To find the formula for the area of a circle, we can use the formula or the area of a parallelogram.

HANDS-ON MINI-LAB

Work with a partner.   compass   straightedge   scissors

Try This

- Draw a circle and several radii that separate the circle into equal-sized sections.

  Let r units represent the length of the radius of the circle. Let C units represent its circumference.

- Cut out each section of the circle.
- Reassemble the sections in the form of a parallelogram.

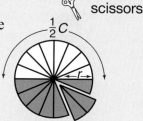

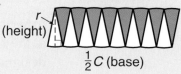

Talk About It

1. What is the height of this "parallelogram"? What is the length of the base?
2. What is the formula for the area of a parallelogram?
3. How could you use this formula to find the area of a circle?

The base of the parallelogram shown in the Mini-Lab is equal to one half of the circumference of the circle $\left(\frac{1}{2}C\right)$ The height of the parallelogram is *r*. Now, use the formula for the area of a parallelogram.

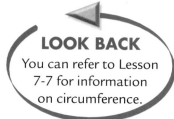

LOOK BACK

You can refer to Lesson 7-7 for information on circumference.

$A = bh$	*Formula for the area of a parallelogram*
$A = \left(\frac{1}{2}C\right)r$	*Substitute $\frac{1}{2}C$ for b and substitute r for h.*
$A = \left(\frac{1}{2} \times 2\pi r\right)r$	*Substitute $2\pi r$ for C. Why?*
$A = \pi r^2$	*Simplify: $\frac{1}{2} \times 2 = 1$, $r \times r = r^2$*

	Words: The area (A) of a circle is equal to pi (π) times the square of the radius (r).
Area of a Circle	**Symbols:** $A = \pi r^2$ **Model:**

Real World APPLICATION

1 **Fashion** Refer to the beginning of the lesson. What is the area of the skirt of the dress?

Explore You are given the diameter of the dress and the diameter of the waist. You want to know the area of the skirt.

Plan Draw a figure to illustrate the skirt. The radius of the dress is $\frac{1}{2} \times 28$ or 14 feet. The radius of the waist is $\frac{1}{2} \times 2$ or 1 foot. To determine the area of the skirt, subtract the area of the waist from the area of a circle with a radius of 14 feet.

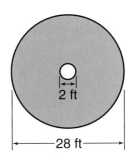

Solve **Area of 28-ft circle**

$A = \pi r^2$

$A = \pi \cdot 14^2$ *r = 14*

[π] [×] 14 [x²]

[=] *615.7521601*

$A \approx 615.8$

Area of 2-ft circle

$A = \pi r^2$

$A = \pi \cdot 1^2$ *r = 1*

[π] [×] 1 [x²]

[=] *3.141592654*

$A \approx 3.1$

The area of the skirt is about $615.8 - 3.1$ or 612.7 square feet.

Examine To estimate the area of the circles, use 3 for π.

$3 \times 14^2 = 588$ $3 \times 1^2 = 3$

An estimate for the area of the skirt is $588 - 3$ or 585 square feet. The answer seems reasonable.

2 Find the length of the radius of a circle if its area is 40 square centimeters.

$A = \pi r^2$ *Formula for the area of a circle*

$40 = \pi r^2$ *Replace A with 40.*

$\dfrac{40}{\pi} = \dfrac{\pi r^2}{\pi}$ *Divide each side by π.*

$12.7 \approx r^2$ *40* [÷] [π] [=] *12.73239545*

$\sqrt{12.7} \approx r$ *Definition of square root*

$3.6 \approx r$ *12.73239545* [2nd] [√] *3.568248232*

The radius is about 3.6 centimeters.

Communicating Mathematics

Read and study the lesson to answer each question.

1. *Explain* how to find the area of a circle if you know its radius.

2. *Draw* a circle with an area less than π square inches.

3. *You Decide* Briana says that the area of a circle with diameter of 20 centimeters is about 314 square centimeters. James says the area is about 1,257 square centimeters. Who is correct? Explain.

HANDS-ON MATH

4. Use a compass to *draw* a circle on grid paper so that the center of the circle is at a place where two grid lines meet and the radius is a whole number. See the example at the right.

   a. Estimate the area of the circle by finding the mean of the inner and outer measures.

   b. Find the area of the circle by using the formula.

   c. Compare the areas found in parts a and b.

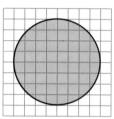

Guided Practice

Find the area of each circle to the nearest tenth.

5.
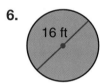
5 m

6.
16 ft

7. radius: 9 in.

8. diameter: 4.6 cm

9. Find the length of the radius of a circle with an area of 28 square meters. Round to the nearest tenth.

10. Find the length of the diameter of a circle with an area of 35 square feet. Round to the nearest tenth.

11. *Architecture* Many buildings have a circular shape.

   a. The Roman Pantheon was completed about 126 A.D. It is 142 feet in diameter. Find the area of the Pantheon.

   b. The Louisiana Superdome is the largest indoor stadium. It is 680 feet in diameter. Find the area of the Superdome.

   c. Compare the area of the Pantheon to the area of the Superdome.

EXERCISES

Find the area of each circle to the nearest tenth.

Practice

12.
4 yd

13.
21cm

14.

24 m

Find the area of each circle to the nearest tenth.

15.
28 in.

16.
35 m

17.
2.4 cm

18. radius: 3 ft

19. diameter: 34 m

20. diameter: 17 cm

21. radius: 11 in.

22. radius: 10.5 cm

23. diameter: 13 yd

Find the length of the radius of each circle given the following areas. Round to the nearest tenth.

24. 14 m²

25. 70 in²

26. 56 ft²

27. Find the length of the diameter of a circle with an area of 42 square centimeters. Round to the nearest tenth.

28. Find the length of the radius of a circle with an area of 63 square meters. Round to the nearest tenth.

29. *Algebra* Let the *x*-axis of a coordinate plane represent the radius of a circle and the *y*-axis represent the area of the circle.

   a. Graph the points that represent circles with radii 0, 1, 2, and 3 units long. Draw a line or curve that goes through each point.

   b. Consider a circle with a radius of 1 unit and a circle with a radius of 2 units. Write a ratio comparing the radii. Write a ratio comparing the areas. Do these ratios form a proportion? Explain.

   c. What happens to the area of a circle when its diameter is doubled? Are the diameters and the areas of the two circles proportional?

Applications and Problem Solving

30. *Coin Minting* Refer to the chart. Find the area of one side of each coin.

31. *Money Matters* Brittany is planning to order a pizza. She can buy a 12-inch pepperoni pizza for $7.99 or a 14-inch pepperoni pizza for $10.49. Which pizza is the better buy?

Diameters of U.S. Coins

Penny 19.05 mm   Nickel 21.21 mm   Dime 17.91 mm   Quarter 24.26 mm

Source: Department of Treasury United States Mint

32. *Critical Thinking* What is the area of the largest circle that will fit inside a square with an area of 36 square centimeters?

Mixed Review

33. *Geometry* Find the area of a trapezoid with bases of 8 centimeters and 20 centimeters and a height of 11 centimeters. *(Lesson 10-5)*

34. *Standardized Test Practice* A cake recipe calls for $2\frac{1}{3}$ cups of sugar and $3\frac{3}{4}$ cups of flour. About how many cups of sugar and flour are in the recipe? *(Lesson 7-1)*

   A 6 cups   **B** 5 cups   **C** 4 cups   **D** 3 cups   **E** Not Here

35. *Algebra* Graph the equation $y = -2x$. *(Lesson 6-7)*

36. Solve $r = -24 \div 8$. *(Lesson 5-7)*

For **Extra Practice,** see page 597.

10-7A Probability and Area Models

A Preview of Lesson 10-7

 inch grid paper

 ruler

 compass

small counters

In this lab, you will investigate the relationship between area and probability.

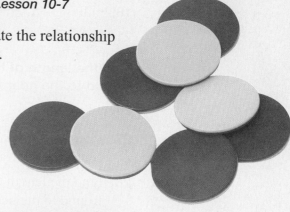

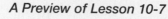 **TRY THIS**

Work in groups of three.

Step 1 On your grid paper, draw a square that has sides 8 inches long. Inside the square, draw a circle with a radius of 2.5 inches.

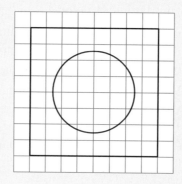

Step 2 Hold 20 counters about 5 inches above the paper and drop them onto the paper.

Step 3 Count the number of counters that landed completely within the square. (This includes those that landed within the circle.) Count the number that landed completely inside the circle. Do not count those that landed on the circle itself. These two numbers make up the first sample. Record your results.

Step 4 Repeat Steps 2 and 3 nine more times, recording your results each time.

Step 5 Add the results of your ten samples to find the total number of counters that fell within the square and the total number that fell within the circle.

Step 6 The probability that a counter will land inside the circle is expressed by the fraction:

$$\text{probability} = \frac{\text{total counters within the circle}}{\text{total counters within the square}}$$

Calculate the experimental probability based on your data.

Step 7 You can use experimental probability to estimate the area of the circle.

$$P(\text{probability}) = \frac{c \ (\text{area of a circle})}{s \ (\text{area of square})}$$

Substitute your experimental probability for P. Calculate the area of the square and substitute it for s. Then solve for c to find the approximate area of the circle.

ON YOUR OWN

1. Count the number of grid squares inside the circle to get an estimate of its area. Since many of the squares are not complete squares, you will need to combine two or three partial squares to get a better estimate of the equivalent number of complete squares. How does this estimate of the area compare with the experimental probability estimate you found in Step 7?

2. Using the radius of 2.5 inches and the formula $A = \pi r^2$, find the area of the circle. How does it compare with your experimental probability estimate?

3. Repeat this activity with a triangle inside your square.

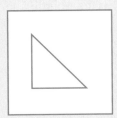

4. Repeat this activity with a trapezoid inside your square.

5. *Look Ahead* Find the probability that a randomly-dropped counter will fall in the shaded region.

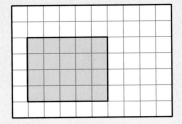

10-7

Integration: Probability
Area Models

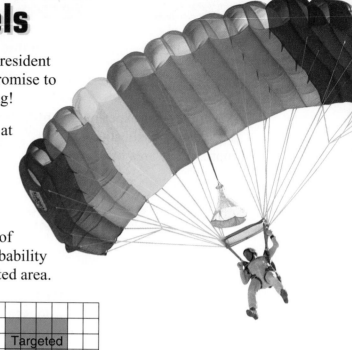

What you'll learn

You'll learn to find probability using area models.

When am I ever going to use this?

Knowing how to find the probability using area models can help you understand the game of darts.

At the age of 72, former president George Bush fulfilled a promise to himself. He went skydiving!

Suppose Mr. Bush landed at random in the rectangular field. What is the probability that he landed in the targeted area?

We can use the definition of probability to find the probability that he landed in the targeted area.

Targeted Area

$$\text{probability} = \frac{\text{number of ways to land in the targeted area}}{\text{number of ways to land in the field}}$$

In this case, the probability can be defined using area.

$$\text{probability} = \frac{\text{area of targeted area}}{\text{area of field}}$$

The area of the targeted area is 12 square units. The area of the field is 40 square units.

$$\text{probability} = \frac{12}{40} \quad \textit{The GCF of 12 and 40 is 4.}$$

$$= \frac{12 \div 4}{40 \div 4} \text{ or } \frac{3}{10}$$

The probability that George Bush landed in the targeted area is $\frac{3}{10}$.

Example

APPLICATION

1 Skydiving A sky diver parachutes at random onto a square field that contains a pond. The field is 150 feet on a side, and the pond has an area of 750 square feet. What is the probability that the diver has a dry landing?

$$\text{area of field} = 150^2 \text{ or } 22{,}500 \text{ square feet}$$

$$\text{dry area of field} = 22{,}500 - 750 \text{ or } 21{,}750 \text{ square feet}$$

$$\text{probability} = \frac{\text{dry area of field}}{\text{area of field}}$$

$$\text{probability} = \frac{21{,}750}{22{,}500} \text{ or } \frac{29}{30}$$

The probability that the diver has a dry landing is $\frac{29}{30}$.

Study Hint

Technology You can use a calculator to express $\frac{29}{30}$ as a decimal.

$\frac{29}{30} = 0.9\overline{6}$ or $96.\overline{6}\%$

An *annulus* is the region bounded by two circles with the same center but different radii.

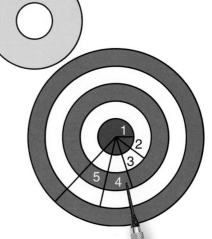

Games A dartboard has four annular rings surrounding a bull's-eye. The circles have radii of 1, 2, 3, 4, and 5 units. Suppose a dart is equally likely to hit any point on the board. Is the dart more likely to hit in the outermost ring or inside the region consisting of the bull's-eye and the two innermost rings?

Step 1 Find the area of the target.

$A = \pi r^2$ *Formula for the area of a circle*

$A = \pi \cdot 5^2$ *Replace r with 5.*

| π | × | 5 | x² | = | 78.53981634 |

The area of the target is about 78.5 square units.

Step 2 Find the area of the outermost ring.

The area of the outermost ring can be found by subtracting the area of a circle with radius of 4 units from the area of the target.

$A = \pi \cdot 5^2 - \pi \cdot 4^2$

| π | × | 5 | x² | − | π | × | 4 | x² | = | 28.27433388 |

The area of the outermost ring is about 28.3 square units.

Step 3 Find the area of the region consisting of the bull's-eye and the two innermost rings.

$A = \pi \cdot 3^2$

| π | × | 3 | x² | = | 28.27433388 |

The area of the region is about 28.3 square units.

In both cases, the probability is about $\frac{28.3}{78.5}$. The dart is as likely to hit in the outermost ring as it is to hit the bull's-eye and the two innermost rings.

Study Hint

Technology You can use a calculator to express $\frac{28.3}{78.5}$ as a decimal. $\frac{28.3}{78.5} \approx 0.361$ or about 36.1%.

CHECK FOR UNDERSTANDING

Communicating Mathematics

Read and study the lesson to answer each question.

1. *Explain* what is meant by a probability of a dry landing is $\frac{29}{30}$ in Example 1. Does the sky diver have a good chance for a dry landing?

2. *Write* the equation you would use to estimate the area of the trapezoid at the right if 4 out of 20 counters landed in the trapezoid.

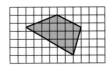

3. Suppose that you are playing a game that uses a spinner and you need the spinner to stop on the red section for you to win the game.

    a. Draw a spinner where the probability of you winning is $\frac{1}{4}$.

    b. Draw a spinner where the probability of you winning is $\frac{2}{7}$.

    c. Draw a spinner where the probability of you winning is 20%.

    d. Write a short paragraph telling which spinner would give you a better chance of winning and why.

Guided Practice **Find the probability that a randomly-dropped counter will fall in the shaded region.**

4. 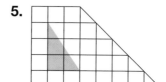   **5.**

6. *Golf* A golfer tees off and the ball lands in the rectangular region at the right. What is the probability that the ball lands on the green?

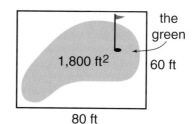

EXERCISES

Practice **Find the probability that a randomly-dropped counter will fall in the shaded region.**

7.    **8.** 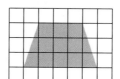   **9.**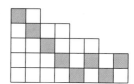

10. 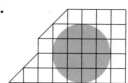   **11.** 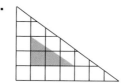   **12.**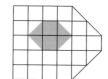

13. Draw a square on a piece of grid paper that is 6 units on a side. Draw a triangle inside the square that has a base of 3 units and a height of 4 units. What is the probability that a randomly-dropped counter will fall in the triangle?

14. **Draw** a square on a piece of grid paper that is 10 units on a side. Draw a trapezoid inside the square so that the probability that a randomly-dropped counter will fall in the trapezoid is 24%.

15. **Geometry** A tangram, a puzzle that originated in China, consists of 7 pieces that form a square as shown. These pieces can be rearranged to form shapes of animals, people, and other objects. Suppose a counter is randomly dropped on the tangram. Find the probability that the counter will land on the small square. (*Hint:* The side of the small square is $\frac{1}{4}$ of the diagonal of the large square.)

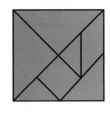

Applications and Problem Solving

16. **Games** A contestant throws a dart at a wall partially covered with balloons. If a balloon is popped, the contestant wins a prize. The wall has an area of 20 square feet. Eight square feet are covered with balloons. What is the probability that a contestant wins?

17. **Computer Technology** A diskette contains 2,847 clusters of storage space. Suppose that 5 clusters are defective. What is the probability that information that needs to be saved will be saved to a portion of the diskette that is not defective?

18. **Working on the** CHAPTER Project Refer to the map you drew on page 407. Suppose a meteor heading toward Earth does not disintegrate as it passes through the atmosphere. What is the probability that the meteor will land in water?

19. **Critical Thinking** Odessa wins a prize by tossing a quarter onto a grid board so that it doesn't touch a line. The sides of the small squares of the grid are 40 millimeters long and the radius of a quarter is 12 millimeters long. What is the probability of winning? (*Hint:* Find the area where the center of the coin could land so that the edges don't touch a line.)

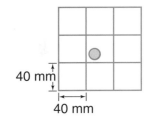

Mixed Review

20. **Architecture** The Connaught Centre building in Hong Kong has 1,748 circular windows. The diameter of each window is 2.4 meters. Find the total area of the glass in the windows. *(Lesson 10-6)*

21. **Standardized Test Practice** Which polygon is a regular polygon? *(Lesson 9-2)*

A   B   C   D

For **Extra Practice**, see page 597.

22. **Algebra** Solve $\frac{15}{32} = \frac{5}{p}$. *(Lesson 8-3)*

 **inter**NET
CONNECTION Chapter Review **For additional lesson-by-lesson review, visit:**
www.glencoe.com/sec/math/mac/mathnet

Vocabulary

After completing this chapter, you should be able to define each term, concept, or phrase and give an example or two of each.

Problem Solving
guess and check (p. 408)

Number and Operation
perfect square (p. 411)
radical sign (p. 411)
square (p. 410)
square root (p. 411)

Geometry
hypotenuse (p. 419)
inner measure (p. 423)
irregular figure (p. 423)
leg (p. 419)
outer measure (p. 423)
Pythagorean Theorem (p. 419)
trapezoid (p. 429)
triangle (p. 428)

Understanding and Using the Vocabulary

Choose the correct term or number to complete each sentence.

1. The number (64, 500) is a perfect square.

2. In a right triangle, the square of the length of the hypotenuse is (equal to, less than) the sum of the squares of the lengths of the legs.

3. A trapezoid is a quadrilateral with exactly one pair of (parallel, perpendicular) sides.

4. $A = \frac{1}{2} h(a + b)$ is the formula for the area of a (triangle, trapezoid).

5. $A = \pi r^2$ is the formula for the area of a (square, circle).

6. Probability can be expressed as a (radical, fraction).

7. The square of 9 is (3, 81).

8. A (square, square root) of 49 is 7.

9. The $\sqrt{}$ symbol is called a (radical, perfect square) sign.

10. The longest side of a right triangle is the (leg, hypotenuse).

11. The legs of a right triangle form a (right angle, hypotenuse).

12. A trapezoid (is, is not) an irregular figure.

In Your Own Words

13. *Explain* how you would estimate the square root of 45.

Objectives & Examples

Upon completing this chapter, you should be able to:

● find squares of numbers and square roots of perfect squares *(Lesson 10-1)*

Evaluate $\sqrt{196}$.

Since $14^2 = 196$, $\sqrt{196} = 14$.

● estimate square roots *(Lesson 10-2)*

Estimate $\sqrt{29}$ to the nearest whole number.

$$25 < \;\; 29 \;\; < 36$$
$$\sqrt{25} < \;\; \sqrt{29} \;\; < \sqrt{36}$$
$$5 < \;\; \sqrt{29} \;\; < 6$$

Since 29 is closer to 25 than 36, the best whole number estimate is 5.

● find length using the Pythagorean Theorem *(Lesson 10-3)*

Find the missing measure.

$$a^2 + b^2 = c^2$$
$$4^2 + 12^2 = c^2$$
$$16 + 144 = c^2$$
$$160 = c^2$$
$$\sqrt{160} = c$$
$$12.6 \approx c$$

4 in. ⌐ *c* in. / 12 in.

The missing measure is about 12.6 inches.

● estimate the area of irregular figures *(Lesson 10-4)*

inner measure: 6
outer measure: 12
mean: $\dfrac{6 + 12}{2} = 9$
The area is about 9 square units.

Review Exercises

Use these exercises to review and prepare for the chapter test.

Find the square of each number.

14. 9   15. 22

16. 43   17. 50

Find each square root.

18. $\sqrt{16}$ 19. $\sqrt{0}$

20. $\sqrt{225}$ 21. $\sqrt{256}$

Estimate each square root to the nearest whole number.

22. $\sqrt{6}$ 23. $\sqrt{37}$

24. $\sqrt{99}$ 25. $\sqrt{48}$

26. $\sqrt{90}$ 27. $\sqrt{410}$

Find the missing measure for each right triangle. Round to the nearest tenth.

28. *a*: 5 ft; *b*: 6 ft

29. *b*: 10 yd; *c*: 12 yd

30. *a*: 12 in.; *b*: 4 in.

31. *a*: 7 m; *c*: 15 m

Estimate the area of each figure.

32. 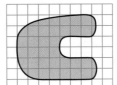   33.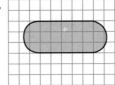

Objectives & Examples

Review Exercises

find the areas of triangles and trapezoids
(Lesson 10-5)

Area of a triangle: $A = \frac{1}{2}bh$

Area of a trapezoid: $A = \frac{1}{2}h(a + b)$

Find the area of the figure.

$$A = \frac{1}{2}h(a + b)$$

$$A = \frac{1}{2}(3)(2 + 8)$$

$$A = 15$$

The area of the figure is 15 cm².

Find the area of each figure to the nearest tenth.

34.
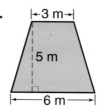

35.
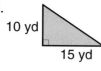

36.
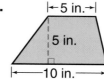

37.

find the area of circles *(Lesson 10-6)*

Area of a circle: $A = \pi r^2$

Find the area of a circle with a radius of 5 inches.

$$A = \pi r^2$$

$$A = \pi \cdot 5^2$$

$$A \approx 78.5$$

The area of the circle is about 78.5 in².

Find the area of each circle to the nearest tenth.

38. radius: 8 ft

39. diameter: 14 mm

40. diameter: 15 yd

41. radius: 25 in.

find probability using area models
(Lesson 10-7)

probability of hitting a white square

$$= \frac{\text{area of white squares}}{\text{total area}}$$

$$= \frac{8}{25}$$

Find the probability that a randomly-dropped counter will fall in the shaded region.

42.

43.

Applications & Problem Solving

44. Guess and Check Admission to the aquarium is $4 for adults, $1.50 for children under 12, and $2 for seniors. Ten people paid a total of $26.50. If the group included 4 adults, how many children and seniors were in the group? *(Lesson 10-1A)*

45. Communication A telephone pole has a wire attached from the top of the pole to a point 30 feet from the base of the pole. If the pole is 20 feet tall, find the length of the wire. Round to the nearest tenth. *(Lesson 10-3)*

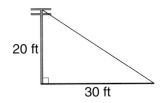

46. Gardening A lawn sprinkler can water a circular area with a radius of 20 feet. Find the area that can be watered with this sprinkler to the nearest tenth. *(Lesson 10-6)*

47. Games A game at the state fair requires that a contestant throw a coin onto a board covered with different-colored squares. If the coin lands on a red square, the contestant wins. The board has an area of 15 square feet. Three square feet are covered with red squares. What is the probability of winning? *(Lesson 10-7)*

Alternative Assessment

● Open Ended

Suppose you want to build a patio that is in the shape of a trapezoid. You want the patio to have an area between 90 and 110 square feet. Draw a plan for the patio on grid paper. Show how to find the exact area of the patio.

Suppose you decide to replace part of the patio with a planter that is in the shape of a right triangle. The part of the patio that remains should still be a trapezoid. Add the planter to your plan. Show how to find the exact area of the planter and the exact area of the patio.

A practice test for Chapter 10 is provided on page 616.

● Completing the CHAPTER Project

Use the following checklist to make sure your map is complete.

☑ The landmasses and bodies of water are the correct size.

☑ The estimates for the area of the landmasses and bodies of water are correct.

☑ A sentence describing the approximate part of Earth that is covered in water is included.

Add any finishing touches that you would like to make your map attractive.

PORTFOLIO Select one of the assignments from this chapter and place it in your portfolio. Attach a note to it explaining why you selected it.

Section One: Multiple Choice

There are eleven multiple choice questions in this section. Choose the best answer. If a correct answer is *not here*, choose the letter for Not Here.

1. Which regular polygon can be used by itself to make a tessellation?

   A pentagon

   B hexagon

   C heptagon

   D nonagon

2. The maximum square footage of a storeroom is 1,500 square feet. What number line shows the range of square footage for this storeroom?

   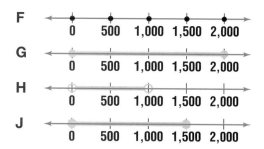

3. What is the probability that a randomly-dropped counter will fall in the shaded region?

   A $\frac{1}{8}$

   B $\frac{1}{4}$

   C $\frac{1}{3}$

   D $\frac{1}{2}$

4. A coffee can contains 1 pound 10 ounces of coffee. How many ounces is this?

   F 18 oz

   G 20 oz

   H 26 oz

   J 42 oz

Please note that Questions 5–11 have five answer choices.

5. How could you calculate the perimeter of an $8\frac{1}{2}$-by-11 piece of paper?

   A Add $8\frac{1}{2}$ and 11.

   B Multiply 2 times $8\frac{1}{2}$ and add to 11.

   C Add $8\frac{1}{2}$ and 11 and multiply the sum by 2.

   D Multiply $8\frac{1}{2}$ and 11.

   E Multiply 2 times 11 and add to $8\frac{1}{2}$.

6. There are 452 Calories in one handful of candy. If a handful of candy is 9 pieces, what is the best estimate of the number of Calories in each piece of candy?

   F 30 Calories

   G 40 Calories

   H 50 Calories

   J 60 Calories

   K 70 Calories

7. Tanya walks 5 kilometers east and 5 kilometers south. To the nearest kilometer, how far is she from her starting point?

   A 25 km

   B 10 km

   C 7 km

   D 5 km

   E 3 km

8. The regular price of a compact disc player is $279.83 without tax. Before tax is added, how much can be saved by buying a compact disc player on sale for $225.65?

   F $61.18

   G $57.18

   H $54.18

   J $54.08

   K Not Here

9. Lyle's job is to pack textbooks into cartons. One day Lyle packed 504 textbooks into 36 cartons. If each carton contains the same number of textbooks, how many textbooks did Lyle put into each carton?

   A 12   B 14

   C 28   D 50

   E 56

10. A wall shaped like a trapezoid needs to be painted. The height of the wall is 12 feet, and the bases are 17 feet and 22 feet. If one can of paint covers 20 square feet, how many cans of paint will be needed to paint the wall?

    F 24 cans   G 21 cans

    H 15 cans   J 12 cans

    K Not Here

11. A fish tank holds 46.2 liters of water. If there should be 2.2 liters of water per fish, how many fish can be placed in the tank?

    A 2 fish   B 20 fish

    C 21 fish   D 25 fish

    E Not here

Test Practice For additional test practice questions, visit:

www.glencoe.com/sec/math/mac/mathnet

Section Two: Free Response

This section contains four questions for which you will provide short answers. Write your answers on your paper.

12. What are the vertices of $\triangle EFG$ after it is reflected over the y-axis?

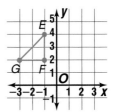

13. Write the expression that represents five more than a number.

14. How many triangles of the shape and size of the shaded triangle can divide into the trapezoid evenly?

15. What is the area of the trapezoid?

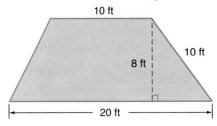

CHAPTER 11

Applying Percents

What you'll learn in Chapter 11

- to estimate percents and solve problems using the percent equation,

- to solve problems by solving a simpler problem,

- to construct and interpret circle graphs,

- to predict actions of a larger group by using a sample, and

- to solve problems involving sales tax, discount, and simple interest.

CHAPTER Project

DON'T TURN THAT DIAL!

In this project, you will listen to a radio station and keep track of the types of programming. You will organize your results in a table and display your data in a circle graph. You will use the results to make predictions about the general format of the radio station. You will present your final results in a report that you will share with the class.

Getting Started

- Work in small groups. Your group should choose a local radio station.
- Choose 6 or 7 types of programming, such as news, commercials, weather reports, station identification, songs, traffic reports, and miscellaneous. For one hour, each person in your group should keep track of when they hear different types of programming and their length.
- Each person in your group should fill out a table like the one below.

Station: (name of station)
Format: (alternative, classic rock, country, jazz, all-news, top 40, classical, all-sports . . .)
Time: (date, day, hour)

Time	Type of Programming	Number of Minutes

Technology Tips

- Use a **calculator** to help you find percents.
- Use **computer software** to make graphs.
- Use a **word processor** to write your report.

 **inter**NET **CONNECTION** **Research** For up-to-date information on radio broadcasting, visit:

www.glencoe.com/sec/math/mac/mathnet

Working on the Project

You can use what you'll learn in Chapter 11 to help you with your report.

Page	Exercise
458	25
463	9
467	10
485	Alternative Assessment

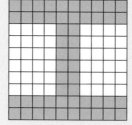

11-1 Percent and Estimation

What you'll learn

You'll learn to estimate percents by using fractions and decimals.

When am I ever going to use this?

Knowing how to estimate with percents will help you find out how much money you'll save when you buy something on sale.

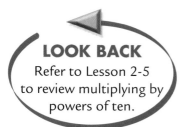

LOOK BACK
Refer to Lesson 2-5 to review multiplying by powers of ten.

In 1996, 14-year-old Subaru Takahashi became the youngest person to sail across the Pacific Ocean alone. He sailed about 47% of the 6,000-mile journey without outside communication after an engine died and the backup systems failed. For how many miles was he without communication?

You can estimate by rounding 47% to 50% and then finding 50% of 6,000.

Method 1	**Method 2**
Use a fraction.	Find 10% and multiply.
50% is the same as $\frac{1}{2}$.	10% is the same as $\frac{1}{10}$ or 0.1.
$\frac{1}{2}$ of 6,000 is 3,000.	10% of 6,000 is 0.1(6,000) or 600.
	Now find 50% or 5 times (10% of 6,000).
	$5 \times 600 = 3,000$

Using either method, the estimate is 3,000. So, for about 3,000 miles, Subaru was without communication with the rest of the world.

MINI-LAB

Work with a partner.   grid paper   marker

You can use area models to estimate percent.

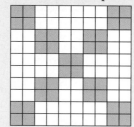
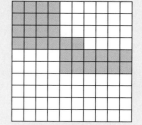

Try This

1. Estimate the percent of the shaded portion of each figure.
2. Count grid squares to find the actual percent shaded.

Talk About It

3. How do the estimates compare with the actual percents?
4. Draw a design on a 10 × 10 grid and shade it. Estimate the percent shaded. Then count to find the exact percent.

1 Estimate 62% of 507.

62% is about 60%, and 60% = $\frac{60}{100}$ or $\frac{3}{5}$; 507 is about 500.

Method 1 Use a fraction.

$\frac{3}{5}$ of 500 is $\frac{3}{5} \times 500$ or 300.

So, 62% of 507 is *about* 300.

Method 2 Find 10% and multiply.

10% of 500 is 0.1(500) or 50.   *10% = 0.1*

Now find 60% or 6 times (10% of 500).

6(50) = 300

So, 62% of 507 is *about* 300.

APPLICATION

Real World

2 **Shopping** Estimate how much money you would save on a $149 coat that is marked 30% off.

30% is about $\frac{1}{3}$ and $149 is about $150.

$\frac{1}{3}$ of 150 is $\frac{1}{3} \times 150$ or 50.

So, you would save about $50.

You can also estimate percents of numbers when the percent is less than 1 or the percent is greater than 100.

3 Estimate 113% of 42.

113% is more than 100%, so 113% of 42 is greater than 42.

113% is *about* 110%.
110% = 100% + 10%
42(100% + 10%) = 42(100%) + 42(10%)
$\qquad\qquad\qquad\quad$ = 42 + 4.2
$\qquad\qquad\qquad\quad$ = 46.2

100% of 42   10% of 42

113% of 42 is *about* 46.

4 Estimate 0.5% of 223.

0.5% is half of 1%.
223 is about 200.
1% of 200 is 0.01 · 200 or 2.   *1% = 0.01*
$\frac{1}{2}$ of 2 is 1.

0.5% of 223 is *about* 1.

Communicating Mathematics

Read and study the lesson to answer each question.

1. *Explain* how area models can be used to estimate a percent.

2. *You Decide* Tanika estimated that she would save about $30 if she bought an $86 dress on sale for 30% off. Is she right? Explain.

HANDS-ON MATH

3. *Draw* a figure or design on a 10×10 grid. Shade $\frac{4}{10}$ of the figure or design. What percent is shaded?

Guided Practice

4. Estimate the percent shaded. Then count to find the exact percent.

Write the fraction, decimal, mixed number, or whole number equivalent of each percent that could be used to estimate.

5. 38%   6. 300%   7. 25%

Estimate.

8. 25% of 18   9. 121% of 56   10. 0.3% of 425

11. *Geology* Granite, often used for stone structures, is 0.8% water. About how many pounds of water are there in a 2,000-pound block of granite?

Practice

Estimate the percent shaded. Then count to find the exact percent.

12.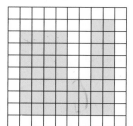

13.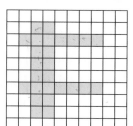

14.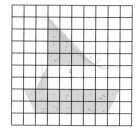

Write the fraction, decimal, mixed number, or whole number equivalent of each percent that could be used to estimate.

15. 87%   16. 200%   17. 13%

18. 43.5%   19. 0.8%   20. 103%

21. 16.97%   22. $\frac{7}{8}$%   23. 350%

Estimate.

24. 16% of 32.6   25. 25% of 408   26. 40% of 62

27. 75% of 125   28. 1% of 89   29. 30.5% of 50

30. $6\frac{1}{2}$% of 236   31. 150% of 52   32. 0.6% of 220

452 Chapter 11 Applying Percents

33. Estimate 20% of $21.99.

34. *Algebra* Estimate 79% of x if $x = 304$.

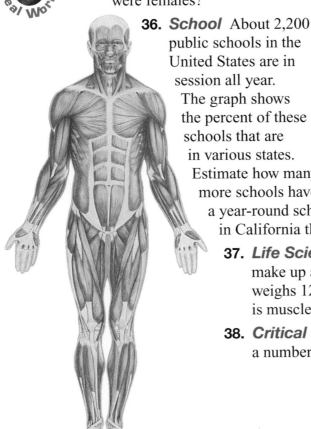

Applications and Problem Solving

35. *Education* In 1997, 70% of applicants to veterinary medical colleges were female. The College of Veterinary Medicine at Colorado State had 758 applicants in 1997. About how many of them might you estimate were females?

36. *School* About 2,200 public schools in the United States are in session all year. The graph shows the percent of these schools that are in various states. Estimate how many more schools have a year-round schedule in California than in Texas.

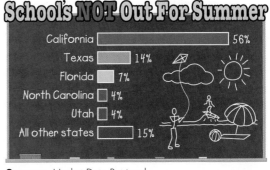

School's NOT Out For Summer

California	56%
Texas	14%
Florida	7%
North Carolina	4%
Utah	4%
All other states	15%

Source: Market Data Retrieval

37. *Life Science* The 639 muscles in your body make up about 40% of your total weight. If a person weighs 120 pounds, about how much of the weight is muscle?

38. *Critical Thinking* How could you find $\frac{1}{4}$% of a number?

Mixed Review

39. *Probability* Find the probability that a randomly-dropped counter will fall in the shaded region. *(Lesson 10-7)*

40. **Standardized Test Practice** Which procedure could be used to find the measure of angle *G*? *(Lesson 9-4)*
  A Add 50° to 180°.
  B Subtract 100° from 180°.
  C Subtract 50° from 90°.
  D Add 50° to 90°.
  E Subtract 180° from 100°.

41. Express 0.08 as a percent. *(Lesson 8-6)*

42. *Algebra* Solve $\frac{3}{5}a = 12$. *(Lesson 7-8)*

For **Extra Practice,** see page 598.

43. Order $\frac{1}{2}, \frac{7}{8}, \frac{1}{16}, \frac{5}{6}$, and $\frac{2}{3}$ from least to greatest. *(Lesson 4-10)*

11-1B Solve a Simpler Problem

A Follow-Up of Lesson 11-1

Jocelyn and Mi-Ling are studying land use in the United States in Earth Science class. Let's listen in!

I wonder how many acres of forests there are in the United States.

The table shows that 20.4% of the land is forests, but how much is that?

Mi-Ling

We need to find 20.4% of 1,940,011,000. We can estimate the number of acres by solving a simpler problem. Round each number to its greatest place value.

Land Cover	Percent
Crops	19.7
Pasture	6.5
Range	20.6
Forest	20.4
Total: 1,940,011,000 acres	

Source: *Statistical Abstract, 1996*

Jocelyn

Think: 20.4% → 20%   *Nearest 10%*

1,940,011,000 → 2,000,000,000   *Nearest billion*

10% of 2,000,000,000 is 200,000,000.

So, 20% of 2,000,000,000 is 400,000,000.

So, about 400 million acres in the United States are forests.

THINK ABOUT IT

Work with a partner.

1. **Think** of another way that Jocelyn and Mi-Ling could have estimated the number of forest acres in the United States.

2. **Find** the approximate number of pasture acres in the United States by **solving a simpler problem**.

3. **Apply** what you have learned to solve the following problem.

   In 1995, the United States imported $743.4 billion in goods from other countries. About 5% of this came from imported clothing. About how much in clothing did the United States import?

For **Extra Practice,** see page 598.

ON YOUR OWN

4. The third step of the 4-step plan for problem solving asks you to *solve* the problem. **Explain** how you can solve problems by solving a simpler problem.

5. *Write a Problem* that you can solve by solving a simpler problem. Solve the problem and explain your answer.

MIXED PROBLEM SOLVING

STRATEGIES

Look for a pattern.
Solve a simpler problem.
Act it out.
Guess and Check.
Draw a diagram.
Make a chart.
Work backward.
Make a list.

Solve. Use any strategy.

6. *Photocopying* Suppose you enlarge a drawing to 120% of its original size on the photocopy machine. If the drawing is 2 inches long and 3 inches wide, what are the dimensions of the copy?

7. *Music* The graph shows how current and former musicians learned to play their instruments.

Learning to Play an Instrument

Private lessons
29%   40%
Lessons at school
26%   30%
Taught self
21%   12%
Took school band/orchestra
17%   17%
Taught by parent/relative
6%   6%
Taught by friend
8%   2%

Men
Women

Note: Could choose more than one.

Source: Gallup for National Association for Music Merchants

Estimate how many women out of a group of 2,493 women musicians learned to play by taking band and/or orchestra in school.

8. *Life Science* The cheetah is the fastest land animal in the world. Its speed is $2\frac{1}{2}$ times that of the fastest human's speed. If the fastest recorded speed for a human is 28 miles per hour, how fast can the cheetah run?

9. *Money Matters* When the Glover family went out for pizza, their bill was $27.97. They wanted to leave a tip of approximately 15%. What is a reasonable estimate of the tip?

10. *Earth Science* Earth's atmosphere exerts a pressure of 14.7 pounds per square inch at the ocean's surface. The pressure increases by 2.7 pounds per square inch for every 6 feet that you descend. Find the pressure at 18 feet below the surface.

11. *Standardized Test Practice* Alisa bought a new stereo. She made a 25% down payment and 12 monthly payments of $45. Which is a reasonable estimate for the total price of the stereo?

 A $590

 B $620

 C $700

 D $840

 E $900

What you'll learn

You'll learn to solve problems by using the percent equation.

When am I ever going to use this?

The percent equation is useful in calculating sales tax, discounts, and commissions.

LOOK BACK

Refer to Lesson 8-9 to review the percent proportion.

Ocean water contains about 3.5% salt. How much dissolved salt is in a 50-gallon tank of ocean water? You could solve this problem by using a percent proportion $\frac{P}{B} = \frac{r}{100}$, where P is the percentage, B is the base, and $\frac{r}{100}$ is the rate. Another method is to write an equation. Let R represent the ratio $\frac{r}{100}$.

$$\frac{P}{B} = \frac{r}{100} \quad \textit{Percent proportion}$$

$$\frac{P}{B} = R \quad \textit{Replace } \frac{r}{100} \textit{ with R.}$$

$$\frac{P}{B} \cdot B = R \cdot B \quad \textit{Multiply each side by B.}$$

$$P = R \cdot B$$

Thus, the percent proportion $\frac{P}{B} = \frac{r}{100}$ can be written as a percent equation $P = R \cdot B$.

Percent Equation	**Words:**	The percentage (P) is equal to the rate (R) times the base (B).
	Symbols:	$P = R \cdot B$

In the percent equation, the rate is usually expressed as a decimal.

In the problem above, the rate R is 3.5%, and the base B is 50.

$P = R \cdot B$

$P = 0.035 \cdot 50 \quad$ *$R = 3.5\% = \frac{3.5}{100}$ or 0.035, B = 50*

$P = 1.75$

So, 50 gallons of ocean water contains about 1.75 gallons of dissolved salt.

Example 1

What number is 24% of 82?   *Estimate: $\frac{1}{4} \cdot 80 = 20$*

$P = R \cdot B$

$P = 0.24 \cdot 82 \quad$ *Replace R with 0.24 and B with 82.*

$P = 19.68$

24% of 82 is 19.68.   *Compare to the estimate.*

2 **Sports** In a recent year, Alex Rodriguez had 215 hits in 601 times at bat. What percent of his times at bat were hits?

Explore   You need to find what percent of 601 is 215.

   Estimate: $\frac{200}{600} = \frac{1}{3}$ or about 33%.

Plan   Use the percent equation $P = R \cdot B$. The percentage, P, is 215, and the base, B, is 601.

Solve $P = R \cdot B$

   $215 = R \cdot 601$ *Replace P with 215 and B with 601.*

   $\frac{215}{601} = R$ *Divide each side by 601.*

   215 ÷ 601 = 0.357737105

   So, about 36% of Mr. Rodriquez's times at bat were hits.

Examine   Comparing the actual answer to the estimate, the answer is reasonable.

3 **42 is 56% of what number?**   *Estimate: 42 is 50% or $\frac{1}{2}$ of 84.*

   $P = R \cdot B$ *Use the percent equation.*

   $42 = 0.56 \cdot B$ *Replace P with 42 and R with 0.56.*

   $\frac{42}{0.56} = B$ *Divide each side by 0.56.*

   42 ÷ .56 = 75

   $B = 75$

   42 is 56% of 75.   *Compare to the estimate.*

CHECK FOR UNDERSTANDING

Communicating Mathematics

Read and study the lesson to answer each question.

1. **Tell** how the percent equation, $P = R \cdot B$, is related to the percent proportion, $\frac{P}{B} = \frac{r}{100}$.

Math Journal

2. **Explain** why the rate is the percent divided by 100.

3. **Write** when it is easier to use the percent equation rather than the percent proportion.

Guided Practice

Write an equation for each problem. Then solve. Round answers to the nearest tenth.

4. 22 is what percent of 50?   5. 30% of what number is 27?

6. 24 is 60% of what number?   7. Find 8% of 38.

8. **Technology** Scientists are trying to find energy-saving alternatives to electrical appliances. A microwave clothes dryer, which reduces drying time by 25%, is currently being developed. If normal drying time is 40 minutes, how much less time would it take a microwave clothes dryer to dry a load of clothes?

Practice

Write an equation for each problem. Then solve. Round answers to the nearest tenth.

9. 16% of 32 is what number?

10. 17 is what percent of 68?

11. 75 is 78% of what number?

12. Find 26% of 119.

13. 45 is what percent of 36?

14. Find 20% of 68.

15. 17 is 40% of what number?

16. 55% of what number is 1.265?

17. 26% of 48 is what number?

18. What percent of 87 is 57?

19. 30 is what percent of 500?

20. 42.5% of what number is 36?

21. Find 5.75% of $69. Round to the nearest cent.

22. *Algebra* If 70% of x is 42, find x.

Applications and Problem Solving

23. *Pets* The table shows the costs of owning a dog over an average 11-year lifespan, not including the initial price of the dog. What percent of the total cost is veterinary bills?

Free(?) to a Good Home

Food	$4,020
Veterinary	$3,930
Grooming, toys, equipment, house	$2,960
Flea and tick treatment	$1,070
Training	$1,220
Other	$1,400

ROVER

Source: American Kennel Club, *USA TODAY* research

24. *Life Science* *Biometrics* is the science of verifying identities by biological characteristics. Researchers have developed a system that checks a person's identity by studying the iris of his or her eye. Currently, there are errors in only 2 to 4 percent of the identities checked by the system. If 1,500 people are tested using this system, about how many of the cases will be accurate?

25. *Working on the* **CHAPTER Project** Refer to your table of data for one hour of radio programming. Find the percent of the total hour for each type of programming.

26. *Critical Thinking* Explain how you can predict when the percentage, P, will be less than, the same as, or greater than the base, B.

Mixed Review

27. Estimate 18% of 40.5. *(Lesson 11-1)*

28. **Standardized Test Practice** A trapezoid has bases of 15 meters and 18 meters and a height of 10 meters. What is the area of the trapezoid? *(Lesson 10-5)*

 A 30 m² **B** 165 m² **C** 60 m² **D** 330 m²

29. *Aviation* A blimp starts a descent from 1,000 feet above the ground. After 10 minutes, the blimp is at 600 feet above the ground. Find the rate of its descent in feet per minute. *(Lesson 8-2)*

For **Extra Practice,** see page 598.

HANDS-ON LAB

COOPERATIVE LEARNING

11-3A Jelly Bean Statistics

A Preview of Lesson 11-3

jelly beans

needles

thread

compass

straightedge

Almost everyone has a favorite flavor of jelly bean. What is yours?

TRY THIS

Work with a partner.

Step 1 Take a survey of the people in your class. Tally responses by flavor in a frequency table. *Refer to Lesson 3-1 to review frequency tables.*

Step 2 Sort the jelly beans to reflect the results of the survey. For example, if there are 3 people whose favorite flavor is cherry (red), you would select 3 red jelly beans, and so on.

Step 3 String the jelly beans with like flavors together.

Step 4 Arrange the jelly beans in a circle. Use a compass to draw a circle the same size.

Step 5 On the circle, mark sections to indicate the separation by flavor.

Step 6 Draw a radius from each mark on the circle to the center.

Step 7 Identify each section by flavor.

ON YOUR OWN

1. Write a short paragraph describing the circle graph. Include a description of the sizes of the sections in relation to each other.

2. Is there a relationship between the number of tally marks and the size of a section by flavor? If so, write a sentence to describe that relationship.

3. Use the percent proportion or percent equation to find the percent represented by each flavor. Label each section by flavor and by the percent it represents.

4. Explain how the ratio of each flavor to the whole is represented on the circle graph.

5. **Reflect Back** The circle graph represents the same information as the frequency table. Discuss the advantages and disadvantages of each.

Integration: Statistics
Making Circle Graphs

Making Circle Graphs

What you'll learn

You'll learn to construct and interpret circle graphs.

When am I ever going to use this?

Magazines and newspapers often use circle graphs to show results of an opinion survey.

Word Wise

circle graph

Do you believe that there is intelligent life on other planets? The results of a recent poll are shown in the table.

You can draw a **circle graph** to show this information. A circle graph is used to compare parts of a whole.

Follow the steps to make a circle graph of the responses of the men.

The Truth is Out There . . .		
Is there life on other planets?	Men	Women
Yes	54%	33%
No	33%	47%
Don't Know	13%	20%

Source: Fox News/Opinion Dynamics poll

Step 1 Find the number of degrees for each part. Use $P = R \cdot B$.

Yes   54% of 360° = 0.54 · 360° = 194.4°
No   33% of 360° = 0.33 · 360° = 118.8°
Don't Know   13% of 360° = 0.13 · 360° = 46.8°

Step 2 Use a compass to draw a circle. Then draw a radius as shown.

Step 3 You can start with the least number of degrees, in this case, 46.8°. Use your protractor to draw an angle of 46.8°.

Step 4 Repeat for another section.

You can measure the last section of a circle graph to verify that the angles have the correct measures.

Step 5 In this case, there is only one section left. Label each section of the graph with the category and percent. Give the graph a title.

Life on Other Planets? What Men Think

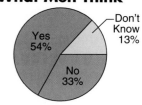

Oceanography The table shows the surface area of the four oceans. Make a circle graph to represent the data.

Ocean Surface Areas	
Ocean	Area (sq mi)
Pacific	64,186,300
Atlantic	33,420,000
Indian	28,350,500
Arctic	5,105,700

• Find the total surface area of the oceans.

Pacific	64,186,300
Atlantic	33,420,000
Indian	28,350,500
Arctic	5,105,700
Total	131,062,500

A ratio is a comparison of two numbers by division.

• Find the ratio that compares each number with the total. Convert the ratio to a decimal. Round to the nearest hundredth.

Pacific $\dfrac{64,186,300}{131,062,500} \approx 0.49$

Atlantic $\dfrac{33,420,000}{131,062,500} \approx 0.25$

Indian $\dfrac{28,350,500}{131,062,500} \approx 0.22$

Arctic $\dfrac{5,105,700}{131,062,500} \approx 0.04$

Study Hint

Technology You can use word processing or spreadsheet software to make circle graphs.

• Find the number of degrees for each section of the graph.

Pacific	$0.49 \cdot 360° = 176.4°$
Atlantic	$0.25 \cdot 360° = 90°$
Indian	$0.22 \cdot 360° = 79.2°$
Arctic	$0.04 \cdot 360° = 14.4°$

The sum of the degrees may not always be 360° due to rounding.

• Make the circle graph.

Ocean Surface Areas

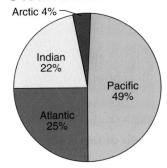

Communicating Mathematics

Read and study the lesson to answer each question.

1. *Tell* how to make a circle graph when you are given the percents of the whole that each category represents.

2. *Explain* what the ratios of angle measures to 360 represent in a circle graph.

3. *You Decide* The table shows the percent of adults surveyed who visited animal attractions in 1996. Could you make a circle graph of the data? If so, explain the steps. If not, explain why not.

Seeing the Animals	
Attraction	**Percent of Adults**
Zoo	28%
Aquarium	17%
Wild Animal Park	10%

Source: Bruskin/Goldring Research

Guided Practice

4. Refer to the table.
   a. Write a ratio that compares each number with the total. Write as a decimal to the nearest hundredth.
   b. Find the number of degrees for each section of the graph. Round to the nearest tenth.
   c. Make a circle graph showing how people record their vacations.

Vacation Memories	
Method	**Number of People**
Camera	667
Camcorder	61
Camera/ Camcorder	131
Neither	151

Source: Opinion Research Corp.

Practice

5. Refer to the table.
   a. Write a ratio that compares each number with the total. Write as a decimal to the nearest thousandth.
   b. Find the number of degrees for each section of the graph. Round to the nearest tenth.
   c. Make a circle graph of the park tourists.

Park Tourists	
Park	**Visitors (thousands)**
A	15,509
B	14,100
C	12,900
D	10,700
E	10,700
F	9,500

6. *Sports* The table shows the percent of total injuries of high school basketball players.
   a. Make a circle graph of girls' injuries.
   b. Make a circle graph of boys' injuries.
   c. Compare and contrast the graphs.

Injury	Girls	Boys
Ankle/Foot	36%	38%
Hip/Leg/Knee	30%	25%
Arm/Hand	11%	12%
Face/Scalp	9%	12%
All Others	14%	13%

Source: National Athletic Trainers' Association

7. *Statistics* Refer to the beginning of the lesson.

   a. Make a circle graph of the responses of women regarding life on other planets.

   b. Compare this circle graph to the one showing the responses of men.

   c. What advantage is there in showing the data in a circle graph instead of a table?

8. *Geography* Students in a New Zealand classroom used a string of beads to make a circle graph showing their ethnicity. A different kind of bead was given to each child depending on his or her ethnicity. Then the string was tied in a circle and divided into segments.

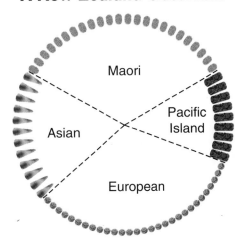

A New Zealand Classroom

Source: *Teaching Statistics,* Autumn

   a. Count the beads in each section of the circle graph.

   b. Find the percent of the class that was in each ethnic group. Round to the nearest tenth.

9. *Working on the* CHAPTER Project Refer to Exercise 25 on page 458.

   a. Make a circle graph of your radio programming data.

   b. Compare your circle graph to the circle graphs of other people in your group. Summarize the similarities and differences.

10. *Critical Thinking* Line graphs are usually best for data that show change over time. When might it be more appropriate to display data in a circle graph?

Mixed Review

11. Find 16% of 74. *(Lesson 11-2)*

12. Standardized Test Practice A brochure is 12 inches wide and 8 inches long. If the only photo on the brochure measures 3 inches by 2 inches, which sentence could be used to find *x,* the amount of space left for information and borders? *(Lesson 1-7)*

   A $x = (12 + 8) - (3 + 2)$ **B** $x = 2(12 + 8) - 2(3 + 2)$

   C $x = 12 \times 8 \times 3 \times 2$ **D** $x = (12 \times 8) - (3 \times 2)$

   E $x = \dfrac{12 \times 8}{3 \times 2}$

For **Extra Practice,** see page 599.

Integration: Statistics
Using Statistics to Predict

What you'll learn

You'll learn to predict actions of a larger group by using a sample.

When am I ever going to use this?

You can use statistics to decide how much of different types of snacks to buy for a party.

Word Wise
population
sample
random

Since 1790, the U.S. government has conducted a census. In a census, every member of a population is contacted by mail or a census taker. The **population** is counted, and other information like annual income, number of people in a household, and ethnic background is gathered.

Surveying every member of a population is very expensive and time consuming. Most of the time pollsters gather information by surveying a **sample**, which is a part of the total population. In order for a sample to be representative, it must be **random**. A random sample will give everyone the same chance of being selected.

Example ① APPLICATION

Marketing One of the reasons that Jay Leno was chosen to promote a snack was to appeal to teenagers. Other celebrities from sports and entertainment were also considered. Would the decision makers have obtained representative results from a survey taken at each location?

a. 25 eighth graders at a middle school basketball game
b. 500 teens at department stores in all parts of the country

a. This sample is not random because it does not represent all teens. The sample is small, and students at a basketball game may be more likely to suggest a basketball player.

b. This sample is representative and fairly large. Teens of different ages and interests visit malls.

If a random sample of the population is surveyed, then the results can be used to make predictions about the entire population.

Example ② APPLICATION

Food A company surveyed people about the type of crust they preferred on their pizza. Use the results to predict how many of the 1,312 students at Morgan Middle School would choose thin crust.

Type of Crust	Percent
Thin	48%
Thick	46%
No Preference	6%

Source: *Pizza Today*

You can use the percent proportion to find the number who would prefer thin crust. Find 48% of 1,312 students.

$$rate \rightarrow \quad \frac{48}{100} = \frac{n}{1,312} \quad \begin{array}{l} \leftarrow \; percentage \\ \leftarrow \; base \end{array}$$

48 ⊠ 1312 ⊡ 100 ⊟ *629.76*

You can predict that about 630 students at Morgan would prefer thin crust pizza.

Communicating Mathematics

Read and study the lesson to answer each question.

1. *Explain* how to use the results of a survey to predict the actions of a population.

2. *Select* a newspaper article that contains a table or graph. Explain how you think the results were found.

Guided Practice

3. *Music* A company randomly surveyed about 1,800 adults and teens to find the age they began playing musical instruments. They estimate that there are 62 million amateur musicians in the United States.

   a. How many amateur musicians in the United States do you predict learned to play before age 5?

   b. About how many amateur musicians learned to play between the ages of 12 and 14?

It's Music To My Ears	
Age	**Percent**
Before 5	3%
5 to 11	65%
12 to 14	21%
After 18	11%

Source: National Assoc. of Music Merchants

EXERCISES

Practice

4. *Entertainment* The table shows the results of a survey of students' favorite TV programs at Trutt Middle School. The school has a total of 840 students.

   a. What was the sample size?

   b. To the nearest percent, what percent of students preferred Program A?

   c. How many students in the school would you expect to say that Program A is their favorite?

   d. Mykia disagreed with the results of the survey and decided to conduct her own. She surveyed all 35 girls in her physical education class. Is Mykia's sample random? Explain.

Program	Number
A	46
B	32
C	28
D	25
E	23
F	21
G	19

interNET CONNECTION

For the latest television programming statistics, visit:

www.glencoe.com/sec/math/mac/mathnet

5. *Careers* Each year, the University of California surveys incoming freshmen on their career intentions.

   a. Do you think this sample is representative of all college students in the United States? Why or why not?

   b. Of the 3,775 freshmen at the University of California during the 1996-1997 school year, how many would you expect to choose a career as an elementary teacher?

Career Goal	Percent
Business executive	7.3%
Elementary teacher	5.5%
Engineer	6.4%
Lawyer	3.4%
Physician	5.7%

Source: University of California

6. *Entertainment* A company asked 6,500 teens in 26 countries about their favorite things to do outside of school.

Teens could choose more than one activity.

a. What percent of teens said they enjoy going to the movies?

b. What percent of teens said they enjoy playing sports?

c. If there were 472 students in your school, how many would you expect to say that they enjoy listening to the radio?

d. Survey the students in your math class about their favorite activities. Are the results similar to these? Explain why or why not.

What Should We Do?	
Activity	**Number**
Watching TV	6,045
Being with friends	6,045
Listening to music	5,915
Listening to radio	5,525
Watching movies at home	5,395
Going to movies	5,200
Going to parties	5,070
Talking on the phone	4,940
Playing sports	4,940

Source: New World Teen Study

Applications and Problem Solving

7. *Gardening* Recently, *Organic Gardening* magazine and the National Gardening Association (NGA) each conducted surveys on gardening habits. *Organic Gardening* used a mail survey of 40,000 households to ask whether anyone in the household did any gardening at all. They found that gardening has gained popularity in the 1990s. In interviews, the NGA asked individuals whether they had done any gardening in the last 12 months. They found that gardening had declined in the 1990s. Why might the results be different for these surveys?

8. *Medicine* Some rural areas of the country have very few doctors. In Mississippi, 0.145% of the population are doctors.

a. Of the 7,900 people in Benton County, Mississippi, how many would you expect to be doctors?

b. Currently, no doctors reside in Benton County. How does this compare with your estimate? Explain your results.

9. *Life* Do you ever wish that life was simpler? In a recent survey, people were asked how complicated they think life will be by the year 2000.

a. If there were 1,000 people in the survey, how many said that life would be much more complicated by 2000?

Life is Not That Simple	
How Complicated by Year 2000?	**Percent**
Much more	31%
Little more	31%
Same	20%
Little less	9%
Much less	5%
Don't know	4%

Source: Claris Corp. for ClarisWorks

b. Of the 185 million adults in the United States, how many would you expect to say they think life will be a little less complicated by 2000?

10. **Working on the** **CHAPTER Project** Refer to the data you collected on radio programming on page 449 and to the circle graph you made in Exercise 9 on page 463. Predict the amount of time the station would spend doing each activity. Use your prediction to make an 8-hour programming schedule for the radio station.

11. **Critical Thinking** A survey of 2,500 teens showed that 45% of girls and 40% of boys are members of the YMCA or YWCA. A marketer has found, based on information obtained from the YMCA and YWCA, that there are 3,959,550 girls and 3,874,000 boys ages 10-14 who are members of a YMCA or YWCA. Estimate how many girls and boys ages 10-14 there are in the United States.

For **Extra Practice,** see page 599.

Mixed Review

12. **Statistics** Refer to the table in Example 2. Make a circle graph showing pizza crust preference. *(Lesson 11-3)*

13. **Standardized Test Practice** The length of your calculator is about — *(Lesson 2-8)*

    A 16 mm.   **B** 16 cm.   **C** 16 m.   **D** 16 km.

CHAPTER 11   Mid-Chapter Self Test

Estimate. *(Lesson 11-1)*

1. 18% of 41

2. 32% of 90

3. 112% of 36

Write an equation for each problem. Then solve. Round answers to the nearest tenth. *(Lesson 11-2)*

4. 24 is what percent of 25?

5. 9% of 72 is what number?

6. Find 36% of 15.

7. 16% of what number is 13.12?

8. The table shows the percent of different kinds of juice sold in the United States. *(Lesson 11-3)*
    a. Make a circle graph of the data.
    b. If a grocery store is ordering 500 cans of frozen juice concentrate, how many of the cans should be orange juice?

Juice	Percent Sold
Orange	56%
Apple	14%
Blends	6%
Grape	5%
Other	19%

Source: Beverage Marketing Corporation

9. **Statistics** Members of the Student Council wanted to know if students thought an end-of-school dance was a good idea. They each asked three of their friends to give their opinion, and they tallied the results. Is this a random sample? Explain. *(Lesson 11-4)*

10. **Entertainment** One hundred people in Houston, ages 13 to 19, are randomly surveyed to find their opinion of their favorite radio station. Sixty-three of them said they liked KLOL-FM. If there are 800,000 people ages 13 to 19 in the listening area, about how many of them would you predict listen to KLOL-FM? *(Lesson 11-4)*

COOPERATIVE LEARNING

11-5A Percent of Change

A Preview of Lesson 11-5

⊞ dot paper

You can use dot paper or a geoboard to help you understand the meaning of percent of increase or percent of decrease.

TRY THIS

Work in groups of three.

Step 1 Make a 2 × 2 square like the one shown at the right.

Step 2 Suppose you want to decrease or increase the area of square A by 25%. Think: 25% = $\frac{1}{4}$. Separate the square into 4 equal parts.

Step 3 Remove 25% or $\frac{1}{4}$ from the original figure to show a decrease of 25%.

Step 4 Add 25% or $\frac{1}{4}$ to the original figure to show an increase of 25%.

ON YOUR OWN

1. Once you showed a 25% decrease in the figure, what percent remained?

2a. If there is a 25% increase in area, find the ratio of the new area to the old area.

b. If there is a 25% decrease in area, find the ratio of the new area to the old area.

c. Write each ratio as a percent.

3. Use the figure at the right to draw an increase of 50% and a decrease of 50%.

4. Draw a 3 × 3 square. Add $33\frac{1}{3}\%$ to this figure to show an increase of $33\frac{1}{3}\%$.

5. *Look Ahead* Refer to the figures at the right. By what percent was the original figure decreased? Explain how you determined the percent.

468 **Chapter 11** Applying Percents

Percent of Change

What you'll learn

You'll learn to find the percent of increase or decrease.

When am I ever going to use this?

You'll find percent of change is often used to summarize growth and decline in population.

Courtney Dann

Courtney Dann of Bellingham, Washington, is the winner of four National Water Ski Championships. Her longest jump is 91 feet, and her goal is to break the girls' jump record of 102 feet. What is the percent of increase from 91 to 102?

You can use the percent proportion to find the percent of increase. Compare the amount of the increase to the original amount.

Step 1 Find the amount of increase. $102 - 91 = 11$

Step 2 Use the percent proportion. $\dfrac{\text{amount of increase}}{\text{original amount}} = \dfrac{r}{100}$

$$\dfrac{11}{91} = \dfrac{r}{100}$$

Step 3 Solve for r. $11 \cdot 100 = 91r$ *Find the cross products.*

$$\dfrac{1{,}100}{91} = \dfrac{91r}{91} \quad \text{\textit{Divide each side by 91.}}$$

$$12.09 \approx r \quad \text{\textit{Use a calculator}}$$

The percent of increase would be about 12%.

HANDS-ON MINI-LAB

Work with a partner.   calculator   ruler

Try This

1. Draw a segment that you estimate to be 25% longer than $\overline{CD}$.

2. Measure the length of $\overline{CD}$ to the nearest centimeter. Use this number as the base B.

3. Measure the length of your segment. Use this as the percentage P.

Talk About It

4. Will 50% of the length of $\overline{CD}$ be greater than, less than, or equal to its length?

5. Will 100% of the length of $\overline{CD}$ be greater than, less than, or equal to its length?

6. Do you think the length of your segment is greater than, less than, or equal to 100% of the length of $\overline{CD}$?

7. Write a proportion or equation to find the percent the length of $\overline{CD}$ is of the length of your segment. Solve.

8. The segment you drew is actually what percent longer than $\overline{CD}$?

You can also find the percent of decrease in a similar way.

APPLICATION

1 **Nutrition** Find the percent of decrease in Calories from Meal A to Meal B.

Find the amount of decrease.

$1,134 - 683 = 451$

Meal A	Meal B
cheeseburger buttered ear of corn ice cream root beer float	veggie burger plain ear of corn frozen yogurt root bear float
Total Calories 1,134	683

Source: *Vitality*

Use the percent proportion.

$$\frac{\text{amount of decrease}}{\text{original amount}} = \frac{r}{100}$$

$$\frac{451}{1,134} = \frac{r}{100}$$

$451 \cdot 100 = 1,134r$ *Find the cross products.*

$\frac{45,100}{1,134} = \frac{1,134r}{1,134}$ *Divide each side by 1,134.*

$39.8 \approx r$

The percent of decrease in Calories is about 40%.

INTEGRATION

2 **Geometry** A loop of string measuring 20 centimeters is formed into a rectangle that has a length of 6 centimeters and a width of 4 centimeters. The loop is then changed to a square with each side measuring 5 centimeters. What is the percent of change in area?

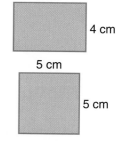

6 cm
4 cm
5 cm
5 cm

Explore   You know the dimensions of both figures. You need to find the area of each figure.

Plan   Use the percent proportion to calculate the percent of change from the area of the rectangle to the area of the square.

Solve   area of rectangle: $A = \ell \cdot w$
$= 6 \cdot 4$ or 24

area of square: $A = 5 \cdot 5$ or 25

$25 - 24 = 1$ *This is a percent of increase since 25 > 24.*

$\frac{1}{24} = \frac{r}{100}$ *Original amount = 24*

$1 \cdot 100 = 24r$

$4.2 = r$

The percent of increase is about 4%.

Examine   Since the difference between the areas is so small, it makes sense that the percent of increase is small.

Communicating Mathematics

Read and study the lesson to answer each question.

1. *Tell* what amount is used as a base in the percent proportion when finding the percent of change.

2. *Determine* whether the percent of increase from 30 to 45 equals the percent of decrease from 45 to 30. Explain.

HANDS-ON

3. If the figure represents 75% of an area, *draw* a diagram to represent 100%.

Guided Practice

Find the percent of change. Round to the nearest whole percent.

4. original: $85
   new: $68

5. original: $456
   new: $500

6. original: 1.6
   new: 0.95

7. *Entertainment* The graph shows the percent of movies that received G, PG, and PG-13 ratings in 1984 and in 1996. Find the percent of change to the nearest whole percent from 1984 to 1996 for each rating. Tell whether it is a percent of increase or a percent of decrease.

   a. G   b. PG   c. PG-13

What's the Rating?

G 2.1%
   2.9%
PG 31.3%
    14.7%
PG-13 7.7%
       16.3%

☐ 1984
☐ 1996

Source: Motion Picture Association of America

Practice

Find the percent of change. Round to the nearest whole percent.

8. original: $4
   new: $6

9. original: $60
   new: $38

10. original: 20.5
    new: 35.5

11. original: 35
    new: 45

12. original: $126
    new: $150

13. original: $30
    new: $24

14. Find the percent of decrease if an item that originally cost $36 goes on sale for $18.

15. Find the percent of change from 87.5 to 112.

16. *Write a Problem* in which the percent of change is 60%.

17. Find the original number if the new number is 16 and the percent of decrease is 68%.

Applications and Problem Solving

18. *Population* The Hispanic population is the fastest-growing minority population in the United States. Currently, there are about 29 million Hispanics in the U.S. There are expected to be more than 41 million by 2010. What is the predicted percent of increase?

19. *Recreation* The graph shows the millions of tax dollars spent on bike paths and walkways in the United States.

Taking the Scenic Route

1992	$94
1993	$150
1994	$211
1995	$222
1996	$325

Source: Bicycle Federation of America

  a. Between which two consecutive years was the percent of increase the greatest?

  b. What was the percent of increase to the nearest whole percent?

20. *Critical Thinking* Find a number such that adding 1 to it represents a percent increase of $33\frac{1}{3}$%, and subtracting 1 from the new total results in a percent decrease of 25%. How many such numbers do you think there are? Justify your answer.

Mixed Review

21. *Statistics* Refer to the table in Exercise 6 on page 466. If there were 890 students in your school, how many would you expect to say they enjoy listening to music? *(Lesson 11-4)*

22. *Geometry* Classify △*KLM* by its angles and by its sides. *(Lesson 9-4)*

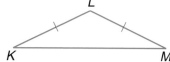

23. *Standardized Test Practice* A piece of wire is 86 inches long. What is the greatest number of 15-inch pieces that can be cut from the wire? *(Lesson 5-7)*

  A 2

  B 4

  C 6

  D 7

  E Not Here

For **Extra Practice,** see page 599.

MATH / IN THE MEDIA

Smart Shopping

A company recently advertised their new heart-shaped cat treats. Although the size of the container stayed the same, the weight of the treats inside went from 4 ounces to 3 ounces. A company representative explained that they lowered the price of the treats from $1.49 to $1.19 to account for the change in weight. However, she also noted that stores are free to set their own price, and it would be wise for people to shop around.

1. Find the percent of decrease in the size and the percent of decrease in the price of the cat treats.

2. Do you think the unadvertised change in size of the treats was fair to consumers? Explain.

MEDIA

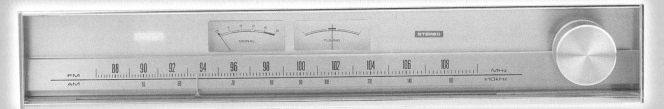

Traci Tong
RADIO PRODUCER/DIRECTOR

Traci Tong is the producer and director of the radio news program *The World* on WGBN in Boston, Massachusetts. Ms. Tong reads, writes, edits, and produces news reports from a large network of international journalists. As in-studio director of the program's daily live broadcast, she ensures that reports are edited to the precise time needed.

To work in radio, a degree in broadcast journalism is usually required. Courses in English, public speaking, foreign languages, mathematics, computers, and electronics are valuable in a field where people work under tight deadlines and accuracy is crucial.

For more information:
Broadcast Education Association
1771 N St., NW
Washington, DC 20036

*inter*CONNECTION
www.glencoe.com/sec/
math/mac/mathnet

Someday I'd like to direct a radio show where important issues are discussed.

Your Turn
Interview a local radio announcer or disc jockey. Make a list of the questions and answers from the interview and write a description of what a job in radio would be like.

What **you'll learn**

You'll learn to solve problems involving sales tax and discount.

When **am I ever going to use this?**

Knowing how to calculate discounts and sales tax will help you determine how much money you're actually spending when you shop.

Recently, tiny video games tucked inside egg-shaped pendants became popular. A toy store in New York City sold nearly 10,000 of the $17.99 toys in one day.

If the sales tax in New York City is $8\frac{1}{4}\%$, what is the total cost of one toy? *Sales tax is the primary way some communities and states raise the money they need to operate.*

One of the following methods can be used to find the total cost.

Method 1 First, find the amount of the sales tax, t.

$$8\frac{1}{4}\% \text{ of } \$17.99 = t$$

.0825 ⊠ 17.99 ⊟ *1.484175*   The sales tax is about $1.49.

Numbers involving money are usually rounded up.

Then add the sales tax to the price of a toy.

$17.99 + \$1.49 = \$19.48$

The total price of a toy is $19.48.

Method 2 First, add the percent of tax to 100%.

$100\% + 8.25\% = 108.25\%$

So, the total price of a toy will be 108.25% of the market price.

Then multiply to find the total cost including tax.

1.0825 ⊠ 17.99 ⊟ *19.474175*

Again, the total price of a toy is $19.48.

These methods can also be used to find the sale price of an item.

Examples
APPLICATION
Real World

① Money Matters A $53 racquet at Sports Galore is on sale for 20% off the regular price. What is the sale price of the racquet?

Method 1 First, find the amount of the discount, d.

20% of $53 = d$

.2 ⊠ 53 ⊟ *10.6*   The discount is $10.60.

Then subtract to find the sale price.

$\$53 - \$10.60 = \$42.40$ The sale price is $42.40.

Method 2 First, subtract the percent of discount from 100%.

$$100\% - 20\% = 80\%$$

So, the sale price of the racquet will be 80% of the regular price.

Then multiply to find the total cost.

.8 ⊠ 53 ⊟ 42.4

Again, the sale price of the racquet is $42.40.

APPLICATION

Real World

②ⓈShopping The advertisement shows the sale price of a new mini portable television. What is the percent of discount?

Find the amount of discount.

$$\$119 - \$89 = \$30$$

Use the percent proportion to find what percent $30 is of $119.

Mini Portable T.V.

Sale Price $89.00

Regular Price $119.00

$$\frac{\text{amount of decrease}}{\text{original amount}} = \frac{r}{100}$$

$$\frac{30}{119} = \frac{r}{100}$$

$$30 \cdot 100 = 119r \quad \textit{Find the cross products.}$$

$$3,000 = 119r$$

$$25.21008403 \approx r \quad \textit{Divide each side by 119.}$$

The percent of discount is about 25%.

CHECK FOR UNDERSTANDING

Communicating Mathematics

Read and study the lesson to answer each question.

1. *Tell* which watch would be cheaper: a $38 watch that is 15% off, or a $50 watch that is 30% off. Explain.

2. *Explain* how you could find the percent of discount of an item that regularly sells for $39, but is marked down to $28. Then find the percent of discount to the nearest whole percent.

Math Journal

3. *Describe* the two methods for finding the total cost of an item if the sales tax is 6%. Which method is more efficient? Explain.

Guided Practice

Find the sales tax or discount to the nearest cent.

4. $17.42 book; $5\frac{1}{2}$% tax

5. $145 chair; 33% discount

Find the total cost or sale price to the nearest cent.

6. $15.99 T-shirt; 20% off   7. $65 video; 7% tax

8. If the regular price of an item is $44 and the sale price is $34, find the percent of discount to the nearest percent.

9. *Money Matters* Jeanelle bought a skirt and a vest on sale. The skirt, originally priced at $28, was 33% off. The vest, originally priced at $21, was 15% off. Which item costs less?

EXERCISES

Practice

Find the sales tax or discount to the nearest cent.

10. $16.58 gloves; $6\frac{1}{2}$% tax   11. $38.50 sweater; 15% off

12. $25 watch; 30% discount   13. $87 radio; 6% tax

14. $36 jeans; $5\frac{1}{2}$% tax   15. $72 in-line skates; 25% discount

Find the total cost or sale price to the nearest cent.

16. $3.49 socks; 35% off   17. $125.59 speakers; 5.75% tax

18. $13.99 CD; 10% discount   19. $64 shoes; 20% off

20. $24 3-D puzzle; $8\frac{1}{2}$% tax   21. $31.65 backpack; $6\frac{1}{2}$% tax

Find the percent of discount to the nearest percent.

22. regular price, $47   23. regular price, $18.99
    sale price, $15   sale price, $13.29

24. Find the total price to the nearest cent if a $69 jacket is on sale for 15% off and the sales tax is 7%.

25. Find the original price of a table if the sale price of $189 was 45% off the original price.

Applications and Problem Solving

26. *Money Matters* The table shows the sneakers that a magazine bought on sale for one of their consumer tests.

    a. Which shoe had the greatest percent of discount?

    b. What was the percent of discount?

Shoe	Original Price	Sale Price
Intimidator	$94	$76
Instapump	$120	$90
L/J Basketball	$90	$37

27. *Employment* Fifteen-year-old Trent Eisenberg owns his own computer company. He offers a 2.5% discount for immediate cash payments. If he collected a $617.77 cash payment for his first job, how much was his original bill without the discount?

28. *Critical Thinking* Find the total percent of change on the final price of an item if the percent of discount is 15% and the sales tax is 5%. Does it matter in which order the discount and the sales tax are applied? Explain.

Mixed Review

29. *Transportation* Sports cars make up about 1% of total car sales. In 1996, 68,203 sports cars were sold. In 1997, about 75,000 sports cars were sold. What was the percent of increase from 1996 to 1997? *(Lesson 11-5)*

30. Standardized Test Practice Which word, written as shown, does not have a vertical line of symmetry? *(Lesson 9-7)*

A	T	B	W	C	H	D	W
	O		H		A		H
	W		I		T		Y
			M		C		
					H		

31. Express 4.8 as a percent. *(Lesson 8-7)*

32. *Advertising* In 1997, packages of a certain cookie claimed "1,000 chips in every bag!" A magazine took six bags and counted every single chip. Here's what the magazine reported. *(Lesson 3-4)*

> The final tally: an average of over 1,100 chips . . . per package! Of course, that doesn't mean every bag measures up – to find that out, we'd have to buy all of these in the world.

Do you think the testers used the mean, median, or mode to summarize their data? Explain.

For **Extra Practice,** see page 600.

Let the Games Begin

Time to Shop

Math Skill
Percent of Discount

Get Ready This game is for 2 or 3 players.   index cards

Get Set Copy the numbers 11, 28, 45, 62, 84, and 98 onto index cards, one number per card. These numbers are the original price of an item. Shuffle the cards and place them facedown in a pile. Label equal sections of two spinners with the digits 0 through 9. Decide which spinner will stand for digits in the tens place and which spinner will stand for digits in the ones place. The number formed by spinning both spinners is the percent of discount.   2 spinners

Go ● One player selects the top card from the pile.

● The player spins both spinners and records the two-digit number. He or she uses this number as the percent of discount, and the number on the card as the original price, and computes the percent of discount. This is the player's score for this turn.

● Continue in this way, taking turns selecting a card, until no cards remain in the pile. The player with the largest total score wins.

 Visit www.glencoe.com/sec/math/mac/mathnet for more games.

What you'll learn

You'll learn to solve problems involving simple interest.

When am I ever going to use this?

You'll use the simple interest formula to find the interest earned on your savings account.

Word Wise

simple interest
principal
rate
time

Mr. Craig borrowed $3,200 from a bank to help pay for a new all-terrain vehicle. If the bank charges him 9.5% interest, how much interest will he pay in a year?

Simple interest is the amount paid for the use of money. The formula for simple interest is $I = prt$, where I is the interest, p is the **principal**, or the amount of money invested or borrowed, r is the annual interest **rate**, and t is the **time** in years.

$I = prt$

$I = 3,200 \cdot 0.095 \cdot 1$ *p = $3,200, r = 9.5% or 0.095, t = 1 year*

$I = 304$

Mr. Craig will pay $304 in interest in a year, in addition to the $3,200 he originally borrowed.

You can also use the formula $I = prt$ to find the simple interest when you deposit money in a savings account. In Example 1, the principal is the amount deposited in the savings account.

Example

APPLICATION

1 **Investing** Cecile has $500 in a savings account that pays 5% simple interest.

   a. How much interest will she earn in 2 years?

   b. How much interest will she earn if she withdraws the money after 9 months?

   a. $I = prt$

     $I = 500 \cdot 0.05 \cdot 2$ *p = $500, r = 5% or 0.05, t = 2 years*

     $I = 50$

     Cecile will earn $50 in interest in 2 years.

If t is given in months, write this as a fraction of one year or as a decimal.

   b. $I = prt$

     $I = 500 \cdot 0.05 \cdot 0.75$ *p = $500, r = 5%,*

     $I = 18.75$ *t = 9 months or 0.75 year*

     Cecile will earn $18.75 in interest in 9 months.

② Finance Mr. Foose bought a watch for $135. He used his credit card, which charges 21% annual interest from the date of purchase. If he does not make any payments or any additional charges, how much would he owe at the end of the first month? Round to the nearest cent.

$$I = prt$$

$$I = 135 \cdot 0.21 \cdot \frac{1}{12} \quad p = \$135, r = 21\%, t = 1 \text{ month or } \frac{1}{12} \text{ year}$$

$$I = 2.3625$$

After 1 month, the interest would be about $2.36. So the total amount owed would be $135 + $2.36 or $137.36.

CHECK FOR UNDERSTANDING

Communicating Mathematics

Read and study the lesson to answer each question.

1. **Explain** how to find the interest on $900 at 4.5% for 1 year.

2. **Tell** how to determine the value of *t* in the formula $I = prt$, if time is given in months.

3. **Write** a few sentences comparing and contrasting principal and interest in borrowing money and in saving money.

Guided Practice

Find the interest to the nearest cent for each principal, interest rate, and time.

4. $2,250, 7%, 3 years

5. $875, 15%, 4 months

Find the interest to the nearest cent on credit cards for each credit card balance, interest rate, and time.

6. $121, 16%, 2 months

7. $5,096, 17%, 2 years

8. **Home Improvement** Mr. Alvarez borrows $5,000 for 3 years at 11.5% simple interest to make home repairs and improvements. How much will he have to pay back, including interest?

EXERCISES

Practice

Find the interest to the nearest cent for each principal, interest rate, and time.

9. $340, 12%, 1.5 years

10. $3,186, 10%, 2 years

11. $4,200, $9\frac{1}{4}$%, 3 years

12. $98.50, $6\frac{1}{2}$%, 16 months

13. $514, 8.75%, 6 months

14. $175.80, 12%, 1.25 years

Find the interest to the nearest cent on credit cards for each credit card balance, interest rate, and time.

15. $839, 21%, 1 year

16. $400, 19%, 6 months

17. $1,200, 19%, 9 months

18. $325, 18.5%, 1 year

19. $672, $15\frac{1}{2}$%, 2 years

20. $1,000, $20\frac{1}{2}$%, 3 months

21. **Write a Problem** that involves a principal amount of $2,600, an interest rate of 17%, and a time of 4 years. Then find the interest.

22. If a principal amount of $600 earned $75 in 2.5 years, find the interest rate.

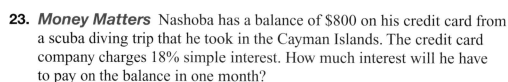

Applications and Problem Solving

23. **Money Matters** Nashoba has a balance of $800 on his credit card from a scuba diving trip that he took in the Cayman Islands. The credit card company charges 18% simple interest. How much interest will he have to pay on the balance in one month?

24. **Finance** Ms. Rollins deposited $700 in a savings account that pays 6.75% annual interest. She also deposited $300 in an account that pays 6% annual interest. If she does not deposit or withdraw any money from her accounts, what will be the total amount in both accounts after 6 months?

25. **Finance** Madeline Morgan has invested $1,500 at 9% annual interest. She has $3,000 more to invest. At what rate must she invest the $3,000 to have a total of $4,950 at the end of the year?

26. **Critical Thinking** Maria Constanza deposits $400 in an account that pays 3.5% annually. At the end of the year, the interest earned is added to the principal. Find the total amount in her account each year for 3 years.

Mixed Review

27. **Money Matters** Find the price to the nearest cent if a $180 electric guitar is on sale for 15% off. *(Lesson 11-6)*

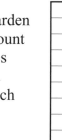

28. **Standardized Test Practice** Ms. Gonzalez has an irregularly shaped garden in her backyard. To determine the amount of mulch she should buy, she estimates the area of the garden. What is a good estimate of this area based on the sketch of her garden? *(Lesson 10-4)*

   A 30 ft²

   B 10 ft²

   C 50 ft²

   D 60 ft²

   E 40 ft²

For **Extra Practice,** see page 600.

29. Subtract $\frac{1}{9}$ from $\frac{5}{12}$. *(Lesson 7-2)*

11-7B Simple Interest

A Follow-Up of Lesson 11-7

computer

spreadsheet software

Simple interest I is calculated by finding the product of the principal p, the rate r, as a decimal, and the time t in years. The formula used is $I = prt$. The new account balance is then found by adding the interest to the principal, or $A = p + I$.

TRY THIS

Work with a partner.

A spreadsheet can be used to generate a simple interest table for various account balances.

Suppose you are the manager of a local bank. Your bank is starting a "Young Savers" program for children. You want to make a table of the interest that children can earn to show how important saving money is. The current rate on the "Young Savers" account is 5%. Use 2 years as the time period. Substitute the values B2 = 5 and C2 = 2 into the spreadsheet to show how much will be in an account with the different starting balances at the end of 2 years.

	A	B	C	D	E
1	Principal	Rate	Time	Interest	New Balance
2					
3	500	= B2/100	= C2	= A3*B3*C3	= A3+D3
4	1000	= B2/100	= C2	= A4*B4*C4	= A4+D4
5	1500	= B2/100	= C2	= A5*B5*C5	= A5+D5
6	2000	= B2/100	= C2	= A6*B6*C6	= A6+D6
7	2500	= B2/100	= C2	= A7*B7*C7	= A7+D7

ON YOUR OWN

Use the spreadsheet to answer each question.

1. Why is the rate in column B divided by 100?

2. What is the account balance after 2 years if the principal is $1,500 and the simple interest rate is 5%?

3. What is the interest earned in 2 years on an account with a principal of $2,000 and an interest rate of 5%?

4. Suppose you wanted to add a new row to the spreadsheet that represents a principal of $3,000. List each of the cell entries (A8, B8, C8, D8, and E8) that you would enter.

5. What entries for cells B2 and C2 would you use to calculate the simple interest on a principal of $1,500 at a rate of 7% for a 9-month period? What is the balance of this account at the end of the 9 months?

Chapter Review For additional lesson-by-lesson review, visit:
www.glencoe.com/sec/math/mac/mathnet

Vocabulary

After completing this chapter, you should be able to define each term, concept, or phrase and give an example or two of each.

Statistics and Probability
circle graph (p. 460)
population (p. 464)
random (p. 464)
sample (p. 464)

Number and Operations
principal (p. 478)
rate (p. 478)
simple interest (p. 478)
time (p. 478)

Problem Solving
solve a simpler problem (p. 454)

Understanding and Using the Vocabulary

State whether each sentence is *true* or *false*. If false, replace the underlined word or number to make a true sentence.

1. A percent is a ratio that compares a number to <u>100</u>.

2. A circle graph is used to <u>compare</u> parts of a whole.

3. There are <u>300°</u> in a circle.

4. When taking a <u>sample</u>, every member of a population is surveyed.

5. When finding a percent of increase, compare the amount of the increase to the <u>new</u> amount.

6. The formula for simple interest is <u>$I = prt$</u>.

7. In order for a sample to be representative, it must be <u>random</u>.

In Your Own Words

8. ***Explain*** two methods for estimating percents.

Objectives & Examples

Upon completing this chapter, you should be able to:

● estimate percents by using fractions and decimals *(Lesson 11-1)*

Estimate 52% of 495.

52% is about 50% or $\frac{1}{2}$.

$\frac{1}{2}$ of 500 is 250.

So, 52% of 495 is *about* 250.

● solve problems by using the percent equation *(Lesson 11-2)*

What number is 16% of 110?

$P = R \cdot B$

$P = 0.16 \cdot 110$ *R = 0.16, B = 110*

$P = 17.6$ 16% of 110 is 17.6.

● construct and interpret circle graphs *(Lesson 11-3)*

Spring: 40%, 144°

Summer: 26%, 93.6°

Autumn: 22%, 79.2°

Winter: 12%, 43.2°

Favorite Season

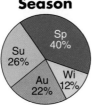

● predict actions of a larger group by using a sample *(Lesson 11-4)*

In a random sample of 100 students at McAuliffe Middle School, 10% have after-school jobs. How many of the 600 students have after-school jobs?

10% of 600 = 0.10 × 600

          = 60 students

Review Exercises

Use these exercises to review and prepare for the chapter test.

Write the fraction, decimal, mixed number, or whole number equivalent of each percent that could be used to estimate.

9. 78%   **10.** 205%

Estimate.

11. 12% of 77   **12.** 88% of 400

13. 149% of 30   **14.** 0.95% of 700

15. 55% of 1,000   **16.** 98% of 1

Write an equation for each problem. Then solve. Round answers to the nearest tenth.

17. 32 is what percent of 50?

18. 65% of what number is 39?

19. Find 62% of 300.

20. 57% of 450 is what number?

21. 15.5% of what number is 108.5?

22. Make a circle graph of favorite soft drinks.

Soft Drink	Percent
Cola	36%
Diet Cola	28%
Root Beer	15%
Lemon Lime	7%
Other	14%

23. Of 20,000 registered voters, the voting preferences of 1,000 are listed in the table.

Candidate	Number of Votes
Chung	220
Addair	390
Armas	310
Undecided	80

How many of the 20,000 voters might you expect to vote for Addair?

Objectives & Examples

Review Exercises

● find the percent of increase or decrease
(Lesson 11-5)

original: $2.75   new: $3.55

difference: $3.55 − $2.75 = $0.80

$\dfrac{0.80}{2.75} = \dfrac{r}{100}$

$80 = 2.75r$ *Find the cross products.*

$\dfrac{80}{2.75} = \dfrac{2.75r}{2.75}$ *Divide each side by 2.75.*

$29.09 \approx r$

The percent of increase is about 29%.

Find the percent of change. Round to the nearest whole percent.

24. original: 44
new: 66

25. original: $22,500
new: $25,400

26. original: $200
new: $180

27. original: $25
new: $33.33

28. Find the percent of increase from 106 miles to 122 miles.

● solve problems involving sales tax and discount *(Lesson 11-6)*

Find the sales tax on a $75 pair of shoes if the tax rate is 6%. Let t represent the sales tax.

$6\% \times \$75 = t$

$0.06 \times 75 = t$

$\$4.50 = t$

Find the sales tax or discount to the nearest cent.

29. $25 shirt; 7% tax

30. $210 bicycle; 15% off

31. $8,000 car; $5\frac{1}{2}\%$ tax

32. $40 sweater; 33% discount

● solve problems involving simple interest
(Lesson 11-7)

Find the interest on $400 at 9% for 3 years.

$I = prt$

$I = 400 \times 0.09 \times 3$ *p = 400, r = 0.09,*

$I = \$108$ *t = 3 years*

Find the interest to the nearest cent for each principal, interest rate, and time.

33. $5,000, 10% 3 years

34. $85, 7.5%, 9 months

35. $2,500, 11%, $1\frac{1}{2}$ years

36. $775, 19%, 30 months

Applications & Problem Solving

37. *Solve a Simpler Problem* Mega Mall had sales of $12.1 million in 1998. Dana's Department Store accounted for 4.9% of the sales. Approximate the sales for Dana's Department Store in 1998. *(Lesson 11-1B)*

38. *Money Matters* Ethan has a credit card balance of $1,000. If he pays off the balance over 2 years at an annual simple interest rate of 18%, how much interest will he pay? *(Lesson 11-7)*

39. *Money Matters* Debi bought a CD player for $115. The sales tax rate was 5%. How much sales tax did she pay? *(Lesson 11-3)*

40. *School* Brenda's attendance during the 1998-99 school year is summarized in the following circle graph. If she was in class on time a total of 167 days, how many total days were in the school year? *(Lessons 11-2 and 11-3)*

School Attendance

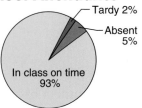

Tardy 2%
Absent 5%
In class on time 93%

Alternative Assessment

● *Open Ended*

You made a circle graph to show the results of a survey taken of 100 students in your school about their favorite cafeteria food. It was published in the school newspaper as shown below.

Favorite Cafeteria Food

Hot Dogs
Pizza
Spaghetti
Hamburgers

What information is missing from the graph? Explain how you can determine this information. Then add it to the graph.

You now find that one of the sections of the graph is too big. What does that tell you about at least one other section of the graph?

● *Completing the* CHAPTER Project

Use the following checklist to make sure your project is complete.

☑ Your data is well-organized and easy to read.

☑ The percents used to describe the data are accurate.

☑ The circle graph is accurate and neat.

☑ The predictions about the radio station programming are logical.

☑ You have a report including a summary of your data.

PORTFOLIO   Select one of the words you learned in this chapter and place the word and its definition in your portfolio. Attach a note explaining why you selected it.

A practice test for Chapter 11 is provided on page 617.

Section One: Multiple Choice

There are nine multiple-choice questions in this section. Choose the best answer. If a correct answer is *not here,* choose the letter for Not Here.

1.

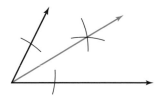

This drawing shows how to —

A construct an angle bisector.

B construct an angle congruent to a given angle.

C construct perpendicular angles.

D construct a segment bisector.

2. Jesse bought 2.4 meters of blue ribbon and 110 centimeters of white ribbon. How many centimeters longer is the blue ribbon than the white ribbon?

F 1.3 cm

G 350 cm

H 107.6 cm

J 130 cm

3. Which best represents a pair of similar figures?

A

B

C

D

Please note that Questions 4–9 have five answer choices.

4. A person's weight on the moon is about $\frac{1}{6}$ of their weight on Earth. About how much would a person weigh on the moon if their weight on Earth is 125 pounds?

F 12 lb

G 20 lb

H 60 lb

J 200 lb

K 600 lb

5. Which procedure could be used to find b in the triangle?

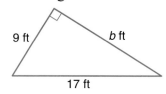

A Add 9 and 17.

B Add 9^2 and 17^2.

C Subtract 9^2 from 17^2.

D Subtract 9 from 17.

E Multiply 9^2 and 17^2.

6. There are 194 Calories in 1 serving of crackers. If a serving consists of 12 crackers, which is the best estimate of the number of Calories in each cracker?

F 10 Calories

G 20 Calories

H 30 Calories

J 60 Calories

K 100 Calories

7. A small sample of students were questioned about their favorite ice cream flavors. The chart shows the results.

Ice Cream	Number of Students
chocolate chip	12
vanilla	6
mint chocolate chip	4
chocolate	3
Total	25

If there are 75 students in the class, what is a good prediction of the number of students who would choose chocolate chip as their favorite ice cream?

A 30   **B** 36
C 45   **D** 60
E 25

8. The temperature rose about 2°F each hour for 9 hours. The beginning temperature was −14°F. What was the temperature after 9 hours?

F 32°F   **G** −12°F
H 4°F   **J** 5°F
K Not Here

9. Tamera had $25.83 in her purse. She spent $6 for lunch. How much money did she have left?

A $20.83   **B** $21.83
C $26.26   **D** $31.83
E Not Here

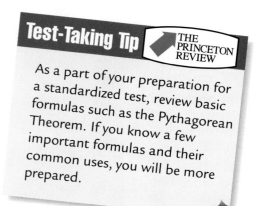

Test-Taking Tip THE PRINCETON REVIEW

As a part of your preparation for a standardized test, review basic formulas such as the Pythagorean Theorem. If you know a few important formulas and their common uses, you will be more prepared.

Section Two: Free Response

This section contains three questions for which you will provide short answers. Write your answers on your paper.

10. The grading scale for a test is shown below.

Wrong Answers	0	1	2	3	4
Score	100	98	95	93	90

Using the pattern in the table, what score would be given for 6 wrong answers?

11. What is the unit price (dollars per ounce) if a 14-ounce can of peaches costs $1.19?

12. Kay spun a spinner 100 times. The results are in the chart.

Color	Number of Spins
Yellow	25
Black	25
Orange	50

If the spinner is divided into four equal sections, how many sections would you expect to be yellow?

Test Practice For additional test practice questions, visit:

www.glencoe.com/sec/math/mac/mathnet

CHAPTER 12

Geometry: Finding Volume and Surface Area

What you'll learn in Chapter 12

- to draw three-dimensional figures when given the top, side, and front views,

- to solve problems by making a model,

- to make a net of the surface area of a solid, and

- to find the volumes and surface areas of rectangular prisms and cylinders.

CHAPTER Project

TURN OVER A NEW LEAF

Leaves come in many different shapes. But one thing that is common to most leaves is that they are flat. Did you ever wonder why? In this project, you will research to find the primary function of a leaf. You will also investigate the relationship between the surface area of a leaf and its volume and summarize your findings in a report.

Getting Started

- Collect a leaf or trace the outline of a picture of a leaf.
- Research leaves and their function.
- Review how to find the area of irregular shapes.

Technology Tips

- Use an **electronic encyclopedia** to do your research.
- Use a **word processor** to write your report.

 **interNET CONNECTION** **Research** For up-to-date information on botany, visit:

www.glencoe.com/sec/math/mac/mathnet

Working on the Project

You can use what you'll learn in Chapter 12 to help you find the relationship between a leaf's surface area and its volume.

Page	Exercise
501	20
513	17
521	Alternative Assessment

COOPERATIVE LEARNING

12-1A Building Three-Dimensional Figures

cubes

A Preview of Lesson 12-1

If you looked at ordinary table salt under a microscope, you would see that salt crystals are cubes. Cubes are examples of three-dimensional figures. They have length, width, and depth.

In this lab, you will use cubes to build other three-dimensional figures.

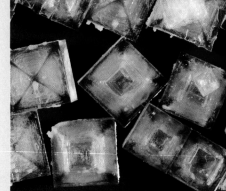

TRY THIS

Work with a partner.

The top view, a side view, and the front view of a stack of cubes are given.

top   side   front

Use cubes to build the three-dimensional figure.
- The top view shows the shape of the base. It is a 3-by-2 rectangle.
- The side view is a 2-by-3 rectangle.
- The front view is a 2-by-2 square.

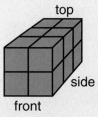

ON YOUR OWN

The top view, a side view, and the front view of three-dimensional figures are shown. Use cubes to build each figure.

1.   top   side   front

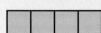  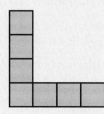

2.

top	side	front

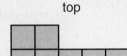 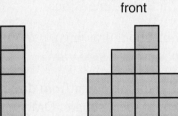

3.

top	side	front

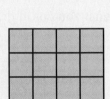

4.

top	side	front

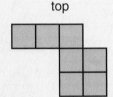 

5. Build a model with cubes and draw the top, side, and front views. Give the drawing of the views to your partner and have him or her build the figure with cubes. Repeat with your partner making the drawings and you building the figure.

6. Share with other groups how you began building the figures.

7. Could you have built the figures in Exercises 1–4 without one of the views? Explain.

8. Is there only one way to build the figures from the drawings given? If no, build another model. If yes, explain why.

9. Build two different models that would look the same from two views, but not the third view. Draw a top view, side view, and front view of each model.

10. *Look Ahead* Describe a real-life situation where it might be necessary for you to draw a top, side, and front view of a three-dimensional figure.

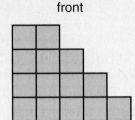

Drawing Three-Dimensional Figures

What you'll learn

You'll learn to draw a three-dimensional figure given the top, side, and front views.

When am I ever going to use this?

Different perspectives of three-dimensional figures are used in art and architecture.

Word Wise

perspective
solids

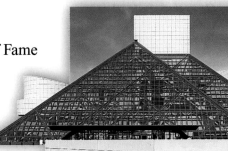

Study the photo of the Rock and Roll Hall of Fame and Museum, which is in Cleveland, Ohio. The main part of the building is a large glass pyramid. In the photograph, it has only two dimensions, width and height. Yet when you look at it, you can visualize the building in three dimensions.

Think about drawing a pyramid in two dimensions. You could draw a top view or a side view.

- If you look down from directly above, you would see a square. Drawing a square would not indicate that the figure has three dimensions.

- Looking at the pyramid directly from the side, you can see a triangle. Again this does not give any indication that this is a view of a pyramid.

However, if you make a drawing that is somewhere between a top and side view, you are able to see that the figure has three dimensions. This view is called a **perspective** view.

You can use isometric dot paper to draw a perspective view.

Example **1** Make a perspective drawing of a figure by using the top, side, and front views of the figure below.

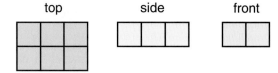

First, sketch a 2-by-3 rectangle for the top. Then, add the front and side views. Finally, add dashed lines to show the hidden edges.

Artists create the illusion of depth in their drawings by using a technique called one-point perspective.

Art The front and side views of an apartment building are shown below. Make a perspective drawing of the building.

front   side

Draw the front of the building like a rectangle. Then, draw lines from the rectangle to a *vanishing point*. Finally, draw the side of the building along the lines to the vanishing point.

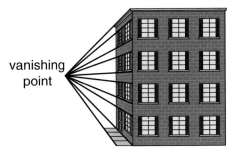

vanishing point

In this chapter, you will study three-dimensional figures called **solids**. Some common solids are shown below.

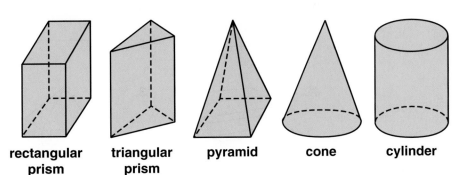

rectangular prism   triangular prism   pyramid   cone   cylinder

---

CHECK FOR UNDERSTANDING

Communicating Mathematics

Read and study the lesson to answer each question.

1. *Draw* a top view and a side view of a cylinder.

2. *Compare and contrast* a pyramid with a cone. How are they alike and how are they different?

3. *Write* about real-world objects that are examples of a triangular prism, a pyramid, a cone, and a cylinder.

Draw a top, a side, and a front view of each figure.

4.

5.

Make a perspective drawing of each figure by using the top, side, and front views as shown. Use isometric dot paper if necessary.

6. top side front

7. top side front

8. *Industrial Technology* A cast iron sleeve has a rectangular hole that extends through its entire length. Draw a top, side, and front view of the sleeve.

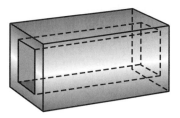

EXERCISES

Practice **Draw a top, a side, and a front view of each figure.**

9. **10.** **11.**

12.  **13.**  **14.**

Make a perspective drawing of each figure by using the top, side, and front views as shown. Use isometric dot paper if necessary.

15. top side front

16. top side front

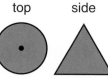

17. top side front

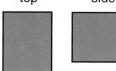

18. top side front

19. Draw a top, a side, and a front view of a piece of furniture.

494   **Chapter 12** Geometry: Finding Volume and Surface Area

20. *Art* Make a drawing of your own that shows one-point perspective.

21. *Architecture* When a building is being designed, an architect provides a set of elevation drawings. These drawings show how the building appears from each side. Draw a set of elevation drawings for your home or school.

22. *Toys* A child's game contains a wood frame with cut-out shapes and some blocks. Name the shape of a block that will fit exactly through both the square and the circle.

23. *Critical Thinking* Draw a three-dimensional figure in which the front and top view has a line of symmetry but the side view does not.

Mixed Review

24. *Money Matters* Maria borrowed $3,500 to help pay for her college tuition. The loan was made at 8% interest for 18 months. Find the amount of interest Maria will pay on the loan. *(Lesson 11-7)*

25. **Standardized Test Practice** Jeremiah walks 7 miles south and 4 miles east. To the nearest tenth of a mile, what is the straight line distance from his starting point? *(Lesson 10-3)*

  A 65.0 mi   **B** 32.5 mi
  C 17.3 mi   **D** 8.1 mi

For **Extra Practice**, see page 600.

26. Find the square root of 196. *(Lesson 10-1)*

MATH IN THE MEDIA

SHOE

1. Which view of the Washington Monument is shown in the comic?

2. Find a photograph of the Washington Monument and draw the remaining views.

PROBLEM SOLVING

12-1B Make a Model

A Follow-Up of Lesson 12-1

Juan and Jessica are designing decorations for the school dance. They want to cover the ceiling of the gymnasium with large geometric shapes. Their math teacher suggests they find out about *Platonic solids* and use them as the basis of their decorations.

Juan

Jessica

I found out that Platonic solids have regular congruent polygons as their faces.

I made a model of one of them — a cube. The faces of a cube are squares.

Here's another one. It's called a tetrahedron. It has four faces that are equilateral triangles.

Let's find as many others as we can. We'll make models of them and show the rest of the decoration committee. Then we can figure out what materials we need to make them.

THINK ABOUT IT

Work with a partner.

1. **Make a model** of a cube and a tetrahedron using straws for the edges and gumdrops for the vertices.

2. **Explain** when making a model is a better strategy than drawing a picture.

3. **Apply** the **make a model** strategy to solve the following problem.

   There are five Platonic solids. One is a cube, and one is a tetrahedron. Find or make a model of the other three.

For **Extra Practice,** see page 601.

ON YOUR OWN

4. *Write a Problem* that can be solved by using the make a model strategy.

5. *Explain* how sculptors and architects use the make a model strategy.

6. *Explain* how Juan and Jessica might use the make a model strategy to plan the decorations for the dance.

MIXED PROBLEM SOLVING

STRATEGIES

Look for a pattern.
Solve a simpler problem.
Act it out.
Guess and check.
Draw a diagram.
Make a chart.
Work backward.

Solve. Use any strategy.

7. *Packaging* A company packages six small books for a children's collection in a decorated 4-inch cube. They are shipped to bookstores in cartons. Twenty cubes fit in a carton with no extra space. What are the dimensions of the carton?

8. *School* Fred, Sarah, and Greg take French, Spanish, and German. No person's language class begins with the same letter as their first name. Sarah's best friend takes French. Which language does each person take?

9. Identical boxes are stacked in the corner of a room as shown below. How many boxes are *not* visible?

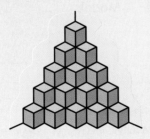

10. During a special on repair work, eight customers lined up outside The Bike Shop with either a bicycle or a tricycle that needed repair. When the owner looked out the window, she counted 21 wheels outside the shop. How many tricycles and bicycles were outside the shop?

11. Jenna spent $1\frac{1}{2}$ hours addressing 50 graduation announcements. At this rate, how long will it take her to address 125 announcements?

12. *Volunteers* Refer to the graph. If you survey 50 teenage volunteers, predict how many volunteer because the cause is important to them.

Teens Help Out

Top reasons they give for volunteering

Feel compassion for needy	84%
Cause important to them	84%
Get new perspective on life	74%
If you help others, others will help you	73%
Is important to people they respect	73%
Looks good on resume	63%

Source: Volunteering and Giving Among Teenagers by Independent Sector

13. **Standardized Test Practice** There are 22 tables for campers in the dining hall at the campground. There are 6 chairs at each of the tables. If 12 chairs in the dining hall are empty, which number sentence could be used to find *N*, the number of campers seated?

A $N = (22 + 6) - 12$

B $N = (22 \times 6) + 12$

C $N = (22 - 12) \times 6$

D $N = (22 \times 6) - 12$

E $N = (22 + 6) + 12$

12-2 Volume of Rectangular Prisms

What you'll learn

You'll learn to find the volume of rectangular prisms.

When am I ever going to use this?

You'll use volume to find the density of an object in science.

Word Wise

rectangular prism
volume

During World War I, some bandages were made from wood pulp, cellulose, and small amounts of cotton. After the war, the company that produced the bandages advertised them as disposable handkerchiefs. The sales increased 400% in only two years.

Today, disposable handkerchiefs, or tissues, are sold in boxes that are shaped like **rectangular prisms**. A rectangular prism is a solid figure that has three sets of parallel congruent sides shaped like rectangles. The **volume** of a solid figure is the measure of the space occupied by it. It is measured in cubic units.

The container at the right has a length of 6 inches, a width of 2 inches, and a height of 4 inches. You can make a model using cubes.

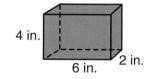

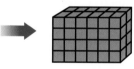

Each layer has 2 · 6 or 12 cubes.

There are 4 layers.

It takes 12 · 4 or 48 cubes to fill the container. The volume of the container is 48 cubic inches.

Study Hint

Reading Math
Prisms have flat surfaces called *faces*. The faces meet to form the *edges* of a prism. The edges meet at corners called *vertices*.

| Volume of a Rectangular Prism | **Words:** The volume (V) of a rectangular prism is found by multiplying the length (ℓ), the width (w), and the height (h). |
| | **Symbols:** $V = \ell wh$ **Model:** |

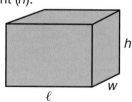

 Example 1

Draw and label a rectangular prism whose length is 7 centimeters, width is 5 centimeters, and height is 10 centimeters. Find its volume.

$V = \ell wh$

$= 7 \cdot 5 \cdot 10$ *Replace ℓ, with 7, w with 5 and h with 10.*

$= 350$

The prism has a volume of 350 cubic centimeters.

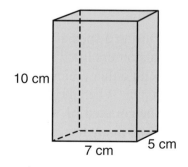

10 cm

7 cm   5 cm

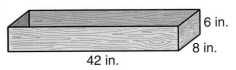

Gardening Charlita wants to buy enough potting soil to fill a window box that is 42 inches long, 8 inches wide, and 6 inches high. If one bag of potting soil contains 576 cubic inches, how many bags should she buy?

Find the volume of the box and then compare it to the volume of potting soil.

Estimate: $40 \times 10 \times 5 = 2,000$

$V = \ell w h$

42 ⊠ 8 ⊠ 6 ⊟ *2016*

42 in.

6 in.

8 in.

Charlita needs 2,016 cubic inches of potting soil. One bag contains 576 cubic inches. Therefore, she needs to buy $2,016 \div 576$ or 4 bags of potting soil.

Study Hint

Reading Math

Volume is expressed in cubic units. So, the abbreviations for volume often use the exponent 3.

cubic inch → in^3

cubic centimeter → cm^3

LOOK BACK

You can refer to Lesson 6-6 to review functions and their graphs.

HANDS-ON MINI-LAB

Work with a partner.

 grid paper   scissors   tape

Try This

1. Cut a piece of grid paper so that it measures 20×20 units. Cut off square corner sections from each corner to make an open box that is $14 \times 14 \times 3$.

2. Fold the paper to make a box. Then tape the corners together.

3. Find the volume of the box. Record your answer as an ordered pair (length of base, volume).

4. Continue making boxes by cutting off square corners from 20×20 grids and finding their volumes until you have made all possible boxes with whole number lengths.

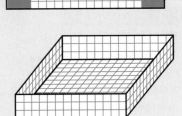

Talk About It

5. Which box has the greatest volume?

6. Graph the ordered pairs (length of base, volume) on a coordinate plane. Describe the graph.

Communicating Mathematics

Read and study the lesson to answer each question.

1. *Write* the abbreviation for cubic meters using an exponent.

2. *Draw* and label a rectangular prism whose length is 5 inches, width is 4 inches, and height is 8 inches.

HANDS-ON MATH

3. *Make models* of three different prisms that have a volume of 24 cubic units.

Guided Practice

Find the volume of each rectangular prism to the nearest tenth.

4.
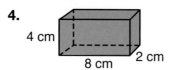
4 cm
8 cm
2 cm

5.
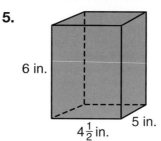
6 in.
$4\frac{1}{2}$ in.
5 in.

6. Find the volume of a rectangular prism whose length is 2.2 centimeters, width is 4.4 centimeters, and height is 5.5 centimeters.

7. *Business* Donna Mendez's new office is 20 feet long, 15 feet wide, and 12 feet high. On average, it costs 9¢ per year to air condition one cubic foot of space. How much will it cost to air condition her office for one month?

Practice

Find the volume of each rectangular prism to the nearest tenth.

8.
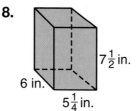
$7\frac{1}{2}$ in.
6 in.
$5\frac{1}{4}$ in.

9.
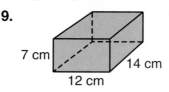
7 cm
14 cm
12 cm

10.
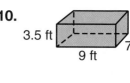
3.5 ft
9 ft
7.2 ft

Family Activity

The capacity of a refrigerator is usually measured in cubic feet. Measure the inside of your refrigerator and see how closely it matches the manufacturer's claims about the capacity.

11.
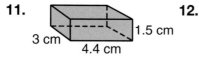
3 cm
1.5 cm
4.4 cm

12.
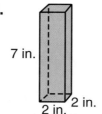
7 in.
2 in.
2 in.

13.
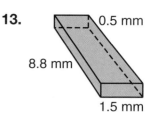
0.5 mm
8.8 mm
1.5 mm

14. Draw and label a rectangular prism whose length is 1 centimeter, width is 8 centimeters, and height is 10 centimeters. Find its volume.

15. A cube has edges that are 7 inches long. Find its volume.

16. Use $V = Bh$ to find the volume of the triangular prism.

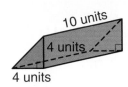

10 units
4 units
4 units

17. *Measurement* How many cubic inches are in a cubic foot?

18. *Swimming* A competition swimming pool is 25 yards long and has 8 lanes that are each 3 yards wide. The pool is filled to a depth of 6 feet.

  a. Express the measurements of the pool in feet.

  b. Find the number of cubic feet of water in the pool.

  c. Each cubic foot of volume contains about 7.5 gallons of water. About how many gallons of water are in the pool?

19. *Landscaping* A landscape architect is designing the outside of a new restaurant. She wants to cover a 40-foot by 12-foot rectangular area with small stones. If she orders 120 cubic feet of stones to spread over the area, how deep will the stones be?

20. *Working on the* **CHAPTER Project** Use the leaf that you collected from page 489.

  a. Trace the outline of the leaf onto centimeter grid paper. Estimate the area of the leaf.

  b. Another formula for the volume of a prism is $V = Bh$, where B is the area of the base and h is the height. Assume the height of your leaf is 0.1 centimeter. Find the volume of the leaf.

21. *Critical Thinking* What is the effect on the volume of a cube if the length of each edge is doubled? tripled? Write an equation to show each relationship.

Mixed Review

22. *Geometry* Make a perspective drawing of a figure by using the top, side, and front views of the figure. *(Lesson 12-1)*

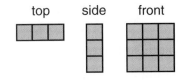

top   side   front

23. *Money Matters* At a sale, Barbara finds a $125 coat marked down to $87.50. What percent of decrease is this? *(Lesson 11-6)*

24. **Standardized Test Practice** The manager of a book store conducted a survey of 1,000 adults to find what kinds of books they like to read. The top five choices are shown at the right. How many adults out of 1,000 who were surveyed like to read biographies? *(Lesson 8-8)*

What People Read	
Mystery/Thriller	25%
Romance	11%
History	7%
Biographies	6%
Religious	6%

  A 6   **B** 60

  C 600   **D** 6,000

For **Extra Practice,** see page 601.

COOPERATIVE LEARNING

12-2B Volume of Pyramids

A Follow-Up of Lesson 12-2

 rice

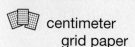 centimeter
grid paper

The newly built entrance to the Louvre Museum in Paris is a glass pyramid. A *pyramid* is a solid figure that has all of its faces, except one, intersecting at a point.

In this lab, you'll investigate the relationship between the volume of a prism and the volume of a pyramid with the same base and height.

TRY THIS

Work with a partner.

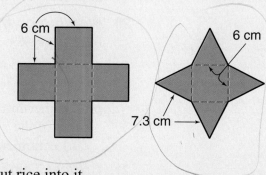

- Copy the two figures at the right onto centimeter grid paper.
- Cut them out and fold on the dashed lines. They will fold into models of an open prism and a pyramid. Tape the edges together to form the models.
- Estimate the ratio of the volume of the prism to the volume of the pyramid.
- Make an opening in the base of the pyramid so you can put rice into it.
- Fill the pyramid with rice. Then pour this rice into the prism. Repeat until the prism is full.

ON YOUR OWN

1. How many pyramids of rice did it take to fill the prism?
2. Compare the heights of the prism and the pyramid.
3. Compare the areas of the bases of each solid.
4. Compare the volume of the prism and the pyramid.
5. Write a formula for the volume of a pyramid.

Volume of Cylinders

The stack of quarters is a model of a **cylinder**. A cylinder is a solid figure that has two congruent, parallel circles as its bases.

To find the value of a stack of quarters, you multiply the value of one quarter, 25¢, by the number of quarters in the stack. You can use a similar method to find the volume of a cylinder.

Cylinder

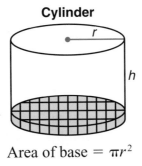

Area of base $= \pi r^2$

The area of the base tells how many unit cubes cover the base. The height tells how many layers there are.

Volume of a Cylinder	**Words:** The volume (*V*) of a cylinder is found by multiplying the area of the base (πr^2) by the height (*h*).	
	Symbols: $V = \pi r^2 h$ **Model:**	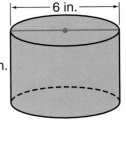

Example ①

Find the volume of a cylinder with a diameter of 6 inches and a height of 4 inches.

The diameter of the cylinder is 6 inches. Therefore, the radius is 3 inches.

Estimate: $3^2 \times 3 \times 4 = 108$

$V = \pi r^2 h$ *Use 3.14 for π.*

$V \approx 3.14 \cdot 3^2 \cdot 4$ *Replace r with 3 and h with 4.*

$V \approx 113.04$

The cylinder has a volume of about 113 cubic inches.

6 in.

4 in.

Catering Mrs. Washington uses a special lemonade pitcher when she caters picnics. It is a cylinder that contains a cylinder on the inside where ice is placed so it does not dilute the lemonade. How many gallons of lemonade can the pitcher hold? (*Hint:* 1 gallon = 231 cubic inches)

Find the volume of the outside cylinder. Then find the volume of the inside cylinder.

Volume of outside cylinder	**Volume of inside cylinder**
The radius is 3.5 inches.	*The radius is 1 inch.*
$V = \pi \cdot 3.5^2 \cdot 10$	$V = \pi \cdot 1^2 \cdot 10$
[π] [×] 3.5 [x²] [×] 10	[π] [×] 1 [x²] [×] 10
[=] *384.8451001*	[=] *31.41592654*
The volume is about 385 cubic inches.	The volume is about 31 cubic inches.

To find the volume of the pitcher, subtract the volume of the inside cylinder from the volume of the outside cylinder.

$385 - 31 = 354$

The pitcher has a volume of about 354 cubic inches. Each gallon of lemonade has a volume of 231 cubic inches. So, the volume of the pitcher is $\frac{354}{231}$ or about 1.5 gallons.

 HANDS-ON **MINI-LAB**

Work in a small group. five cylinder-shaped objects

Try This

centimeter ruler

1. Based on your observations alone, rank the cylinders in order from least to greatest volume.

2. Estimate the volume of each cylinder. Then rank them in order from least to greatest volume based on your estimates.

3. Measure the height and radius of each cylinder. Calculate the volume of each cylinder and rank them in order from least to greatest volume based on your calculations.

Talk About It

4. How close were your estimates to your actual calculations?

5. How precise did you need to be when you measured the cylinders?

Communicating Mathematics

Read and study the lesson to answer each question.

1. *Tell* how the formula for the volume of a cylinder is similar to the formula for the volume of a prism.

2. *Write a Problem* in which you find the volume of a cylinder.

3. *You Decide* Refer to the cylinders at the right. Latisha thinks the volumes of the two cylinders are equal. Cleveland thinks they are not equal. Who is correct? If they are not equal, tell which cylinder has the greater volume.

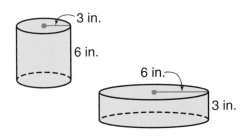

Guided Practice

Find the volume of each cylinder to the nearest tenth.

4.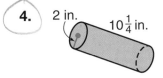

5.

6. *Food* A can of potato chips is 8 inches high and has a radius of 1.5 inches. Find the volume of the can.

EXERCISES

Practice

Find the volume of each cylinder to the nearest tenth.

7.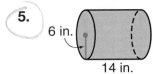

8.

9.

10.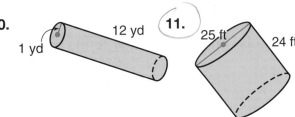

11.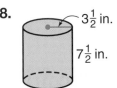

12.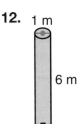

13. The diameter of a cylinder is 5 feet, and the height is 8.25 feet. Find the volume of the cylinder.

14. Use the formula $\pi r h^2$ for the volume of a right circular cylinder.

   a. What is the ratio of the volume of such a cylinder to the volume of one having twice the height but the same radius?

   b. What is the volume of such a cylinder to the volume of one having the same height but twice the radius?

15. *Energy* A pipe is 100 feet long and has an inside diameter of 0.5 foot. Find the number of cubic feet of oil that it can hold.

16. *Baking* A rectangular cake pan is 13 inches by 9 inches by 2 inches. A round cake pan has a diameter of 8 inches and a height of 2 inches. Which will hold more batter, the rectangular pan or two round pans?

17. *Measurement* Firewood is usually sold by a unit of measure known as a cord. A cord is a stack of wood that is 8 feet long, 4 feet wide, and 4 feet high. Suppose a tree has a diameter of 2 feet. Find the height of the tree trunk that would produce about 1 cord of firewood.

18. *Critical Thinking* Design an activity to find the relationship between the volume of a cylinder and the volume of a cone with the same base and height. Describe the relationship. (*Hint:* See Hands-On Lab 12-2B on page 502.)

Mixed Review

19. *Geometry* Find the volume of a rectangular prism with a length of 6 meters, a width of 4.9 meters, and a height of 5.2 meters. *(Lesson 12-2)*

20. *Standardized Test Practice* A 20-foot ladder is leaning against a building. The base of the ladder is 4 feet from the base of the wall. To the nearest tenth of a foot, how far up the wall does the ladder reach? *(Lesson 10-3)*

A 19.2 ft **B** 19.6 ft **C** 29.6 ft **D** 31.4 ft

21. *Patterns* Draw the next two figures in the pattern. *(Lesson 1-6)*

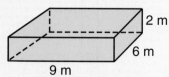

For **Extra Practice,**
see page 601.

CHAPTER 12

Mid-Chapter Self Test

1. Draw the top, front, and side views of a cylinder. *(Lesson 12-1)*

Find the volume of each rectangular prism or cylinder to the nearest tenth.
(Lessons 12-2 and 12-3)

2.

2 m
6 m
9 m

3. 3 in.

3 in.

4.

7 ft
$2\frac{1}{2}$ ft

5. The members of student council are selling popcorn at home basketball games. They can use a rectangular box that is 5 inches long, 2 inches wide, and 8 inches high, or they can use a cylinder-shaped bag that has a 4-inch diameter and is 6 inches high. Which choice holds more popcorn? *(Lessons 12-2 and 12-3)*

BIOCHEMISTRY

Dr. Eloy Rodriguez
BIOCHEMIST

Dr. Rodriguez, a professor at Cornell University in Ithaca, New York, works in a new field of chemistry called *zoopharmacognosy*. He studies how some animals choose to eat plants with healing substances that cure them of their illnesses. He determines what substances occur in these plants and tries to reproduce them in the laboratory. He hopes that someday these synthetic substances can be used to cure human illnesses.

Someday, I'd like to find a cure for cancer.

Most careers in biochemistry require a minimum of four years of college and a Bachelor's degree. Courses in English, foreign language, mathematics, chemistry, physics, biology, social studies, and humanities will prepare students to work in different communities, countries, and cultures while gathering data for their research.

For more information:
Botanical Society of America
Department of Botany
The Ohio State University
1735 Neil Avenue
Columbus, OH 43210

interNET CONNECTION
www.glencoe.com/
sec/math/mac/mathnet

Your Turn
Interview a botanist. Ask what plants they are currently studying and what they hope to learn from their study of such plants. Write a report based on the information gathered in your interview. Request, or find, photos or drawings of the plants being studied. Be sure to include these pictures in your report.

COOPERATIVE LEARNING

12-4A Nets and Surface Area

A Preview of Lesson 12-4

- dot paper
- scissors
- tape

Imagine that you cut a cardboard box along its edges, open it up, and lay it flat. The result is a two-dimensional figure called a *net*. Nets can help you see the regions or faces that make up the surface of the figure.

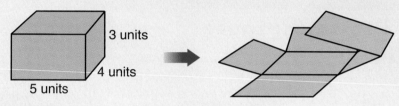

TRY THIS

Work with a partner.

To make a net of the prism shown above, follow these steps.

- Start with the base of the prism. Draw a rectangle that is 5 units long and 4 units wide on dot paper.

- Then visualize unfolding the solid along the edges. Draw the rectangles for the front, back, and sides of the prism.

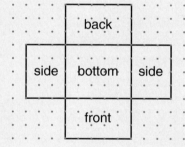

- Finally, draw the top of the prism. *This is only one of several possible nets that you could draw.*

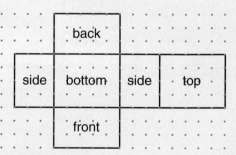

ON YOUR OWN

1. The net shown above is made of rectangles. How many rectangles are in the net?

2. Explain how you can find the total area of the rectangles.

3. The *surface area* of a prism is the total area of its net. Find the surface area of the prism.

Draw a net for each figure. Then cut out the net, fold it, and tape it together to form the three-dimensional figure. Find the surface area of each figure.

4.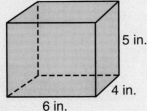
5 in.
4 in.
6 in.

5.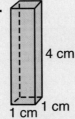
4 cm
1 cm
1 cm

6.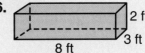
2 ft
3 ft
8 ft

7.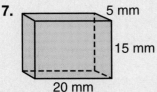
5 mm
15 mm
20 mm

8.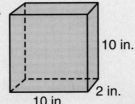
10 in.
10 in.
2 in.

9.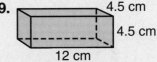
4.5 cm
4.5 cm
12 cm

10. A tetrahedron is a three-dimensional figure made of equilateral triangles. Draw a net for a tetrahedron.

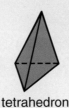

tetrahedron

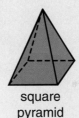

square pyramid

11. Draw a net for a square pyramid.

12. All of the faces of a cube are congruent squares.

  a. Copy and complete the table to find the surface area of each cube. Use nets if necessary.

Dimensions of Each Face	Area of Each Face	Surface Area
1 unit by 1 unit		
2 units by 2 units		
3 units by 3 units		
4 units by 4 units		
5 units by 5 units		

  b. What is the effect on the surface area of a cube if its dimensions are doubled? tripled?

  c. Using the pattern in the table, write an equation to find the surface area of a cube whose length is s units.

13. *Look Ahead* Let ℓ represent the length, w represent the width, and h represent the height of a prism. Write an equation that shows how to find the surface area if you know the length, width, and height of a rectangular prism.

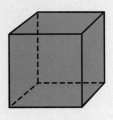
h
w
ℓ

Surface Area of Rectangular Prisms

What you'll learn

You'll learn to find the surface area of rectangular prisms.

When am I ever going to use this?

You'll use surface area to find how much wrapping paper you'll need to wrap a present.

Word Wise

surface area

Did you ever wonder why houses in New England are often two-story houses that resemble a cube, but houses in the southern part of the U.S. are often one-story ranch houses? Blame it on the weather.

Builders in the north want to build energy-efficient homes that keep the heat inside in winter. Builders in the south want to build open, well-ventilated houses to let the cool breezes in all year long.

What does this have to do with geometry? It has to do with **surface area**. Surface area is the sum of the areas of all of the outside surfaces of a three-dimensional figure.

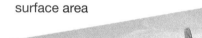

HANDS-ON MINI-LAB

Work with a partner. 8 small cubes

In this Mini-Lab, you will construct prisms with small cubes. The volume of each small cube is 1 cubic unit, and the area of each face is 1 square unit.

Try This

1. Arrange the eight cubes into a prism. Count the number of faces that are on the outside of the prism. This number is the surface area of the prism. Don't forget the faces on the bottom.

2. Draw a sketch of your prism and write the surface area beside it.

3. Repeat Steps 1-2 and make other prisms with the eight cubes.

Talk About It

4. Which prism has the least surface area? the greatest?

5. When you heat a home in the winter, some heat is lost through the walls and roof. What shape house would you want to minimize wall and roof area?

6. Suppose you want lots of air and light in your house. What shape house would you build to maximize wall and roof area?

Example 1

Find the surface area of the rectangular prism.

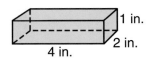

The net shows that there are six faces.

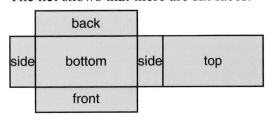

Notice there are three pairs of congruent faces — the front and back, the top and bottom, and the two sides.

Use the formula $A = \ell w$ to find each area.

Face	Dimensions	Area
front	$\ell = 4, h = 1$	$A = 4 \times 1$ or 4
back	$\ell = 4, h = 1$	$A = 4 \times 1$ or 4
top	$\ell = 4, w = 2$	$A = 4 \times 2$ or 8
bottom	$\ell = 4, w = 2$	$A = 4 \times 2$ or 8
side	$w = 2, h = 1$	$A = 2 \times 1$ or 2
side	$w = 2, h = 1$	$A = 2 \times 1$ or 2
		TOTAL 28

The surface area of the prism is 28 square inches.

	Words: The surface area of a rectangular prism equals the sum of the areas of the faces.
Surface Area of a Rectangular Prism	**Symbols:** surface area $= 2\ell w + 2\ell h + 2wh$
	Model: 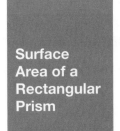

Example 2

Manufacturing Find the surface area of a rectangular shipping box that is 15 inches by 10 inches by 12 inches.

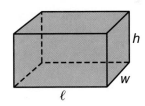

Replace ℓ with 15, w with 10, and h with 12.

surface area $= 2\ell w + 2\ell h + 2\, wh$

surface area $= 2 \times 15 \times 10 + 2 \times 15 \times 12 + 2 \times 10 \times 12$

surface area $= 300 + 360 + 240$ *Multiply first. Then add.*

surface area $= 900$

The surface area of the box is 900 square inches.

LOOK BACK
You can refer to Lesson 1-2 to review the order of operations.

Communicating Mathematics

Read and study the lesson to answer each question.

1. **Explain** how to find the surface area of the rectangular prism at the right.

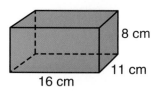

2. **Tell** why surface area is measured in square units even though the figure is three-dimensional.

3. **Write a Problem** about a situation in which you might find the surface area of a rectangular prism.

Guided Practice

Find the surface area of each rectangular prism to the nearest tenth.

4.
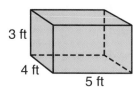

5.
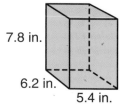

6. Find the surface area of a rectangular prism with a length of 5 centimeters, width of 10 centimeters, and height of 7 centimeters.

7. **Packaging** What is the least amount of wrapping paper that is needed to wrap a gift box that is 2 feet by 1 foot by 3 feet?

Practice

Find the surface area of each rectangular prism to the nearest tenth.

8.
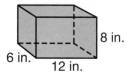

9.
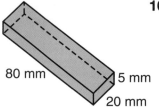

10.
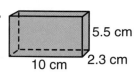

11.
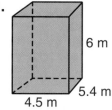

12.
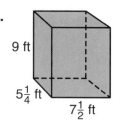

13.
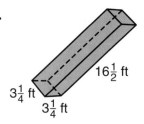

14. Each face of a cube has an area of 8 square inches. What is the surface area of the cube?

Applications and Problem Solving

15. **Algebra** Write a formula for the surface area of a cube in which each edge measures x units.

16. **Physical Science** When you make fruit-flavored drink mix, you dissolve sugar into water. Granulated sugar dissolves faster than a sugar cube.

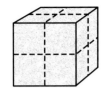

   a. Suppose the length of each edge of a sugar cube is 1 centimeter. Find the surface area of the cube.

   b. Imagine cutting the cube in half horizontally and vertically. Find the total surface area of the eight cubes.

   c. Make a conjecture as to why granulated sugar dissolves faster than a sugar cube.

17. **Working on the** CHAPTER **Project** Refer to Exercise 20 on page 501.

   a. What is the surface area of your leaf?

   b. Find the ratio of the surface area of your leaf to its volume.

18. **Critical Thinking** Draw two prisms such that one has a greater surface area and the other has a greater volume.

Mixed Review

19. **Geometry** Find the volume of a cylinder having a radius of 4 centimeters and a height of 6.5 centimeters. *(Lesson 12-3)*

20. **Standardized Test Practice** A football player has made 80% of the field goals he has attempted in his career. If he attempts 5 field goals in a game, how many would he be expected to make? *(Lesson 8-8)*

   A 5   **B** 4   **C** 3   **D** 2

For **Extra Practice**, see page 602.

Shape-Tac-Toe

Get Ready   This game is for two players.    one index card

   two cubes   ten 2-color counters

Get Set   One player draws a game board on the index card like the one shown. The second player labels the six faces of each cube with these words.

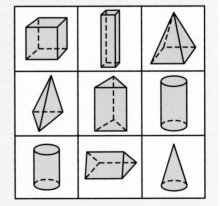

First Cube	Second Cube
prism	circular base
pyramid	triangular base
cylinder	square base
cone	two parallel bases
flat surfaces	one base
curved surface	congruent faces

Go ● The first player rolls both cubes and places a counter on any one shape that matches the conditions on the cubes. Players alternate turns.

   ● The first player to cover three shapes in a row wins.

*inter***NET** CONNECTION Visit www.glencoe.com/sec/math/mac/mathnet for more games.

12-5 Surface Area of Cylinders

What you'll learn

You'll learn to find the surface area of cylinders.

When am I ever going to use this?

You'll use surface area to find the amount of paint needed to cover a cylindrical container.

Isn't it great to have a cold soft drink on a hot day! Now you can have one without the hassle of carrying a bulky ice chest or finding a refrigerated vending machine. A recently-invented can contains your favorite soft drink and can chill it anytime. Just turn the can over and press the button on the bottom of the can. In about 90 seconds, the temperature inside the can drops by 30°F, and you have a cold soft drink.

Soft drink cans are in the shape of cylinders. You can find the surface area of a cylinder by finding the area of all of the surfaces.

HANDS-ON MINI-LAB

Work with a partner.   soft drink can   grid paper   scissors

Try This

• Trace the top and bottom of the can on grid paper. Cut out the shapes.

• Cut a long rectangle from grid paper. The width of the rectangle should be the same as the height of the can. Wrap the grid paper around the side of the can. Cut off the excess paper so the edges just meet.

Talk About It

1. Make a net of the cylinder.

2. Describe the shapes in the net.

3. Tell how the length of the rectangle is related to the radii of the circles.

4. Explain how to find the surface area of the cylinder.

The Mini-Lab suggests that you can open a cylinder and lay it flat in the same way you did with a rectangular prism.

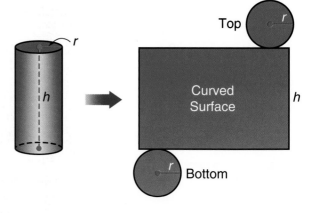

Example 1 **Find the surface area of a 12-ounce soft drink can.**

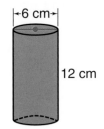

- First, find the area of the top and bottom. Use the formula for the area of a circle, $A = \pi r^2$. *The radius of the circle is 3 centimeters.*

| π | × | 3 | x² | = | *28.27433388* |

The area of one circle is about 28 square centimeters. So, the area of the top and bottom of the can is about 2×28 or 56 square centimeters.

- Now calculate the area of the curved surface. When unfolded, the curved surface has the shape of a rectangle. The width of the rectangle is the height of the can, and the length is the circumference of the base.

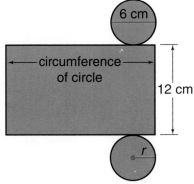

$A = \ell w$

$A = 2\pi r \times w$ *Replace ℓ with $2\pi r$, the formula for circumference.*

$A = 2\pi r \times h$ *Replace w with h.*

$A = 2 \times \pi \times 3 \times 12$ *Replace r with 3 and h with 12.*

| 2 | × | π | × | 3 | × | 12 | = | *226.1946711* |

The area of the curved surface is about 226 square centimeters.

- Finally, add the area of the curved surface to the area of the two circles.

$226 + 56 = 282$

The surface area of the soft drink can is about 282 square centimeters.

Surface Area of a Cylinder	**Words:**	The surface area of a cylinder equals the sum of the areas of the circular bases ($2\pi r^2$) and the area of the curved surface ($2\pi rh$).
	Symbols:	surface area $= 2\pi r^2 + 2\pi rh$
	Model:	

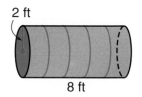
2 ft
8 ft

Agriculture A cylindrical gasoline storage tank on Mr. Baker's farm needs to be painted. He has almost one gallon of paint leftover from another painting job. If one gallon of paint covers 350 square feet, does Mr. Baker have enough paint for this job?

First, find the surface area of the tank.

surface area $= 2\pi r^2 + 2\pi rh$

surface area $= (2 \times \pi \times 2^2) + (2 \times \pi \times 2 \times 8)$ *Replace r with 2 and h with 8.*

2 [×] [π] [×] 2 [x²] [+] 2 [×] [π] [×] 2 [×] 8 [=] *125.6637061*

The surface area of the tank is about 126 square feet. Since one gallon of paint will cover 350 square feet and Mr. Baker has almost one gallon of paint, he probably has enough paint.

CHECK FOR UNDERSTANDING

Communicating Mathematics

Read and study the lesson to answer each question.

1. **Explain** how you use the circumference of the base of a cylinder to help you find the surface area of the cylinder.

2. **Choose** the situation that represents the surface area of a cylinder.
   a. the amount of tomato juice that is inside a can
   b. the amount of sheet metal needed to make the can

HANDS-ON MATH

3. **Estimate** the surface area of three different cans of food. Then measure each can, find the surface area, and rank them in order from greatest surface area to least surface area.

Guided Practice

Find the surface area of each cylinder to the nearest tenth.

4.
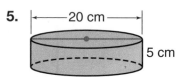
4 m
7 m

5.
20 cm
5 cm

6. The radius of a cylinder is 5 feet, and its height is 4 feet. Find the surface area.

EXERCISES

Practice

Find the surface area of each cylinder to the nearest tenth.

7.
8 mm
4 mm

8.
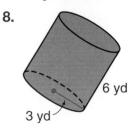
6 yd
3 yd

9.
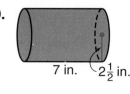
7 in. $2\frac{1}{2}$ in.

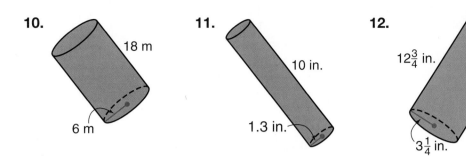

10. 18 m / 6 m

11. 10 in. / 1.3 in.

12. $12\frac{3}{4}$ in. / $3\frac{1}{4}$ in.

13. Find the surface area of a cylinder whose height is 14 inches and whose base has a diameter of 16 inches.

14. Find the surface area of a cylinder whose height is 12 inches and whose base has a circumference of 37.88 inches.

15. *Write a Problem* in which you need to find the surface area of a cylinder.

Applications and Problem Solving

16. *Design* A can of vegetables is 5 inches high, and its base has a radius of 2 inches. How much paper is needed to make the label on the can?

17. *Manufacturing* The three metal containers below each hold about 1 liter of liquid. Find the surface area of each container.

a. 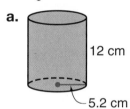 12 cm / 5.2 cm

b. 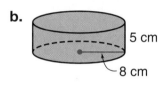 5 cm / 8 cm

c. 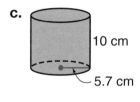 10 cm / 5.7 cm

d. If you were in charge of manufacturing the containers, which container would you choose so that you would use the least amount of metal?

For **Extra Practice,** see page 602.

18. *Critical Thinking* If you double the height of a cylinder, will its surface area also double? Explain your reasoning.

Mixed Review

19. **Standardized Test Practice** The surface area of a cube is 294 mm². What is the length of one side of the cube? *(Lesson 12-4)*

   A 7 mm   **B** 17 mm   **C** 21 mm   **D** 27 mm

20. Divide $4\frac{2}{5}$ by $\frac{1}{2}$. *(Lesson 7-9)*

21. *Entertainment* Refer to the graph. *(Lesson 6-6)*

   a. Graph the ordered pairs (year, ticket sales) on a coordinate plane.

   b. Write a statement that describes the trend in movie ticket sales.

Ticket Sales per Film	(million dollars)
1992	$32.5
1993	$32.0
1994	$29.4
1995	$23.5
1996	$24.6

Source: Motion Picture Association of America, 1997

interNET
CONNECTION Chapter Review **For additional lesson-by-lesson review, visit:**
www.glencoe.com/sec/math/mac/mathnet

Vocabulary

After completing this chapter, you should be able to define each term, concept or phrase and give an example or two of each.

Geometry	Problem Solving
cylinder (p. 503)	make a model (p. 496)
net (p. 508)	
perspective (p. 492)	
Platonic solids (p. 496)	
pyramid (p. 502)	
rectangular prism (p. 498)	
solids (p. 493)	
surface area (p. 510)	
volume (p. 498)	

Understanding and Using the Vocabulary

Choose the correct term or number to complete each sentence.

1. When a figure is drawn so you are able to see that it has three dimensions, the view is called a(n) (isometric, perspective) view.

2. A (rectangular prism, rectangle) is a solid figure that has three sets of parallel congruent sides.

3. The (volume, surface area) of a solid figure is the measure of the space occupied by it.

4. Volume is measured in (square, cubic) units.

5. The volume of a rectangular prism is found by (adding, multiplying) the length, the width, and the height.

6. A (cylinder, prism) is a solid figure that has two congruent, parallel circles as its bases.

7. The volume of a cylinder with a radius of 4 inches and a height of 2 inches is about (100.5, 25.1) cubic inches.

8. Surface area is the sum of the (areas, volumes) of all of the outside surfaces of a three-dimensional figure.

9. The surface area of a prism with a length of 6 meters, width of 2 meters, and a height of 1 meter is (20, 40) square meters.

In Your Own Words

10. *Explain* how to find the surface area of a cylinder.

Objectives & Examples

Upon completing this chapter, you should be able to:

● draw a three-dimensional figure when given the top, side, and front views *(Lesson 12-1)*

Make a perspective drawing by using the top, side, and front views.

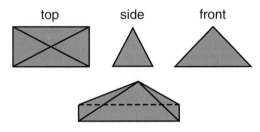

top   side   front

Review Exercises

Use these exercises to review and prepare for the chapter test.

Draw a top, a side, and a front view of the figure.

11.

Make a perspective drawing of each figure by using the top, side, and front views as shown. Use isometric dot paper if necessary.

top   side   front

12.

13.

● find the volume of rectangular prisms *(Lesson 12-2)*

Find the volume of a rectangular prism with a length of 5 inches, a width of 3 inches, and a height of 2 inches.

$V = \ell wh$
$V = 5 \times 3 \times 2$
$V = 30$

2 in.
3 in.
5 in.

The volume is 30 cubic inches.

Find the volume of each rectangular prism to the nearest tenth.

14.

6.3 mm
1.2 mm   2.5 mm

15.

6 ft
$4\frac{1}{2}$ ft $6\frac{1}{4}$ ft

16.

7 cm
4 cm   3.5 cm

17.
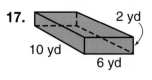
2 yd
10 yd
6 yd

18. Find the volume of a rectangular prism whose length is 9 inches, width is 5 inches, and height is 3 inches.

Objectives & Examples

find the volume of cylinders *(Lesson 12-3)*

Find the volume of a cylinder with a radius of 4 centimeters and a height of 8 centimeters.

$V = \pi r^2 h$

$V \approx 3.14 \times 4^2 \times 8$

$V \approx 401.92$

The volume is about 402 cubic centimeters.

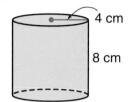

Review Exercises

Find the volume of each cylinder to the nearest tenth.

19. 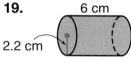 6 cm / 2.2 cm

20. 18 mm 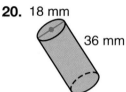 36 mm

21. The diameter of a cylinder is 3.4 inches, and the height is 5.2 inches. Find the volume of the cylinder.

---

find the surface area of rectangular prisms *(Lesson 12-4)*

Find the surface area of a rectangular prism with a length of 3 centimeters, a width of 8 centimeters, and a height of 2 centimeters.

surface area $= 2\ell w + 2\ell h + 2wh$

surface area $= 2 \times 3 \times 8 + 2 \times 3 \\ \times 2 + 2 \times 8 \times 2$

surface area $= 48 + 12 + 32$

surface area $= 92$

The surface area is 92 square centimeters.

Find the surface area of each rectangular prism to the nearest tenth.

22. 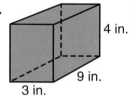 4 in. / 9 in. / 3 in.

23. 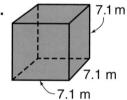 7.1 m / 7.1 m / 7.1 m

24. Find the surface area of a rectangular prism with a length of 4 meters, width of 3 meters, and height of 4 meters.

---

find the surface area of cylinders *(Lesson 12-5)*

Find the surface area of a cylinder whose height is 6 millimeters and whose radius is 2 millimeters.

surface area $= 2\pi r^2 + 2\pi rh$

surface area $= 2 \times \pi \times 2^2 + 2 \times \pi \times 2 \times 6$

surface area ≈ 100.48

The surface area is about 100.5 mm².

Find the surface area of each cylinder to the nearest tenth.

25. 8 cm / 20 cm   **26.** 3 m / 9 m

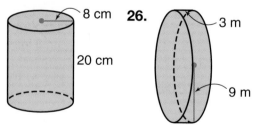

27. Find the surface area of a cylinder whose height is 3.2 inches and whose base has a diameter of 4.6 inches.

Applications & Problem Solving

28. *Make a Model* A large cube is made up of 27 small cubes. The outside of the cube is painted blue. If the large cube is taken apart, how many small cubes would have none of their sides painted? *(Lesson 12-1B)*

29. *Manufacturing* A cereal box has a length of 11 inches, a height of 14 inches, and a depth of 1.5 inches. What is the volume of the box? *(Lesson 12-2)*

30. *Pottery* In his art class, Arturo made a vase in the shape of a cylinder. The diameter is 5 inches, and the height is 10 inches. Find the maximum volume of water the vase can hold. *(Lesson 12-3)*

31. *Pets* The Totally-Pets Company wants to make a pet carrier that has a length of 2.5 feet, a height of 1 foot, and a width of 1.25 feet, as shown below. How much plastic is needed to make this carrier? *(Lesson 12-4)*

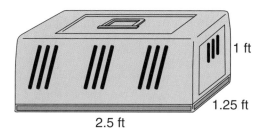

1 ft

1.25 ft

2.5 ft

Alternative Assessment

● *Open Ended*

Suppose you are given an empty oatmeal canister that measures 11 inches tall with a 6-inch diameter to use for a class project. You are told to cover the canister with wrapping paper, using the smallest possible amount of paper. How can you determine the amount of paper you will need?

Suppose you found a sheet of wrapping paper at home that is $12\frac{1}{2}$ inches wide by 19 inches long. Is the paper big enough to cover the canister?

● *Completing the* CHAPTER Project

Use the following checklist to make sure your report is complete.

☑ You have included a drawing of your leaf.

☑ Your report includes a discussion of the main function of a leaf and an explanation why leaves have a large surface area compared to their volume.

☑ You have used the terms *ratio, volume,* and *surface area* in your report.

 PORTFOLIO Review the items in your portfolio. Make a table of contents of the items, noting why each item was chosen. Replace any items that are no longer appropriate.

A practice test for Chapter 12 is provided on page 618.

Section One: Multiple Choice

There are nine multiple choice questions in this section. Choose the best answer. If a correct answer is *not here*, choose the letter for Not Here.

1. A CD holder is shaped like a rectangular prism. What is the volume of the box if it measures 6 inches by 6 inches by 12 inches?

A 24 in³

B 48 in³

C 72 in³

D 432 in³

2. Which figure has exactly four faces?

F

G

H

J

3. Montega worked 15 hours last week. He earned $5 per hour. Which equation can be used to find his total earnings?

A $x = 15 \div 5$

B $5x = 15$

C $x = 15 \times 5$

D $15x = 5$

4. The top, side, and front views of a three-dimensional figure are given. What is the figure?

top   side   front

F cone

G cylinder

H rectangular prism

J sphere

Please note that Questions 5–9 have five answer choices.

5. In Fairfield County, about 49% of the registered voters actually vote. If a district in the county has about 600 registered voters, about how many should vote in the next election?

A less than 290

B between 290 and 310

C between 310 and 320

D between 320 and 330

E more than 330

6. Last year, a business owner donated 10% of her profits to a shelter for the homeless. If her business is selling T-shirts and she makes a profit of $9.95 for each shirt, what other information is needed to find how much money she donated?

F the cost of making one T-shirt

G the number of weeks per year

H the total number of T-shirts sold

J the number of shirts in each of her stores

K the selling price of each shirt

7. Alicia bought a new guitar. She made a 40% down payment and will make 12 monthly payments of $30 each. What is a reasonable total price for the guitar?

A $400   **B** $500
C $600   **D** $700
E $800

8. The seventh grade class collected money each year for the last three years to make a donation to the student council.

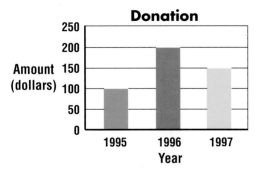

Donation

Amount (dollars)

According to the graph, which is the average yearly amount of money collected over the three-year period?

F $150   **G** $200
H $250   **J** $350
K $450

9. To the nearest cubic inch, find the volume of the cylinder.

A 7.5 in³
B 30 in³
C 35 in³
D 45 in³
E Not Here

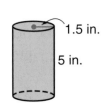

1.5 in.

5 in.

Test-Taking Tip

THE PRINCETON REVIEW

Many problems can be solved without much calculating if you estimate carefully. Look to see which answer choices are *not* reasonable for the information given in the problem.

Section Two: Free Response

This section contains six questions for which you will provide short answers. Write your answers on your paper.

10. Colleen wants to buy a pair of shoes that cost $54.98. She must also pay a sales tax of 6%. To the nearest cent, what is the total cost of the shoes?

11. What is the reciprocal of $3\frac{3}{4}$?

12. Suppose you want to find $12\frac{1}{2}\%$ of a number using a calculator. What decimal can you enter for $12\frac{1}{2}\%$?

13. Three out of five students usually buy their lunch from the cafeteria. If there are 500 students in the school on a given day, about how many lunches should be prepared?

14. What is the prime factorization of 45?

15. The figure below shows a rectangular swimming pool. How much water is in the pool when it is filled to a depth of 5 feet?

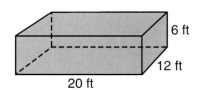

6 ft

12 ft

20 ft

THE PERFECT PACKAGE

Did you ever buy a product because of the way it was packaged? Manufacturers try to design packages that are economical to produce yet appeal to consumers. Most of the products you buy are packaged. Foods are often in boxes or cans, bars of soap are wrapped in paper, and cologne may be in a bottle that is inside a box. How much material does it take to make a package? How much does a package hold?

What You'll Do

In this investigation, you will find the volume and surface area of some packages from products you use. You will also design your own package.

Materials

 packages from a supermarket or department store

 calculator   cardboard or poster board

ruler

Procedure

1. Work individually. Find two packages from products such as an empty vegetable can or a box from a tube of toothpaste. Measure each package and find the volume and surface area.

2. Work with a partner. Make a table to record the measurements, volume, and surface area of the packages.

3. Choose a product for which to design a package. Design two different packages—one should be a prism and the other should be a cylinder. The packages should have about the same volume. Explain how you determined the dimensions of the two containers so that their volumes were about the same.

4. Use cardboard or poster board to make your packages and then find their surface area.

5. Make a display that includes the packages, the surface area and volume, and scale drawings of patterns for the packages.

Technology Tips

- Use a **spreadsheet** to calculate the surface areas and volumes.

- Use a **word processor** to prepare your display.

- Use **geometry software** to help you design a pattern.

Making the Connection

Use the data collected about packages to help in these investigations.

Language Arts

Decide which container would be best for your product, the prism or the cylinder. Prepare a presentation using the facts about volume and surface area to convince your class which package would be best.

Science

Investigate the history of packaging. Research plastic and polystyrene and tell the impact that these materials have on the environment. Which materials are safest for the environment?

Art

Make a logo or design that would help sell your product. Use color and present the logo or design on the container you made.

Go Further

- Investigate the ratio of height to width of packages that are rectangular prisms. Describe what you find.

- Search for unique packages that are not cylinders or rectangular prisms. Find the volumes and surface areas of these containers. Make your own unique packages.

🖳 interNET
CONNECTION For current information of packaging, visit:
www.glencoe.com/sec/math/mac/mathnet

 You may want to place your work on this investigation in your portfolio.

Exploring Discrete Math and Probability

What you'll
learn in Chapter 13

- to find and compare experimental and theoretical probabilities,
- to solve problems by acting them out,
- to count outcomes by using a tree diagram or the Counting Principle,
- to find the probability of independent and dependent events, and
- to find the number of permutations or combinations of a group of objects.

CHAPTER Project

ADVANCE TO *GO* AND COLLECT $200

In this project, you will analyze the use of probability in your favorite board game or television game show. You will also design your own game that combines skill and chance.

Getting Started

- Choose a game to analyze.
- Outline a plan for designing your own board game.

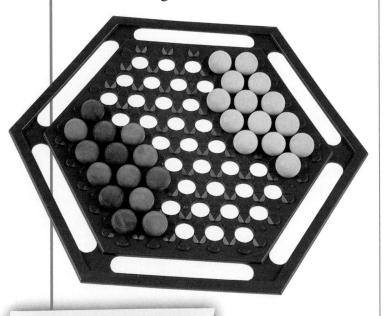

Technology Tips

- Surf the **Internet** for information about games.
- Use a **word processor** to write the rules of your game.

interNET **CONNECTION** **Research** **For up-to-date information on board games, visit:**

www.glencoe.com/sec/math/mac/mathnet

Working on the Project

You can use what you'll learn in Chapter 13 to help you analyze and design your games.

Page	Exercise
533	22
545	18
557	Alternative Assessment

PROBLEM SOLVING

13-1A Act it Out

A Preview of Lesson 13-1

Do you dread pop quizzes? Most students don't like them. But if you do your homework, you'll be more likely to answer the questions correctly and get a good grade. Dave and Selena just finished taking a pop quiz that had 10 true-false questions. Let's listen in as they talk about it.

Sometimes I think tossing a coin would be a good way to answer the questions on a true-false test. If the coin showed tails, I'd write T, and if it showed heads, I'd write F.

Selena

That seems risky. Why don't we do an experiment to act it out? We could see if tossing a coin is a good strategy.

Dave

OK—let's experiment with a 10-point quiz. Let's suppose the correct answers are T, F, F, F, T, T, F, T, F, T.

I'll toss a coin 10 times for each "quiz" and you record the answers. Then we'll circle the ones that are correct for each trial.

Answers	T	F	F	F	T	T	F	T	F	T	Number Correct
Trial 1	Ⓣ	T	Ⓕ	Ⓕ	F	Ⓣ	Ⓕ	Ⓣ	T	F	6
Trial 2	Ⓣ	T	Ⓕ	T	Ⓣ	F	Ⓕ	Ⓣ	T	F	5
Trial 3	F	T	Ⓕ	T	Ⓣ	Ⓣ	Ⓕ	F	Ⓕ	Ⓣ	6
Trial 4	Ⓣ	Ⓕ	T	T	F	Ⓣ	Ⓕ	Ⓣ	Ⓕ	F	6

THINK ABOUT IT

Work with a partner.

1. **Compute** the average score for the four trial quizzes shown above.

2. If a passing grade is 70%, **decide** whether tossing a coin is a good strategy to use when taking true-false tests.

3. **Explain** whether the results of this experiment would be the same if it were repeated.

4. **Apply** the **act it out** strategy to solve this problem.

   A baseball card manufacturer is holding a contest. Each package of cards contains a puzzle piece. If you collect all 6 different pieces, you win two tickets to a major league game. There is an equally-likely chance of getting a different puzzle piece each time. How many packages of cards would you need to buy to win the contest?

For **Extra Practice,** see page 602.

ON YOUR OWN

5. *Make a model* using a spinner, that will act out a basketball player making free throws, if she usually succeeds in making 75% of her free throws.

6. *Explain* an advantage of using the act it out strategy to solve a problem.

7. *Look Ahead* Suppose Mavis, Miguel, and Carianne each left one book in the school library. When they returned to their homeroom, their teacher randomly handed out the three books to the students. Act out this situation 20 times. For how many of these times did Carianne receive the same book she was reading in the library?

MIXED PROBLEM SOLVING

STRATEGIES

Look for a pattern.
Solve a simpler problem.
Act it out.
Guess and check.
Draw a diagram.
Make a chart.
Work backward.

Solve. Use any strategy.

8. *Money Matters*
Mikael received a birthday gift of money. He loaned $5 to his friend Tequisha and spent half of the remaining money. The next day he received $10 from his uncle. After spending $9 at the movies, he still had $11 left. How much money did he receive for his birthday?

9. *Geometry* The length of a rectangle is 8 inches longer than its width. What are the length and width of the rectangle if the area is 84 square inches?

10. *Patterns* This pattern below is known as Pascal's Triangle. Find the pattern and complete the 6th and 7th rows.

```
1st row ──────────→ 1
2nd row ─────────→ 1    1
3rd row ───────→ 1    2    1
4th row ─────→ 1    3    3    1
5th row ──→ 1    4    6    4    1
```

11. *Earth Science* Mauna Kea, a Hawaiian mountain, is 3.35×10^4 feet tall when its height is measured from the ocean floor. Mt. Everest, the highest mountain above sea level, is about 29,000 feet. Which mountain is taller?

12. *Life Science* About 190 million years ago, a giant lizard-like dinosaur called an apatosaurus roamed Earth. It weighed about 30 tons. Today, the largest living land animal is the African elephant. It weighs about 16,500 pounds. How many more tons did the brontosaurus weigh than today's elephant?

13. *Patterns* Complete the pattern.
100, 98, 94, __?__, 80, __?__ .

14. *Standardized Test Practice* About 48% of the registered voters actually vote in an election. In a district with 900 registered voters, which is a good estimate of the number who will vote in the next election?

A less than 300

B between 300 and 450

C between 450 and 600

D between 600 and 750

E more than 750

Theoretical and Experimental Probability

What you'll learn

You'll learn to find and compare experimental and theoretical probabilities.

When am I ever going to use this?

Probability is used in weather forecasting, games, and business.

Word Wise

outcome
sample space
theoretical probability
experimental probability

If you are one of the estimated 480 million people who have played the most popular board game, you know that skill helps you win the game. But there is also an element of chance in the game. Chance makes games more exciting.

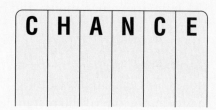

Let the Games Begin

Take a Chance

Get Ready   This game is for two, three, or four players.

   2 number cubes

Get Set   Each player makes a score sheet like the one below.

Go   Each letter of the word CHANCE represents a different round of the game. The object of the game is to get the greatest number of points during the six rounds.

C	H	A	N	C	E

- The first player rolls the number cubes and records the sum of the numbers in the first C column, unless a "1" is rolled on either of the cubes.

- If a "1" is rolled, all of the player's points in the first C column are wiped out, and his or her turn is over. If "double 1" is rolled, all of the points in the previous rounds are also wiped out.

- If a "1" is *not* rolled, the player chooses to roll again or to stop and keep the points he or she already has. Then it is the next player's turn. After each player has a turn, the first round is complete. Continue in this manner for the next rounds.

- The winner is the player with the most points at the end of six rounds.

inter NET CONNECTION   Visit www.glencoe.com/sec/math/mac/mathnet for more games.

In the game on page 530, you tried to guess the probability that a "1" would be rolled. There are many possible results, or **outcomes**, when you roll two number cubes. The list of all possible outcomes is called the **sample space**.

1, 1	1, 2	1, 3	1, 4	1, 5	1, 6	*(1, 6) means that the*
2, 1	2, 2	2, 3	2, 4	2, 5	2, 6	*first number cube is*
3, 1	3, 2	3, 3	3, 4	3, 5	3, 6	*a 1 and the second*
4, 1	4, 2	4, 3	4, 4	4, 5	4, 6	*number cube is a 6.*
5, 1	5, 2	5, 3	5, 4	5, 5	5, 6	
6, 1	6, 2	6, 3	6, 4	6, 5	6, 6	

LOOK BACK

Refer to Lesson 4-8 to review how to find the probability of simple events.

There are 36 possible outcomes and 11 ways of rolling at least one "1". The **theoretical probability** of rolling a "1" is $\frac{11}{36}$. Theoretical probability is the ratio of the number of favorable outcomes to the total number of possible outcomes. It tells us that, in the long run, a "1" should occur about 11 times in every 36 rolls.

Example

APPLICATION

1

Games Find the theoretical probability of rolling "double 1" with a pair of number cubes. *A "double 1" is given by (1, 1).*

$$P(\text{"double 1"}) = \frac{\text{number of ways rolling "double 1"}}{\text{number of possible outcomes}} \text{ or } \frac{1}{36}.$$

The probability of rolling "double 1" is $\frac{1}{36}$ or about 3%.

Study Hint

Reading Math

P("double 1") is read as *the probability of rolling "double 1".*

You can also conduct an experiment by rolling two number cubes many times and recording the number of times "double 1" occurs. This will give you an **experimental probability**. Experimental probability is based on frequencies obtained in an experiment.

MINI-LAB

Work with a partner.

 2 number cubes

Try This

- Roll two number cubes 100 times. Record each "double 1".
- Compute the experimental probability of a "double 1" occurring.

Talk About It

1. How does your experimental probability compare to the probabilities of those in other groups?
2. Compare your experimental probability to the theoretical probability. If the numbers are different, explain why.
3. Combine your results with other groups to find the experimental probability for the entire class. How does it compare to the theoretical probability?

The frequency table shows the results of an experiment in which one coin was tossed. Find the experimental probability of tossing heads for this experiment.

Outcome	Tally	Frequency
Heads	ⅢⅢ Ⅲ Ⅱ	12
Tails	Ⅲ Ⅲ	8

$$\frac{\text{number of times heads occur}}{\text{number of possible outcomes}} = \frac{12}{12 + 8} \text{ or } \frac{12}{20}$$

The experimental probability of tossing heads is $\frac{12}{20}$ or 60%.

CHECK FOR UNDERSTANDING

Communicating Mathematics

Read and study the lesson to answer each question.

1. *Explain* the difference between theoretical probability and experimental probability.

2. *Tell* whether the experimental probability of an event is always the same. Explain your reasoning.

HANDS-ON MATH

3. *Give an example* of an event with a theoretical probability of $\frac{1}{6}$.

Guided Practice

Refer to the sample space on page 531. Find each theoretical probability.

4. a sum of 7

5. 3 on the first cube and 5 on the second

6. Find the theoretical probability of choosing a girl's name at random from 20 boy's names and 10 girl's names.

7. Ayani spins a spinner like the one at the right 30 times. It lands on red 16 times. What is the experimental probability of spinning red?

8. *Bowling* Sonia averages 3 strikes for every 10 frames of bowling. What is the probability she will get a strike in the first frame of her next game?

EXERCISES

Practice

Refer to the sample space on page 531. Find each theoretical probability.

9. the same number on both cubes

10. a sum less than 4

11. a sum of 1

12. a sum less than 15

13. the first number greater than the second

14. both numbers odd

There are 4 red marbles, 5 green marbles, 6 yellow marbles, and 3 blue marbles in a bag. Suppose you select one marble at random.

15. Find $P(\text{red})$.

16. Find $P(\text{green})$.

17. Find $P(\text{yellow or blue})$.

18. Conduct an experiment in which you use small slips of paper. Write "red" on four slips, "green" on five slips, "yellow" on six slips, and "blue" on three slips. Without looking, choose a slip of paper and record the color. Replace the slip and repeat this another nine times.

   a. Find the experimental probability of choosing yellow or blue slip.

   b. How does the experimental probability compare to the theoretical probability you found in Exercise 17?

19. *Geography*
A state is chosen at random from the United States. Find each probability.

   a. The state borders on the Gulf of Mexico.

   b. The state begins with the letter "T".

20. *Games* Parcheesi is a game that was first played in India. In this game, two dice are rolled. You can put a game piece into play if a total of 5 is shown on the dice or if a 5 is shown on at least one of the dice. What is the probability that a game piece can be put into play in one roll of the dice?

21. *Advertising* The results of an Arbitron survey of radio stations in Fayette County, Kentucky, are shown in the chart. Both WVLK-92.9 FM and WWYC-100.1 FM are country music stations. What is the probability that a person chosen at random who is tuned into one of these five stations is tuned into a country station? Express your answer to the nearest whole percent.

Station	Number of Listeners
WVLK-92.9 FM	115,000
WKQQ-98.1 FM	81,000
WMXL-94.5 FM	76,000
WGKS-96.9 FM	54,100
WWYC-100.1 FM	54,000

Source: Arbitron, December

22. *Working on the* **CHAPTER Project** Make a list of the ways probability is used in the game you have chosen.

23. *Critical Thinking* Twenty red cards are added to a stack of cards with an unknown number of black cards. The stack is then shuffled to thoroughly mix the cards, and the first 20 cards on the stack are turned over. Five of these cards are red. Does this imply that the number of black cards in the stack was 60? Explain.

24. **Standardized Test Practice** Gilberto is covering a cylindrical can with paper for a school project. The can has a height of 15 inches and a radius of 4 inches. How much paper will he need to cover the can completely? *(Lesson 12-5)*

   A 477.5 in²   **B** 375.2 in²   **C** 104.5 in²   **D** 60.0 in²

25. Find 35% of 20 using the percent equation. *(Lesson 11-2)*

26. *Algebra* Solve $5t = 125$. *(Lesson 6-2)*

Lesson 13-1 Theoretical and Experimental Probability   **533**

13-2 Tree Diagrams

What you'll learn

You'll learn to use tree diagrams to count outcomes.

When am I ever going to use this?

You can use tree diagrams to analyze games of chance.

Word Wise

tree diagram
fair game

Recently, there were over 5 million families in the United States that had three children under age 18. What is the probability that a family with three children has all boys?

First, find the number of possible outcomes. You can make a list or draw a diagram. The **tree diagram** below shows that each child might be a boy (B) or a girl (G).

First Child	Second Child	Third Child	Outcomes
		B	BBB
	B	G	BBG
B		B	BGB
	G	G	BGG
		B	GBB
	B	G	GBG
G		B	GGB
	G	G	GGG

There are eight possible outcomes, but only one has all boys. So the probability that a family with three children has all boys is $\frac{1}{8}$ or 12.5%.

Example
APPLICATION

Making a tree diagram is a good way to determine a sample space.

Inventory The Sports Store carries one brand of basketball tank-top in four sizes: M, L, XL, and XXL. Each size comes in two colors: gold (G) and black (B). Make a tree diagram that shows all of the possible combinations of size and color.

Size	Color	Outcomes
M	G	MG
	B	MB
L	G	LG
	B	LB
XL	G	XLG
	B	XLB
XXL	G	XXLG
	B	XXLB

There are eight possible combinations of size and color.

You can use tree diagrams to analyze whether or not games are fair. In a **fair game**, players of equal skill have the same chance of winning.

 **MINI-LAB**

Work with a partner. 3 counters marker

Try This

- Mark one side of a counter A. Mark the other side B.

  Mark one side of another counter A. Mark the other side C.

  Mark one side of the last counter B. Mark the other side C.

- Player 1 tosses the counters. If two counters show the same letter, Player 1 wins. Otherwise, Player 2 wins. Record who wins.

- Repeat the counter toss experiment 19 more times, recording who won each time.

Talk About It

1. Find the experimental probability that Player 1 and Player 2 each won.

2. Make a tree diagram to show all the possible outcomes for this game.

3. Find the theoretical probability that Player 1 and Player 2 each won. Does this explain the results of your game?

4. Is this game a fair game? Explain your reasoning.

CHECK FOR UNDERSTANDING

Communicating Mathematics

Read and study the lesson to answer each question.

1. *Tell* how to use a tree diagram to list outcomes.

2. *Write a Problem* that can be solved by using the tree diagram at the right.

   hot dog ⎯ cola
   ⎯ diet cola
   ⎯ root beer

HANDS-ON MATH

3. *Determine* whether the following game for two players is fair.

   - Toss three pennies.
   - If exactly two pennies match, Player 1 wins.
   - Otherwise, Player 2 wins.

   hamburger ⎯ cola
   ⎯ diet cola
   ⎯ root beer

Guided Practice

For each situation, make a tree diagram to show the sample space. Then give the total number of outcomes.

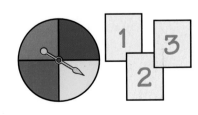

4. spinning the spinner and choosing a card

5. tossing a coin and rolling a number cube

6. *Travel* Rosa packed a red shirt, a white shirt, blue jeans, and black jeans for a weekend trip. Make a tree diagram to show all of the possible shirt/jeans selections she can make.

Practice

For each situation, make a tree diagram to show the sample space. Then give the total number of outcomes.

7. tossing a penny and tossing a dime

8. choosing a card with a letter, then choosing a card with a number from the choices shown at the right

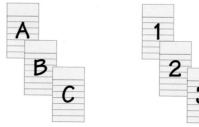

9. choosing a wall or portable phone, in either black, beige, red, or white

10. choosing a bicycle having 10 speeds, 18 speeds, or 21 speeds and either red, blue, green, or white in color

11. rolling a number cube, tossing a coin, and choosing a card from among cards marked W, X, Y, and Z

12. choosing cereal, French toast, or pancakes and choosing orange, apple, or grapefruit juice

Applications and Problem Solving

13. *Geography* The ZIP codes for Virginia addresses all begin with a 2, which identifies its national area. The second and third digits identify the geographic area. The second digit in Virginia is either a 0, 2, 3, or 4. The third digit can be any digit from 0 to 9. How many geographic areas are possible in Virginia?

14. *Population* Refer to the beginning of the lesson. What is the probability that a family with three children has at least two girls? at least two boys?

15. *Critical Thinking* A softball tournament between the Jaguars and the Bandits is won when one team wins two of three games. Construct a tree diagram to show all the possible outcomes for the tournament.

Mixed Review

16. **Standardized Test Practice** Poloma has four pennies in her wallet. The dates on the pennies are 1998, 1995, 1989, and 1988. If she picks one penny from her wallet without looking, what is the probability that it will have a date in the 1980s? *(Lesson 13-1)*

   A $\frac{1}{4}$ **B** $\frac{1}{2}$ **C** $\frac{2}{3}$ **D** $\frac{3}{4}$

17. *Geometry* Draw top, side, and front views of each three-dimensional figure. *(Lesson 12-1)*

   a.   **b.**   **c.**

                       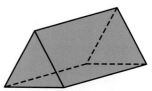

For **Extra Practice,** see page 603.

18. *Geometry* What is the diameter of a circle whose radius is 13 feet? *(Lesson 7-7)*

DESIGN

Chris Haney, John Haney, and Scott Abbott

INVENTORS

Canadian entrepreneurs Chris Haney, John Haney, and Scott Abbott dreamed up the idea for a board game in 1979. They started their own company to develop, manufacture, and distribute 1,100 copies of the game. Within fifteen years, 60 million copies of the popular game had been sold.

Independent toy designers are important sources of new game and toy ideas. To become a game and toy inventor or designer, you must be creative and have a good business sense. Other careers in the toy and game industry include toy buyers, who place orders for toys to be sold in stores, and toy assemblers or machine operators, who help manufacture toy products.

For more information:
Toy Manufacturers of America, Inc.
200 Fifth Ave., Suite 740
New York, NY 10010

interNET CONNECTION
www.glencoe.com/sec/
math/mac/mathnet

Who comes up with ideas for games?

Your Turn

Suppose you have a great new game idea. Make a plan for producing and selling your game.

The Counting Principle

What you'll learn

You'll learn to use multiplication to count outcomes.

When am I ever going to use this?

The Counting Principle can be used to find the number of possible phone numbers in an area code.

Word Wise

Counting Principle

Colonel Mustard, in the kitchen, with the wrench … . With words like these, players of the popular board game *Clue* accuse a suspect of murder. Players can choose from six suspects, nine rooms, and six weapons. How many different ways are there to make an accusation?

The partial tree diagram helps to show the number of outcomes in the sample space.

6 Suspects	9 Rooms	6 Weapons
Colonel Mustard	kitchen	rope
Professor Plum	study	lead pipe
Mrs. Peacock	library	knife
Miss Scarlet	hall	wrench
Mrs. White	lounge	candlestick
Mr. Green	dining room	revolver
	ballroom	
	conservatory	
	billiard room	

Each suspect has nine possible rooms. So, there are 6 × 9 or 54 possible suspect/room outcomes. Finally, each of these outcomes has six different weapons to choose from. So, there are 54 × 6 or 324 possible ways to make an accusation.

In this situation, you used multiplication instead of a tree diagram to find the number of possible outcomes in the sample space.

The Counting Principle	If event *M* can occur in *m* ways and is followed by event *N* that can occur in *n* ways, then the event *M* followed by *N* can occur in *m* × *n* ways.

Example ➊ APPLICATION

Retail Sales The Jean Scene sells women's jeans in sizes 3, 5, 7, 9, 11, 13, and 15. Each size comes in slim fit, regular fit, and relaxed fit. There are also three possible lengths: petite, regular, and tall. How many different kinds of jeans are there?

$$\underbrace{7}_{number\ of\ sizes} \times \underbrace{3}_{number\ of\ fits} \times \underbrace{3}_{number\ of\ lengths} = \underbrace{63}_{number\ of\ kinds}$$

There are 63 different kinds of jeans at the Jean Scene.

Real World APPLICATION

② Travel Air Canada offers nonstop flights from several cities in the U.S. to its Canadian hubs. Once in Canada, travelers can fly to destinations across Canada. The figure shows the routes that go through Vancouver, British Columbia. How many different routes are possible from U.S. cities through Vancouver with destinations in Canada?

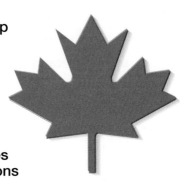

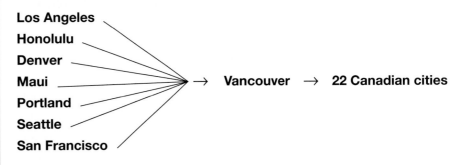

Los Angeles
Honolulu
Denver
Maui → **Vancouver** → **22 Canadian cities**
Portland
Seattle
San Francisco

routes from United States   *routes from Vancouver*   *total routes*
$$7 \qquad \times \qquad 22 \qquad = \qquad 154$$

There are 154 different routes.

APPLICATION **③ Braille** Many people who are visually impaired use a code called Braille to be able to read. Each character in Braille lies in a 3 × 2 cell of raised dots or blank spaces. Some examples are shown at the right. The cell consisting of six blank spaces is not used. How many different cells are possible?

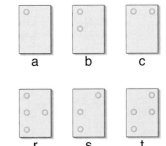

Explore	You know that Braille consists of raised dots or blank spaces. There are six positions in each cell. You need to find how many cells are possible.
Plan	There are two choices for each of the six positions. You can use the Counting Principle to find the total number of groupings.
Solve	$2 \cdot 2 \cdot 2 \cdot 2 \cdot 2 \cdot 2 = 64$

Remember, there is no cell with six blank spaces. Therefore, the total number of cells is $64 - 1$ or 63.

Examine	There should be enough cells for 26 letters, 10 digits, and about 10 punctuation marks. So, 63 is a reasonable answer.

Communicating Mathematics

Read and study the lesson to answer each question.

1. *Calculate* the number of ways of selecting one shirt/tie combination from among 3 different shirts and 4 different ties.

2. *Draw* a tree diagram to verify the results of Exercise 1.

3. *Write* a problem in which it is easier to use the Counting Principle than a tree diagram.

Guided Practice

Use the Counting Principle to find the total number of outcomes in each situation.

4. choosing a car from five different exterior colors, four different interior colors, and automatic or standard transmission

5. making a sandwich with raisin bread, whole wheat bread, white bread, or a bagel and choosing peanut butter, cream cheese, or jelly

6. *Travel* There are three highways connecting Tomville and Greensburg. There are two roads connecting Greensburg to North Huntingdon. How many ways are there to drive from Tomville to North Huntingdon?

Tomville   Greensburg   North Huntingdon

EXERCISES

Practice

Use the Counting Principle to find the total number of outcomes in each situation.

7. spinning the spinners shown at the right

8. tossing a penny, a nickel, a dime, and a quarter

9. choosing the first two characters for a license plate if it begins with a letter of the alphabet and is followed by a digit

10. choosing a way to drive from Milton to Harper's Township

Milton   Westwood   Morgantown   Harper's Township

11. choosing a dinner with one entrée, one salad, and one dessert

12. choosing a 4-digit Personal Identification Number (PIN) if the numbers can be repeated

Entrée	Salad	Dessert
▸ ham ◂	▸ potato ◂	▸ ice cream ◂
▸ beef ◂	▸ tossed ◂	▸ pie ◂
▸ turkey ◂	▸ cole slaw ◂	▸ cake ◂

Applications and Problem Solving

13. *Advertising* The Bowl 'N' Ladle Restaurant advertises that you can have a different lunch every day of the year. It offers 13 different kinds of soups and 24 different kinds of sandwiches. If the restaurant is open every day of the year, is its claim valid? Explain.

14. *Games* Refer to the beginning of the lesson. What is the probability that the murder was committed by Miss Scarlet, in the kitchen, with the knife?

15. *Critical Thinking* How many outcomes are there if you toss
   a. one coin?   **b.** two coins?   **c.** three coins?   **d.** *n* coins?

Mixed Review

16. *School* Jewel must take a science class, a math class, and an English class next year. She can choose from two science classes, three math classes, and two English classes. Make a tree diagram to show all of the possible schedules she can arrange. *(Lesson 13-2)*

17. **Standardized Test Practice** What number should come next in the pattern 3, 7, 15, 31, 63? *(Lesson 4-3)*

   A 73   **B** 84   **C** 106   **D** 127

For **Extra Practice**, see page 603.

Let the Games Begin

Cherokee Butterbean Game

Math Skill
Probability

Get Ready   This game is for two, three, or four players.
It is a variation of a traditional Cherokee game.

   ⚬ 6 dry lima beans   ◡ bowl   ✎ marker

Get Set   Color one side of each bean with the marker.

Go   The first player places the beans in the bowl, gently tosses the beans into the air, and catches them in the bowl. Points are scored as follows.

- If all of the beans land with the unmarked sides up, score 6 points.

- If all of the beans land with the marked side up, score 4 points.

- If exactly one bean lands with the marked or unmarked side up, score 2 points.

- If three beans are marked and three beans are unmarked, score 1 point.

If a toss scores points, the player takes another turn. If a toss does not score any points, it is the next player's turn.

The winner is the player with the most points after a given number of rounds.

 Visit www.glencoe.com/sec/math/mac/mathnet for more games.

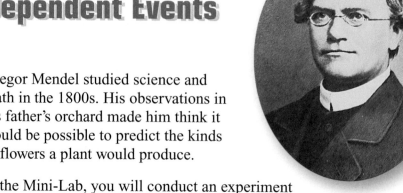

13-4 Independent and Dependent Events

What you'll learn

You'll learn to find the probability of independent and dependent events.

When am I ever going to use this?

You'll use probability of two events when you study genetics in life science.

Word Wise

compound events
independent events
dependent events

Gregor Mendel studied science and math in the 1800s. His observations in his father's orchard made him think it would be possible to predict the kinds of flowers a plant would produce.

In the Mini-Lab, you will conduct an experiment to predict how many white flowers will result from a cross between two parent plants. Each parent has an equal chance of passing on a gene for a red or white flower.

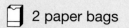

HANDS-ON MINI-LAB

Work with a partner.   📄 2 paper bags   🫘 50 red beans

🫘 50 white beans

Try This

- Place 25 red beans and 25 white beans in bag 1 and 25 red beans and 25 white beans in bag 2.
- Without looking, remove one bean from each bag. The two beans represent the cross between the two flowers. Note the color combination and return the beans to their respective bags.
- Make a frequency table and record the color combination each time you remove two beans. Repeat 99 times.
- Count and record the total number of red/red, red/white, white/red, and white/white combinations.

Talk About It

1. Estimate the probability of each result.
   a. red/red   **b.** red/white   **c.** white/red   **d.** white/white
2. Suppose the red color is *dominant* over the white color. That means a red/white and a white/red plant will be red. The only way to have a white flower is with white/white. Estimate the probability of getting a white flower.

In the experiment, you removed one bean from each bag. This is a **compound event**. Choosing a bean from bag 1 did not affect choosing a bean from bag 2. These events are called **independent events**. You can analyze the experiment with a tree diagram.

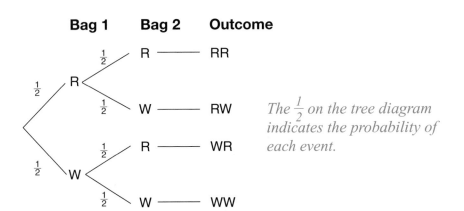

Bag 1	Bag 2	Outcome

There are four equally-likely outcomes. So, $P(\text{white/white}) = \frac{1}{4}$.

You can also multiply to find the probability of two independent events.

$$\underbrace{P(\text{white from bag 1})}_{\frac{1}{2}} \times \underbrace{P(\text{white from bag 2})}_{\frac{1}{2}} = \underbrace{P(\text{white/white})}_{\frac{1}{4}}$$

Probability of Independent Events	The probability of two independent events can be found by multiplying the probability of one event by the probability of the second event.

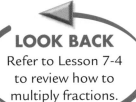

LOOK BACK

Refer to Lesson 7-4 to review how to multiply fractions.

Example 1

A green number cube and a red number cube are rolled. Find the probability that an odd number is rolled on the green number cube and a multiple of 3 is rolled on the red number cube.

There are three odd numbers. So, $P(\text{odd}) = \frac{3}{6}$ or $\frac{1}{2}$.

There are two multiples of 3. So, $P(\text{multiple of 3}) = \frac{2}{6}$ or $\frac{1}{3}$.

So, $P(\text{odd and multiple of 3}) = \frac{1}{2} \cdot \frac{1}{3}$ or $\frac{1}{6}$.

If the result of one event affects the result of a second event, the events are called **dependent events**.

Example 2

APPLICATION

Game Shows On a popular television game show, contestants can win a car if they draw the five digits of the price of the car before they draw three strikes. Once a digit or strike is drawn, it is not replaced. Find the probability that a contestant draws two strikes in the first two draws.

$P(\text{strike on first draw}) = \frac{3}{8}$ ← *There are 3 strikes.*
← *There is a total of 8 outcomes— 5 digits and 3 strikes.*

The result of the first draw affects the probability of the second draw.

$P(\text{strike on second draw}) = \frac{2}{7}$ ← *There are 2 strikes left.*
← *There are 7 digits or strikes left.*

(continued on the next page)

Lesson 13-4 Independent and Dependent Events **543**

To find the probability of 2 strikes in a row, multiply the probability of a first strike and a second strike.

$$P(\text{strike, then strike}) = \frac{3}{8} \cdot \frac{2}{7}$$

$$= \frac{3}{\underset{4}{8}} \cdot \frac{2^{1}}{7} \text{ or } \frac{3}{28}$$

So, the probability is $\frac{3}{28}$ or about 11%.

CHECK FOR UNDERSTANDING

Communicating Mathematics

Read and study the lesson to answer each question.

1. *Give an example* of two events that are dependent.

2. *Write a Problem* about the marbles. Then tell whether the events are independent or dependent.

3. *You Decide* Jared thinks that the results of rolling a number cube twice are independent events. Angie thinks they are dependent events. Who is correct? Explain.

Guided Practice

Tell whether the event is *independent* or *dependent*.

4. choosing a card from a hat and then choosing a second card without replacing the first one

5. A coin is tossed and a number cube is rolled. Find the probability of getting heads and a multiple of 2.

6. A bag contains 10 white, 8 blue, and 6 red marbles. Two marbles are drawn, but the first marble is not replaced. Find $P(\text{both blue})$.

7. *Games* An American game involving five dice is similar to a Puerto Rican dice game called Generala. In both games, players try to roll five of a kind. In Generala, five of a kind in one roll is called a Big General and wins the game automatically. Find the probability of throwing a Big General.

EXERCISES

Practice

Tell whether each event is *independent* or *dependent*.

8. selecting a name from the Chicago telephone book and a name from the Houston telephone book

9. tossing a coin twice

10. choosing a President, Vice President, and Secretary from three members of student council

11. A blue number cube and a yellow number cube are rolled. Find the probability that an odd number is rolled on the blue number cube and a multiple of 6 is rolled on the yellow number cube.

12. A wallet contains five $5 bills, three $10 bills, and two $20 bills. Two bills are selected without the first selection being replaced. Find $P(\$10,$ then $\$10)$.

13. A wallet contains five $5 bills, three $10 bills, and two $20 bills. Three bills are selected without each selection being replaced. Find P($5, then $10, then $20).

14. A blue number cube and a green number cube are rolled. Find the probability that a multiple of 2 is rolled on the blue number cube and a multiple of 3 is rolled on the green number cube.

Applications and Problem Solving

15. *Game Shows* Refer to Example 2 on page 543. Find the probability that a contestant draws three strikes in the first three draws.

16. *Baseball* Toshiro hit 9 home runs during his last 100 times at bat during the baseball season. What is the probability that he will hit home runs in his next two times at bat?

17. *Critical Thinking* Make a tree diagram of all the possible outcomes of three successive spins of the spinner shown.

   a. How many paths in the tree diagram represent two red spins and one blue?

   b. Suppose the spinner is designed so that for each spin there is a 40% probability of spinning red and a 20% chance of spinning blue. What is the probability of spinning two reds and one blue?

18. *Working on the* CHAPTER Project For the board game that you are designing, describe how you could include a situation that involves the probability of two events. Then explain whether these events are independent or dependent.

Mixed Review

For **Extra Practice**, see page 604.

19. **Standardized Test Practice** Susanna has 3 sweaters, 5 blouses, and 6 skirts that coordinate. How many different outfits can Susanna make if each outfit consists of a sweater, a blouse, and a skirt? *(Lesson 13-3)*
   A 105   **B** 90   **C** 45   **D** 14   **E** Not Here

20. Find $\sqrt{144}$. *(Lesson 10-1)*

CHAPTER 13

Mid-Chapter Self Test

1. Out of 30 rolls of a number cube, Allison rolls a 4 three times. What is the experimental probability of rolling a 4? *(Lesson 13-1)*

2. Find the theoretical probability of rolling a 4 on a number cube. *(Lesson 13-1)*

3. Dion has a choice of two juices (orange or apple) and three cereals (wheat, rice, or corn) for breakfast. Make a tree diagram to show all the possible outcomes. *(Lesson 13-2)*

4. Find the total number of outcomes if you toss a penny and spin the spinner at the right. *(Lesson 13-3)*

5. *Traffic Planning* One traffic light is red 60% of the time. The next traffic light is red 50% of the time. If the lights operate independently of one another, find P(both red). *(Lesson 13-4)*

13-5A Exploring Permutations

A Preview of Lesson 13-5

3 index cards

How many different ways are there to arrange your school schedule if you take math, science, and language arts?

TRY THIS

Work with a partner.

1 You can make an organized list.

- Write math, science, and language arts on the index cards.

- Choose one of the three subjects as the first class of the day. Choose one of the remaining two subjects for the second class. The third class is the card that remains.

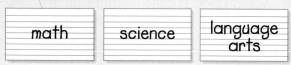

- Record this arrangement of classes.

- Change the order of the last two classes. Record this arrangement.

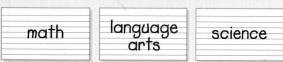

- Continue rearranging the cards until you have found all of the possible arrangements.

2 You can also make a tree diagram.

- Copy and complete the tree diagram shown below.

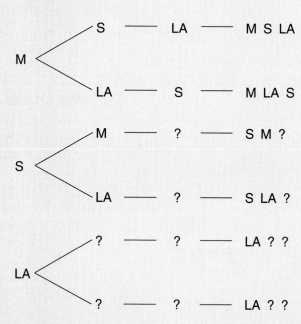

ON YOUR OWN

1. When you first started to make your list or tree diagram, how many choices did you have for your first class?

2. Once your first class was selected, how many choices did you have for the second class?

3. Once the first two were selected, how many choices did you have for the third class?

4. How many different arrangements were possible?

5. Explain how you can use the Counting Principle to find the number of arrangements.

6. Suppose you also take social studies. How many arrangements are possible?

Permutations

What you'll learn

You'll learn to find the number of permutations of a set of objects.

When am I ever going to use this?

You can use permutations to find how many different ways you can arrange a collection of CDs.

Word Wise

permutation
factorial

Have you ever noticed that the characters in Peanuts often discuss mathematics?

PEANUTS

A **permutation** is an arrangement, or listing, of objects in which order is important. Let's help Peppermint Patty overcome her math anxiety and find the number of ways to arrange 9 books on a shelf.

Peppermint Patty has 9 different choices for placing the first book on the shelf. Once that book is placed, she then has 8 choices for the second book. This continues until she has 1 choice for the final book.

The following expression shows the number of permutations.

$$9 \cdot 8 \cdot 7 \cdot 6 \cdot 5 \cdot 4 \cdot 3 \cdot 2 \cdot 1 = 362,880$$

There are 362,880 ways to arrange 9 books on a shelf.

The expression $9 \cdot 8 \cdot 7 \cdot 6 \cdot 5 \cdot 4 \cdot 3 \cdot 2 \cdot 1$ can be written as 9!, which is read "nine **factorial**." In general, $n!$ is the product of all the counting numbers beginning with n and counting backward to 1.

Study Hint

Technology If your calculator has an x! key, you can use it to compute 9!.

Example 1 Compute 5!.

$$5! = 5 \cdot 4 \cdot 3 \cdot 2 \cdot 1$$
$$= 120 \qquad \text{Therefore, } 5! = 120.$$

Some arrangements involve only part of a group.

Example 2
CONNECTION

Geography The flag of Mali has three vertical stripes that are green, yellow, and red. The flag of Italy is very similar. It has three vertical stripes that are green, white, and red. How many different flags can be made from the colors green, white, yellow, and red if each flag has three vertical stripes?

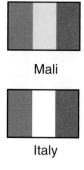

Mali

Italy

For the first vertical stripe, there are four possible choices. After that, there are three possible choices. Finally, there are two possible choices for the third stripe.

$$4 \cdot 3 \cdot 2 = 24$$

There are 24 different flags with three vertical stripes that can be made from green, yellow, white, and red. *To check the answer, you could use colored pencils and actually draw each possible flag.*

CHECK FOR UNDERSTANDING

Communicating Mathematics

Math Journal

Read and study the lesson to answer each question.

1. *Define* factorial.

2. *List* all of the permutations of the digits 2, 4, and 6.

3. *Write a Problem* in which you need to find the number of permutations of three objects.

Guided Practice

Find the value of each expression.

4. 3!

5. $5 \cdot 4 \cdot 3$

6. At a pet show, first, second, and third prizes will be awarded to Cookie, Charlie, and Max. In how many ways can the prizes be awarded?

7. A license plate begins with three letters. If the alphabet contains 26 letters, how many different permutations of these letters can be made if no letter is used more than once?

8. *Music* The pentatonic scale has five notes, C#, D#, F#, G#, and A#. These are the black keys on a piano. How many different five-note sequences can be written if each note is used only once?

Practice

Find the value of each expression.

9. 4!　　　　　　**10.** 6!　　　　　　**11.** $6 \cdot 5 \cdot 4$

12. $9 \cdot 8 \cdot 7 \cdot 6$　　　**13.** 2!　　　　　　**14.** seven factorial

15. In how many ways can the starting five players of a basketball team stand in a row for a team picture?

16. How many different three-letter "words" can be formed from the letters E, N, and D if no letter may be used more than once? (*Hint:* "Words" means any arrangement of letters, not just English words.)

17. In how many ways can a president, treasurer, and a secretary be chosen from among 8 candidates?

18. There are five finalists for the science fair, and trophies will be given to the first three finishers. How many ways are there for the three trophy winners to be selected?

Applications and Problem Solving

19. *Sports* A softball team has three power hitters. The coach wants to place them in the 3rd, 4th, and 5th positions in the batting order. How many different ways can these players be placed in those positions?

20. *Parades* If there are 50 floats in Pasadena's Rose Parade, how many ways can a first-place and second-place trophy be awarded?

21. *Critical Thinking* Suppose Della has forgotten her locker combination. There are three numbers in the combination, and each number is different. The numbers on the locker go from 0 to 35. If she tries one combination every 10 seconds, how long will it take her to test all of the possible combinations?

Mixed Review

22. *Probability* John is late for work 15% of the time. Diane is late for work 20% of the time. Find the probability that they will both be late for work on the same day. *(Lesson 13-4)*

23. **Standardized Test Practice** Amelia performed a probability experiment by spinning a spinner 20 times. The results are shown in the chart.

Color	Red	Green	Blue
Number	5	10	5

If the spinner is divided into four equal sections, how many sections would you expect to be colored blue? *(Lesson 13-1)*

A 1　　　　　**B** 2　　　　　**C** 3　　　　　**D** 4

24. *Algebra* Evaluate $6x - 3(x - y)$ if $x = 8$ and $y = 2$. *(Lesson 1-3)*

For **Extra Practice**, see page 604.

COOPERATIVE LEARNING

13-6A Exploring Combinations

A Preview of Lesson 13-6

4 index cards

The student council at Heritage Middle School is planning to sell pizzas with two different toppings. The choices are shown below.

> # Pizzas for Sale!
>
> pepperoni   sausage
>
> mushroom   green pepper

How many different combinations are possible?

TRY THIS

Work with a partner.

- Write the names of the four pizza toppings on the index cards.
- To make a pizza, select any pair of cards. Make a list of all the different combinations that are possible. Note that the order of the toppings is not important.

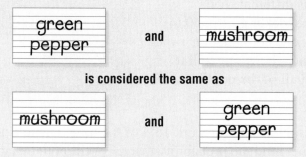

green pepper **and** mushroom

is considered the same as

mushroom **and** green pepper

ON YOUR OWN

1. How many different combinations are possible with two toppings?

2. How many different pizzas could be made with two toppings if the order of the toppings *was* important?

3. How many ways are there to arrange two toppings if order is important?

4. How are the answers to Exercises 1, 2, and 3 related?

5. Repeat Exercises 1–4 to find how many three-topping pizzas are possible from a list of five toppings.

Combinations

You'll learn to find the number of combinations of a set of objects.

When am I ever going to use this?

You'll use combinations when you choose a committee of four from a larger group.

Word Wise

combination

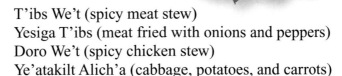

Suppose you are invited to an Ethiopian restaurant and the following items on the menu attract your attention.

T'ibs We't (spicy meat stew)
Yesiga T'ibs (meat fried with onions and peppers)
Doro We't (spicy chicken stew)
Ye'atakilt Alich'a (cabbage, potatoes, and carrots)

In how many ways can you choose two different items to sample? *This problem will be solved in Example 1.*

Sometimes, order is *not* important. In this case, choosing T'ibs We't and Yesiga T'ibs is the same as choosing Yesiga T'ibs and T'ibs We't. An arrangement, or listing, of objects in which order is not important is called a **combination**.

Example ①

Food Refer to the beginning of the lesson. In how many ways can two items be chosen from a list of four items?

Let's call the menu items A, B, C, and D. First, list *all* the arrangements of A, B, C, and D taken two items at a time.

AB	AC	AD	BC	BD	CD
BA	CA	DA	CB	DB	DC

From the list, count only the different arrangements. Arrangements AB and BA are the same in this case.

AB   AC   AD   BC   BD   CD

So, there are six ways to choose two items from a list of four items.

Permutations and combinations are related. You can find the number of combinations of objects by dividing the number of permutations of the entire set by the number of ways each smaller set can be arranged.

Example ②

In how many ways can you choose a committee of three people from a group of six people?

There are $6 \cdot 5 \cdot 4$ permutations of three people chosen from six. There are 3! or $3 \cdot 2 \cdot 1$ ways to arrange the three people.

$\frac{6 \cdot 5 \cdot 4}{3 \cdot 2 \cdot 1} = \frac{120}{6}$ or 20   There are 20 ways to choose the committee.

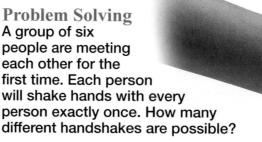

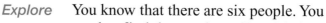

Example

INTEGRATION

③ Problem Solving

A group of six people are meeting each other for the first time. Each person will shake hands with every person exactly once. How many different handshakes are possible?

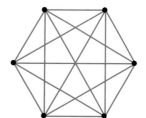

Explore   You know that there are six people. You need to find the number of handshakes.

Plan   Find the number of ways that two people can be chosen from a group of six people. In this case, order is not important.

Solve   *There are 6 · 5 ways to choose 2 people.* $\rightarrow$ $\dfrac{6 \cdot 5}{2 \cdot 1} = \dfrac{30}{2}$
There are 2 · 1 ways to arrange 2 people. $\rightarrow$

There are 15 possible handshakes. $= 15$

Examine   Make a diagram in which each person is represented by points. Draw line segments between two points to represent the handshakes. There are 15 line segments.

CHECK FOR UNDERSTANDING

Communicating Mathematics

Read and study the lesson to answer each question.

1. ***Explain*** why a combination lock really should be called a permutation lock.

2. ***Tell*** whether there are more permutations of a set of objects taken three at a time or more combinations of the objects taken three at a time.

3. ***You Decide*** Malik thinks that choosing four CDs from a group of 20 to take on a trip is a permutation. Miyoki thinks it is a combination. Who is correct? Explain your reasoning.

Guided Practice

Tell whether each problem represents a *permutation* or a *combination*. Then solve the problem.

4. In how many ways can four cars line up for a race?

5. In how many ways can four swimmers for a team be chosen from six swimmers?

6. ***Food*** Yogi's Yogurt offers a choice of chocolate syrup, butterscotch syrup, strawberries, pineapple, coconut, and peanuts as toppings for its yogurt. In how many different ways can Brandon choose two different toppings for his yogurt?

Practice

Tell whether each problem represents a *permutation* or a *combination*. Then solve the problem.

7. Six students remain in a game of musical chairs. If two chairs are removed, how many different groups of four students can remain?

8. In how many ways can three clarinet players be seated in the first, second, and third seats in the orchestra?

9. In how many ways can three wooden carvings be displayed from a collection of 10?

10. Given six different toppings from which to choose, how many different four-topping pizzas are possible?

11. Any four people on a committee of seven can make a decision for the committee. How many groups of four people are there?

12. In how many ways can five people be seated in a row of five chairs?

Applications and Problem Solving

13. ***Lottery*** Each ticket in a lottery has five one-digit numbers, 0 – 9 on it. No two tickets have the same five numbers. The Lottery Commission will announce the combination of five numbers that will win the Grand Prize. You win if your ticket has those five numbers, in any order. What are your chances of winning?

14. ***School*** Raul, Debbie, Julio, Terese, Yoki, and Tim have completed their mathematics project. Their teacher will choose three students at random to present the project to the class. Debbie, Terese, and Yoki hope they will be chosen together to present the project. What are their chances of being chosen together?

15. ***Geometry*** Eight points are marked on a circle. How many different line segments can be drawn between any two of the points?

16. ***Critical Thinking*** There are 504 ways in which three students can be selected first, second, and third place in the science fair. How many students are competing?

Mixed Review

17. ***Sailing*** When Mr. Rivera purchased his sailboat, it came with six different-colored flags to be used for sending signals. The specific signal depended on the order of the flags. How many different three-flag signals can he send? *(Lesson 13-5)*

18. **Standardized Test Practice** Which figure has exactly 5 faces? *(Lesson 12-1)*

   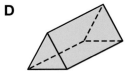

A   B   C   D

For **Extra Practice**, see page 604.

Vocabulary

After completing this chapter, you should be able to define each term, concept, or phrase and give an example or two of each.

Statistics and Probability
combination (p. 551)
compound events (p. 542)
Counting Principle (p. 538)
dependent events (p. 543)
experimental probability (p. 531)
factorial (p. 547)
fair game (p. 535)
independent events (p. 542)
outcome (p. 531)
permutation (p. 547)
sample space (p. 531)
theoretical probability (p. 531)
tree diagram (p. 534)

Problem Solving
act it out (p. 528)

Understanding and Using the Vocabulary

Choose the correct term to complete the sentence.

1. The set of all possible outcomes for an experiment is called the (sample space, combination).

2. The Counting Principle counts the number of possible outcomes using the operation of (addition, multiplication).

3. The ratio of the number of times an event occurs to the number of trials done is called the (theoretical, experimental) probability.

4. When the outcome of one event influences the outcome of a second event, the events are called (independent, dependent).

5. A (permutation, combination) is an arrangement of objects in which order is important.

6. In a(n) (fair game, independent event), players of equal skill have the same chance of winning.

7. A tree diagram can be used to find the number of (combinations, outcomes).

In Your Own Words

8. *Explain* the difference between a permutation and a combination.

Objectives & Examples

Upon completing this chapter, you should be able to:

● find and compare experimental and theoretical probabilities *(Lesson 13-1)*

If 1 marble is drawn from a bag containing 8 red and 2 green marbles, the theoretical probability that it is green is $\frac{2}{10}$ or $\frac{1}{5}$.

● use tree diagrams to count outcomes *(Lesson 13-2)*

If a family has two children, there are 4 possible outcomes.

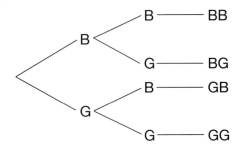

● use multiplication to count outcomes *(Lesson 13-3)*

There are 2 possible outcomes each time a family has a child. If a family has 2 children, there are 2 × 2 or 4 outcomes.

Review Exercises

Use these exercises to review and prepare for the chapter test.

A bowl contains the names of 25 students. Ten are 10 years old, ten are 11 years old, and five are 12 years old. One name is selected at random. Find each probability.

9. *P*(10 years old)

10. *P*(12 years old)

11. *P*(at least 11 years old)

For each situation, make a tree diagram and list the outcomes. Then give the total number of outcomes.

12. choosing a red, blue, or white shirt with either black or gray lettering

13. tossing a coin and choosing a card from six cards numbered from 1 to 6

Use the Counting Principle to find the total number of outcomes in each situation.

14. rolling 2 number cubes

15. selecting a car from 3 styles, 3 interior colors, and 3 exterior colors

Objectives & Examples

Review Exercises

● find the probability of independent and dependent events *(Lesson 13-4)*

Find the probability of spinning a 2 on the first spinner and a 3 on the second spinner.

$$P(2) = \frac{1}{4} \qquad P(3) = \frac{1}{3}$$

$$P(2, \text{then } 3) = \frac{1}{4} \cdot \frac{1}{3} \text{ or } \frac{1}{12}$$

A bag contains 4 green, 6 white, and 8 blue counters. Two counters are randomly drawn.

16. Find P(white, white) if the first counter drawn is replaced.

17. Find P(white, white) if the first counter drawn is not replaced.

A box contains 5 blue, 4 red, and 3 yellow marbles. Two marbles are randomly drawn.

18. Find P(blue, red) if the first marble drawn is not replaced.

19. Find P(blue, red) if the first marble drawn is replaced.

● find the number of permutations of a set of objects *(Lesson 13-5)*

In how many ways can a president and vice-president be chosen from among 4 candidates?

┌── *choices for president*

$4 \cdot 3 = 12$

└── *choices for vice-president*

Find the value of each expression.

20. 6!   21. 4!

22. In how many ways can five basketball players be placed in three positions?

23. The pole vault competition has 10 people in it. In how many ways can first-, second-, and third-place ribbons be awarded?

● find the number of combinations of a set of objects *(Lesson 13-6)*

In how many ways can you choose two items from a menu with 4 items on it?

$$\frac{4 \cdot 3}{2 \cdot 1} = \frac{12}{2} \text{ or } 6$$

24. Given eight different toppings to choose from, how many three-topping pizzas are possible?

25. Any five people on a committee of nine can make a decision for the committee. How many groups of five people are there?

Applications & Problem Solving

26. *Act It Out* How many times do you need to spin this spinner to spin all six numbers? *(Lesson 13-1A)*

27. *Shopping* One catalog offers a jogging suit in two colors, blue and red. It comes in sizes S, M, L, XL, and XXL. How many possible jogging suits can be ordered? *(Lesson 13-3)*

28. *Raffle* In how many ways can the grand-prize ticket, second-prize ticket, and third-prize ticket be selected from the 30 raffle tickets sold? *(Lesson 13-5)*

29. *Games* Andrea has 10 different games that she keeps in her bedroom. She wants to select two of these games to play in the evening. In how many ways can Andrea select the two games? *(Lesson 13-6)*

30. *Clothes* A drawer contains two blue socks and four black socks. Without looking, you choose one sock, and then another sock without replacing the first one. What is the probability that you choose two black socks? *(Lesson 13-3)*

Alternative Assessment

● ***Open Ended***

Mia and her friends are creating a game for her party. They are labeling each side of a six-sided cube with numbers from 1 to 4, using each number at least once. They want to label the faces of the cube so that the probability of rolling a 3 or a 4 is $\frac{1}{2}$ and the probability of not rolling a 1 is $\frac{2}{3}$. How should they label the sides of the cube?

Mia decides to label a second six-sided number cube with numbers from 1 to 4, using each number at least once. She wants to label the faces of this second cube so that the probability of getting a sum of 4 by rolling both cubes is $\frac{1}{6}$. How should she label the sides of the second cube?

A practice test for Chapter 13 is provided on page 619.

● ***Completing the***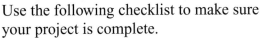

Use the following checklist to make sure your project is complete.

☑ You have listed several examples that describe how probability is used in the board game you have chosen.

☑ The design of your own board game includes a complete set of rules for the game, and you have explained the role probability plays in your game.

● *PORTFOLIO* Select one of the vocabulary words you learned in this chapter and place the word and its definition in your portfolio. Attach a note explaining why you selected it.

Section One: Multiple Choice

There are nine multiple-choice questions in this section. Choose the best answer. If a correct answer is *not here,* choose the letter for Not Here.

1. There are four flavors of yogurt, six kinds of toppings, and four different kinds of syrup. How many different combinations of yogurt, topping, and syrup can be ordered if you can choose one of each?

   A 14

   B 24

   C 48

   D 96

2. A cube has a surface area of 144 square centimeters. How can you find the surface area of one face?

   F Divide 144 by 4.

   G Divide 144 by 6.

   H Divide 144 by 8.

   J Divide 144 by 10.

3. What is the theoretical probability of choosing a vowel from the word MATHEMATICS?

   A $\frac{4}{7}$ B $\frac{4}{11}$

   C $\frac{7}{11}$ D $\frac{4}{9}$

4. Suppose you need 0.65 liter of water for a science experiment, but the container is measured in milliliters. How much water do you need?

   F 6,500 milliliters

   G 650 milliliters

   H 65 milliliters

   J 0.00065 milliliter

Please note that Questions 5–9 have five answer choices.

5. People were asked to choose their favorite type of movie. The graph shows the results of the survey.

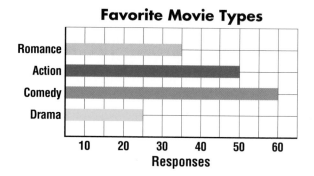

Favorite Movie Types

What was the total number of people who were surveyed?

   A 60

   B 75

   C 90

   D 170

   E 270

6. The Girls Scouts placed a bank in the school office to collect money for a service project. At the end of the week, the bank contained 32 quarters, 27 nickels, and 105 dimes. Which number sentence could be used to find M, the total amount of money in the bank?

   F $M = (32 \times 0.25) + (27 \times 0.05) + (105 \times 0.10)$

   G $M = (32 + 0.25) \times (27 + 0.05) \times (105 + 0.10)$

   H $M = (32 \div 0.25) \times (27 \div 0.05) \times (105 \div 0.10)$

   J $M = (0.25 + 0.10 + 0.05) \times 86$

   K $M = (32 \div 0.25) + (27 \div 0.05) + (105 \div 0.10)$

7. If $592 was made in sales of sweatshirts, what was the total amount of sales to the nearest dollar?

Sales of Items

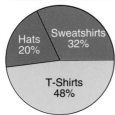

- Hats 20%
- Sweatshirts 32%
- T-Shirts 48%

A $1,850   **B** $1,798

C $185   **D** $180

E Not Here

8. Cali bought a sweater priced at $60. If the sales tax is 9%, which is a good estimate for the total purchase price of the jacket, including sales tax?

F $55   **G** $65

H $70   **J** $75

K $80

9. If one chocolate bar contains $\frac{3}{4}$ of a pound of chocolate, how many chocolate bars can be made from $6\frac{1}{2}$ pounds of chocolate?

A $4\frac{7}{8}$ **B** $5\frac{3}{4}$

C $8\frac{2}{3}$ **D** $9\frac{2}{3}$

E Not Here

Test Practice For additional test practice questions, visit:

www.glencoe.com/sec/math/mac/mathnet

Section Two: Free Response

This section contains five questions for which you will provide short answers. Write your answers on your paper.

10. Find 2.36×100.

11. At the school carnival, 10 prizes were written on slips of paper. There were 3 markers, 2 movie passes, 4 pencils, and 1 five-dollar bill. If a slip of paper is selected without looking, what is the probability that it will be a pencil?

12. What is the length of the hypotenuse of the triangle?

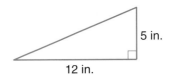

5 in.

12 in.

13. A store advertised a backpack for 25% off the regular price. Express 25% as a fraction.

14. A deck is shaped like a square with an area of 400 square feet. What is the length of one side of the deck?

Student Handbook
Table of Contents

Basic Skills

Adding and Subtracting Decimals

1. 0.132
 − 0.021

2. 3.78
 + 0.21

3. 12.3
 − 0.847

4. 5.86
 − 1.51

5. 128.01
 − 39.117

6. 14.7
 + 351.82

7. 42.3
 + .81

8. 13.2
 + 12.8

9. 342.9
 − 0.18

10. 282.45
 − 111.3

11. 100
 − 0.48

12. 12.888
 − 4.996

13. 42.07
 − 38.78

14. 80.05
 − 79.06

15. 104.98
 − 0.12

16. 304.999
 + 1.1

17. 82.23
 + 0.88

18. 898.667
 − 0.56

19. 13
 − 0.324

20. 0.42
 + 0.68

21. 9.1 − 5.625

22. 0.48 + 2.901

23. 5.8 + 3.92

24. 38.63 + 38.63

25. 8 − 2.54

26. 16.354 − 0.2

27. 0.125 + 0.78

28. 8.2 − 6.9

29. 1.245 + 3.842

30. 3.2 + 1.23

31. 0.889 − 0.3

32. 22.22 + 1.475

33. 10 − 0.25

34. 33.16 − 0.08

35. 1.254 + 0.5

36. 44.698 − 14.903

37. 10 − 0.005

38. 722.86 + 0.024

39. 100.211 + 8.004

40. 86.124 + 32.822

41. 6.9 + 1.1

42. 75 − 0.24

43. 822.003 + 0.22

44. 84.98 + 0.129

45. 3.2 − 1.3

46. 35.009 + 3.6

47. 800.972 + 0.4

48. 125.011 − 2.344

49. 0.8 + 1.2

50. 0.1 + 0.2

Basic Skills

Multiplying Decimals

1. 5.08
$\underline{\times\ 0.19}$

2. 15.8
$\underline{\times\quad 11}$

3. 1.6
$\underline{\times 1.6}$

4. 3.5
$\underline{\times 1.5}$

5. 88
$\underline{\times 2.5}$

6. 99
$\underline{\times 1.1}$

7. 0.042
$\underline{\times\quad 6}$

8. 16.8
$\underline{\times\ 2.2}$

9. 33
$\underline{\times 1.2}$

10. 1.25
$\underline{\times 1.33}$

11. 1.23
$\underline{\times\ 8.8}$

12. 18.9
$\underline{\times\ 0.8}$

13. 100
$\underline{\times\ .04}$

14. 4.32
$\underline{\times 1.23}$

15. 3.6
$\underline{\times 1.2}$

16. 8.2
$\underline{\times 3.9}$

17. 0.6
$\underline{\times\ 2}$

18. 3.2
$\underline{\times 0.8}$

19. 43.2
$\underline{\times 0.13}$

20. 68
$\underline{\times 1.9}$

21. 2×0.3

22. 0.4×3.8

23. 9×0.5

24. 0.3×1.2

25. 1.88×1.11

26. $33 \times .03$

27. 0.003×482

28. 5×0.9

29. 0.4×16

30. 1.23×3

31. 0.8×1

32. 36×0.46

33. 0.5×1.6

34. 200×0.004

35. 0.7×18

36. 18×0.04

37. 12.2×12.4

38. 4.3×2.8

39. 800×0.8

40. 125×1.29

41. 8×0.3

42. 4×0.4

43. 0.3×2.3

44. 0.23×0.2

45. 380×0.125

46. 0.004×2

47. 38.3×29.1

48. 0.44×0.5

49. 34.2×80.1

50. 42×0.17

Basic Skills

Dividing Decimals

1. $0.3 \overline{)9.81}$

2. $12 \overline{)0.12}$

3. $3.2 \overline{)5.76}$

4. $0.22 \overline{)0.0132}$

5. $0.04 \overline{)0.008}$

6. $3.18 \overline{)0.636}$

7. $0.2 \overline{)8.24}$

8. $82.3 \overline{)823}$

9. $49.92 \overline{)803.712}$

10. $100 \overline{)0.01}$

11. $13.8 \overline{)131.1}$

12. $10.81 \overline{)363.216}$

13. $74.9 \overline{)5.992}$

14. $0.5 \overline{)85}$

15. $1.9 \overline{)38.57}$

16. $100 \overline{)8.235}$

17. $64.80 \overline{)25920}$

18. $19.2 \overline{)4.416}$

19. $8.43 \overline{)0.02529}$

20. $12.02 \overline{)24.04}$

21. $812 \div 0.4$

22. $0.34 \div 0.2$

23. $1680.042 \div 44.2$

24. $90.175 \div 2.5$

25. $39.95 \div 799$

26. $88.8 \div 444$

27. $613.8 \div 66$

28. $2445.3 \div 33$

29. $500 \div 0.10$

30. $44.82 \div 45$

31. $0.01197 \div 3.99$

32. $2.232 \div 0.036$

33. $14.4 \div 0.12$

34. $4.6848 \div 0.366$

35. $2.475 \div 0.03$

36. $45 \div 0.09$

37. $180 \div 0.36$

38. $97.812 \div 1.1$

39. $23 \div 0.023$

40. $20.24 \div 2.3$

41. $0.004 \div 0.0002$

42. $10.557 \div 0.23$

43. $485.76 \div 8.8$

44. $493.19 \div 33.1$

Basic Skills

Adding and Subtracting Fractions

1. $\dfrac{17}{5} - \dfrac{2}{5}$

2. $\dfrac{1}{10} + \dfrac{1}{6}$

3. $\dfrac{1}{32} + \dfrac{2}{3}$

4. $\dfrac{4}{5} - \dfrac{1}{3}$

5. $\dfrac{3}{7} + \dfrac{1}{4}$

6. $\dfrac{2}{25} + \dfrac{10}{30}$

7. $\dfrac{17}{20} - \dfrac{3}{10}$

8. $\dfrac{10}{11} + \dfrac{1}{2}$

9. $\dfrac{9}{11} - \dfrac{2}{11}$

10. $\dfrac{3}{4} - \dfrac{1}{2}$

11. $\dfrac{10}{11} - \dfrac{1}{2}$

12. $\dfrac{1}{2} - \dfrac{1}{3}$

13. $\dfrac{6}{7} - \dfrac{2}{3}$

14. $\dfrac{4}{5} - \dfrac{1}{5}$

15. $\dfrac{14}{45} - \dfrac{3}{10}$

16. $\dfrac{3}{8} - \dfrac{2}{8}$

17. $\dfrac{7}{100} + \dfrac{3}{100}$

18. $\dfrac{3}{10} - \dfrac{1}{5}$

19. $\dfrac{8}{14} + \dfrac{2}{5}$

20. $\dfrac{3}{10} + \dfrac{5}{16}$

21. $\dfrac{7}{22} - \dfrac{1}{11}$

22. $\dfrac{3}{4} - \dfrac{1}{4}$

23. $\dfrac{18}{50} - \dfrac{1}{25}$

24. $\dfrac{6}{13} + \dfrac{4}{13}$

25. $\dfrac{6}{7} - \dfrac{1}{7}$

26. $\dfrac{5}{46} + \dfrac{4}{23}$

27. $\dfrac{1}{12} + \dfrac{1}{6}$

28. $\dfrac{1}{99} + \dfrac{3}{99}$

29. $\dfrac{5}{90} - \dfrac{1}{30}$

30. $\dfrac{8}{15} + \dfrac{4}{15}$

31. $\dfrac{4}{5} - \dfrac{1}{25}$

32. $\dfrac{1}{6} + \dfrac{2}{6}$

33. $\dfrac{11}{12} - \dfrac{1}{2}$

34. $\dfrac{1}{3} + \dfrac{2}{3}$

35. $\dfrac{1}{6} + \dfrac{2}{3}$

36. $\dfrac{1}{3} + \dfrac{4}{9}$

37. $\dfrac{3}{4} + \dfrac{1}{12}$

38. $\dfrac{1}{20} - \dfrac{1}{100}$

39. $\dfrac{1}{8} + \dfrac{2}{3}$

40. $\dfrac{99}{100} - \dfrac{24}{25}$

Basic Skills

Multiplying Fractions

1. $\frac{1}{3} \times \frac{1}{6}$

2. $\frac{3}{4} \times \frac{2}{3}$

3. $\frac{10}{5} \times \frac{5}{6}$

4. $\frac{1}{2} \times \frac{4}{5}$

5. $120 \times \frac{3}{10}$

6. $144 \times \frac{3}{12}$

7. $\frac{6}{11} \times \frac{121}{3}$

8. $\frac{4}{5} \times \frac{20}{2}$

9. $\frac{1}{4} \times 30$

10. $\frac{1}{3} \times \frac{1}{2}$

11. $\frac{3}{8} \times \frac{10}{21}$

12. $\frac{18}{3} \times \frac{2}{9}$

13. $\frac{15}{7} \times \frac{3}{2}$

14. $\frac{4}{100} \times \frac{5}{6}$

15. $\frac{11}{17} \times \frac{1}{2}$

16. $\frac{12}{13} \times \frac{39}{6}$

17. $\frac{6}{7} \times \frac{42}{3}$

18. $\frac{3}{19} \times \frac{19}{27}$

19. $\frac{3}{40} \times \frac{4}{9}$

20. $36 \times \frac{1}{6}$

21. $\frac{18}{25} \times \frac{10}{90}$

22. $\frac{1}{3} \times \frac{3}{4}$

23. $\frac{1}{8} \times 64$

24. $\frac{3}{2} \times \frac{4}{9}$

25. $\frac{48}{50} \times \frac{3}{2}$

26. $\frac{102}{121} \times \frac{11}{51}$

27. $\frac{1}{2} \times \frac{4}{7}$

28. $\frac{80}{90} \times \frac{9}{10}$

29. $\frac{1}{5} \times 25$

30. $\frac{8}{9} \times \frac{81}{4}$

31. $\frac{64}{2} \times \frac{2}{18}$

32. $5 \times \frac{3}{4}$

33. $\frac{36}{38} \times \frac{76}{12}$

34. $\frac{32}{37} \times \frac{1}{8}$

35. $\frac{40}{55} \times \frac{5}{10}$

36. $9 \times \frac{1}{3}$

37. $\frac{72}{80} \times \frac{3}{4}$

38. $\frac{18}{32} \times \frac{22}{46}$

39. $\frac{3}{8} \times 24$

40. $\frac{82}{85} \times \frac{15}{16}$

Basic Skills

Dividing Fractions

1. $\dfrac{15}{7} \div \dfrac{3}{2}$

2. $5 \div \dfrac{3}{4}$

3. $\dfrac{42}{3} \div \dfrac{7}{6}$

4. $\dfrac{3}{19} \div \dfrac{27}{19}$

5. $\dfrac{6}{11} \div \dfrac{3}{121}$

6. $\dfrac{4}{5} \div \dfrac{2}{20}$

7. $\dfrac{3}{40} \div \dfrac{9}{4}$

8. $9 \div \dfrac{1}{3}$

9. $\dfrac{2}{9} \div \dfrac{3}{18}$

10. $\dfrac{1}{6} \div \dfrac{1}{36}$

11. $\dfrac{1}{3} \div \dfrac{3}{4}$

12. $\dfrac{80}{9} \div \dfrac{10}{9}$

13. $\dfrac{10}{90} \div \dfrac{25}{18}$

14. $\dfrac{15}{7} \div \dfrac{2}{3}$

15. $\dfrac{48}{50} \div \dfrac{2}{3}$

16. $\dfrac{1}{8} \div 64$

17. $\dfrac{12}{13} \div \dfrac{6}{39}$

18. $\dfrac{4}{100} \div \dfrac{6}{5}$

19. $\dfrac{1}{2} \div \dfrac{5}{4}$

20. $\dfrac{4}{9} \div \dfrac{2}{3}$

21. $\dfrac{64}{4} \div \dfrac{18}{4}$

22. $\dfrac{11}{17} \div 2$

23. $\dfrac{3}{12} \div \dfrac{1}{144}$

24. $\dfrac{1}{4} \div \dfrac{1}{30}$

25. $5 \div \dfrac{4}{3}$

26. $\dfrac{3}{10} \div \dfrac{1}{120}$

27. $\dfrac{10}{5} \div \dfrac{6}{5}$

28. $\dfrac{1}{3} \div \dfrac{1}{6}$

29. $\dfrac{3}{4} \div \dfrac{2}{3}$

30. $24 \div \dfrac{3}{8}$

31. $\dfrac{1}{2} \div \dfrac{5}{4}$

32. $\dfrac{1}{5} \div 25$

33. $\dfrac{36}{38} \div \dfrac{12}{76}$

34. $\dfrac{8}{9} \div \dfrac{4}{81}$

35. $\dfrac{110}{121} \div \dfrac{50}{11}$

36. $\dfrac{1}{3} \div 2$

37. $\dfrac{1}{3} \div \dfrac{1}{9}$

38. $24 \div \dfrac{2}{3}$

39. $\dfrac{1}{25} \div 5$

40. $\dfrac{3}{8} \div \dfrac{21}{10}$

Extra Practice

Lesson 1-1 *(Pages 4–7)*
Use the four-step plan to solve each problem.

1. The Gonzales family rode their bicycles for 10 miles to the campground. The ride back was along a different route for 13 miles. How many miles did they ride in all?

2. A farmer planted 405 acres of land with 71,685 corn plants. How many plants were planted per acre?

3. A group of 251 people is eating dinner at a school fundraiser. If each person pays $4.00 for their meal, how much money is raised?

4. When Marcy calls home from college, she talks ten minutes per call for 3 calls per week. How many minutes does she call in a 15-week semester?

Lesson 1-2 *(Pages 8–10)*
Evaluate each expression.

1. $14 - 5 + 7$

2. $12 + 10 - 5 - 6$

3. $50 - 6 + 12 + 4$

4. $12 - 2 \cdot 3$

5. $16 + 4 \times 5$

6. $5 + 3 \times 4 - 7$

7. $2 \times 3 + 9 \times 2$

8. $6 \cdot 8 + 4 \div 2$

9. $7 \times 6 - 14$

10. $8 + 12 \times 4 \div 8$

11. $13 - 6 \times 2 + 1$

12. $80 \div 10 \times 8$

13. $1 + 2 + 3 + 4$

14. $1 \cdot 2 \cdot 3 \cdot 4$

15. $6 + 6 \times 6$

16. $14 - 2 \times 7 + 0$

17. $156 - 6 \times 0$

18. $30 - 14 \cdot 2 + 8$

Lesson 1-3 *(Pages 12–15)*
Evaluate each expression if $a = 3$, $b = 4$, and $c = 12$.

1. $a + b$

2. $c - a$

3. $a + b + c$

4. $b - a$

5. $c - a \times b$

6. $a + 2 \times b$

7. $b + c \div 2$

8. ab

9. $a + 3b$

10. $a + c \div 6$

11. $25 + c \div b$

12. abc

13. $2(a + b) \div 7$

14. $2c \div b$

15. $144 - abc$

16. $2ab$

17. $c \div a + 10$

18. $9b \div 3$

19. $2b - a$

20. ac

Lesson 1-4 *(Pages 17–20)*

Write each power as a product of the same factor.

1. 13^4

2. 9^6

3. $2^3 \cdot 3^2$

4. x^5

5. 169^3

6. $13,410^2$

Write each product using exponents.

7. $2 \cdot 2 \cdot 2 \cdot 2 \cdot 2$

8. $6 \cdot 6 \cdot 6 \cdot 7 \cdot 7$

9. $9 \cdot 9 \cdot 9 \cdot 9 \cdot 9 \cdot 9 \cdot 10$

10. $k \cdot k \cdot k \cdot \ell \cdot \ell \cdot \ell$

11. $14 \cdot 14 \cdot 6$

12. $3 \cdot 3 \cdot 3 \cdot 3 \cdot y \cdot y$

Evaluate each expression.

13. 5^6

14. 17^3

15. 2^{12}

16. $3^5 \cdot 2^3$

17. $6^4 \cdot 3$

18. $2^2 \cdot 3^2 \cdot 4^2$

19. 176^2

20. $6 \cdot 4^3$

21. five squared

22. 2 to the fifth power

23. 4 cubed

Lesson 1-5 *(Pages 21–23)*

Solve each equation.

1. $b + 7 = 12$

2. $a + 3 = 15$

3. $s + 10 = 23$

4. $9 + n = 13$

5. $20 = 24 - n$

6. $4x = 36$

7. $2y = 10$

8. $15 = 5h$

9. $j \div 3 = 2$

10. $14 = w - 4$

11. $24 \div k = 6$

12. $b - 3 = 12$

13. $c \div 10 = 8$

14. $y \div 2 = 8$

15. $6 = t \div 5$

16. $42 = 6n$

17. $14 + m = 24$

18. $g - 3 = 10$

19. $7 + a = 10$

20. $3y = 39$

21. $\frac{f}{2} = 12$

22. $16 = 4v$

23. $81 = 80 + a$

24. $9 = \frac{72}{x}$

Lesson 1-6 *(Pages 24–27)*

Draw the next two figures that continue each pattern.

1.

2.

3.

4.

Lesson 1-7 *(Pages 30–33)*
Find the area of each rectangle or parallelogram.

1.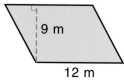
7 cm
2 cm

2.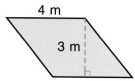
4 m
3 m

3.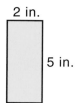
2 in.
5 in.

4.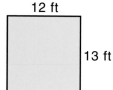
9 m
12 m

5.
12 ft
13 ft

6.
4 ft
3 ft

7. rectangle: ℓ, 19 m; w, 6 m

8. parallelogram: b, 15 m; h, 12 m

9. rectangle: ℓ, 8 m; w, 5 m

Lesson 1-7B *(Pages 34–35)*
Solve.

1. A trip from Cleveland to Columbus is 120 miles. If 5 gallons of gasoline are used, how many miles per gallon is this?

2. A national debate tournament had 2,673 students registered. The tournament director had to assign 336 students per hotel. To how many different hotels did the director have to assign the students?

3. During a political campaign, each of 125 persons donated $100. How much money was donated?

4. Marnie is planting a garden that is 18 feet long and 12 feet wide. How many feet of fencing will be needed to enclose the entire garden?

5. If a bag of fertilizer feeds 75 square feet of garden, how many bags will Marnie need for the garden in Problem 4?

Lesson 2-1 *(Pages 44–46)*
Replace each ● with <, >, or = to make a true sentence.

1. 0.36 ● 0.63

2. 1.74 ● 1.7

3. 4.03 ● 4.003

4. 0.06 ● 0.066

5. 10.5 ● 10.05

6. 3.0 ● 3

7. 5.632 ● 5.623

8. 0.423 ● 0.5

9. 2.020 ● 2.202

10. 0.93 ● 0.9

11. 0.205 ● 0.025

12. 0.46 ● 0.49

13. 13.100 ● 13.1

14. 6.25 ● 6.20

15. 9.99 ● 9.099

16. 0.030 ● 0.03

17. 0.062 ● 0.62

18. 1.14 ● 1.09

19. 10.1 ● 100.0

20. 0.02 ● 0.002

21. 2.101 ● 2.11

Lesson 2-2 (Pages 47–49)

Round each number to the place indicated.

1. 5.64; tenth   **2.** 0.2625; hundredth   **3.** 0.45695; thousandth

4. 6.249; tenth   **5.** 0.00263; thousandth   **6.** 758.997; hundredth

Round each number to the underlined place-value position.

7. 32.65<u>8</u>2   **8.** <u>0</u>.025   **9.** 1.00<u>4</u>9   **10.** 9.<u>2</u>5

11. 67.4<u>9</u>2   **12.** 25.<u>1</u>9   **13.** 26.<u>9</u>6   **14.** 4.0<u>0</u>65

15. 26.96<u>6</u>6   **16.** 1.<u>2</u>499999   **17.** 2.0<u>1</u>2   **18.** 1<u>6</u>.569

Lesson 2-3 (Pages 50–53)

Estimate by rounding.

1. 0.245
     + 0.256

2. 2.45698
     − 1.26589

3. 0.5962
     + 1.2598

4. 17.985
     − 9.001

5. 8.5
     × 9.1

6. 12.9568
     × 6.1563

7. 9.652
     ×   6.2

8. 25.49862
     ×   4.2136

9. 3)11.75   **10.** 4.1)16.123   **11.** 2.7)29.5   **12.** 8.14)81.27

Estimate by clustering.

13. 1.12 + 0.9865 + 1.023 + 0.89 + 0.99 + 1.03569

14. 82.1 + 79.3 + 81.5 + 79 + 80 + 81.256

Lesson 2-3B (Pages 54–55)

Solve.

1. Mr. Ludwig eats food that has 3,115 calories in an average day. When he multiplied the number of calories he eats per day by the number of days in a week, the calculator showed 218,050. Is this answer reasonable? Explain.

2. You need to buy 3 cans of tomato soup at 79¢ each, a box of crackers at $1.89, and a gallon of orange juice at $2.39. Should you take $10.00 with you to the store or will $5 be enough?

3. When Nia received her work schedule for the week, she multiplied the 28 hours she was scheduled to work by her wage of $5.15 per hour. Her calculator showed that her pay for the week should be $112.75. Is this answer reasonable?

4. A newspaper article reported that a running back ran for 182 yards in 17 carries during the football game. Jamison determined that this was about 10 yards per carry. Is his answer reasonable?

5. There were three recycling centers set up for the Wyandot Junior High aluminum can drive. One center collected 867 pounds of cans, the second collected 1,236 pounds of cans, and the third collected 1,827 pounds of cans. The Junior High reported that they had collected 4,000 pounds of cans. Is this answer reasonable?

Lesson 2-4 (*Pages 56–59*)
Multiply.

1. 9.6×10.5

2. 3.2×0.1

3. 10.5×9.6

4. 5.42×0.21

5. 7.42×0.2

6. 0.001×0.02

7. 0.6×542

8. 6.7×5.8

9. 3.24×6.7

10. 9.8×4.62

11. 7.32×9.7

12. 0.008×0.007

13. 0.0001×56

14. 4.5×0.2

15. 9.6×2.3

16. 5.63×8.1

17. 10.35×9.1

18. 28.2×3.9

19. 102.13×1.221

20. 2.02×1.25

21. 8.37×89.6

Lesson 2-5 (*Pages 61–63*)
Multiply mentally.

1. 1.2×10

2. 0.23×100

3. $1.235 \times 1,000$

4. 1.2×10^3

5. 3.97×0.1

6. 3.56×10^2

7. 123.92×0.01

8. 95.23×10^1

9. 76.425×0.1

10. $1.0056 \times 10,000$

11. 4.7×10^0

12. 9.6×10^0

Solve each equation.

13. $1.47 \times 10^3 = x$

14. $y = 0.82 \times 1,000$

15. $m = 2.8 \times 0.1$

16. $h = 15.23 \times 10^4$

17. $53.7 \times 0.01 = y$

18. $0.9 \times 10^0 = w$

Lesson 2-6 (*Pages 66–69*)
Without finding or changing each quotient, change each problem so that the divisor is a whole number.

1. $3.6 \div 0.6$

2. $0.36 \div 0.4$

3. $82 \div 0.4$

4. $4.876 \div 0.72$

5. $0.009 \div 0.03$

6. $21.8 \div 0.005$

Divide.

7. $7.2 \div 0.4$

8. $0.76 \div 0.5$

9. $8.4 \div 0.8$

10. $0.23\overline{)2.76}$

11. $3.2\overline{)8.32}$

12. $0.003\overline{)0.018}$

13. $51 \div 0.8$

14. $1.826 \div 8.3$

15. $702 \div 6.5$

16. Round the quotient of 6.51 and 0.8 to the nearest tenth.

17. What is $31.76 divided by 0.7, rounded to the nearest cent?

Lesson 2-7 (Pages 70–73)
Write each repeating decimal using bar notation.

1. 0.3333333....

2. 0.121212....

3. 4.3151515....

4. 7.023023023....

5. 0.544444....

6. 18.75484848....

Express each fraction or mixed number as a decimal. If the decimal is a repeating decimal, use bar notation.

7. $\frac{16}{20}$

8. $\frac{25}{100}$

9. $1\frac{7}{8}$

10. $\frac{1}{6}$

11. $\frac{11}{40}$

12. $5\frac{13}{50}$

13. $\frac{55}{300}$

14. $\frac{18}{12}$

Lesson 2-8 (Pages 74–76)
Complete.

1. 400 mm = _____ cm

2. 4 km = _____ m

3. 660 cm = _____ m

4. 0.3 km = _____ m

5. 30 mm = _____ cm

6. 84.5 m = _____ km

7. _____ m = 54 cm

8. 18 km = _____ cm

9. _____ mm = 45 cm

10. 4 kg = _____ g

11. 632 mg = _____ g

12. 4,497 g = _____ kg

13. _____ mg = 21 g

14. 61.2 mg = _____ g

15. 61 g = _____ mg

16. _____ mg = 0.51 kg

17. 0.63 kg = _____ g

18. _____ kg = 563 g

Lesson 2-9 (Pages 77–79)
Write each number in scientific notation.

1. 720

2. 7,560

3. 892

4. 1,400

5. 91,200

6. 51,000

7. 145,000

8. 90,100

9. 123,000,000,000

10. 4,500

11. 20,000

12. 1,700,000

13. 961

14. 10,000,000

15. 820,000,000

16. 52,000

17. 680,000

Lesson 3-1 *(Pages 88–91)*

Choose an appropriate scale and interval for each set of data. Make a frequency table.

1.

Length of Time Walking (min.)			
15	30	15	45
45	30	30	60
30	60	15	30
45	45	60	15

2.

Number of Raisins Eaten in Preschool			
40	49	45	49
42	41	45	41
45	41	41	40
41	43	40	41

Lesson 3-2A *(Pages 92–93)*

Use the circle graph to solve each problem.

1. What part of the budget is spent on printing?

2. How much more money is spent on photography than design?

3. If the total budget is increased by $200 next year, about how much could be spent on printing?

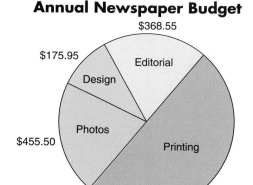

Annual Newspaper Budget

Lesson 3-2 *(Pages 94–97)*

1. Darlene's quiz scores in science have been steadily going up since her parents hired a tutor for her. Based on the graph below, predict what Darlene's score will be on the next quiz.

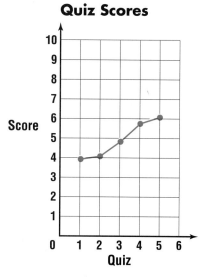

2. A balloon maker asked 100 kids what their favorite color of balloon is. The graph below shows their responses. What color of balloon should he make the most of?

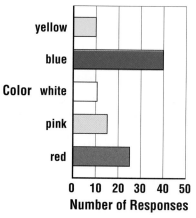

Favorite Color of Balloon

Lesson 3-3 *(Pages 98–101)*

Make a line plot for each set of data.

1. 25, 26, 27, 25, 28, 27, 21, 26, 28, 25

2. 110, 210, 156, 174, 125, 196, 165, 185

3. 600, 650, 700, 600, 625, 675, 450, 650

4. 0.5, 0.6, 0.1, 0.4, 0.8, 0.6, 0.7, 0.5

5. 3, 4, 5, 2, 6, 1, 2, 4, 3, 6, 9, 1, 2, 3, 4

Lesson 3-4 *(Pages 102–105)*

Find the mean, mode(s), and median for each set of data. Round answers to the nearest tenth.

1. 1, 5, 9, 1, 2, 5, 8, 2

2. 2, 5, 8, 9, 7, 6, 3, 5

3. 1, 2, 1, 2, 2, 1, 2

4. 12, 13, 15, 12, 12, 11

5. 256, 265, 247, 256

6. 957, 562, 462, 847, 721

7. 46, 54, 66, 54, 46, 66

8. 81, 82, 83, 84, 85, 86, 87

Lesson 3-5 *(Pages 108–111)*

Write the stems that would be used in a stem-and-leaf plot for each set of data. Then make the stem-and-leaf plot.

1. 23, 15, 39, 68, 57, 42, 51, 52, 41, 18, 29

2. 5, 14, 39, 28, 14, 6, 7, 18, 13, 28, 9, 14

3. 189, 182, 196, 184, 197, 183, 196, 194, 184

4. 71, 82, 84, 95, 76, 92, 83, 74, 81, 75, 96

Lesson 3-6 *(Pages 114–117)*

Use the box-and-whisker plot to answer each question.

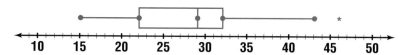

1. What is the median?
2. What is the upper quartile?
3. What is the lower quartile?
4. What is the upper extreme?
5. What is the lower extreme?
6. What is the interquartile range?
7. Name any outliers.
8. What fraction of the data fall between 22 and 29?

Lesson 3-7 *(Pages 119–121)*

Which graph could be misleading? How are the graphs misleading?

1. Both graphs show pounds of grapes sold to Westview School in one week.

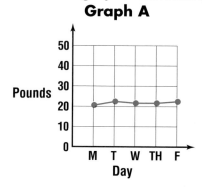

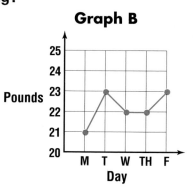

2. Both graphs show commissions made by Mai-Lin for a four-week pay period.

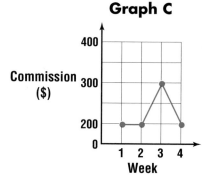

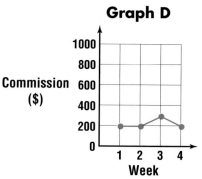

Lesson 4-1 *(Pages 133–136)*

Determine whether the first number is divisible by the second number.

1. 279; 3
2. 1,240; 6
3. 3,250; 5
4. 835; 4
5. 5,550; 10
6. 315; 9
7. 777; 6
8. 4,214; 3
9. 3,012; 2
10. 244; 4
11. 984; 6
12. 1,000; 5

Determine whether each number is divisible by 2, 3, 4, 5, 6, 9, or 10.

13. 453
14. 2,225
15. 504
16. 4,300
17. 672
18. 8,240
19. 111
20. 6,232
21. 999
22. 5,200
23. 3,217
24. 804

Lesson 4-2 *(Pages 138–141)*

Determine whether each number is *composite* or *prime*.

1. 32

2. 417

3. 5,212

4. 2,111

5. 71

6. 1,005

7. 239

8. 3,215

Use a factor tree to find the prime factorization of each number.

9. 81

10. 525

11. 245

12. 1,120

13. 750

14. 2,400

15. 914

16. 975

Use your calculator to find the prime factors of each number. Then write the prime factorization of each number.

17. 423

18. 972

19. 144

20. 72

Lesson 4-3 *(Pages 142–145)*

Describe the pattern in each sequence. Identify the sequence as *arithmetic*, *geometric*, or *neither*. Then find the next three terms.

1. 5, 9, 13, 17, ...

2. 3, 6, 12, 24, ...

3. 10, 15, 25, 40, ...

4. 4.5, 5.4, 6.3, 7.2, ...

5. 90, 91, 94, 99, ...

6. 0.3, 0.4, 0.5, ...

7. 8, 24, 72, 216, ...

8. 16, 17, 19, 22, ...

Create a sequence using each rule. Provide four terms for the sequence beginning with the given number. State whether the sequence is *arithmetic*, *geometric*, or *neither*.

9. Add 9 to each term; 1.

10. Multiply each term by 4; 3.

11. Multiply each term by 0.2; 6.

12. Add 16 to each term; 14.

13. Add 1 to the first term, 11 to the second term, 111 to the third term, and so on; 40.

Lesson 4-4A *(Pages 148–149)*

Solve.

1. Jo sees that every third Fibonacci number is even. (F3 = 2, F6 = 8, F9 = 34). Find F30.

2. Are there any perfect squares (for example: 4 × 4 = 16) or perfect cubes (for example: 4 × 4 × 4 = 64) in the first 12 terms of the Fibonacci sequence? If so, name them.

3. Make a list to find the number of 4-person teams that can be formed from a 5-member group. Use the letters A, B, C, D, E to represent the people.

4. Study the pattern below. List the next three numbers in the pattern.

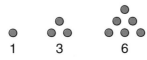

Lesson 4-4 (Pages 150–153)
Find the GCF of each set of numbers.

1. 12, 16   **2.** 63, 81   **3.** 225, 500   **4.** 37, 100

5. 240, 32   **6.** 640, 412   **7.** 36, 81   **8.** 350, 140

9. 72, 170   **10.** 255, 51   **11.** 48, 72

12. 86, 200   **13.** 24, 56, 120   **14.** 48, 60, 84

15. 32, 80, 96   **16.** 49, 14, 70   **17.** 6, 8, 12

18. 33, 55, 77   **19.** 27, 15, 300   **20.** 45, 150, 225

Lesson 4-5 (Pages 154–157)
Express each fraction or ratio in simplest form.

1. $\frac{14}{28}$ **2.** 15:25   **3.** $\frac{100}{300}$ **4.** 14:35

5. 9:51   **6.** $\frac{54}{56}$ **7.** 75:90   **8.** $\frac{24}{40}$

9. 180:270   **10.** $\frac{312}{390}$ **11.** 240:448   **12.** $\frac{71}{82}$

13. $\frac{333}{900}$ **14.** 85:255   **15.** $\frac{84}{128}$ **16.** 64:96

Lesson 4-6 (Pages 158–160)
Write a percent to represent the shaded area.

1. 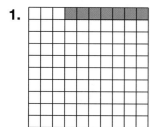   **2.** 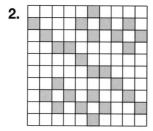   **3.**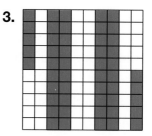

Express each ratio as a percent.

4. 39 out of 100   **5.** $\frac{23}{100}$ **6.** 17:100

7. $72 per $100   **8.** 4 to 100   **9.** 98 in 100

Lesson 4-7 *(Pages 161–164)*

Express each fraction as a percent.

1. $\frac{1}{2}$ 2. $\frac{6}{6}$ 3. $\frac{8}{10}$ 4. $\frac{3}{5}$

5. $\frac{1}{4}$ 6. $\frac{21}{25}$ 7. $\frac{16}{100}$ 8. $\frac{13}{20}$

9. $\frac{3}{10}$ 10. $\frac{17}{20}$ 11. $\frac{9}{10}$ 12. $\frac{9}{50}$

Express each percent or decimal as a fraction in simplest form.

13. 28%   14. 0.10   15. 18%   16. 0.02

17. 0.09   18. 36%   19. 49%   20. 3%

21. 0.25   22. 0.98   23. 0.88   24. 50%

Lesson 4-8 *(Pages 165–168)*

The spinner shown is equally likely to stop on each of its regions numbered 1 to 8. Find the probability that the spinner will stop on each of the following.

1. an even number   2. a prime number

3. a factor of 12   4. a composite number

5. a number less than 5   6. a factor of 36

A package of balloons contains 5 green, 3 yellow, 4 red, and 8 pink balloons. If you reach in the package and choose one balloon at random, what is the probability that you will select each of the following? Express each ratio as a fraction in simplest form and as a percent.

7. a red balloon   8. a green balloon   9. a pink balloon

10. a yellow balloon   11. a red or yellow balloon

Lesson 4-9 *(Pages 169–171)*

Find the LCM of each set of numbers.

1. 4, 9   2. 6, 16   3. 3, 8, 14   4. 24, 36

5. 48, 84   6. 12, 18, 28   7. 8, 9   8. 49, 56

9. 42, 66   10. 15, 39   11. 32, 80, 96   12. 56, 64

13. 24, 42   14. 250, 80   15. 26, 169   16. 5, 18, 45

17. 11, 22, 33   18. 56, 14, 70   19. 16, 24   20. 13, 14

Lesson 4-10 *(Pages 172–175)*

Find the LCD for each pair of fractions.

1. $\frac{3}{8}, \frac{2}{3}$

2. $\frac{5}{9}, \frac{7}{12}$

3. $\frac{4}{9}, \frac{8}{15}$

4. $\frac{11}{24}, \frac{17}{42}$

5. $\frac{12}{36}, \frac{15}{42}$

6. $\frac{25}{27}, \frac{43}{81}$

7. $\frac{32}{64}, \frac{15}{48}$

8. $\frac{2}{6}, \frac{14}{15}$

Replace each ● with <, >, or = to make a true sentence.

9. $\frac{7}{9}$ ● $\frac{3}{5}$

10. $\frac{14}{25}$ ● $\frac{3}{4}$

11. $\frac{8}{24}$ ● $\frac{20}{60}$

12. $\frac{5}{12}$ ● $\frac{4}{9}$

13. $\frac{18}{24}$ ● $\frac{10}{18}$

14. $\frac{4}{6}$ ● $\frac{5}{9}$

15. $\frac{11}{49}$ ● $\frac{12}{42}$

16. $\frac{5}{14}$ ● $\frac{2}{6}$

Lesson 5-1 *(Pages 184–186)*

Write an integer for each situation.

1. a gain of 14 points

2. a $25 withdrawal

3. six degrees below zero

4. a loss of 3 pounds

5. a loss of 20 yards

6. a profit of $16

Write the integer represented by the point for each letter. Then find its opposite and its absolute value.

7. A

8. B

9. C

10. D

11. E

12. F

Lesson 5-2 *(Pages 188–190)*

Replace each ● with < or > to make a true sentence.

1. 7 ● -7

2. -8 ● 4

3. -4 ● -9

4. -3 ● 0

5. 8 ● 10

6. -5 ● -4

7. 6 ● -7

8. -12 ● -13

Order the integers from least to greatest.

9. $-2, -8, 4, 10, -6, -12$

10. $19, -19, -21, 32, -14, 18$

11. $18, 23, 95, -95, -18, -23, 2$

12. $46, -48, -47, -52, -18, 12$

Lesson 5-3 (Pages 191–194)

Name the *x*-coordinate and the *y*-coordinate for each point labeled at the right. Then tell in which quadrant each point lies.

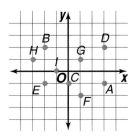

1. A
2. B
3. C
4. D
5. E
6. F
7. G
8. H
9. I

On graph paper, draw a coordinate plane. Then graph and label each point.

10. $N(-4, 3)$
11. $K(2, 5)$
12. $W(-6, -2)$
13. $X(5, 0)$
14. $Y(4, -4)$
15. $M(0, -3)$
16. $Z(-2, 0.5)$
17. $S(-1, -3)$

Lesson 5-4 (Pages 197–200)

Solve each equation.

1. $a = -4 + 8$
2. $14 + 16 = b$
3. $-7 + (-7) = h$
4. $g = -9 + (-6)$
5. $-18 + 11 = d$
6. $k = -36 + 40$
7. $42 + (-18) = f$
8. $-42 + 29 = r$
9. $m = 18 + (-32)$

Evaluate each expression if *a* = 6, *b* = −2, and *c* = −6.

10. $-96 + a$
11. $b + (-5)$
12. $c + (-32)$
13. $a + 98$
14. $-120 + b$
15. $-120 + c$
16. $5 + b$
17. $a + b$
18. $c + a$

Lesson 5-5 (Pages 202–205)

Solve each equation.

1. $3 - 7 = y$
2. $-5 - 4 = w$
3. $a = -6 - 2$
4. $r = 8 - 13$
5. $6 - (-4) = b$
6. $12 - 9 = x$
7. $-2 - 23 = c$
8. $z = 63 - 78$
9. $a = 0 - (-14)$

Evaluate each expression if *k* = −3, *p* = 6, and *n* = 1.

10. $55 - k$
11. $p - 7$
12. $n - 15$
13. $n - 12$
14. $-51 - p$
15. $k - 21$
16. $n - k$
17. $-99 - k$
18. $p - k$

Lesson 5-6 *(Pages 207–209)*
Solve each equation.

1. $5(-2) = d$
2. $a = 6(-4)$
3. $4(21) = y$
4. $-11(-5) = c$
5. $x = -6(5)$
6. $a = -50(0)$
7. $-5(-5) = z$
8. $-4(8) = q$
9. $b = 3(-13)$

Evaluate each expression if $a = -5$, $b = 2$, $c = -3$, and $d = 4$.

10. $-2d$
11. $6a$
12. $3ab$
13. $-12d$
14. $-4b^2$
15. $-5cd$
16. a^2
17. $13ab$
18. $-6ac$

Lesson 5-7A *(Pages 210–211)*
Solve.

1. Kit received an e-mail message from a Simon in England. After 10 minutes, she forwarded the message to 3 of her friends. After 10 more minutes, each of those friends forwarded the message to 3 more people. If the message was forwarded like this every 10 minutes, how many people received Simon's e-mail message after 40 minutes?

2. A display of laundry detergent boxes at Mike's Market is stacked in the shape of a pyramid. There are 2 boxes in the first row, 4 in the second row, 6 in the next row, and so on. The display contains 10 rows of boxes. How many boxes are in the display?

3. State the pattern and find the next three terms in the sequence 8, 12, 18, 27, ...

4. Sheri has decided to start an exercise program. She plans to begin by running 4 laps and then doubling her number of laps. Write a sequence showing the number of laps she runs each day for one week. If each lap is $\frac{1}{4}$ mile, is her plan reasonable? Why or why not?

Lesson 5-7 *(Pages 212–214)*
Solve each equation.

1. $a = 4 \div (-2)$
2. $16 \div (-8) = x$
3. $-14 \div (-2) = c$
4. $d = 32 \div 8$
5. $g = 18 \div (-3)$
6. $h = -18 \div 3$
7. $8 \div (-8) = y$
8. $t = 0 \div (-1)$
9. $-25 \div 5 = k$
10. $c = -14 \div (-7)$
11. $-32 \div 8 = m$
12. $n = -56 \div (-8)$
13. $-81 \div 9 = y$
14. $81 \div (-9) = w$
15. $x = 81 \div 9$
16. $q = -81 \div (-9)$
17. $18 \div (-2) = a$
18. $-55 \div 11 = c$
19. $25 \div (-5) = r$
20. $x = -21 \div 3$
21. $-42 \div (-7) = y$
22. $y = -121 \div (-11)$

Lesson 5-8 *(Pages 215–217)*

Graph each figure and its transformation. Write the ordered pairs for the vertices of the new figure.

1. $\triangle ABC$ with vertices $A(-4, 3)$, $B(2, -1)$, and $C(0, 5)$ translated 3 units left and 4 units down.

2. $\triangle DEF$ with vertices $D(5, 2)$, $E(-1, -1)$, and $F(3, 4)$ reflected over the x-axis.

3. $\triangle GHI$ with vertices $G(0, 7)$, $H(5, 0)$, and $I(-2, -4)$ translated 2 units right and 3 units up.

4. $\triangle JKL$ with vertices $J(-4, -4)$, $K(4, -4)$, and $L(0, 0)$ reflected over the y-axis.

5. Rectangle $PQRS$ with vertices $P(3, 5)$, $Q(-4, 5)$, $R(-4, -1)$, and $S(3, -1)$ translated 1 unit down and 4 units left.

Lesson 6-1 *(Pages 228–231)*

Solve each equation. Check your solution.

1. $r - 3 = 14$
2. $t + 3 = 21$
3. $s + 10 = 23$
4. $7 + a = -10$
5. $14 + m = 24$
6. $-9 + n = 13$
7. $s - 2 = -6$
8. $x - 1.3 = 12$
9. $y + 3.4 = 18$
10. $0.013 + h = 4.0$
11. $6 + f = 71$
12. $7.2 + g = 9.1$
13. $z - 12.1 = 14$
14. $w - 0.1 = 0.32$
15. $v - 18 = 13.7$
16. $s + 1.3 = 18$
17. $t + 3.43 = 7.4$
18. $x + 7.4 = 23.5$
19. $p + 3.1 = 18$
20. $q - 2.17 = 21$
21. $w - 3.7 = 4.63$
22. $m - 4.8 = 7.4$

Lesson 6-1B *(Pages 232–233)*

Solve.

1. I'm thinking of a number. If I multiply it by 7 and add 23, the result is 107. What is the number?

2. Sam is planning a luncheon. He goes to the grocery store and buys a ham for $24.98 and a vegetable tray for $17.49. There is no tax on food. He gives the cashier one bill and receives less than $10 in change. What was the denomination of the bill Sam gave the cashier?

3. Maya is two years older than her sister Jana. Jana is 5 years older than her brother Trevor, who is 9 years younger than his brother Trent. If Trent is 17 years old, how old is Maya?

4. A can of evaporated milk weighs 15 ounces. Mrs. Martinez uses half of the milk to make pumpkin pudding. The can and the milk that is left weigh 9 ounces. How much does the can weigh?

5. A parking garage in New York City charges $3 for the first two hours and then $0.75 for each additional hour. Walter parks his car in the garage at 9:00 A.M. and when he returns must pay a $5.25 parking fee. What time did Walter return?

Lesson 6-2 *(Pages 234-237)*
Solve each equation. Check your solution.

1. $2m = 18$

2. $-42 = 6n$

3. $72 = 8k$

4. $-20r = 20$

5. $420 = 5s$

6. $325 = 25t$

7. $-14 = -2p$

8. $18q = 36$

9. $40 = 10a$

10. $100 = 20b$

11. $416 = 4c$

12. $45 = 9d$

13. $0.5m = 3.5$

14. $1.8 = 0.6x$

15. $0.4y = 2$

16. $1.86 = 6.2z$

17. $-8x = 24$

18. $8.34 = 2r$

Lesson 6-3 *(Pages 239-241)*
Solve each equation. Check your solution.

1. $3x + 6 = 6$

2. $2r - 7 = -1$

3. $-10 + 2d = 8$

4. $2b + 4 = -8$

5. $5w - 12 = 3$

6. $5t - 4 = 6$

7. $2q - 6 = 4$

8. $2g - 3 = -9$

9. $15 = 6y + 3$

10. $3s - 4 = 8$

11. $18 - 7f = 4$

12. $13 + 3p = 7$

13. $7.5r + 2 = -28$

14. $4.2 + 7z = 2.8$

15. $-9m - 9 = 9$

16. $32 + 0.2c = 1$

17. $5t - 14 = -14$

18. $-0.25x + 0.5 = 4$

19. $5w - 4 = 8$

20. $4d - 3 = 9$

21. $2g - 16 = -9$

22. $4k + 13 = 20$

23. $7 = 5 - 2x$

24. $8z + 15 = -1$

25. $92 - 16b = 12$

26. $14e + 14 = 28$

27. $1.1j + 2 = 7.5$

28. $4r + 3 = 25$

Lesson 6-4 *(Pages 242–245)*
Write each phrase as an algebraic expression.

1. six less than p

2. twenty more than c

3. the quotient of a and b

4. Ann's age plus 6

5. x increased by twelve

6. $1,000 divided by z

7. 3 divided into y

8. the product of 7 and m

9. the difference of f and 9

10. twenty-six less q

11. 19 decreased by z

12. two less than x

Write each sentence as an algebraic equation.

13. Three times a number less four is 17.

14. The sum of a number and 6 is 5.

15. Twenty more than twice a number is -30.

16. The quotient of a number and -2 is -42.

17. Four plus three times a number is 18.

18. Five times a number minus 15 is 92.

Lesson 6-5 (Pages 246–248)
Solve each inequality. Graph the solution on a number line.

1. $x + 2 > -3$

2. $x + 2.9 \leq 9.1$

3. $8t \geq 24$

4. $v - 3 < -3$

5. $6y \geq -12$

6. $a + 3 \leq -2$

7. $k - 5 < -2$

8. $q - 3 \leq 14$

9. $c - 4 \leq -2$

10. $n + 2 > -5$

11. $j + 1.2 > 4.8$

12. $4x < 40$

13. $2y \leq 10$

14. $g + 8 < 10$

15. $2 + b > 4$

16. $3m \geq 9$

17. $2y \leq 6$

18. $w - 6 \geq 4$

Lesson 6-6 (Pages 249–252)
Graph the ordered pairs in each table on a coordinate plane. Then write a sentence describing each relationship as a function.

1. In the nitrogen family of elements, the atomic number and the atomic mass are given.

Element Name	Atomic Number	Atomic Mass
Nitrogen	7	14
Phosphorus	15	31
Arsenic	33	75
Antimony	51	122
Bismuth	82	209

2. The mass for different numbers of pennies is given.

Number of pennies	1	2	3	4	6
Mass of pennies (g)	5.1	6.2	9.3	12.4	13.6

Lesson 6-7 (Pages 254–257)
Copy and complete each table. Then graph the ordered pairs.

1.

x	2x	y
2		
1		
0		
-1		

2.

x	3x + 1	y
1		
0		
-1		
-2		

3.

x	-2x - 3	y
0		
1		
2		
3		

4.

x	-0.5x - 1	y
2		
4		
6		
8		

Graph each equation.

5. $y = 3x$

6. $y = 2x + 3$

7. $y = -x$

8. $y = 4x + 2$

9. $y = 0.5x + 2$

10. $y = -x + 3$

11. $y = 0.25x + 6$

12. $y = -3x + 6$

13. $y = 2x + 7$

14. $y = -5x + 1$

15. $y = 13 + x$

16. $y = 5 - 0.5x$

17. $y = x - 6$

18. $y = 5x + 1.5$

19. $y = 16 - 4x$

20. $y = 4x + 5$

Lesson 7-1 *(Pages 268–271)*
Round each fraction to 0, $\frac{1}{2}$, or 1.

1. $\frac{3}{8}$ **2.** $\frac{1}{9}$ **3.** $\frac{6}{7}$ **4.** $\frac{7}{12}$ **5.** $\frac{1}{6}$ **6.** $\frac{10}{12}$

Round to the nearest whole number.

7. $5\frac{7}{8}$ **8.** $3\frac{7}{12}$ **9.** $7\frac{1}{10}$ **10.** $2\frac{5}{12}$ **11.** $2\frac{4}{9}$ **12.** $8\frac{3}{4}$

Estimate.

13. $\frac{3}{7} + \frac{6}{8}$ **14.** $\frac{3}{9} + \frac{7}{8}$ **15.** $\frac{1}{8} + \frac{8}{9}$

16. $3\frac{1}{8} + 7\frac{6}{7}$ **17.** $4\frac{2}{3} + 6\frac{7}{8}$ **18.** $3\frac{2}{3} \times 2\frac{1}{3}$

19. $\frac{4}{5} \times 3$ **20.** $9\frac{7}{8} - 6\frac{2}{3}$ **21.** $\frac{3}{7} - \frac{1}{15}$

Lesson 7-2 *(Pages 272–275)*
Add or subtract. Write each sum or difference in simplest form.

1. $\frac{5}{11} + \frac{9}{11}$ **2.** $\frac{5}{8} - \frac{1}{8}$ **3.** $\frac{7}{10} + \frac{7}{10}$

4. $\frac{9}{12} - \frac{5}{12}$ **5.** $\frac{2}{9} + \frac{1}{3}$ **6.** $\frac{1}{2} + \frac{3}{4}$

7. $\frac{1}{4} - \frac{3}{12}$ **8.** $\frac{3}{7} + \frac{6}{14}$ **9.** $\frac{1}{4} + \frac{3}{5}$

10. $\frac{4}{9} + \frac{1}{2}$ **11.** $\frac{5}{7} - \frac{4}{6}$ **12.** $\frac{3}{4} - \frac{1}{6}$

13. $\frac{3}{5} + \frac{3}{4}$ **14.** $\frac{2}{3} - \frac{1}{8}$ **15.** $\frac{9}{10} + \frac{1}{3}$

16. $\frac{8}{15} + \frac{2}{9}$ **17.** $\frac{6}{7} + \frac{6}{9}$ **18.** $\frac{3}{7} + \frac{3}{4}$

19. $\frac{5}{7} + \frac{5}{9}$ **20.** $\frac{7}{8} + \frac{5}{6}$ **21.** $\frac{5}{8} + \frac{3}{4}$

Lesson 7-3 *(Pages 276–279)*
Add or subtract. Write each sum or difference in simplest form.

1. $2\frac{1}{3} + 1\frac{1}{3}$ **2.** $5\frac{2}{7} - 2\frac{3}{7}$ **3.** $6\frac{3}{8} + 7\frac{1}{8}$

4. $2\frac{3}{4} - 1\frac{1}{4}$ **5.** $5\frac{1}{2} - 3\frac{1}{4}$ **6.** $2\frac{2}{3} + 4\frac{1}{9}$

7. $7\frac{4}{5} + 9\frac{3}{10}$ **8.** $3\frac{3}{4} + 5\frac{5}{8}$ **9.** $10\frac{2}{3} + 5\frac{6}{7}$

10. $17\frac{2}{9} - 12\frac{1}{3}$ **11.** $6\frac{5}{12} + 12\frac{5}{12}$ **12.** $7\frac{1}{4} + 15\frac{5}{6}$

13. $6\frac{1}{8} + 4\frac{2}{3}$ **14.** $7 - 6\frac{4}{9}$ **15.** $8\frac{1}{12} + 12\frac{6}{11}$

16. $7\frac{2}{3} + 8\frac{1}{4}$ **17.** $12\frac{3}{11} + 14\frac{3}{13}$ **18.** $21\frac{1}{3} + 15\frac{3}{8}$

19. $19\frac{1}{7} + 6\frac{1}{4}$ **20.** $9\frac{2}{5} - 8\frac{1}{3}$ **21.** $18\frac{1}{4} - 3\frac{3}{8}$

Lesson 7-3B *(Pages 280–281)*
Solve.

1. A fishbowl holds $2\frac{1}{2}$ gallons of water. If there is $\frac{2}{3}$ gallon of water in the bowl, how many more gallons are needed to fill the bowl?

   A $2\frac{5}{6}$ gal   **B** $\frac{1}{3}$ gal

   C $3\frac{1}{6}$ gal   **D** $1\frac{5}{6}$ gal

2. Ryla bought apples, 2 for $0.79; oranges, 3 for $2.49; and grapes, 1 pound for $1.29. Choose the best estimate for the amount of change she will get from $10.

   A $0.40   **B** $4.60

   C $5.40   **D** $6.20

3. A taxi charges $1.25 for the first 0.5 mile and $0.50 for each additional 0.25 mile. Choose the best estimate for the cost of a 10-mile taxi ride.

   A $18.25   **B** $20.25

   C $22.25   **D** $24.25

4. Jeremiah runs 15.5 miles every week. He ran 3 miles on Monday, 4.25 miles on Tuesday, and 5 miles on Thursday. How many more miles does he have to run this week?

   A 3.25   **B** 4.25

   C 2.75   **D** 3.75

Lesson 7-4 *(Pages 284–287)*
Multiply. Write each product in simplest form.

1. $\frac{2}{3} \times \frac{3}{5}$ 2. $\frac{1}{6} \times \frac{2}{5}$ 3. $\frac{4}{9} \times \frac{3}{7}$ 4. $\frac{5}{12} \times \frac{6}{11}$

5. $\frac{3}{8} \times \frac{8}{9}$ 6. $\frac{3}{5} \times \frac{1}{12}$ 7. $\frac{2}{5} \times \frac{5}{8}$ 8. $\frac{7}{15} \times \frac{3}{21}$

9. $\frac{5}{6} \times \frac{15}{16}$ 10. $\frac{6}{14} \times \frac{12}{18}$ 11. $\frac{2}{3} \times \frac{3}{13}$ 12. $\frac{4}{9} \times \frac{1}{6}$

13. $3 \times \frac{1}{9}$ 14. $5 \times \frac{6}{7}$ 15. $\frac{3}{5} \times 15$ 16. $3\frac{1}{2} \times 4\frac{1}{3}$

17. $3\frac{5}{8} \times 4\frac{1}{2}$ 18. $\frac{4}{5} \times 2\frac{3}{4}$ 19. $6\frac{1}{8} \times 5\frac{1}{7}$ 20. $2\frac{2}{3} \times 2\frac{1}{4}$

Lesson 7-5 *(Pages 289–291)*
Complete.

1. 4,000 lb = _____ T   2. 5 T = _____ lb   3. 2 lb = _____ oz

4. 12,000 lb = _____ T   5. $\frac{1}{4}$ lb = _____ oz   6. 6 lb 2 oz = _____ oz

7. 3 gal = _____ pt   8. 24 fl oz = _____ c   9. 8 pt = _____ c

10. 10 pt = _____ qt   11. $2\frac{1}{4}$ c = _____ fl oz   12. 12 pt = _____ c

13. 4 gal = _____ qt   14. 4 qt = _____ fl oz   15. 4 pt = _____ c

16. 9 lb = _____ oz   17. 15 qt = _____ gal   18. 6 lb = _____ oz

19. 2 gal = _____ fl oz   20. 3 T = _____ lb   21. 18 qt = _____ pt

Lesson 7-6 *(Pages 292–295)*

Find the perimeter of each figure.

1.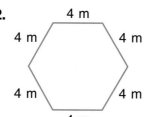
8 yd

3 yd

2.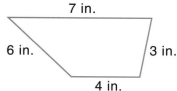
4 m

4 m   4 m

4 m   4 m

4 m

3.
7 in.

6 in.   3 in.

4 in.

4. rectangle: $\ell = 7\frac{1}{2}$ inches

     $w = 4\frac{3}{4}$ inches

5. rectangle: $\ell = 8\frac{1}{3}$ feet

     $w = 2\frac{1}{6}$ feet

Lesson 7-7 *(Pages 297–300)*

Find the circumference of each circle to the nearest tenth. Use $\frac{22}{7}$ or 3.14 for π.

1.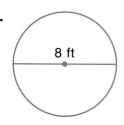
8 ft

2.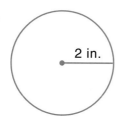
2 in.

3.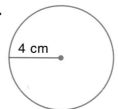
4 cm

4.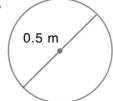
0.5 m

5. $r = 1.5$ in.

8. $d = 1$ m

11. $d = 1.5$ in.

6. $d = \frac{2}{3}$ cm

9. $r = 6$ cm

12. $d = 2$ yd

7. $r = 4$ yd

10. $r = 1$ m

13. $r = 0.5$ cm

Lesson 7-8 *(Pages 301–304)*

Name the property shown by each statement.

1. $\frac{4}{5} \times \frac{2}{3} = \frac{2}{3} \times \frac{4}{5}$

2. $\frac{3}{10} \times 3\frac{1}{3} = 1$

3. $\frac{24}{27} \times 1 = \frac{24}{27}$

4. $\left(\frac{1}{2} + \frac{3}{4}\right) + \frac{5}{6} = \frac{1}{2} + \left(\frac{3}{4} + \frac{5}{6}\right)$

5. $\frac{2}{3} \times \left(\frac{1}{2} + \frac{5}{6}\right) = \frac{2}{3} \times \frac{1}{2} + \frac{2}{3} \times \frac{5}{6}$

6. $\frac{2}{3} \times \frac{3}{2} = 1$

Solve each equation. Write the solution in simplest form.

7. $\frac{a}{13} = 2$

8. $3 \times 1\frac{2}{5} = p$

9. $\frac{8}{9}x = 24$

10. $\frac{3}{8}r = 36$

11. $\frac{3}{4}t = \frac{1}{2}$

12. $16 = \frac{h}{4}$

13. $\frac{1}{5} \times 10 = b$

14. $\frac{k}{2.1} = 0.7$

Lesson 7-9 *(Pages 305–307)*
Divide. Write each quotient in simplest form.

1. $\frac{2}{3} \div \frac{3}{2}$

2. $\frac{3}{5} \div \frac{2}{5}$

3. $\frac{7}{10} \div \frac{3}{8}$

4. $\frac{5}{9} \div \frac{2}{5}$

5. $4 \div \frac{2}{3}$

6. $8 \div \frac{4}{5}$

7. $9 \div \frac{5}{9}$

8. $\frac{2}{7} \div 2$

9. $\frac{1}{14} \div 7$

10. $\frac{2}{13} \div \frac{5}{26}$

11. $\frac{4}{7} \div \frac{6}{7}$

12. $\frac{7}{8} \div \frac{1}{3}$

13. $15 \div \frac{3}{5}$

14. $\frac{9}{14} \div \frac{3}{4}$

15. $\frac{8}{9} \div \frac{5}{6}$

16. $\frac{4}{9} \div 36$

17. $\frac{3}{5} \div \frac{2}{3}$

18. $\frac{8}{9} \div \frac{4}{5}$

19. $\frac{3}{4} \div \frac{15}{16}$

20. $6 \div \frac{1}{5}$

21. $\frac{5}{8} \div 2$

Lesson 8-1 *(Pages 317–320)*
Express each ratio as a fraction in simplest form.

1. 45 to 15

2. 64:128

3. 12 weeks out of 15

4. 14 to 49

5. 125:25

6. 18 to 81

7. 33 minutes:60 minutes

8. 16:40

9. 120 to 180

10. 32:64

11. 10 ft to 8 yd

12. 90 to 100

Tell whether the ratios are equivalent. Show your answer by simplifying.

13. 14 to 77 and 8 to 44

14. $\frac{48}{16}$ and $\frac{1}{3}$

15. 65:13 and 500:100

16. 72 to 90 and 20 to 16

17. 250:100 and 5:2

18. $\frac{32}{2}$ and $\frac{3}{48}$

19. 8 hours to 5 days and 24 hours to 15 days

Lesson 8-2 *(Pages 321–324)*
Express each rate as a unit rate.

1. $240 for 4 days

2. 250 people in 5 buses

3. 500 miles in 10 hours

4. 18 cups for 24 pounds

5. 32 people in 8 cars

6. 3 dozen for $4.50

7. 245 tickets in 5 days

8. 12 classes in 4 semesters

9. 60 people in 4 rows

10. 48 ounces in 3 pounds

11. 20 people in 4 groups

12. 1.5 pounds for $3.00

13. 45 miles in 60 minutes

14. $5.50 for 10 disks

15. 360 miles for 12 gallons

16. $8.50 for 5 yards

17. 24 cups for $1.20

18. 160 words in 4 minutes

19. $60 for 5 books

20. $24 for 6 hours

Lesson 8-3 *(Pages 325–328)*
Solve each proportion.

1. $\dfrac{4}{9} = \dfrac{x}{3}$

2. $\dfrac{12}{m} = \dfrac{15}{10}$

3. $\dfrac{36}{90} = \dfrac{16}{t}$

4. $\dfrac{g}{32} = \dfrac{8}{64}$

5. $\dfrac{5}{14} = \dfrac{10}{a}$

6. $\dfrac{k}{18} = \dfrac{5}{3}$

7. $\dfrac{120}{150} = \dfrac{p}{20}$

8. $\dfrac{15}{w} = \dfrac{60}{4}$

9. $\dfrac{81}{90} = \dfrac{y}{20}$

10. $\dfrac{14}{s} = \dfrac{8}{4}$

11. $\dfrac{h}{3} = \dfrac{36}{9}$

12. $\dfrac{44}{8} = \dfrac{150}{t}$

13. $\dfrac{42}{8} = \dfrac{36}{d}$

14. $\dfrac{125}{v} = \dfrac{35}{5}$

15. $\dfrac{u}{72} = \dfrac{2}{4}$

16. $\dfrac{45}{80} = \dfrac{j}{3}$

17. $\dfrac{3}{7} = \dfrac{21}{d}$

18. $\dfrac{3}{10} = \dfrac{z}{36}$

Lesson 8-4A *(Pages 330–331)*
Solve.

1. Thirty-two basketball teams are participating in a single-elimination tournament; that means that if a team loses one game it is eliminated. How many games will the winning team have played?

2. After a student council meeting, each of the seven members shook hands with each other. How many handshakes were there in all?

3. A shuttle bus at Cedar Point Amusement Park holds 28 passengers. It starts out empty and picks up 1 passenger at the first stop, 2 passengers at the second stop, 3 at the third stop and so on. After how many stops will the bus be full?

4. Madrina mails a recipe to five of her friends. Each of the five friends mails the recipe to five of their friends and so on. How many recipes are in the fourth mailing?

5. Rene is arranging chairs in a meeting room. Each row has 1 more chair than the last so that no chair is directly behind another. If there are 6 chairs in the first row, how many will be in the fifth row?

Lesson 8-4 *(Pages 332–335)*
On a map, the scale is 1 inch:50 miles. For each map distance, find the actual distance.

1. 5 inches

2. 12 inches

3. $3\frac{1}{2}$ inches

4. $2\frac{3}{8}$ inches

5. $\frac{4}{5}$ inch

6. $6\frac{3}{4}$ inches

7. $2\frac{5}{6}$ inches

8. 8 inches

On a scale drawing, the scale is $\frac{1}{2}$ inch:2 feet. Find the dimensions of each room in the scale drawing.

9. 14 feet by 18 feet

10. 32 feet by 6 feet

11. 3 feet by 5 feet

12. 20 feet by 30 feet

13. 8 feet by 15 feet

14. 25 feet by 80 feet

Lesson 8-5 *(Pages 336–338)*
Express each fraction as a percent.

1. $\frac{14}{25}$ **2.** $\frac{28}{50}$ **3.** $\frac{14}{20}$ **4.** $\frac{9}{12}$

5. $\frac{4}{6}$ **6.** $\frac{3}{8}$ **7.** $\frac{7}{10}$ **8.** $\frac{17}{17}$

9. $\frac{9}{16}$ **10.** $\frac{80}{125}$ **11.** $\frac{8}{9}$ **12.** $\frac{3}{16}$

Express each percent as a fraction in simplest form.

13. 32%   **14.** 18.5%   **15.** 89%   **16.** 72%

17. $52\frac{1}{4}\%$ **18.** $33\frac{1}{3}\%$ **19.** 11%   **20.** 1%

21. 28%   **22.** 55%   **23.** $26\frac{1}{4}\%$ **24.** $3\frac{1}{3}\%$

Lesson 8-6 *(Pages 339–341)*
Express each decimal as a percent.

1. 0.03   **2.** 0.16   **3.** 0.1   **4.** 0.5

5. 0.08   **6.** 0.98   **7.** 0.666   **8.** 0.31

9. 0.76   **10.** 0.725   **11.** 0.07   **12.** 0.8

Express each percent as a decimal.

13. 42%   **14.** 100%   **15.** 8%   **16.** 20%

17. 35%   **18.** 3%   **19.** 62%   **20.** 1%

21. 50%   **22.** 7.5%   **23.** $2\frac{1}{2}\%$ **24.** 87.5%

Lesson 8-7 *(Pages 342–345)*
Express each percent as a decimal.

1. 125%   **2.** 0.045%   **3.** 895%   **4.** 0.000075%

5. 200%   **6.** 0.001%   **7.** 0.01345%   **8.** 555%

Express each number as a percent.

9. $4\frac{1}{4}$ **10.** $7\frac{9}{10}$ **11.** 3.245   **12.** 0.003

13. 25   **14.** 16.74   **15.** $2\frac{3}{5}$ **16.** 900

Replace each ● with <, >, or = to make a true sentence.

17. 3.25 ● 325%   **18.** 2,000% ● 2   **19.** 45 ● 4.5%   **20.** 245% ● 2.45

21. $24 \times \frac{1}{4}$ ● $24 \times 25\%$ **22.** $16 \times 1\frac{1}{3}$ ● $133\frac{1}{3}\% \times 16$

Lesson 8-8 (Pages 346–348)

Find each number. Round to the nearest tenth if necessary.

1. 5% of 40 is what number?
2. What number is 10% of 120?
3. Find 12% of 150.
4. Find 12.5% of 40.
5. What number is 75% of 200?
6. Find 13% of 25.3.
7. 250% of 44 is what number?
8. What number is 0.5% of 13.7?
9. Find 600% of 7.
10. Find 1.5% of $25.
11. Find 81% of 134.
12. What number is 43% of 110?
13. What number is 61% of 524?
14. Find 100% of 3.5.
15. 20% of 58.5 is what number?
16. Find 45% of 125.5
17. What number is 23% of 500?
18. Find 80% of 8.
19. 90% of 72 is what number?
20. What number is 32% of 54?

Lesson 8-9 (Pages 349–351)

Find each number. Round to the nearest tenth if necessary.

1. What number is 25% of 280?
2. 38 is what percent of 50?
3. 54 is 25% of what number?
4. 24.5% of what number is 15?
5. What number is 80% of 500?
6. 12% of 120 is what number?
7. Find 68% of 50.
8. What percent of 240 is 32?
9. 99 is what percent of 150?
10. Find 75% of 1.
11. What number is $33\frac{1}{3}$% of 66?
12. 50% of 350 is what number?
13. What percent of 450 is 50?
14. What number is $37\frac{1}{2}$% of 32?
15. 95% of 40 is what number?
16. Find 30% of 26.
17. 9 is what percent of 30?
18. 52% of what number is 109.2?
19. What number is 65% of 200?
20. What number is 15.5% of 45?

Lesson 9-1 (Pages 362–365)

Classify each angle as *acute*, *obtuse*, *right*, or *straight*.

1.
2.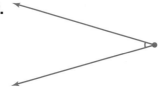
3.

4. 65° angle
5. 24° angle
6. 110° angle
7. 112° angle
8. 90° angle
9. 97° angle

Lesson 9-2 *(Pages 370–373)*

Determine which figures are polygons. If the figure is a polygon, name it and tell whether it is a regular polygon. If the figure is *not* a polygon, explain why.

1.
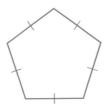

2.

3.
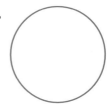

4.
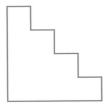

5.
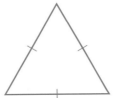

6.
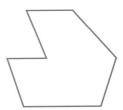

Lesson 9-3 *(Pages 376–379)*

Find the value of *x* in each pair of similar polygons.

1.
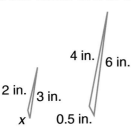
4 in. 6 in.
2 in. 3 in.
x 0.5 in.

2.
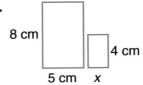
8 cm 4 cm
5 cm x

3.
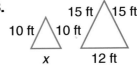
15 ft 15 ft
10 ft 10 ft
x 12 ft

Lesson 9-4 *(Pages 382–385)*

Classify each triangle by its angles and its sides.

1.
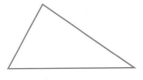

2.

3.
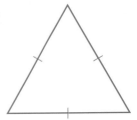

Name every quadrilateral that describes each figure. Then underline the name that best describes the figure.

4.

5.

6.

Lesson 9-4B *(Pages 386–387)*
Solve.

1. Pam, Bob, and Chi each have a collection. One collects stamps, one coins, and the other pins. The coin collector showed Bob and Pam his collection last Saturday. Pam does not collect pins. Who collects stamps?

2. Ana, Iris, and Oki each have a pet. The pets are a fish, a cat, and a bird. Ana is allergic to cats. Oki's pet has 2 legs. Whose pet is a fish?

3. Regular polygons Q, R, and S are a hexagon, a square, and an octagon but not necessarily in that order. Polygon Q and S have the same number of letters in their names. Each angle of polygon S measures less than 135°. Classify the polygons.

4. A number is divisible by three if the sum of its digits is divisible by three. Is 92,742 divisible by 3?

5. A banker, a cook, and a farmer are named Benjamin, Carl, and Fernando. No one's job starts with the same letter as his name. Benjamin bought eggs from the farmer. Who is the banker?

Lesson 9-5 *(Pages 388–391)*
Determine whether each polygon can be used by itself to make a tessellation. Verify your results by finding the number of angles at a vertex. The sum of the measures of the angles of each polygon is given.

1. triangle; 180°   2. square; 360°   3. dodecagon; 1800°

4. Regular hexagons and triangles tessellate. Determine how many of each polygon you need at each vertex. Draw a sketch of the tessellation.

Lesson 9-6 *(Pages 392–394)*
Complete the pattern unit for each translation. Then draw the tessellation.

1.

2.

3.
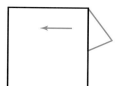

Lesson 9-7 *(Pages 395–397)*
Copy each figure. Draw all lines of symmetry.

1.

2.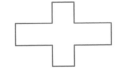

3.

Lesson 10-1A *(Pages 408–409)*
Solve.

1. Josie arranged square tables, each seating 4 people, into one long rectangular table so that her 12 dinner guests could eat together. How many tables did she use?

2. Mariko is thinking of two whole numbers. When she adds them together, the sum is 123. When she subtracts the lesser number from the greater number, their difference is 69. What are the numbers?

3. Seth is the oldest of four children. Each of his sisters is 3 years older than the next oldest sibling. The combined age of Seth and his three sisters is 46. None of the children is over the age of 20. How old is Seth?

4. The length of a rectangle is 6 inches longer than its width. What are the length and width if the area of the rectangle is 216 square inches?

5. Three consecutive integers have a sum of 33 and a product of 1,320. What are the integers?

Lesson 10-1 *(Pages 410–414)*
Find the square of each number.

1. 6	**2.** 12	**3.** 7
4. 15	**5.** 20	**6.** 14
7. 24	**8.** 1	**9.** 11
10. 40	**11.** 25	**12.** 9

Find each square root.

13. $\sqrt{49}$	**14.** $\sqrt{64}$	**15.** $\sqrt{169}$
16. $\sqrt{324}$	**17.** $\sqrt{900}$	**18.** $\sqrt{225}$
19. $\sqrt{2,500}$	**20.** $\sqrt{81}$	**21.** $\sqrt{289}$
22. $\sqrt{576}$	**23.** $\sqrt{8,100}$	**24.** $\sqrt{676}$

Lesson 10-2 *(Pages 415–417)*

Estimate each square root to the nearest whole number.

1. $\sqrt{15}$

2. $\sqrt{35}$

3. $\sqrt{112}$

4. $\sqrt{75}$

5. $\sqrt{27}$

6. $\sqrt{249}$

7. $\sqrt{88}$

8. $\sqrt{1,500}$

9. $\sqrt{612}$

10. $\sqrt{340}$

11. $\sqrt{495}$

12. $\sqrt{264}$

13. $\sqrt{350}$

14. $\sqrt{834}$

15. $\sqrt{3,700}$

16. $\sqrt{298}$

17. $\sqrt{101}$

18. $\sqrt{800}$

19. $\sqrt{58}$

20. $\sqrt{750}$

21. $\sqrt{1,200}$

22. $\sqrt{1,000}$

23. $\sqrt{5,900}$

24. $\sqrt{999}$

25. $\sqrt{374}$

26. $\sqrt{512}$

27. $\sqrt{3,750}$

28. $\sqrt{255}$

29. $\sqrt{83}$

30. $\sqrt{845}$

31. $\sqrt{200}$

32. $\sqrt{500}$

33. $\sqrt{10,001}$

Lesson 10-3 *(Pages 419–422)*

Use the Pythagorean Theorem to find the length of each hypotenuse given the lengths of the legs. Round to the nearest tenth.

1. 4 ft, 6 ft

2. 12 cm, 25 cm

3. 15 yd, 24 yd

4. 8 mm, 11 mm

Find the missing measure for each right triangle. Round to the nearest tenth.

5. a: 14 cm; c: 18 cm

6. b: 15 ft; c: 24 ft

7. a: 5 yd; b: 8 yd

Given the lengths of the sides of a triangle, determine whether each triangle is a right triangle. Write *yes* or *no*.

8. 6 mm, 8 mm, 10 mm

9. 12 ft, 15 ft, 20 ft

10. 300 m, 400 m, 500 m

Lesson 10-4 *(Pages 423–426)*

Estimate the area of each figure.

1.

2.
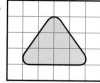

3.

4.

5.

6.

Lesson 10-5 *(Pages 428–431)*

Find the area of each triangle to the nearest tenth.

1.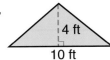

2. base: 5 in.
height: 9 in.

3.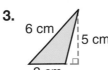

4. base: 12 cm
height: 8 cm

Find the area of each trapezoid to the nearest tenth.

5.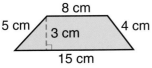

6. bases: 3 cm, 8 cm
height: 12 cm

7.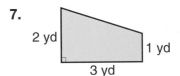

8. bases: 10 ft, 15 ft
height: 12 ft

9.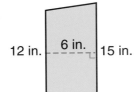

10. bases: 5 m, 9 m
height: 10 m

Lesson 10-6 *(Pages 432–435)*

Find the area of each circle to the nearest tenth.

1. radius, 8 in.

2.

3. diameter, 5 ft

4.

5. radius, 24 cm

6.

7. diameter, 2.3 m

8.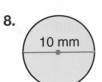

Find the length of the radius of each circle given the following areas. Round to the nearest tenth.

9. 15 cm^2 **10.** 24 ft^2 **11.** 125 in^2 **12.** 36 yd^2

13. 100 m^2 **14.** 200 mm^2 **15.** 72 ft^2 **16.** 142 in^2

Lesson 10-7 *(Pages 438–441)*

Find the probability that a randomly-dropped counter will fall in the shaded region.

1.   **2.**   **3.**   **4.**

5.   **6.**   **7.**   **8.**

Lesson 11-1 *(Pages 450–453)*

Write the fraction, decimal, mixed number, or whole number equivalent of each percent that could be used to estimate.

1. 28%

2. 99%

3. 450%

4. 0.09%

5. $\frac{3}{4}$%

6. 65.5%

7. $15\frac{3}{5}$%

8. 39.45%

9. $8\frac{1}{2}$%

10. 48.2%

11. 0.009%

12. 287%

Estimate.

13. 50% of 37

14. 18% of 90

15. 60.5% of 60

16. 300% of 245

17. 0.7% of 200

18. 1% of 48

19. 7% of 24

20. 400% of 13

21. $5\frac{1}{2}$% of 100

22. 40.01% of 16

23. 70% of 300

24. 35% of 35

Lesson 11-1B *(Pages 454–455)*

Solve.

1. The Oakland Coliseum has a capacity of 48,621 people. For one afternoon baseball game, 38,824 tickets were sold. About what percent of the stadium was full?

2. Olympic stadium in Montreal has a capacity of 59,511. For a baseball game the stadium was about 75% full. About how many people attended the game?

3. There were 22,306 fans at a baseball game in Fenway Park. The ballpark was about $\frac{2}{3}$ full. What is the approximate capacity of Fenway Park?

4. Of the people Joaquin surveyed, 60% had eaten a meal in a restaurant in the past two weeks. If Joaquin surveyed 150 people, how many had eaten a meal in a restaurant in the past two weeks?

Lesson 11-2 *(Pages 456–458)*

Write an equation for each problem. Then solve. Round answers to the nearest tenth.

1. 12% of what number is 50?

2. Find 45% of 50.

3. 38 is what percent of 62?

4. $28\frac{1}{2}$% of 64 is what number?

5. 5% of what number is 12?

6. 80 is what percent of 90?

7. $66\frac{2}{3}$% of what number is 40?

8. Find 46.5% of 75.

9. 90 is what percent of 95?

10. Find 22% of 22.

11. 16% of what number is 2?

12. 75 is what percent of 300?

13. 75% of 80 is what number?

14. Find 60% of 45.

15. What number is 55.5% of 70?

16. 80.5% of what number is 80.5?

Lesson 11-3 *(Pages 460–463)*

Use the information in the following charts to make a circle graph.

1.

Car Sales	
Style	Percent
Sedan	45
Station Wagon	22
Pickup Truck	9
Sports Car	13
Compact Car	11

2.

Favorite Flavor of Ice Cream	
Flavor	Percent
Vanilla	28
Chocolate	35
Strawberry	19
Mint Chip	12
Coffee	6

Lesson 11-4 *(Pages 464–467)*

The table shows the results of a survey of students' favorite cookie flavors at Bush Middle School. The school has 328 students.

1. What was the sample size?

2. To the nearest percent, what percent of students preferred peanut butter cookies?

3. How many students in the school would you expect to say that sugar cookies are their favorite?

Cookie	Number
Chocolate chip	49
Peanut butter	12
Oatmeal	10
Sugar	8
Raisin	3

Lesson 11-5 *(Pages 469–472)*

Find the percent of change. Round to the nearest whole percent.

1. original: $75
   new: $50

2. original: 450
   new: 675

3. original: 3.25
   new: 2.95

4. original: $5.75
   new: $6.25

5. original: 180
   new: 160

6. original: 32.5
   new: 44

7. original: 1.5
   new: 1.0

8. original: 450
   new: 400

9. original: $1,500
   new: $1,200

10. original: 750
    new: 600

11. original: $65
    new: $75

12. original: 380
    new: 320

13. original: 0.75
    new: 1.0

14. original: $3.95
    new: $4.25

15. original: 350
    new: 420

16. original: 500
    new: 100

Lesson 11-6 *(Pages 474–477)*

Find the sales tax or discount to the nearest cent.

1. $45 sweater; 6% tax
2. $18.99 CD; 15% off
3. $39 shoes; $5\frac{1}{2}$% tax
4. $199 ring; 10% off
5. $29 shirt; 7% tax
6. $55 plant; 20% off

Find the total cost or sale price to the nearest cent.

7. $19 purse; 25% off
8. $150 clock; 5% tax
9. $2 notebook; 15% off
10. $145 coat; $6\frac{1}{4}$% tax
11. $89 radio; 30% off
12. $300 table; $\frac{1}{3}$ off

Find the rate of discount to the nearest percent.

13. regular price, $45
    sale price, $40
14. regular price, $250
    sale price, $200
15. regular price, $89
    sale price, $70

Lesson 11-7 *(Pages 478–480)*

Find the interest to the nearest cent for each principal, interest rate, and time.

1. $2,000, 8%, 5 years
2. $500, 10%, 8 months
3. $750, 5%, 1 year
4. $175.50, $6\frac{1}{2}$%, 18 months
5. $236.20, 9%, 16 months
6. $89, $7\frac{1}{2}$%, 6 months
7. $800, 5.75%, 3 years
8. $5,500, 7.2%, 4 years
9. $245, 6%, 13 months

Find the interest to the nearest cent on credit cards for each credit card balance, interest rate, and time.

10. $750, 18%, 2 years
11. $1,500, 19%, 16 months
12. $300, 9%, 1 year
13. $4,750, $19\frac{1}{2}$%, 30 months
14. $2,345, 17%, 9 months
15. $689, 12%, 2 years
16. $390, 18.75%, 15 months
17. $1,250, 22%, 8 months
18. $3,240, 18%, 14 months

Lesson 12-1 *(Pages 492–495)*

Draw a top, a side, and a front view of each figure.

1.

2.

3.

4.

5.

6.

Lesson 12-1B *(Pages 496–497)*

Solve.

1. Mark wants to make a pyramid-shaped display of basketballs for his sports shop. Each basketball comes in a 10-inch cubic box. Mark starts with a base six boxes wide and six boxes long. He decreases each dimension by one box for each layer, how many basketballs will he need for his display?

2. How many different rectangular prisms can be formed with 12 cubes?

3. How many 9-inch by 6-inch by 2-inch paperback books can be shipped in a carton 18-inches by 12-inches by 12-inches?

4. Wes, Yoki, Gabe, and Beth each live in a different colored house. The houses are white, green, blue, and yellow. No person's house color starts with the same letter as his or her name. Gabe and Yoki stop at the white house each morning to pick up their friend. Who lives in the white house?

Lesson 12-2 *(Pages 498–501)*

Find the volume of each rectangular prism to the nearest tenth.

1. length, 1.5 in.
   width, 3 in.
   height, 6 in.

2. length, 4.5 cm
   width, 6.75 cm
   height, 2 cm

3. length, 3 ft
   width, 10 ft
   height, 2 ft

4. length, 16 mm
   width, 0.7 mm
   height, 12 mm

5. length, 18 cm
   width, 23 cm
   height, 15 cm

6. length, $3\frac{1}{2}$ ft
   width, 10 ft
   height, 6 ft

7. length, 25 mm
   width, 32 mm
   height, 10 mm

8. length, 12 in.
   width, $5\frac{1}{2}$ in.
   height, $3\frac{3}{8}$ in.

9.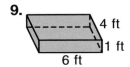

10.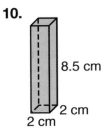

11.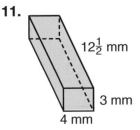

12.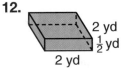

Lesson 12-3 *(Pages 503–506)*

Find the volume of each cylinder to the nearest tenth.

1. radius, 6 in.
   height, 3 in.

2. radius, 8.5 cm
   height, 3 cm

3. diameter, 16 yd
   height, 4.5 yd

4. diameter, 3.5 mm
   height, 2.5 mm

5. radius, 8 ft
   height, 10 ft

6. diameter, 12 m
   height, 4.75 m

7. radius, 6 cm
   height, 12 cm

8. diameter, $\frac{5}{8}$ in.
   height, 4 in.

9.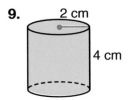

10.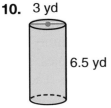

11.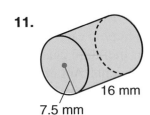

12.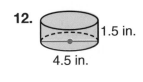

Lesson 12-4 *(Pages 510–513)*

Find the surface area of each rectangular prism to the nearest tenth.

1. length, 8 ft
   width, 6.5 ft
   height, 7 ft

2. length, $4\frac{1}{2}$ cm
   width, 10 cm
   height, $8\frac{3}{4}$ cm

3. length, 9.4 yd
   width, 2 yd
   height, 5.2 yd

4. length, 20 mm
   width, 15 mm
   height, 25 mm

5.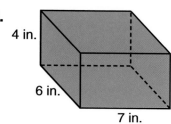

6.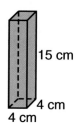

7.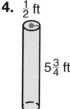

Lesson 12-5 *(Pages 514–517)*

Find the surface area of each cylinder to the nearest tenth.

1.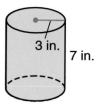

2.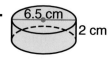

3. 1.5 m, 6 m

4. $\frac{1}{2}$ ft, $5\frac{3}{4}$ ft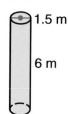

5. height, 6 cm
   radius, 3.5 cm

6. height, $5\frac{1}{2}$ in.
   diameter, 3 in.

7. height, 16.5 mm
   diameter, 18 mm

8. height, 22 yd
   radius, 10.5 yd

Lesson 13-1A *(Pages 528–529)*

Apply the act it out strategy to solve each problem.

1. A fast food restaurant is giving away sports pins with each children's meal. There are 4 different pins: baseball, football, basketball, and soccer. If the pins are given out randomly, how many times would you need to purchase a children's meal to be sure you get one of each of the pins?

2. How many times do you need to toss a die to get all 6 numbers?

3. Suppose Kaley, Marla, and Sophie each left one book in the school library. When they returned to their literature class, their teacher randomly handed out the three books to the students. Act out this situation 25 times. For how many of these times did Sophie receive the same book she was reading in the library?

Lesson 13-1 *(Pages 530–533)*

Refer to the sample space below. Find each theoretical probability.

1, 1	1, 2	1, 3	1, 4	1, 5	1, 6
2, 1	2, 2	2, 3	2, 4	2, 5	2, 6
3, 1	3, 2	3, 3	3, 4	3, 5	3, 6
4, 1	4, 2	4, 3	4, 4	4, 5	4, 6
5, 1	5, 2	5, 3	5, 4	5, 5	5, 6
6, 1	6, 2	6, 3	6, 4	6, 5	6, 6

1. a sum of 4

2. a sum less than 3

3. the first number equals the second

4. a sum of 6

5. both numbers are even

6. both numbers are prime

Lesson 13-2 *(Pages 534–536)*

Make a tree diagram and list the outcomes. Then give the total number of outcomes.

1. rolling 2 number cubes

2. choosing an ice cream cone from waffle, plain, or sugar and a flavor of ice cream from chocolate, vanilla, or strawberry

3. making a sandwich from white, wheat, or rye bread, cheddar or swiss cheese and ham, turkey, or roast beef

4. flipping a penny twice

5. choosing one math class from algebra and geometry and one foreign language class from French, Spanish, or Latin

Lesson 13-3 *(Pages 538–541)*

Use the Counting Principle to find the total number of outcomes in each situation.

1. choosing a local phone number if the exchange is 234 and each of the four remaining digits is different

2. choosing a way to drive from Millville to Westwood if there are 4 roads that lead from Millville to Miamisburg, 2 roads that connect Miamisburg to Hathaway, and 4 highways that connect Hathaway to Westwood

3. tossing a quarter, rolling a number cube, and tossing a dime

4. spinning the spinners shown below

  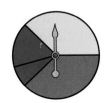

Lesson 13-4 *(Pages 542–545)*
Find each probability.

1. Two evenly-balanced nickels are flipped. Find the probability that one head and one tail result.

2. A wallet contains four $5 bills, two $10 bills, and eight $1 bills. Two bills are selected without the first selection being replaced. Find $P(\$5, \text{then } \$5)$.

3. Two chips are selected from a box containing 6 blue chips, 4 red chips, and 3 green chips. The first chip selected is not replaced before the second is drawn. Find $P(\text{red, then green})$.

4. A blue die and a red die are rolled. Find the probability that an odd number is rolled on the blue die and a multiple of 3 is rolled on the red die.

Lesson 13-5 *(Pages 547–549)*
Find the value of each expression.

1. 3!
2. 0!
3. 6!
4. $P(5, 3)$
5. $P(6, 6)$
6. $P(10, 2)$
7. $P(5, 0)$
8. $P(3, 2)$

Solve.

9. How many different five-digit zip codes can be formed if no digit can be repeated?

10. Eight runners are competing in a 100-meter sprint. In how many ways can the gold, silver, and bronze medals be awarded?

11. In a lottery for which 30 tickets were sold (all to different people), in how many ways can the grand prize, second prize, and third prizes be awarded?

Lesson 13-6 *(Pages 551–553)*
Tell whether each situation represents a *permutation* or *combination*.

1. Six people remaining in a game of musical chairs.

2. First, second, and third place awards of three students who are finalists in a writing competition.

Solve.

3. List all of the possible combinations of Andrew, Jonathan, Megan, Rebecca, and Jeffrey taken 2 at a time. If all of the combinations are equally likely, find the probability that the combination chosen will consist of 2 males.

4. List all the combinations of the digits 1, 3, 5, 7, 9 taken 3 at a time. If all of the combinations are equally alike, find the probability that the combination chosen will contain a 1.

Mixed Problem Solving

Solve using any strategy.

1. *Money Matters* Cara needs to buy five different colored pens for a writing project. She has $12.00. Does she have enough money if each pen costs $1.98?

2. *Patterns* Study the pattern below.
$$1 = 1$$
$$1 + 3 = 4$$
$$1 + 3 + 5 = 9$$
List the next 3 lines in the pattern. What is the pattern of the sums?

3. *Money Matters* The You-Rent-It Auto Company charges $30 a day or $165 a week to rent a compact car. Choose the best estimate of how much money the weekly rate saves the customer who rents a car for 7 days.
   a. $25
   b. $35
   c. $45
   d. $55

4. Coach Williams wants to schedule a round-robin tournament for his chess club where every student plays every other student. If there are seven club members, how many games should the coach schedule?

5. *Geometry* A fence is put around a square dog run whose area is 400 feet. Enough is left over to fence a square garden whose area is 25 square feet. What is the minimum amount of fencing used?

6. *School* Dan, Nan, and Fran have lockers next to each other. Nan rides the bus with the person whose locker is at the right. Dan's locker is not next to Nan's locker. Who has the locker at the left?

7. *Money Matters* Andrea has $2.80 worth of quarters and dimes in her pocket. If the number of quarters equals the number of dimes, how many quarters does she have?

8. Logan used blocks to build a "fort." The blocks were cubes and were stacked five high. The top, side, and front views were all squares. How many blocks did Logan need to build the fort?

Mixed Problem Solving

Solve using any strategy.

1. *Patterns* The NCAA basketball tournament starts with 64 teams. After the first round, there are 32 teams left; after the second round there are 16 teams left, and so on. Complete the pattern until there is only one team left. How many rounds does it take to determine a winner?

2. *Nutrition* The average American eats 28 pounds of cheese a year, and the population of the U.S. was 248,709,873 in 1990. The product manager at Franklin Dairy reported that about 2 billion pounds of cheese were sold in 1990. Is this reasonable?

3. An elevator starts with four passengers and stops at three floors. How many possible ways can the passengers get off the elevator?

4. *Geometry* How many diagonals does a 20-sided polygon have?

5. *Statistics* Use the bar graph to predict which age group will be increasing in numbers over the next century.

Percent Distribution of the Population by Age

Source: U.S. Bureau of the Census

6. *School* Ms. Bosco's class started a six-day problem-solving contest. Each day, only students who correctly solved the previous day's problem are allowed to participate. There are 32 students in the class. If only half of the students get a problem correct on any day, how many students can participate on the fourth day?

7. *Money Matters* The sale price of a sweater was $19 after an additional $5 was taken off the sweater that was already marked 50% off. What was the original price of the sweater?

8. *Sports* American football is played on a rectangular field that is 160 feet by 120 yards including the end zones. What is a reasonable estimate of the area of the playing field in square feet?

1. *Transportation* The Lewis Middle School Band is planning a bus trip to a band competition. There are 138 members in the band, and each bus will hold 32 people. How many buses are needed for the trip?

2. Name the operation to be done first in the expression $26 + 12 \div 3$.

Evaluate each expression.

3. $10 \cdot 2 + 3$

4. $7 - 5 + 3$

5. $(3 + 6) \div (2 + 1)$

6. $7 + 12 \div 4 - 2$

7. 5^2

8. $4 \cdot 2 + 3^3$

9. *Entertainment* An adult movie ticket costs $7.25, and a child's ticket costs $5.50. If two adults and three children go to the movie, find the total cost.

Evaluate each expression if $j = 4$, $k = 7$, and $m = 12$.

10. $\frac{m}{j}$

11. $3k - m$

12. $4j - m + 2k$

13. j^3

14. $k^2 - 2j$

15. $2(j + k)$

16. Write b^5 as a product of the same factor.

17. Write $11 \cdot 11 \cdot 11 \cdot 11 \cdot 11 \cdot 11 \cdot 11$ using exponents.

Solve each equation.

18. $6a = 72$

19. $\frac{x}{8} = 100$

20. $p + 13 = 25$

21. $35 - m = 19$

22. Draw the next two figures that continue the pattern.

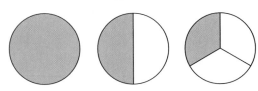

Find the area of each rectangle or parallelogram.

23. rectangle: ℓ, 12 in.; w, 7 in.

24. parallelogram: b, 50 cm; h, 20 cm

25. *Animals* Heather left both her dog and cat at a kennel for 3 nights. The kennel charges $8 per night for the dog and $5 per night for the cat. Compute Heather's total bill.

CHAPTER

2

Test

Order each set of decimals from least to greatest.

1. 12.6, 4.3, 8.7, 4, 12.06

2. 0.07, 0.7, 0.71, 1.07, 1.71

Round each number to the underlined place-value position.

3. 13.2$\underline{7}$5

4. 0.0$\underline{7}$6

5. 1$\underline{2}$,436

6. $\underline{0}$.995

7. *Sports* The average attendance at football games last season was 2,176.34. Round this number to the nearest whole number.

Estimate.

8. 27.34 + 12.95

9. 236.95 − 107.07

10. 23.6 × 2.95

11. 11.1)$\overline{142.6}$

12. 91.6 × 7.999

13. *Money Matters* During a two-week period, Hiroko worked 89.7 hours. Her hourly wage is $6.85. Estimate the amount she earned during that period.

Multiply or divide.

14. 0.3 × 8

15. 5.5 × 0.004

16. 6.7 × 1,000

17. 0.0047 × 10^5

18. 3.003 × 100,000

19. 0.4)$\overline{4.8}$

20. 0.24)$\overline{0.0072}$

21. 0.016)$\overline{64}$

22. *Pets* Phillip's cat Felix weighed 0.95 pound at birth. On his first birthday, Felix weighed 9.2 times his birth weight. What was Felix's weight on his first birthday?

Express each fraction or mixed number as a decimal. If the decimal is a repeating decimal, use bar notation.

23. $\frac{5}{8}$

24. $\frac{1}{9}$

25. $5\frac{1}{20}$

Express each number in scientific notation.

26. 23,000

27. 632

28. 518,000,000

Complete.

29. 1.62 L = __?__ mL

30. 243 g = __?__ kg

31. 0.09 km = __?__ mm

32. *Food* To mix a punch, Andrea starts with 4 liters of ginger ale and adds 2,650 milliliters of cranberry juice.

   a. How many liters are in the punch bowl when the punch is complete?

   b. How many 200-mL servings of punch can be served?

33. *Money Matters* Booker bought a 3-pound bag of apples for $2.89, a 2-pound bag of carrots for $1.79, and 4 avocados for $0.99 each. Should he expect to pay about $5 or $10 at the checkout?

CHAPTER 3 Test

The French test grades were 95, 76, 82, 90, 71, 76, 79, 82, 95, 85, 93, 71, and 63.

1. Find the range.

2. Choose an appropriate scale and an interval for a frequency table.

3. Make a frequency table of the data.

4. *Sports* The number of girls participating in the summer soccer league has increased during the last five years. Predict the number of girls participating in the year 2000.

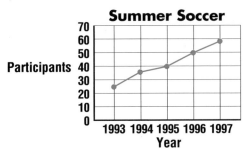

5. *Employment* The line graph shows the percent of women holding jobs outside the home from 1975 to 1995. In which year were the greatest number of women working outside the home?

6. Use the graph to predict the percent of women who will hold jobs outside the home in the year 2000.

7. Name the outliers on the line plot.

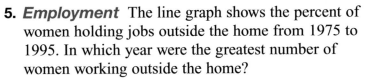

8. *Families* The Hanna family has five children whose ages are 20, 20, 19, 17, and 26. Make a line plot of the ages.

Find the mean, mode(s), and median for each set of data.

9. 4, 6, 11, 7, 4, 11, 4

10. 12.4, 17.9, 16.5, 10.2

Make a stem-and-leaf plot for each set of data.

11. 37, 59, 26, 42, 57, 53, 31, 58.

12. 46¢, 59¢, 42¢, 69¢, 55¢, 48¢, 66¢, 43¢

13. Refer to the box-and-whisker plot.
    a. What is the median?
    b. What is the upper quartile and lower quartile?
    c. What is the interquartile range?

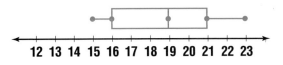

The prices of 15 pairs of shoes at a local shoe store are $34, $28, $32, $28, $25, $69, $25, $75, $30, $29, $32, $28, $27, $30, and $26.

14. Find the mean, mode(s), and median. Round to the nearest cent.

15. Which average might the manager prefer to quote to a cost-conscious customer?

CHAPTER TEST

1. Determine whether 639 is divisible by 2, 3, 4, 5, 6, 9, or 10.

2. Write the prime factorization of 250.

Identify each sequence as *arithmetic*, *geometric*, or *neither*. Then find the next three terms.

3. 9, 15, 21, 27, 33, . . .

4. 2, 6, 18, 54, 162, . . .

5. *Savings* In January, Elena deposited $50 in her savings account. She plans to increase the amount she deposits by $5 each month. How much will Elena deposit in April?

Find the GCF of each set of numbers.

6. 36, 54

7. 52, 100

Express each fraction or ratio in simplest form.

8. $\frac{33}{55}$

9. 24:64

Express each fraction or ratio as a percent.

10. $\frac{28}{70}$

11. 34:100

Express each percent or decimal as a fraction in simplest form.

12. 0.32

13. 73%

The spinner at the right is equally likely to stop on each of the regions. Find the probability that the spinner will stop on each of the following.

14. a prime number

15. a factor of 24

Find the LCM of each set of numbers.

16. 8, 28

17. 14, 21, 27

18. *Life Science* One type of cicada emerges from hibernation every 17 years. Another type emerges every 13 years. If both types came out of hibernation one year, in how many years would this happen again?

Replace each ● with <, >, or = to make a true sentence.

19. $\frac{5}{8}$ ● $\frac{12}{20}$

20. $\frac{12}{15}$ ● $\frac{9}{12}$

Write the integer represented by each letter. Then find its opposite and its absolute value.

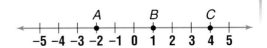

1. A

2. B

3. C

Replace each ● with $<$ or $>$ to make a true sentence.

4. -9 ● 6

5. 0 ● -3

6. **Earth Science** On the same day, the thermometer registered 5° below 0 in Cleveland, 2° above 0 in Columbus, and 2° below 0 in Cincinnati. Which city was the coldest?

On graph paper, draw a coordinate plane. Then graph and label each point.

7. $P(6, -3)$

8. $B(0, -4)$

9. $T(-5, 1)$

Solve each equation.

10. $g = 5 + (-3)$

11. $-9 + (-3) = m$

12. $r = -4 + 4$

13. $k = 11 - 15$

14. $-7 - (-2) = s$

15. $b = -3 - 4$

16. $c = -5(-3)$

17. $h = 12(-2)$

18. $(-7)^2 = p$

19. $q = 90 \div (-3)$

20. $(-25) \div 5 = t$

21. $m = (-72) \div (-9)$

Evaluate each expression if $a = 8$, $b = -3$, $c = 2$, and $d = -8$.

22. $-10 + a$

23. $b + d$

24. $d - c$

25. $23 - b$

26. $-9a$

27. d^2

28. $5ac$

29. $a \div c$

30. **Games** Byron is playing a popular board game and has $250 left. He lands on Boardwalk and needs $200 more than he has to pay rent. How much is the rent on Boardwalk?

31. **Allowance** Ming is supposed to receive an allowance of $25 each month. However, for each day he forgets to take out the garbage, his allowance decreases by $2. Ming forgets to take the garbage out three times during October. Find the amount of his allowance for October.

Graph each triangle and its transformation. Write the ordered pairs for the vertices of the new triangle.

32. $\triangle ABC$ with vertices $A(-4, 2)$, $B(3, 4)$, and $C(-1, 6)$ translated 2 units right and 4 units down

33. $\triangle KLM$ with vertices $K(2, 2)$, $L(4, -1)$, and $M(0, 0)$ reflected over the y-axis

Solve each equation. Check your solution.

1. $12 + t = 32$

2. $m + 7 = 2$

3. $2.9 = w + 1.7$

4. $12e = -120$

5. $14.7 = 3.5d$

6. $s - 5.9 = 12.1$

7. $6x + 4 = 10$

8. $-2b - 5 = 25$

9. $2.6 = 0.2n - 4$

10. $-14s + 5 = 33$

11. $3f - 15 = 6$

12. $0.42 = 0.17 + 0.5j$

13. *Physical Science* During a chemical reaction, 3 milliliters of the original chemical evaporates, leaving 2.6 milliliters in the test tube. Write and solve an equation to find a, the amount of the chemical in the test tube before the reaction occurs.

14. *Money Matters* Ann prices a sweater in two stores. It is $29 in one store and $34 in the other. Write and solve an equation to find p, the price difference between the two stores.

Write each phrase as an algebraic expression.

15. 26 less than x

16. t increased by 23

Write each sentence as an algebraic equation.

17. The product of 7 and a number is 35.

18. A number increased by 125 is 315.

19. Twice a number less 5 is 19.

Solve each inequality. Graph the solution on a number line.

20. $y + 5 < 13$

21. $3n \le 21$

22. $4c - 3 > 17$

23. Graph the ordered pairs in the table on a coordinate plane. Then write a sentence describing the relationship as a function.

Month	1	2	3	4
Enrollment at Small Tots Day Care	40	42	42	44

Graph each equation.

24. $y = -2x - 1$

25. $y = 3x + 1$

Estimate.

1. $1\frac{1}{3} + \frac{6}{7}$

2. $10\frac{4}{5} - 4\frac{1}{8}$

3. $5\frac{1}{8} \div \frac{5}{6}$

Add, subtract, or multiply. Write each sum, difference, or product in simplest form.

4. $\frac{1}{10} + \frac{2}{5}$

5. $\frac{5}{6} - \frac{2}{9}$

6. $\frac{4}{9} \times \frac{3}{8}$

7. $2\frac{1}{4} \times 6$

8. $2\frac{1}{7} - \frac{3}{7}$

9. $\frac{3}{5} + \frac{5}{9}$

10. $5\frac{1}{3} + 4\frac{3}{4}$

11. $\frac{6}{7} \times \frac{5}{8}$

12. $15\frac{4}{9} - 5\frac{5}{12}$

Complete.

13. 4 lb = ___?___ oz

14. 20 qt = ___?___ gal

15. How many fluid ounces are in 3 cups?

16. Find the perimeter of a rectangle with a length of $6\frac{1}{2}$ yards and a width of $3\frac{1}{4}$ yards.

Find the circumference of each circle to the nearest tenth.

17. $d = 2.8$ cm

18. $r = \frac{1}{3}$ yd

Solve each equation.

19. $w = 5 \times 2\frac{2}{5}$

20. $\frac{3}{4}b = \frac{1}{3}$

Divide. Write each quotient in simplest form.

21. $\frac{3}{5} \div \frac{9}{10}$

22. $2\frac{5}{12} \div \frac{5}{6}$

23. $3\frac{1}{3} \div 1\frac{1}{6}$

24. *Money Matters* Willie found a jacket he likes that costs $60. If he waits until the store has its annual $\frac{1}{3}$-off sale, how much can he save?

25. *Cooking* A recipe calls for 2 cups of milk. Miriam wants to double this recipe. How many *quarts* of milk will she need?

Express each ratio as a fraction in simplest form.

1. $\frac{20}{24}$

2. 35:15

3. 42 out of 60 days

4. Tell whether 21:35 and 15:30 are equivalent.

Express each rate as a unit rate.

5. 24 cards for $4.80

6. 330 miles on 15 gallons of gas

7. *Population Density* In 1994, Washington, D.C. had a population of 567,094 people and an area of 67 square miles. To the nearest whole number, how many people per square mile were there in Washington, D.C. in 1994?

Solve each proportion.

8. $\frac{2}{3} = \frac{x}{42}$

9. $\frac{9}{m} = \frac{12}{36}$

10. *Physical Fitness* Alyssa swims 3 laps in 12 minutes. At this same rate, how many laps will she swim in 10 minutes?

Find the distance between each pair of cities, given the map distance and scale.

11. Indianapolis, Indiana and Columbus, Ohio; $3\frac{3}{4}$ inches; 1 inch:49 miles

12. Pensacola and Tallahassee, Florida; $3\frac{1}{2}$ inches; $\frac{1}{2}$ inch:50 km

Express each fraction as a percent.

13. $\frac{3}{8}$

14. $\frac{51}{200}$

15. $2\frac{1}{2}$

Express each decimal as a percent.

16. 0.65

17. 0.079

18. 0.0091

Express each percent as a fraction in simplest form and as a decimal.

19. 47%

20. 80%

21. 450%

Find each number. Round to the nearest tenth if necessary.

22. What number is 35% of 650?

23. Find 69% of 2,398.

24. 82 is what percent of 415?

25. 458 is 105% of what number?

CHAPTER 9 Test

Classify each angle as *acute*, *obtuse*, *right*, or *straight*.

1.

2. 135°

3. *Architecture* Classify the angle made by a wall and the ceiling of your classroom.

4. Classify the angles at the right as complementary, supplementary, or neither.

Determine which figures are polygons. If the figure is a polygon, name it and tell whether it is a regular polygon. If the figure is *not* a polygon, explain why.

5.  **6.** **7.**   **8.**

Tell whether each pair of polygons is similar. Justify your answer.

9.
5 in. 7.5 in.
2 in. 3 in.

10.
12 mm
6 mm
18 mm
20 mm

11. *Geography* Dan wants to draw a map on the chalkboard that is similar to a map in his book. The outline of the map is a rectangle that is 6 inches wide and 8 inches long. If he draws the longer side 36 inches long, how wide should he draw the map?

Classify each triangle by its angles and by its sides.

12.  **13.**

Name every quadrilateral that describes the figure. Then underline the name that best describes it.

14.  **15.** 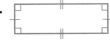 **16.**

17. Determine if a regular heptagon, whose angle measures total 900°, can be used by itself to make a tessellation.

Complete the pattern unit for each translation. Then draw the tessellation.

18.   **19.**

20. Draw a figure that has no lines of symmetry.

1. Amanda is thinking of two numbers. When she adds them together, the sum is 107. When she subtracts the lesser number from the greater number, their difference is 17. What are the numbers?

Find the square of each number.

2. 8   3. 34

Find each square root.

4. $\sqrt{81}$ 5. $\sqrt{400}$

6. *Physical Fitness* Every morning, Elisa jogs around a square region that has an area of 9 square miles. How far does Elisa jog each morning?

Estimate each square root to the nearest whole number.

7. $\sqrt{24}$ 8. $\sqrt{120}$

Find the missing measure for each right triangle. Round to the nearest tenth.

9. *a*: 5 m; *b*: 4 m   10. *b*: 12 in.; *c*: 14 in.   11. *a*: 5 yd; *c*: 10 yd

12. *Traveling* Jerome gets lost while driving to a new vacation spot. After looking at a map, he sees that he is 16 miles too far east and 8 miles too far north. What is the straight-line distance to Jerome's destination? Round to the nearest tenth.

Estimate the area of each figure.

13. 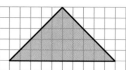   14.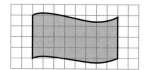

Find the area of each figure.

15. triangle: base: 15 ft   16. trapezoid: bases: 5 km, 10 km
           height: 4 ft   height: 6 km

Find the area of each circle to the nearest tenth.

17. radius: 14 cm   18. diameter: 25 in.

19. *Recreation* A circular playground at a park covers an area of 1,000 square feet. Find the length of the diameter of the playground to the nearest tenth.

20. *Games* To win a carnival game, you must throw a dart at a board and hit one of the several cards on the board. The board is 7 feet by 4 feet. If the cards cover 3 square feet, what is the probability that a randomly-thrown dart will hit a card?

Estimate.

1. 18% of 246

2. 145% of 81

Write an equation for each problem. Then solve. Round answers to the nearest tenth.

3. Find 15% of 60.

4. 75 is what percent of 50?

5. 50% of what number is 335.8?

6. What percent of 48 is 9?

7. The table shows the result of a survey of students' favorite types of fiction at Haskell Middle School. Find the number of degrees for each section if you make a circle graph of the data.
   a. Mystery   **b.** Science Fiction

Type of Fiction	Number of Students
Mystery	24
Science Fiction	8
Sports	30
Romance	38

8. Refer to the table above. If Haskell Middle School has a total of 500 students, how many would you expect to choose each of the following types of fiction as their favorite?
   a. Sports   **b.** Romance

Find the percent of change. Round to the nearest whole percent.

9. original: $60
   new: $75

10. original: 145
   new: 216

11. Find the percent of decrease if an item that originally cost $54 goes on sale for $48.

12. One pocket calendar can display 42 characters on its screen. A newer model of the pocket calendar can display 64 characters. Find the percent of increase in displayed characters.

Find the sales tax or discount to the nearest cent.

13. $19.99 book, 20% off

14. $2,200 computer, $6\frac{1}{2}$% sales tax

15. $35.49 jeans, 33% discount

16. $85 boots, 5.75% tax

Find the interest to the nearest cent for each principal, interest rate, and time.

17. $1,250, 11%, 3 years

18. $2,000, 11.75%, 18 months

19. *Education* The University of North Carolina presently charges $1,091 per year for tuition for residents of North Carolina. If tuition increases 7% for the coming academic year, what will the tuition be next year?

20. *Money Matters* Jorge borrows $4,500 to buy a new motorcycle. His loan is for 4 years at an annual interest rate of 10%. Find the total amount Jorge will pay for his motorcycle.

Draw a top, a side, and a front view of each figure.

1.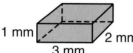

2.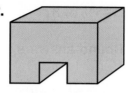

3.

4.

Find the volume of each rectangular prism or cylinder to the nearest tenth.

5. 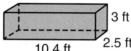 1 mm 2 mm 3 mm

6. 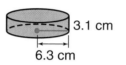 3 ft 10.4 ft 2.5 ft

7. 3.1 cm 6.3 cm

8. 1.5 yd 2 yd

9. A rectangular prism has a length of 1 inch, a width of 6 inches, and a height of 9 inches. Find the volume of the prism.

10. Find the volume of a cube with sides 4 centimeters long.

11. Find the volume of a cylinder with a height of 5 feet and a diameter of 12 feet.

12. *Drinking Straws* The standard-size drinking straw has a radius of $\frac{1}{8}$ inch and a height of $7\frac{3}{4}$ inches. What is the maximum volume of liquid that can be contained in the straw at any given time?

Find the surface area of each rectangular prism or cylinder to the nearest tenth.

13. 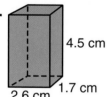 4.5 cm 2.6 cm 1.7 cm

14. 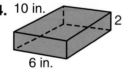 10 in. 2 in. 6 in.

15. 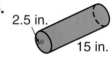 3.7 m 4 m

16. 2.5 in. 15 in.

17. Find the surface area of a rectangular prism with a length of 3.6 centimeters, a width of 2.1 centimeters, and a height of 8 centimeters.

18. *Packaging* What is the least amount of paper needed to wrap a box that is 12 inches by 20 inches by 2 inches?

19. Find the surface area of a cylinder whose height is $\frac{1}{3}$ yard and whose base has a diameter of 14 yards.

20. *Woodworking* Sheri built a jewelry box as a class project. It is 14 inches long, 12 inches wide, and 8 inches high. She would like to paint the outside of the box. What is the surface area of the box?

Test

1. Gary is making a sandwich. He has two kinds of bread and four kinds of meat. If he only uses one kind of bread and one kind of meat, how many different sandwiches can Gary make?

2. A coin is tossed three times. Make a tree diagram to show all the possible outcomes.

Use the Counting Principle to find the total number of outcomes in each situation.

3. choosing a three-digit security code
4. rolling 4 number cubes

5. choosing a pair of slacks from three pairs, a shirt from five shirts, and a pair of shoes from two pairs

The spinner at the right has an equal chance of landing on each number. Find each probability.

6. P(odd number)
7. P(1 or 7)

8. P(number greater than 2)

The entry forms for a contest are in a container. There are forms for 100 students of which 30 are sixth graders, 35 are seventh graders, and 35 are eighth graders. One entry form is selected. Find each probability.

9. P(a sixth grader)
10. P(not an eighth grader)
11. P(at least a seventh grader)

Tell whether the events are independent or dependent.

12. rolling a pair of number cubes and getting a 5 on the first cube and a sum of 10 for both cubes

13. having black hair and owning a black car

14. The two fire engines in a small town operate independently. The probability that each engine is available when needed is 0.95. Find the probability that both engines are available at one time.

Find the value of each expression.

15. 5!
16. 7!

17. In how many ways can a president, vice president, and secretary be chosen from among eight candidates?

18. A menu has 15 different items. In how many ways can you choose three items from the menu?

19. *Sports* Eight runners are participating in a 200-meter race. In how many ways can the gold, silver, and bronze medals be awarded?

20. *School Government* From a student council consisting of nine students, three are selected to represent the student body at a school board meeting. In how many ways can these students be selected?

Getting Acquainted with the Graphing Calculator

When some students first see a graphing calculator, they think, "Oh, no! Do we *have* to use one?", while others may think, "All right! We get to use these neat calculators!" There are as many thoughts and feelings about graphing calculators as there are students, but one thing is for sure: a graphing calculator *can* help you learn mathematics. Keep reading for answers to some frequently asked questions.

What is it?
So what is a graphing calculator? Very simply, it is a calculator that draws graphs. This means that it will do all of the things that a "regular" calculator will do, *plus* it will draw graphs of equations.

What does it do?
A graphing calculator can do more than just calculate and draw graphs. For example, you can program it and work with data to make statistical graphs and computations. If you need to generate random numbers, you can do that on the graphing calculator. If you need to find the absolute value of a number, you can do that too. It's really a very powerful tool, so powerful that it is often called a pocket computer.

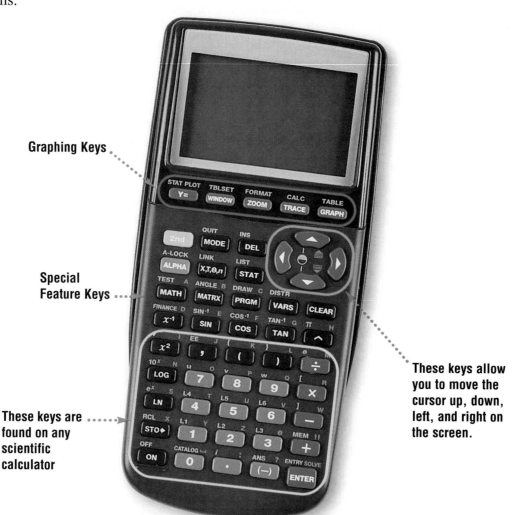

Graphing Keys

Special Feature Keys

These keys allow you to move the cursor up, down, left, and right on the screen.

These keys are found on any scientific calculator

Basic Keystrokes

- The yellow commands written above the calculator keys are accessed with the [2nd] key, which is also yellow. Similarly, the green characters above the keys are accessed with the [ALPHA] key, which is also green. In this text, commands that are accessed by the [2nd] and [ALPHA] keys are shown in brackets. For example, [2nd] [QUIT] means to press the [2nd] key followed by the key below the yellow [QUIT] command.

- [2nd] [ENTRY] copies the previous calculation so you can edit and use it again.

- [2nd] [QUIT] will return you to the home (or text) screen.

- Negative numbers are entered using the [(-)] key, not the minus sign, [–].

- [2nd] [OFF] turns the calculator off.

Order of Operations

As with any scientific calculator, the graphing calculator observes the order of operations.

Example	Keystrokes	Display
4 + 13	4 [+] 13 [ENTER]	4 + 13 17
5^3	5 [∧] 3 [ENTER]	5 ∧ 3 125
4 (9 + 18)	4 [(] 9 [+] 18 [)] [ENTER]	4(9 + 18) 108
$\sqrt{24}$	[2nd] [√] 24 [ENTER]	√ (24 4.8989 79486

Programming

Programming features allow you to write and execute a series of commands for tasks that may be too complex or cumbersome to perform otherwise. Each program is given a name. Commands begin with a colon (:), which the calculator enters automatically, followed by an expression or an instruction.

When you press [PRGM], you see three menus: EXEC, EDIT, and NEW. EXEC allows you to execute a stored program, EDIT allows you to edit or change a program, and NEW allows you to create a program.

- To begin entering a new program, press [PRGM] [▶] [▶] [ENTER].

- You do not need to type each letter using the [ALPHA] key. Any command that contains lowercase letters should be entered by choosing it from a menu. Check your user's guide to find any commands that are unfamiliar.

- After a program is entered, press [2nd] [QUIT] to exit the program mode and return to the home screen.

- To execute a program, press [PRGM]. Then use the down arrow key to locate the program name and press [ENTER] twice, or press the number or letter next to the program name followed by [ENTER].

- If you wish to edit a program, press [PRGM] [▶] and choose the program from the menu.

- To immediately re-execute a program after it is run, press [ENTER] when Done appears on the screen.

- To stop a program during execution, press [ON] or [2nd] [QUIT].

While a graphing calculator cannot do everything, it can make some things easier and help your understanding of math. To prepare for whatever lies ahead, you should try to learn as much as you can. Who knows? Maybe one day you will be designing the next satellite or building the next skyscraper with the help of a graphing calculator!

Getting Acquainted with Spreadsheets

What do you think of when people talk about computers? Maybe you think of computer games or using a word processor to write a school paper. But a computer is a powerful tool that can be used for many things.

One of the most common computer applications is a spreadsheet program. Here are answers to some of the questions you may have if you're new to using spreadsheets.

What is it?

You have probably seen tables of numbers in newspapers and magazines. Similar to those tables, a spreadsheet is a table that you can use to organize information. But a spreadsheet is more than just a table. You can also use a spreadsheet to perform calculations or make graphs.

Why use a spreadsheet?

The advantage a spreadsheet has over a simple calculator is that when a number is changed, the entire spreadsheet is recalculated and the new results are displayed. So with a spreadsheet, you can see patterns in data and investigate what happens if one or more of the numbers is changed.

How do I use a spreadsheet?

A spreadsheet is organized into boxes called *cells*. The cells are named by a letter, that identifies the column, and a number, that identifies the row. In the spreadsheet below, cell C4 is highlighted.

	A	B	C
1	Width	Length	Area
2	3	4	12
3	2	10	20
4	5	12	60
5	8	14	112

To enter information in a spreadsheet, simply move the cursor to the cell you want to access and click the mouse. Then type in the information and press Enter.

How do I enter formulas?

If you want to use the spreadsheet as a calculator, begin by choosing the cell where you want the result to appear.

- For a simple calculation, type = followed by the formula. For example, in the spreadsheet above, the formula in cell C2 is entered as "=A2*B2." *Notice that * is the symbol for multiplication in a spreadsheet.*

- Sometimes you will want similar formulas in more than one cell. First type the formula in one cell. Then select the cell and click the copy button. Finally select the cells where you want to copy the formula and click the paste button.

- Often it is useful to find the sum or average of a row or column of numbers. The spreadsheet allows you to choose from several functions like this instead of entering the formula manually. To enter a function, click the cell where you want the result to appear. Then click on the = button above the cells. A list of formulas will appear to the left. Click the down arrow button and choose your function. The spreadsheet will enter a range, which you may alter. For example, to find the average of row 2 of the spreadsheet below, the function chooses to find the average of cells B2, C2, and D2.

	A	B	C	D	E
1	Student	Test 1	Test 2	Test 3	Average
2	Kathy	88	85	91	88
3	Ben	86	89	92	89
4	Carmen	92	86	92	90
5	Anthony	80	88	87	85

The formula for cell E2 is =(B2+C2+D2)/3.

Spreadsheet software is one of the most common tools used in business today. You should try to learn as much as you can to prepare for your future. Who knows? Maybe you'll use what you're learning today as a company president tomorrow!

Selected Answers

CHAPTER 1
Problem Solving, Algebra, and Geometry

Pages 6–7   Lesson 1-1

1. You need to determine how all the facts are related and what strategy to use to solve the problem.   **5.** 58 tables   **7.** Sample answer: They both have 4 congruent sides; one has right angles and the other does not.   **9.** Sample answer: Assuming it takes 1 second to say a number, it would take about 11.5 days.   **11.** Sample answer: bicycling and in-line skating would burn $708 + 600$ or 1,308 calories

Pages 9–10   Lesson 1-2

1. addition   **3.** Tia; you should divide first.
5. subtraction   **7.** 16   **9.** 13   **11.** multiplication
13. subtraction   **15.** subtraction   **17.** 23   **19.** 28
21. 8   **23.** 6   **25.** 20   **27.** 14   **29.** 83
31. $2(14 - 9) - (17 - 14) = 7$
33. $64 \div (8 + 24) - 1 = 1$
35a. $2 \cdot 5 + 1 \cdot 4 + 3 \cdot 3$ **35b.** $23   **37.** Sample answer: For 150 pounds, sleep for 8 hours, in-line skate for 2 hours, and swim for 1 hour; $8 \times 90 + 2 \times 600 + 1 \times 497 = 2,417$ **39.** A

Page 11   Lesson 1-3A

1. 7   **3.** 9   **5.** Sample answer: The cup represents the variable or unknown quantity.

Pages 14–15   Lesson 1-3

1. Sample answer: Numbers have a constant value, while variables represent many different values.
3a. Sample answer: The perimeter is two more than the number of triangles.   **3b.** $n + 2$ **5.** 10   **7.** 20
9. 4   **11.** 9   **13.** 12   **15.** 17   **17.** 11   **19.** 16
21. 36   **23.** 2   **25.** 3   **27.** 16   **29.** 1,000
31a.

Number of hours	1	2	3	4	5
Amount earned	3	6	9	12	15

31b. $3n$ **31c.** $24   **33.** $246   **35.** C

Page 15   Mid-Chapter Self Test

1. about 498   **3.** 35   **5.** about 4 quarts

Page 16   Lesson 1-3B

1. 48   **3.** 72   **5.** 6   **7.** Sample answer: Store the value of $4 + 8 \times 2$ in a and the value of $15 - 4 + 2$ in b. Then find ab.

Pages 18–20   Lesson 1-4

1. $6 \cdot 6 = 36$ **3.** Sample answer: Using exponents is more convenient and saves space.   **5.** $z \cdot z \cdot z$
7. x^6 **9.** 16   **11.** 35   **13.** $2 \cdot 2 \cdot 2 \cdot 2$ **15.** $4 \cdot 4 \cdot 4 \cdot 4 \cdot 4$ **17.** $m \cdot m \cdot m \cdot m$ **19.** 12^2 **21.** 15^4
23. n^2 **25.** 49   **27.** 21   **29.** 23   **31.** 90   **33.** 54
35. 90   **37.** 16   **39.** false   **41a.** $1^3, 2^3, 3^3, 4^3, 5^3$
41b. Sample answer: A number, n, taken to the third power is the same as the volume of a cube whose edge is n units long.   **43.** Sample answer: $3^3 = 27$, $3^2 = 9$, $3^1 = 3$, $3^0 = 1$; Each power is three times greater than the next.   **45.** 110   **47.** false

Pages 22–23   Lesson 1-5

1. Sample answer: Find the value of the variable that makes a true sentence.   **3.** Latisha; $343 \div 7 = 49$ **5.** 41   **7.** 80   **9.** 18   **11.** 8
13. 4   **15.** 7   **17.** 18   **19.** 56   **21.** 86   **23.** 72
25. 10   **27.** 6   **29.** 63   **31.** 55   **33.** 143
35. 173   **37.** 7.5 hours   **39.** 84 centimeters
41. They are equal.   **43.** 32   **45.** 9

Pages 26–27   Lesson 1-6

1. Small parts of the fern look like the entire fern.
3.    **7.**

9. 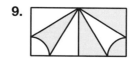   **13.** 75

Pages 28–29   Lesson 1-7A

1. 15 square units   **3.** 16 square units
5. 10 square units   **7.** 36 square units
9. 135 square units   **11.** They have the same area.
13. 8 square units   **15.** The base and height of the parallelogram are the same as the length and width of the rectangle.   **17.** 32 square units   **19.** 60 square units

Pages 32–33   Lesson 1-7

1.

3. Since the length and width are each doubled, the area increases by 2 × 2 or 4 times.   **5.** 40 cm²
7. 30 ft²   **9.** 14 m²   **11.** 80 ft²   **13.** 72 ft²
15. 216 cm²   **17.** 375 ft²   **19.** 132 yd²   **21.** 7 yd
23. 1,960 ft²; from 1,200 to 1,999 ft²
25. Sample answer:

27. A

Pages 34–35   Lesson 1-7B

1. Sample answer: You can use the exact answer as a basis when you estimate other quantities.
3. Sample answer: If no advantage is obtained by having an exact answer then estimation is an acceptable method.   **5.** Sample answer: You can compare your answer against your estimate to determine whether your answer is reasonable.
7. Sample answer: An estimate is close to 1,999 square feet, which is the dividing point for two categories.   **9a.** yes   **9b.** 126 ft²
11. about $16 × 12 or $192
13. yes; 3,000 + 3,000 + 4,000 + 3,000 = 13,000

Pages 36–39   Study Guide and Assessment

1. true   **3.** false, squared   **5.** true   **7.** true
9. false, 15 square inches   **11.** true
13. 420 miles   **15.** 5,932 books   **17.** 82   **19.** 89
21. 10   **23.** 3   **25.** 40   **27.** 78   **29.** 729
31. 225   **33.** 64   **35.** 9   **37.** 47   **39.** 64
41.

43. 45 square yards

45. 168 square meters
47. 12 square feet   **49.** no   **51.** 63°F

Pages 40–41   Standardized Test Practice

1. C   **3.** C   **5.** C   **7.** D   **9.** B   **11.** Sample answer: 240 ÷ 8 = 30; about 30 hours

13. 604,800 seconds   **15.** 20 days

**CHAPTER 2
Applying Decimals**

Pages 45–46   Lesson 2-1

1. Sample answer: 0.30 < 0.50   **3.** Sample answer: You can add a zero to the right of a decimal without changing the value. So, 0.4 = 0.40.   **5.** <   **7.** <
9. chocolate jimmies   **11.** >   **13.** =   **15.** <
17. <   **19.** <   **21.** >   **23.** 0.087   **25.** 6.55, 6.505, 6.5, 6.05   **27a.** visiting friends, reading
27b. teenagers: 17.7, 4.4, 1.3, 1.2, 1.1, 0.9; unmarried adults: 14.2, 7.8, 1.9, 0.7, 0.7, 0.5
27c.

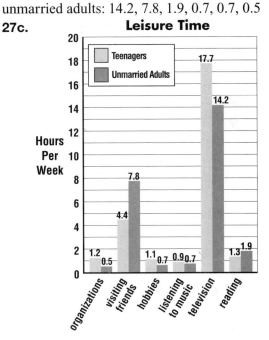

29. 0.999, 4.63, 20.8, 175   **31.** B   **33.** 26

Pages 48–49   Lesson 2-2

1. 14.4   **3.** Tomas; 12.0 is expressed in tenths.
5. 0.25   **7.** 0.79   **9.** 1.57   **11.** 0.22   **13.** 10
15. 16.4   **17.** 9.128   **19.** 0.45   **21.** 1.70
23. 15.5   **25.** 0.8   **27.** 60
29.

5.0 ─────── 5.67 ──────── 6.0

31a. 11.9 lb   **31b.** 5.9 lb   **33.** 8, 8.75, 9.15, 9.5
35. 29

Pages 51–53   Lesson 2-3

1. Sample answer: to help you catch errors in entering the numbers   **5.** 30 − 20 = 10
7. 42 ÷ 7 = 6   **9.** 3(20) = 60

11. about 1 million   **13.** $30 - 20 = 10$
15. $63 \div 9 = 7$ **17.** $30 - 8 = 22$
19. $40 \div 20 = 2$ **21.** $72 \div 8 = 9$
23. $30 \times 80 = 2{,}400$ **25.** $36 \div 12 = 3$
27. $7 \times 5 = 35$ **29.** $4(40) = 160$
31. $4(200) = 800$ **33.** $5(50) = 250$ **35.** about
340 million   **37.** about 70 million   **39.** Sample
answer: Using 15 miles per gallon, he will use 100
÷ 15 or about 6 gallons of gasoline. At $1.25 per
gallon, he will pay about $7.50.   **41.** <
43. $500 + 5n$

Pages 54–55   Lesson 2-3B
5. Sample answer: Estimate before you calculate.
9. 10 cars   **11.** about 300 times   **13.** $9   **15.** C

Pages 58–59   Lesson 2-4
3. 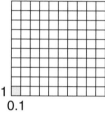   **5.** 0.077

7. 0.1845   **9.** 0.35 km   **11.** 1.14   **13.** 19.728
15. 0.009   **17.** 0.27   **19.** 0.0736   **21.** 0.000805
23. 4.05   **25.** 0.152   **27.** 0.1053   **29.** 6,423
pesetas   **31a.** VCR, 33.6; video game, 27.0; cable,
23.1; computer, 16.1; cellular phone, 12.3; on-line
service, 6.0   **33.** Always; the rectangle will always
be a portion of the full model.   **35.** 1.0   **37.** C

Page 59   Mid-Chapter Self Test
1. 0.28   **3.** 350   **5.** 0.0025 km

Pages 62–63   Lesson 2-5
1. Move the decimal point 2 places to the right;
$x = 237.8$ **5.** 4,600   **7.** 20,310   **9.** 0.045   **11.** 5
13. 2,780   **15.** 5,490   **17.** 9.25   **19.** 0.16
21. 560   **23.** 1,123,000   **25.** 0.8   **27.** 930
29. 2.53   **31a.** 0.72 million   **31b.** 7.92 million
33. 1.2   **35.** 0.01

Pages 64–65   Lesson 2-6A
1. $0.14 \div 0.2 = 0.7$ **3.** $0.16 \div 0.4 = 0.4$
5. 0.6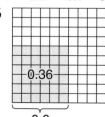

7. 1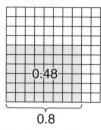

9. 0.6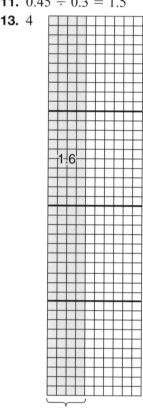

11. $0.45 \div 0.3 = 1.5$
13. 4

15. 0.7

Pages 68–69   Lesson 2-6
1. Sample answer: Yes; dividing both the dividend
and divisor of $35 \div 0.5$ by 10 results in $3.5 \div 0.05$.
3. $3.6 \div 4$ **5.** $5{,}040 \div 56$ **7.** 1.5   **9.** 0.12
11. 62.61 times   **13.** $2{,}940 \div 84$ **15.** $8.2 \div 4$
17. $26 \div 13$ **19.** $14.88 \div 31$ **21.** 3.5   **23.** 0.35
25. 0.3   **27.** 0.48   **29.** 4.6   **31.** 0.088   **33.** 140
35. 0.65   **37.** $1.54   **39a.** $0.26, $0.24
39b. the large box   **39c.** Sample answer: You only
want small servings.   **39d.** Two regular boxes cost
$4.58, which is less than $4.99.   **41.** 9.2 meters per
second   **43.** 283   **45.** $20 \times 6 = 120$ **47.** 10

Pages 71–73   Lesson 2-7

1. Divide the numerator by the denominator.
3. Kim; $0.\overline{5} = 0.55555555...$ **5.** $6.\overline{34}$ **7.** 0.8
9. $0.\overline{7}$ **11.** $0.\overline{4}$ **13.** $1.\overline{12}$ **15.** $13.\overline{245}$ **17.** $0.8\overline{3}$
19. $3.01\overline{523}$ **21.** 0.55   **23.** $0.8\overline{3}$ **25.** 0.032
27. 3.875   **29.** =   **31.** =   **33.** >   **35.** 0.27
37a. 3.141, 3.143; yes   **37b.** 3.160, Archimedes'
39. Sample answer: 0.7, $0.\overline{71}$ **41.** 0.108   **43.** 81

Page 76   Lesson 2-8

1. You should divide because you are changing from a smaller unit to a larger unit.   **3.** 75,000
5. 0.923   **7.** 8.2   **9.** 1,240 g   **11.** 67,100
13. 23.4   **15.** 0.132   **17.** 46,000   **19.** 0.567
21. 8,100   **23.** 0.047   **25.** 90 mL   **27a.** Mercury – 0.38 cm; Venus – 0.95cm; Mars – 0.53 cm; Jupiter – 11.19 cm; Saturn – 9.46 cm; Uranus – 4.01 cm; Neptune 3.88 cm; Pluto – 0.18 cm   **27b.** Mercury – 0.387 cm; Venus – 0.723 cm; Mars – 1.524 cm; Jupiter – 5.203 cm; Saturn – 9.529 cm; Uranus – 19.191 cm; Neptune – 30.061 cm; Pluto – 39.529 cm   **29.** $0.\overline{59}$ **31.** 0.1145

Pages 78–79   Lesson 2-9

1. It is more convenient.   **3.** Alma; 24.59 is not less than 10.   **5.** 8.3×10^3 **7.** 5.2×10^4
9. 1.264×10^8 **11.** 7.5×10^3 **13.** 4.07×10^4
15. 6×10^2 **17.** 2.3×10^4 **19.** 3.2×10^7
21. 5.7×10^5 **23.** 8.08×10^3 **25.** 2.71×10^7
27. 5.58×10^5 **29.** =   **31.** <   **33.** >
35. 3.4×10^6, 3,400,000   **37.** 3.0×10^6; Chicago is one of the largest cities in the U.S., and a population of 3,000,000 is reasonable.
39. 5.95×10^{10} cans   **41.** 1,010 g   **43.** C
45. 13

Pages 80–83   Study Guide and Assessment

1. less   **3.** sum   **5.** gram   **7.** $0.5 \times 0.3 = 0.15$
9. $0.2\overline{3}$ **11.** Sample answer: One centimeter is one-hundredth of a meter.   **13.** 0.06, 0.159, 1.4, 1.59, 15.91   **15.** 15.0, 15.99, 16, 16.03, 16.3
17. 13.27   **19.** 0.1   **21.** 57.20   **23.** $40 \times 3 = 120$ **25.** $12 \times 3 = 36$ **27.** $150 \div 5 = 30$
29. 0.26   **31.** 22.725   **33.** 13,700   **35.** 0.0637
37. 10   **39.** 0.004   **41.** 2.7   **43.** 0.4   **45.** 0.375
47. $0.\overline{5}$ **49.** 0.027   **51.** 0.0033   **53.** 160
55. 0.043   **57.** 6×10^3 **59.** 29.13, 29.97, 30.22, 30.53, 31.01   **61.** 200 geysers

Pages 84–85   Standardized Test Practice

1. D   **3.** D   **5.** B   **7.** C   **9.** D   **11.** D
13. $905   **15.** 9.2 cm   **17.** 65 miles per hour

CHAPTER 3
Statistics: Analyzing Data

Pages 89–91   Lesson 3-1

1. Range: find the difference between the greatest number and the least number. Scale: include all numbers of the data set, plus numbers that are higher or lower than the set, to get an appropriate scale. Interval: choose the number of categories that you want and divide the scale by that number.
3. Tatanka is correct. A frequency table includes all the numbers in a set of data.
5. 19; Sample answer: 0-19, 5

Interval	Tally	Frequency
0-4	|||	3
5-9	|||	3
10-14	|	1
15-19	||	2

7a. 62
7b. Sample answer: 1-70, 10

Reign	Tally	Frequency
1-10	||||	4
11-20	||	2
21-30	|	1
31-40	|	1
41-50	|	1
51-60	|	1
61-70	|	1

9–13. Sample scales and intervals are given.
9. 100; 20-120, 20   **11.** 19; 0-20, 2   **13.** 7.5; 14-22, 1   **15–17.** Sample answers are given.

15.

Number of Books	Tally	Frequency
0	|||	3
1	|||| |	6
2	|||| |||	8
3	|||	3
4	|	1
5	||	2
6	|	1

0–6, 1

17.

Rainfall (in.)	Tally	Frequency
7	|||	3
8	|||	3
9	||	2
10	||||	4

7–10, 1

19.

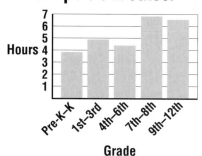

0 5 10 15 20 25 30 35 40 45 50

21a.

Year	Tally	Frequency
1900-1919	|||	3
1920-1939	||||	4
1940-1959	⊪	5
1960-1979	⊪ |	6
1980-1999	||||	4

21b. 1960-1979   **25.** C

Pages 92–93   Lesson 3-2A

1. Sample answer: The table gives information about each kind of bike. The graph shows the relationship between rating and price.   **3.** 2 games
5. Using a graph can help you analyze whether a solution is correct.   **7.** 1.5 in.   **9.** 11   **11a.** about 3 in.   **11b.** Sample answer: May

Pages 96–97   Lesson 3-2

1. Graphs often show trends over time.   **5.** Sample answer: 3.5 million people

7a.   **Computers in School**

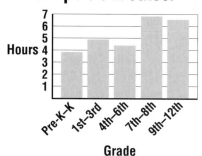

Hours (1–7) vs Grade (Pre-K–K, 1st–3rd, 4th–6th, 7th–8th, 9th–12th)

7b. Students in grades 7-8 and 9-12 spend the most time using computers, so the new computer software should be designed for students in these grades.
9. Sample answer: Wally World probably will not catch up to Valley World in attendance, since Valley World's attendance appears to be increasing faster than Wally World's.   **11.** 29; sample answer: 18-48, 2   **13.** C

Pages 100–101   Lesson 3-3

1. Sample answer: It's easier to see trends over time.

3.

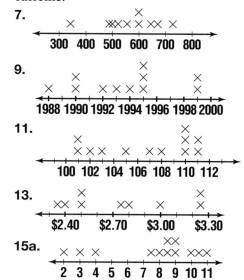

15 20 25 30 35 40 45

5a.

10 15 20 25 30 35 40 45 50 55 60

5b. 35; Many of the items contain 35 mg of caffeine.

7.

300 400 500 600 700 800

9.

1988 1990 1992 1994 1996 1998 2000

11.

100 102 104 106 108 110 112

13.

$2.40 $2.70 $3.00 $3.30

15a.

2 3 4 5 6 7 8 9 10 11

15b. A majority of the players lose between 7.5 and 9 pounds during the game.   **17a.** They both can show the number of times data occur.   **17b.** A line plot shows points of data. A bar graph can show intervals of data.   **19.** B

Pages 104–105   Lesson 3-4

1a. List the data in ascending order. Choose the middle number.   **1b.** List the data in ascending order. Find the mean of the two middle numbers.
3. Erica; the data do not all have to be the same for the mean, median, and mode to be equal. For example, in the set 7, 8, 8, 9, the mean, median, and mode all equal 8.   **5.** 8.95, none, 9.05   **7.** The mode, because it appears most often.   **9.** 93, 90 and 94, 93   **11.** $65\frac{1}{2}$, 65, 65   **13.** 1,780; 1,755 and 1,805; 1,780   **15.** 90.95, 95, 95   **17.** Sample answer: Mean; this represents the average size of all families.   **19.** mean: $331.0 million; median: $310.9 million

21.

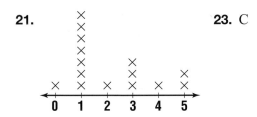

23. C

Page 105 Mid-Chapter Self Test
1. 13; Sample answer: 10-24, 2   **3.** Sample answer: 25 million bales   **5.** 46.9, 45, 46

Page 106 Lesson 3-4B
1. Sample answer: You would probably not get exactly the same results, but the average student would probably be similar.   **3a.** Sample answer: The median; it averages the data.   **3b.** Sample answer: Line plots; they show the distribution graphically.

Pages 109–111 Lesson 3-5
1. Both a stem-and-leaf plot and a bar graph show the frequency of data occurring. However, a stem-and-leaf plot shows individual data values and a bar graph shows only a bar for each interval with the length representing the number of data in the interval.

5. stems: 1, 2, 3

Stem	Leaf
1	3 3 5
2	4 8
3	0 1 2 2 5 6 8 8 8 $1|3 = 13$

7a. 0, 1, 2, 3, 4, 5, 6, 7, 8, 9, 10

7b.

Stem	Leaf
0	1 3 4 5 6 7 7
1	0 0 0 2 2 2 2 5 5 8
2	0 0 0 0 5
3	
4	0
5	
6	
7	
8	
9	
10	0 $1|2 = 12$ years

7c. 1 year, 100 years   **7d.** 10-19 years

9. stems: 0, 1, 2, 3

Stem	Leaf
0	1 2 7 8 9
1	1 2 4 8 8 9
2	1 2
3	1 $0|7 = 7$

11. stems: 1, 2, 3, 4, 5, 6, 7, 8, 9

Stem	Leaf
1	5 8
2	6 7
3	6 7 9
4	4 9
5	6 8
6	1 8
7	5
8	
9	0 $1|8 = 18$

13a.

Top Ranked	Stem	Lower Ranked
8 7	1	3 5 7 8
9 9 8 5 5 5 0	2	0 2 2 6
0 0 0	3	0 0 1 8
$8|1 = \$18$		$1|3 = \$13$

Sample answer: There are more lower ranked jeans available at lower prices. But there are several types of jeans of top quality available at lower prices.

13b.

Male	Stem	Female
7 5 3	1	7 8 8
9 5 5 2 2 0	2	0 5 6 8 9
0 0 0	3	0 0 1 8
$5|1 = \$15$		$1|7 = \$17$

Sample answer: It appears that jeans for females are somewhat more expensive than jeans for males.

15a.

Stem	Leaf
0	N E N N E E E
1	E N N N N E E N
2	E N N
3	
4	
5	E

15b. You gain information about how the pasta consumption of European Union countries compares to pasta consumption of nonmember countries. You lose information about the exact amounts of pasta eaten.   **15c.** This plot is similar to a line plot written horizontally. However, a line plot shows exact data values; this plot shows intervals of 10.
17. 16

Page 113 Lesson 3-6A
1. 221   **3.** 3 or 4

Pages 116–117   Lesson 3-6

1. Sample answer: You can see the range of each quartile, and you know that one-fourth of the data fall in each quartile.   **3.** Sample answer: All of the data do not need to be displayed; it doesn't show all the frequency.   **5a.** 54.5   **5b.** 58   **5c.** 51   **5d.** 69   **5e.** 42   **5f.** 7   **5g.** 40.5, 68.5

5h.
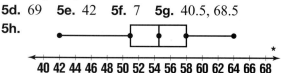

7.

9.
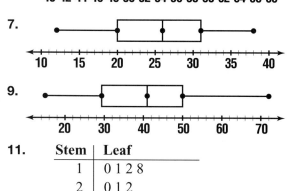

11.

Stem	Leaf
1	0 1 2 8
2	0 1 2
3	4 5 8
4	
5	6 $1|8 = 18$

Page 118   Lesson 3-6B

1. Sample answer: The mean best describes this set of data because the data are centered and evenly distributed.   **3a.** Answers will vary depending on class's data. Prediction should reflect this data.   **3b.** Sample answer: The mean is very close to the prediction.   **3c.** Sample answer: The teacher grabbed more kernels than most of the students. Just data from her alone would not be sufficient data to predict for the rest of the adults.

Pages 120–121   Lesson 3-7

1. Sample answer: Outlier may distort the data; data may be inaccurate; data may be incomplete.   **5a.** Graph B   **5b.** Graph B could be misleading because of the change in the vertical scale.   **7.** Either; accept answers students can justify.   **9.** Sample answer:

Comparable Cost

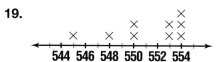

Better Quality

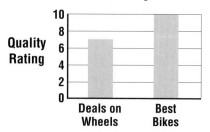

11. No, for example, the median of the set 9, 10, 11, 12, 100 is the same as the median of the set 9, 10, 11, 12, 13.   **13.** E

Pages 122–125   Study Guide and Assessment

1. e   **3.** g   **5.** b   **7.** k   **9.** d   **11.** Find the least and greatest numbers. Draw a vertical line and determine the stems. Write these numbers from least to greatest to the left of the line. Write the leaves, which are the last digits, from least to greatest to the right of the line, next to the corresponding stems. Include an explanation.   **13.** 475; Sample answer: 50 – 600; 50   **15.** Sample answer: 2 – 6; 1

Number of People	Tally	Frequency
2	||	2
3	₩₩	5
4	₩₩ ₩₩	10
5	₩₩	5
6	||	2

17.
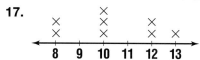

19.

21. 4, 3, 3   **23.** 84.3, none, 86
25. 5, 6, 7, 8, 9

Stem	Leaf
5	3
6	0 1
7	5 7 8 8
8	3 5 7 7 9
9	0 1 2 9 $5|3 = 53$

27.

Seattle	Stem	Olympia
6	4	5 8
2 3 4	5	1 3 6
9 8 2 2 1 0	6	0 2 5 6 8
8 7 2 0	7	2 3 7 9
5 0	8	0 4

$6|4 = 46°F$ $4|5 = 45°F$

29a. 81.6, none, 89   **29b.** The mean; six of the scores were well above 81.6.   **31.** $5.02, $4.90, $4.95

Pages 126–127   Standardized Test Practice

1. C   **3.** C   **5.** A   **7.** B   **9.** 0   **11.** mode

CHAPTER 4
Using Number Patterns, Fractions, and Percents

Page 132   Lesson 4-1A

1. Numbers on back are factors of the numbers on the front.   **3.** 2, 3, 5, 7, 11, 13, 17, 19, 23, 29   **5.** 24 and 30   **7.** 1, 4, 9, 16, 25, 36, 49; perfect square numbers because there is an odd number of factors.

Pages 134–136   Lesson 4-1

1. Sample answer: 15 is not divisible by 4, because a rectangle cannot be formed with 4 as the length of one side.   **3.** Rectangles should have dimensions $1 \times 18, 2 \times 9, 3 \times 6$; 7 is not a factor of 18.   **5.** no   **7.** 2, 4   **9.** 2, 3, 4, 6, 9   **11.** no   **13.** yes   **15.** no   **17.** yes   **19.** no   **21.** 2, 3, 5, 6, 9, 10   **23.** none   **25.** 2, 4, 5, 10   **27.** 2   **29.** yes   **31.** 3, 4, 6, 12, 15, 20, 30, 60   **33.** 252   **35.** $42,100 is considerably higher than all the other salaries.   **37.** 6.35×10^5

Page 137   Lesson 4-1B

1. 96, 108, 36   **3.** In column A, beginning with A2, enter the numbers to be tested. In E1, enter the number 228. In E2, enter A2/E1; in E3, enter A3/E1, and so on.   **5.** Place the numbers 2, 3, 4, 5, 6, 7, 8, 9, and 10 in row 1. Place the numbers 1 through 100 in column A.

Pages 140–141   Lesson 4-2

1. The sum of the digits is divisible by 3, so 3 is a factor of 387.   **3.** Sample answer: You can use divisibility rules to find factors of a number. If the number has a factor other than 1 and itself, then it is not prime.   **5.** prime   **7.** $2^2 \times 3 \times 11$ **9.** $2^5 \times 3^2$ **11.** 3×5^3 **13.** composite   **15.** composite   **17.** composite   **19.** prime   **21.** composite   **23.** $2^5 \times 3$ **25.** $2^4 \times 3^2$ **27.** $2 \times 3^2 \times 5 \times 11$ **29.** $2^2 \times 5^2 \times 17$ **31.** 2^6 **33.** $2 \times 3^3 \times 5$ **35.** $2^4 \times 5 \times 11$ **37.** $5^2 \times 59$ **39.** 3^2 **41a.** Sample answer: $2 \times 2 \times 6, 1 \times 3 \times 8, 1 \times 4 \times 6$, or $1 \times 1 \times 24$ **41b.** Sample answer: $2 \times 3 \times 6, 2 \times 2 \times 9, 3 \times 3 \times 4, 1 \times 6 \times 6, 1 \times 2 \times 18$, or $1 \times 4 \times 9$ **41c.** 5 ways   **41d.** $1 \times 1 \times 17$ **41e.** Sample answer: 32; $2 \times 4 \times 4$; it takes the least amount of space.   **43a.** 4, 9, 16, 25, . . .   **43b.** a square number   **43c.** 21, 28; $15 + 21 = 36$ or $6^2, 21 + 28 = 49$ or 7^2 **45.** A   **47.** 48 mph

Pages 144–145   Lesson 4-3

1. 1, 3, 7, 13, 21, 31; $+2 \ +4 \ +6 \ +8 \ +10$ **3.** Add 5; arithmetic; 28, 33, 38.   **5.** Multiply by 0.5; geometric; 0.25, 0.125, 0.0625.   **7.** 35, 105, 315, 945; geometric   **9.** $125, $62.50, $31.25   **11.** Multiply by 3; geometric; 324, 972, 2,916.   **13.** Add 2 more than was added to the previous term; neither; 220, 230, 242.   **15.** Multiply by 3; geometric; 405, 1,215, 3,645.   **17.** Each term has one more digit, and the digits are 1 greater than in the previous term; neither; 55,555, 666,666, 7,777,777.   **19.** The terms are $1^3, 2^3, 3^3, 4^3, . . .$; neither; 125, 216, 343.   **21.** 7, 35, 175, 875; geometric   **23.** 5, 5.4, 5.8, 6.2; arithmetic   **25.** 70, 7, 0.7, 0.07; geometric   **27.** 3, 3, 6, 18; neither   **29.** 54, 486   **31.** 1, 3, 6, 10, 15, 21, 28, 36; add 2 to the first term, add 3 to the second term, add 4 to the third term, and so on.   **33a.** 39   **33b.** 8,748   **35.** Sample answer: scale = 0-40; interval = 5; range = 33

37. 10

Pages 146–147   Lesson 4-3B

1. geometric   **3.** 14 folds is 32.768 in., 15 folds is 65.536 in., and 16 folds is 131.072 in.   **5.** The terms in the sequences would increase faster.

Pages 148–149   Lesson 4-4A

1. 89, 144, 233   **3.** 13 pairs; the pattern is the

Fibonacci sequence. **7.** Yes; 250,000 stades is approximately 25,000 miles. This is close to the actual measure of 24,901 miles. **9.** 186 miles **11.** about 25 Skybabies

Pages 152–153   Lesson 4-4

1. Identify the common prime factors, 2 and 7, and find their product; 14. **3.** 4 **5.** 2 **7.** 15 **9.** 5 **11.** 5a **13.** 28 **15.** 2 **17.** 2 **19.** 3 **21.** 18 **23.** 10 **25.** 10 **27.** 24 **29.** 12 **31.** 1 **33.** 13 **35.** Sample answer: 26, 52 and 52, 78 **37.** 4, 9 **41.** C **43.** 18

Page 153   Mid-Chapter Self Test

1. 2, 5, 10 **3.** 3, 5, 9 **5.** $2 \times 3 \times 5 \times 11$ **7.** Add 5; arithmetic; 29, 34, 39. **9.** Multiply by 2; geometric; 384, 768, 1,536.

Pages 155–157   Lesson 4-5

1. You cannot cancel out the units digits in the numerator and the denominator. Divide 16 and 36 by the GCF 4 to get $\frac{4}{9}$, not $\frac{1}{3}$. **3.** $\frac{5}{6}$ **5.** $\frac{2}{5}$ **7.** Sample answer: $\frac{4}{18}, \frac{6}{27}$ **9.** $\frac{7}{9}$ **11.** $\frac{61}{102}$ **13.** 1:2 **15.** 16:35 **17.** 11:50 **19.** 2:5 **21.** $\frac{2}{3}$ **23.** $\frac{4}{5}$ **25.** Sample answer: $\frac{6}{8}, \frac{9}{12}$ **27.** Sample answer: $\frac{10}{18}, \frac{15}{27}$ **29.** 8:13 **31.** 13:4 **33a.** $\frac{8}{17}$ **33b.** $\frac{6}{85}$ **33c.** $\frac{2}{17}$ **35.** Multiply the numerator and the denominator of a fraction by 13. (Both numbers must be greater than 7.) Sample answer: $\frac{8 \times 13}{9 \times 13} = \frac{104}{117}$ **37.** $-17°, -3°$ **39.** 18,700

Pages 159–160   Lesson 4-6

1. Sample answer: a ratio that compares a number to 100 **3.** **5.** 60%

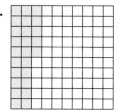

7. 34% **9.** 77% **11.** 23% **13.** 18% **15.** 50% **17.** 57% **19.** 38.4% **21.** 1% **23.** 15% **25.** 36% **27.** 61% **29.** 40% **31.** 2 groups of 18, 3 groups of 12, 4 groups of 9, 6 groups of 6, 9 groups of 4, 12 groups of 3, or 18 groups of 2 **33.** D

Pages 162–164   Lesson 4-7

1. 36%, $\frac{9}{25}$ **3.** Juliana; $0.250 = \frac{1}{4}$ and $0.025 = \frac{1}{40}$ **5.** 65% **7.** 90% **9.** $\frac{11}{50}$ **11.** $\frac{4}{5}$ **13.** 74% **15.** 10% **17.** 91% **19.** 30% **21.** 75% **23.** 55% **25.** $\frac{1}{4}$ **27.** $\frac{4}{5}$ **29.** $\frac{9}{10}$ **31.** $\frac{8}{25}$ **33.** $\frac{1}{10}$ **35.** $\frac{1}{50}$ **37.** 20% **39.** 75% **41a.** $\frac{3}{10}$ **41b.** 0.3

43.

Company	Decimal	Percent
A	0.02	2
B	0.10	10
C	0.23	23
D	0.42	42
E	0.10	10
F	0.03	3
G	0.05	5
H	0.02	2
I	0.08	8
Other	0.00	0

45. 47%

47.

Stem	Leaf
0	2 5 9
1	0 2 3 6 7
2	3 5 5
3	1 $2|3 = 23$

49. 48 miles

Pages 167–168   Lesson 4-8

1. $\frac{2}{5}$ or 40% **3.**

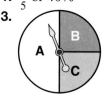

5. $\frac{3}{10}$ **7.** $\frac{3}{14} \approx 21.4\%$ **9.** $\frac{1}{20}$ **11.** $\frac{3}{20}$ **13.** $\frac{2}{5}$ **15.** $\frac{4}{25} = 16\%$ **17.** $\frac{6}{25} = 24\%$ **19.** $\frac{12}{25} = 48\%$ **21.** $\frac{1}{5}$ **23.** $\frac{1}{14}$ **25.** $\frac{1}{6} \approx 16.7\%$ **27.** No; the chance of rolling a 5 $\left(\frac{4}{36}\right)$, 6 $\left(\frac{5}{36}\right)$, 7 $\left(\frac{6}{36}\right)$, or 8 $\left(\frac{5}{36}\right)$ gives Marvin $\frac{20}{36}$ chance of winning a point. The probability of Naomi winning is only $\frac{16}{36}$. **29.** A

Pages 170–171   Lesson 4-9

1. A common multiple should have all the prime

factors of each number; $2^3 \times 3^2 = 72$ **3.** Sample answer: List several multiples of each number. To do the problem mentally, you could find the multiples of 9 and see which ones were multiples of both 3 and 5; 45.   **5.** 45   **7.** 24   **9.** 225   **11.** 60 **13.** 30 **15.** 900   **17.** 24   **19.** 60   **21.** 784   **23.** 102 **25.** 3,750   **27.** 336   **29.** 20   **31.** 12 weeks **33.** Sample answer: {2, 3, 5}   **35.** 18   **37.** $2.10

Pages 174–175   Lesson 4-10

3. $\frac{5}{7}$ and $\frac{4}{9}$; you can see from the graph that $\frac{5}{7}$ of the circle is a greater part than $\frac{4}{9}$ of the circle. So, $\frac{5}{7} > \frac{4}{9}$.   **5.** 65   **7.** 24   **9.** $<$ **11.** $>$ **13.** 45 **15.** 16   **17.** 8   **19.** 36   **21.** 24   **23.** 68   **25.** $<$ **27.** $<$ **29.** $>$ **31.** $=$ **33.** $<$ **35.** $>$ **36.** $<$ **37.** 126   **39.** students in math   **41.** 5-6 hours **43.** 42   **45.**

```
            ×
   ×   × ×     ××      × ×
───┼───┼───┼───┼───┼───┼───┼──→
   65  70  75  80  85  90  95
```

47. B

Pages 176–179   Study Guide and Assessment

1. false, $2^3 \times 3$ **3.** true   **5.** false, $\frac{2}{3}$ **7.** yes **9.** no   **11.** 5   **13.** $2^3 \times 5^3$ **15.** $2 \times 5^2 \times 19$ **17.** $2^5 \times 3$ **19.** $2^4 \times 5^2 \times 7$ **21.** Add 5; arithmetic; 41, 46, 51.   **23.** Add 3 more than was added to the previous term; neither; 63, 84, 108. **25.** 9   **27.** 84   **29.** 8   **31.** $\frac{4}{5}$ **33.** 2:3   **35.** $\frac{7}{11}$ **37.** 56%   **39.** 49%   **41.** 56%   **43.** 40%   **45.** $\frac{3}{20}$ **47.** $\frac{29}{50}$ **49.** $\frac{1}{2} = 50\%$ **51.** $\frac{1}{4} = 25\%$ **53.** 30 **55.** 80   **57.** 3,969   **59.** 12   **61.** $<$ **63.** 2 **65.** 97.3%

Pages 180–181   Standardized Test Practice

1. B   **3.** A   **5.** B   **7.** B   **9.** E   **11.** arithmetic; 62, 74, 86   **13.** 0.3   **15.** $\frac{2}{5}$

CHAPTER 5
Algebra: Using Integers

Pages 185–186   Lesson 5-1

1. It has a decimal part.   **3.** Keisha; -3 and 7 are not the same distance from 0.   **5.** $+5$ **7.** -76 **9.** 7, -7, 7   **11.** $+9$ **13.** -120 **15.** $+1600$ **17.** -4 **19.** 0   **21.** $+15$ **23.** $-1, 1, 1$ **25.** 0, 0, 0   **27.** 7, -7, 7   **29.** 0   **31.** 3,212; $-8,685$

33a. 7   **33b.** -7 **33c.** odd number of times, the result is negative; even number of times, the result is positive   **35.** $>$ **37.** 3, 3, 3

Pages 189–190   Lesson 5-2

1.
```
←──┼───┼───┼───┼───┼───┼──→
  -6  -5  -4  -3  -2  -1
```
3. $>$ **5.** $>$ **7.** 15   **9.** $>$ **11.** $>$ **13.** $>$ **15.** $>$ **17.** $-59, -43, -3, 0, 5, 11$ **19.** 7 **21.** $-35,800; -20,000; -13,000; -4,000; -1,000$ **23a.** Mt. Elbert   **23b.** Granite Peak, Wheeler Peak, Mt. Hood **25a.**

City	Latitude (nearest degree)	Low Temp. (°F)
Honolulu, HI	21	56
San Diego, CA	33	43
Atlanta, GA	34	13
Nashville, TN	36	9
Denver, CO	40	-7
New York, NY	41	6
Minneapolis, MN	45	-11
Bismarck, ND	47	-28
Seattle, WA	48	22
Fairbanks, AK	65	-48

25b. Generally, as the latitude increases, the low temperature decreases.   **27.** A   **29.** 0.003 g

Pages 193–194   Lesson 5-3

1. quadrant II   **5.** (1, 0), x-axis **7-9.**   **11.** (2, 2), I

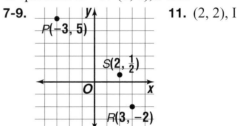

13. $(5, -2)$, IV   **15.** $(0, -4)$, y-axis **17.** $(-1, 4)$, II   **19.** $(-2, -2)$, III **21-27.**

29. quadrant IV   **31a.** $(3, 3)$ **31b.** $(-3, -1)$
31c. $(-2, -2)$ **31d.** $(-2, 3)$ **33a.** Honolulu
$(21, 56)$, San Diego $(33, 43)$, Atlanta $(34, 13)$,
Nashville $(36, 9)$, Denver $(40, -7)$, New York $(41,$
$6)$, Minneapolis $(45, -11)$, Bismarck $(47, -28)$,
Seattle $(48, 22)$, Fairbanks $(65, -48)$

33b.

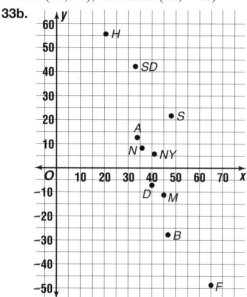

35. $-5, -1, 3, 5$ **37.** $\$0.05$

Page 195   Lesson 5-3B

1.

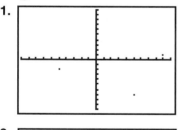

3.

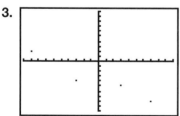

Page 196   Lesson 5-4A

1. $-7 + 2 = -5$
3. $-4 + 1 = -3$

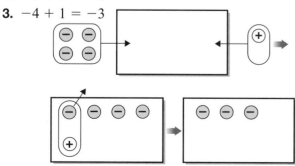

5. negative

Pages 199–200   Lesson 5-4

1. Not always; there may be 1 or 2 positive integers,
but they have a smaller absolute value than the
negative integer(s).
3. Sample answer:

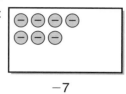

-7

5. zero   **7.** -9 **9.** 7   **11.** -4 **13.** $-56°F$
15. negative   **17.** positive   **19.** negative
21. zero   **23.** 4   **25.** 6   **27.** 13   **29.** 1
31. -19 **33.** -17 **35.** 2   **37.** 11   **39.** 0
41. -86 **43.** 9 yards gained   **45.** Marsha did not
owe Yori any money, because she did not initially
borrow any money.   **47.** 21, 21

49.

Stem	Leaf
0	3 7 9 9
1	0 3 4 5
2	4
3	1 $2|4 = 24$

Page 201   Lesson 5-5A

1. $-7 - (-3) = -4$
3. $-4 - (-1) = -3$

5. $-6 - 1 = -7$

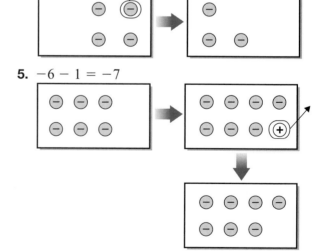

Pages 203–205   Lesson 5-5

1. $a + b$ **3.** Ellen; $15 - 24 = -9$, but
$24 - (15) = 9$.   **5.** -13 **7.** -6 **9.** 23   **11.** 4
13a. 506 years   **13b.** 1,300 years   **15.** 35
17. 10   **19.** -1 **21.** -53 **23.** -32 **25.** 40
27. 38   **29.** 32   **31.** -19 **33.** -8 **35.** -3

37. 4 **39.** −7 **41.** −2 **43.** 221°

45a. Honolulu (21, 38), San Diego (33, 47), Atlanta (34, 89), Nashville (36, 90), Denver (40, 106), New York (41, 96), Minneapolis (45, 112), Bismarck (47, 126), Seattle (48, 74), Fairbanks (65, 136)

45b.

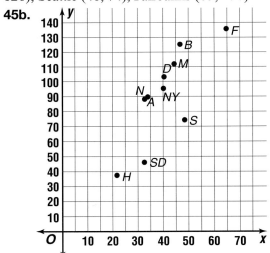

47. A **49.** 3, 5, 9 **51.** 0.375

Page 205 Mid-Chapter Self Test

1. +7 **3.** > **5.** > **7.** (−1, −4), III **9.** −9

Page 206 Lesson 5-6A

1. $2 \times 3 = 6$

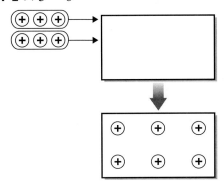

3. $-2 \times (-3) = 6$

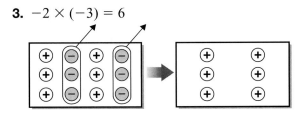

5. $4 \times 0 = 0$

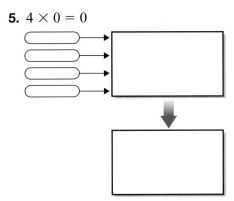

7. −27

Pages 208–209 Lesson 5-6

1. One is positive and one is negative.

3. $-3 \times (-4) = 12$

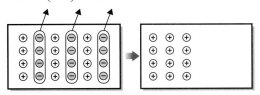

5. 35 **7.** −108 **9.** 36 **11.** 24 **13.** −78
15. 9 **17.** −56 **19.** −20 **21.** 16 **23.** −63
25. −100 **27.** −126 **29.** −24 **31.** −225
33. −78 **35.** −100 **37a.** $-2(14) = x$
37b. −28 or 28 cubic meters removed **39.** One of the integers is negative or all three integers are negative. If all three integers are positive, or if two of the integers are negative, the product is positive. If all three integers are negative, or if one of the integers is negative, the product is negative.
41. −8, −4, −3, 0, 1, 4, 6 **43.** 0.75

Pages 210–211 Lesson 5-7A

1. Multiply by 3; 486; 1,458; 4,374. **3.** 70 boxes
7. 5, 10, 20, 40, 80, 160, 320; no, because 320 min. $= 5\frac{1}{3}$ hr, so she will be exercising more than 5 hours each day by the 7th day. **9.** 24 pegs **11.** 120 feet

Pages 213–214 Lesson 5-7

1. $-12 \div 4 = -3; -12 \div (-3) = 4$ **3.** 4
5. −6 **7.** −3 **9.** −32 **11.** −2 **13.** −5
15. 14 **17.** 12 **19.** −9 **21.** 3 **23.** −7
25. 220 **27.** −20 **29.** 9 **31.** 81 **33.** 1
35. −27 **37.** $y = 24 \div 6$ **39.** −30, −15, −10, −6, −5, −3, −2, −1, 1, 2, 3, 5, 6, 10, 15, 30
41. > **43a.** 115.2 **43b.** 225.6

Pages 216–217　Lesson 5-8

1.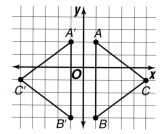

3. Sample answer: The diamonds are both translations and reflections of each other.

5. $A'(-1, 2)$, $B'(-1, -4)$, $C'(-5, -1)$

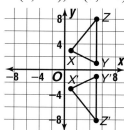

7. on the lid: reflections; on the side: translations and reflections　**9.** translation

11. $X'(1, -3)$, $Y'(5, -1)$, $Z'(5, -8)$

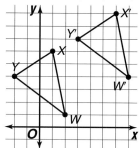

13. $W'(7, 4)$, $X'(6, 9)$, $Y'(3, 7)$

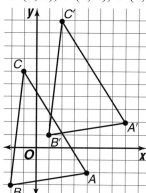

15. $J'(7, 1)$, $K'(2, 5)$, $L'(5, 7)$

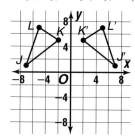

17a. translation　**17b.** $(-2, -4)$

19.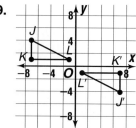

$\triangle JKL$ was reflected over the y-axis and then over the x-axis. This is a double reflection.
21. $2 \times 3^2 \times 5 \times 7$

Pages 218–221　Study Guide and Assessment

1. negative　**3.** 7　**5.** origin　**7.** quadrants
9. y-coordinate　**11.** zero　**13.** A reflection is a transformation where a figure is flipped. A translation is a transformation where a figure is slid.
15. $+14$　**17.** -5　**19.** $-1, 1, 1$　**21.** $4, -4, 4$
23. $>$　**25.** $>$　**27.** $-13, -11, 0, 5, 8, 10$
29. $(1, 3)$, I　**31.** $(-2, -3)$, III
33-37. 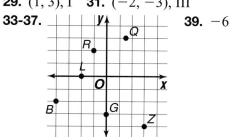　**39.** -6

41. -13　**43.** 0　**45.** 24　**47.** 8　**49.** 2
51. 9　**53.** -25　**55.** 24　**57.** -5　**59.** 6
61. -2　**63.** $A'(7, 2)$, $B'(1, 1)$, $C'(2, 10)$

65. -18　**67.** 4 yards lost　**69.** $x = \$327 \div 3$

Pages 222–223　Standardized Test Practice

1. D　**3.** C　**5.** C　**7.** A　**9.** D　**11.** 7　**13.** 86
15. $\frac{13}{20}$

CHAPTER 6
Algebra: Exploring Equations and Functions

Pages 226–227   Lesson 6-1A

1. 5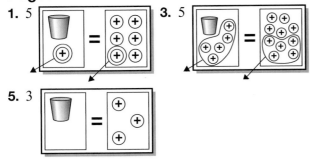

3. 5

5. 3

7. Sample answer: The forces on each arm of the seesaw are equal; each side of an equation has the same value.

9. −5

11. −3

13. −2

15. Sample answer: A zero pair has a value of 0. So, adding or subtracting a zero pair is like adding or subtracting zero.

Pages 230–231   Lesson 6-1

1.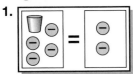

3.
$$x - 6 = 3$$
$$x - 6 + 6 = 3 + 6$$
$$x = 9$$

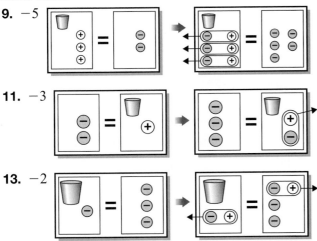

5. 3   **7.** 2   **9.** −8   **11.** 555 feet   **13.** 21
15. −8   **17.** 25   **19.** 8   **21.** −12   **23.** −43
25. 1.7   **27.** 8.5   **29.** 11   **31.** −4

33. $a + 50 + 75 = 180$; 55°   **35.** Sample answer:
$x - 5 = -7, x + 4 = 2$ **37.** 16   **39.** The one very high salary increases the mean.

Pages 232–233   Lesson 6-1B

1. 1952   **3.** Sample answer: There are too many guesses you might make; it would take less time to work backward.   **5.** Sample answer: Since 13 was added to x to get 25, subtract 13 from 25. The solution is 12.   **7.** Sample answer: Subtract 120 from 364.   **9.** 2 letters and 10 postcards   **11.** 1894
13. 55   **15.** B

Pages 236–237   Lesson 6-2

1. $3x = -15$; −5

3. 6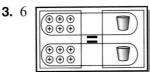

5. −6   **7.** 12   **9.** 6   **11.** 7   **13.** −17
15. 2   **17.** −49   **19.** 5   **21.** 15   **23.** 4.7
25. 5.4   **27.** 18   **29.** −5   **31.** 50 inches
33. 100 seconds   **35.** 7.6   **37.** 1

Page 238   Lesson 6-3A

1. 2

3. 2

5. 0

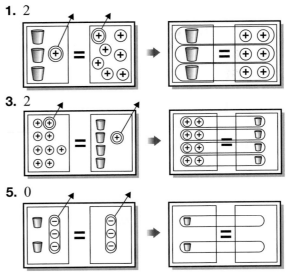

7. It is made up of two operations and takes two steps to solve.

SELECTED ANSWERS

Pages 240–241   Lesson 6-3

1. -2

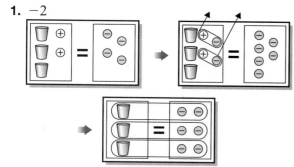

3. Hector; the first step is to create zero pairs.
5. -4 **7.** -4 **9.** 2.1 **11.** 2 **13.** -3 **15.** 3
17. 2.4 **19.** -5 **21.** 1 **23.** -4 **25.** 0 **27.** 4
29. 3 **31.** 5 balloons
33. 3

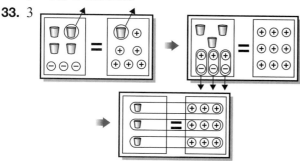

35. D   **37.** 9.8×10^3

Pages 244–245   Lesson 6-4

1. c   **3.** Sample answers: the age of a person 3
years older; the person's age in 3 years   **5.** $p - 8$
7. $-9 + n$ **9.** $20n = 120$ **11.** $m + 10.1 = 33.0$;
$m = 22.9$ **13.** $x \cdot 2$ or $2x$ **15.** $9 + x$ **17.** $a \div 3$
or $\frac{a}{3}$ **19.** $2a$ **21.** $s + 10$ **23.** $\frac{n^2}{10}$
25. $8n = -64$ **27.** $7n - 5 = 37$ **29.** $s + 15 =$
220 **31.** $8 + 2a = 60$ **33.** $p + 0.2 = 1.9$;
$p = 1.7$ **35.** $1.5m + 10 = 19$; $m = 6$ **37.** -13
39. 18 **41.** 243

Page 245   Mid-Chapter Self Test

1. 30 **3.** 5.2 **5.** 12 **7.** 1 **9.** -4

Page 248   Lesson 6-5

1. $x \le -1$
3.

$$\begin{array}{c} 0\ 1\ 2\ 3\ 4\ 5\ 6\ 7\ 8 \end{array}$$

5. $a \ge -2$,

$$\begin{array}{c} -5\ -4\ -3\ -2\ -1\ 0\ 1\ 2\ 3 \end{array}$$

7. $x \le -6$,

$$\begin{array}{c} -8\ -6\ -4\ -2\ 0\ 2 \end{array}$$

9. $x > -7$,

$$\begin{array}{c} -10\ -8\ -6\ -4\ -2 \end{array}$$

11. $d \ge 4$,

$$\begin{array}{c} 0\ 1\ 2\ 3\ 4\ 5\ 6\ 7\ 8\ 9 \end{array}$$

13. $p > -2$,

$$\begin{array}{c} -5\ -4\ -3\ -2\ -1\ 0\ 1\ 2 \end{array}$$

15. $r < 8$,

$$\begin{array}{c} -2\ -1\ 0\ 1\ 2\ 3\ 4\ 5\ 6\ 7\ 8 \end{array}$$

17. $y < 3.2$,

$$\begin{array}{c} 0\ 1\ 2\ 3\ 4\ 5\ 6\ 7\ 8 \\ 3.2 \end{array}$$

19. $a > 7$,

$$\begin{array}{c} 3\ 4\ 5\ 6\ 7\ 8\ 9\ 10\ 11 \end{array}$$

21. $5n > 60$, $n > 12$ **23.** $a \ge 18$
25. $0.06 < s < 2$ **27.** 11

Pages 251–252   Lesson 6-6

1. a relationship between two quantities
3. Sample answer: It would take longer for each
student, so the graph would be above the original
graph.
5. The population
increases each year.

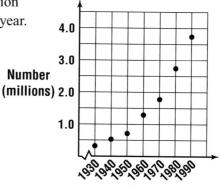

7a.

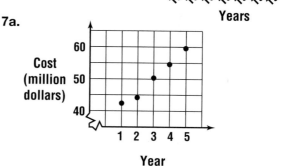

7b. Production costs have increased each year.
9. As you grow older, you are more likely to snore
and less likely to talk in your sleep.   **11.** Between
12 and 14; the line connecting the points is steeper
for these years.   **13.** 55

Page 253 Lesson 6-7A

3. Extend the line.

Pages 256–257 Lesson 6-7

1. d **3.** Grace; $2(1) - 1 \neq -1$

5.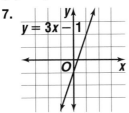

7.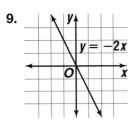

9.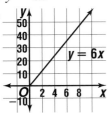

11a.

Hours	Earnings
3	18
5	30
7	42
9	54

11b. $y = 6x$

11c.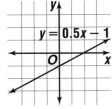

13.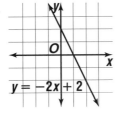

15.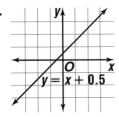

17.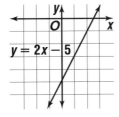

19.

21.

23.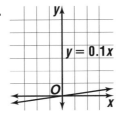

25. $y = 2x$ **27.** $x + y = 10$ **29a.** Sample answer: Emily is 6 years older than Jared.

29b. $y = x + 6$ **29c.** 16 years

29d.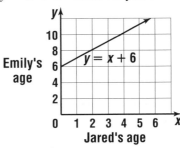

31a. 4.3 lb/in^2

31b.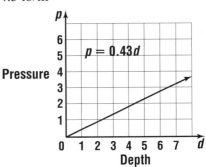

31c. about 12 or 13 lb/in^2 **33.** No; the number of members of the band can only be represented with whole numbers. **35.** 12 **37.** B

Pages 258–261 Study Guide and Assessment

1. false, variable **3.** true **5.** true **7.** false, 7
9. false, -6 **11.** If you add the same number to each side of an equation, the two sides will still be equal. **13.** -38 **15.** 63 **17.** 1.5 **19.** 4 **21.** 9
23. -3 **25.** -1 **27.** -2 **29.** -10 **31.** -4
33. $s - 13$ **35.** $14n = 56$
37. $g \leq 2$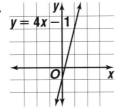

39. $m \geq -3$

41. As the length of a side of a square increases, the area also increases.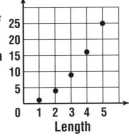

43.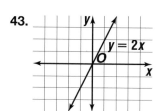

45.

$y = 2x$

$y = 3x + 2$

47. Jeff: 27, Fina: 9, Mario: 12, Danielle: 16
49a. $y = 12.50x$ **49b.** $500

Pages 262–263 Standardized Test Practice

1. D **3.** B **5.** A **7.** C **9.** B **11.** E **13.** 3
15. $E = 6(20)$ **17.** -1

CHAPTER 7
Applying Fractions

Pages 270–271 Lesson 7-1

1a. $\frac{6}{11}, \frac{1}{2}$ **1b.** $\frac{6}{7}, 1$ **3.** 0 **5.** $\frac{1}{2}$ **7.** 9
9. $\frac{1}{2} + 1 = 1\frac{1}{2}$ **11.** $1 \times 11 = 11$ **13.** $5 \times 3 =$
15 **15.** Sample answer: $1 \times 24 = 24$ cups **17.** $\frac{1}{2}$
19. 1 **21.** 0 **23.** 1 **25.** 3 **27.** 4 **29.** 6
31. 7 **33–49.** Sample answers are given.
33. $\frac{1}{2} - 0 = \frac{1}{2}$ **35.** $1 \div 1 = 1$ **37.** $5 - 3 = 2$
39. $\frac{1}{2} \times 18 = 9$ **41.** $12 - 2 = 10$
43. $22 \div 2 = 11$ **45.** $\frac{1}{2} \times \frac{1}{2} = \frac{1}{4}$ **47.** $2 + 1 + 6$
$= 9$ **49.** $1 \times 250 = 250$ pounds **53.** Sample
answer: (1, 4), (2, 7), (3, 10), (4, 13) **55.** 23
57. $<$

Pages 274–275 Lesson 7-2

1. The units of measure must be the same.
3.

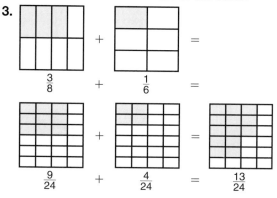

$\frac{3}{8}$ + $\frac{1}{6}$ =

$\frac{9}{24}$ + $\frac{4}{24}$ = $\frac{13}{24}$

5. $\frac{2}{3}$ **7.** $\frac{1}{4}$ **9.** $1\frac{7}{18}$ **11.** $\frac{3}{8}$ tank **13.** $\frac{11}{15}$
15. $1\frac{8}{35}$ **17.** $\frac{1}{6}$ **19.** $\frac{19}{30}$ **21.** $\frac{26}{45}$ **23.** $1\frac{1}{6}$
25. $1\frac{16}{45}$ **27.** $\frac{7}{24}$ **29.** $1\frac{11}{18}$ **31.** $\frac{21}{44}$ **33.** $\frac{3}{5}$

35. No; commutative property does not hold for subtraction. **37.** 2 **39.** A

Pages 277–279 Lesson 7-3

1.

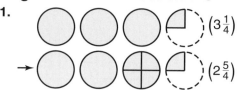

$\left(3\frac{1}{4}\right)$

$\left(2\frac{5}{4}\right)$

3. Renaming of mixed numbers is used when the fraction in the first mixed number is less than the fraction in the second. Renaming of whole numbers is used when the digit in the first number is less than the digit in the same place value in the second number. **5.** 7 **7.** $8\frac{1}{3}$ **9.** $1\frac{3}{4}$ **11.** $212\frac{4}{5}$ carats
13. 3 **15.** 8 **17.** 11 **19.** 13 **21.** $4\frac{2}{3}$ **23.** $4\frac{1}{2}$
25. $29\frac{23}{40}$ **27.** $3\frac{7}{9}$ **29.** $11\frac{13}{24}$ **31.** $2\frac{1}{6}$ **33.** $2\frac{11}{20}$
35. $3\frac{1}{18}$ **37.** No, the ammonia and vinegar make
$2\frac{5}{6}$ cups. A $\frac{1}{2}$-quart pan holds only 2 cups.
39. 20 ft **41.** Sample answers: $(-1, 10), (0, 7),$
$(1, 4), (2, 1)$ **43a.** $\frac{1}{25}$ **43b.** $\frac{13}{100}$ **43c.** $\frac{7}{50}$

Pages 280–281 Lesson 7-3B

1. Sample answer: Use estimation, work backward, draw a diagram. **3.** D **7.** $4\frac{1}{6}$ c **9.** no; $2\frac{5}{6} + 3\frac{1}{2}$
> 6 **11.** Sample answer: about 98,000 thousand

Page 283 Lesson 7-4A

1. $\frac{1}{2} \times \frac{1}{3} = \frac{1}{6}$ $\leftarrow \frac{1}{2} \rightarrow$

$\frac{1}{3}$

3. $3 \times \frac{1}{2} = \frac{3}{2}$ or $1\frac{1}{2}$

$\leftarrow$ 3 $\rightarrow$

$\frac{1}{2}$

5. $1\frac{1}{3} \times \frac{1}{4} = \frac{4}{12}$ or $\frac{1}{3}$

$\leftarrow$ $1\frac{1}{3}$ $\rightarrow$

$\frac{1}{4}$

7. $\frac{20}{6}$ or $3\frac{1}{3}$

Pages 285–287   Lesson 7-4

1. Rename each mixed number as an improper fraction. Multiply the numerators and the denominators. Simplify.
3. $\frac{4}{5} \times \frac{1}{4} = \frac{4}{20}$ or $\frac{1}{5}$

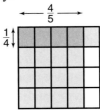

5. $\frac{2}{7}$ **7.** $\frac{1}{4}$ **9.** $4\frac{4}{15}$ **11.** $6\frac{13}{20}$ ft   **13.** $\frac{3}{32}$
15. $1\frac{3}{5}$ **17.** $\frac{1}{2}$ **19.** $\frac{3}{10}$ **21.** $\frac{2}{7}$ **23.** $1\frac{9}{16}$
25. 33   **27.** 6   **29.** $42\frac{1}{6}$ **31.** $7\frac{1}{3}$ **33.** $\frac{1}{4}$
35. $1,500\frac{3}{4}$ mi   **37.** $\frac{2}{3}, \frac{2}{5}$ **39.** $r > -8$
41. 52, 47, 36, 27, 13, 0, $-2, -3, -6, -14$

Page 287   Mid-Chapter Self Test

1. $\frac{1}{2} + 1 = 1\frac{1}{2}$ **3.** $\frac{1}{2} \times 14 = 7$ **5.** $1\frac{7}{36}$ **7.** $6\frac{1}{3}$
9. $32\frac{2}{15}$

Page 288   Lesson 7-4B

1. 72 sq units; 64 sq units   **3.** You can multiply the area of each stage by $\frac{8}{9}$ to get the area of the next stage.   **5a.** $\frac{3}{4}$ sq units   **5b.** $\frac{9}{16}$ sq units
5c. $\frac{27}{64}$ sq units

Pages 290–291   Lesson 7-5

1. Division; pints are larger than cups.   **3.** 80
5. 9   **7.** 24   **9.** 3.5 lb   **11.** 10,000   **13.** 6
15. 4   **17.** 48   **19.** 5   **21.** 40   **23.** 9   **25.** $\frac{1}{4}$
27. 2   **29.** $\frac{3}{8}$ gal   **31.** 52 fl oz   **33.** $7\frac{3}{4}$ T
35. No, the recipe makes 9 cups of punch and the pitcher holds only 2 quarts, or 8 cups.   **37.** D
39. 80

Pages 293–295   Lesson 7-6

3. Sample answer: wallpaper border around bedroom   **5.** 56 cm   **7.** 10.4 yd   **9.** 300 ft
11. 18.8 ft   **13.** 50 mi   **15.** 110 m   **17.** $21\frac{3}{4}$ in.
19. 24 m   **21.** 52.2 cm   **23.** $2\frac{3}{16}$ ft   **25.** $2\frac{3}{4}$ in.
27. $3\frac{1}{2}$ in.   **29.** 132 ft   **31.** Sample answer: 4 in., 4 in., 4 in., $3\frac{5}{6}$ in.   **33.** 99.9   **35.** A

Pages 299–300   Lesson 7-7

3. 29.8 km   **5.** 9.4 yd   **7.** 32.3 cm   **9.** 31.4 ft
11. 38.3 m   **13.** 3.1 mi   **15.** 44 m   **17.** 38.9 cm

19. $2\frac{18}{35}$ or 2.5 yd   **21.** 55 ft   **23a.** 201.0 in.
23b. 5.3 times   **25.** 4.7 ft   **27.** 2π **29.** A
31. $\frac{49}{100}, \frac{7}{20}, \frac{4}{25}$

Pages 303–304   Lesson 7-8

1. No; $4 \cdot 20 \neq 5$.   **3a.** $80 + 20 + 276 = 100 + 276 = 376$; commutative   **3b.** $-26 + 26 + 54 = 0 + 54 = 54$; commutative, additive inverse, identity (+)   **3c.** $\left(-\frac{5}{9}\right)(1) = -\frac{5}{9}$; identity ($\times$)   **3d.** $(-7)(3)(-15)(0) = 0$; additive inverse, multiplicative prop. of zero   **5.** Identity (+)   **7.** $\frac{5}{22}$ **9.** 32
11. 11   **13.** commutative (+)   **15.** identity ($\times$)
17. multiplication (=)   **19.** $\frac{3}{2}$ or $1\frac{1}{2}$ **21.** $\frac{1}{12}$
23 36   **25.** 24   **27.** 112   **29.** $24\frac{1}{2}$ **31.** $\frac{4}{5}$
33. Yes; there is enough turkey for 16 people.
35. 420   **37.** D   **39.** 20; Sample answer: 20-50; 5

Pages 306–307   Lesson 7-9

1. Rename $1\frac{1}{9}$ as $\frac{10}{9}$, and multiply by the inverse of $\frac{8}{9}$.   **3.** Omar; each person will get $2\frac{1}{4} \div 4 = \frac{9}{16}$ or 0.5625 lb.   **5.** 12   **7.** $5\frac{1}{3}$ **9.** $\frac{4}{15}$ **11.** $\frac{7}{16}$
13. $2\frac{2}{5}$ **15.** $1\frac{1}{4}$ **17.** 2   **19.** $\frac{3}{8}$ **21.** $\frac{4}{15}$ **23.** $1\frac{2}{3}$
25. $1\frac{3}{4}$ **27.** $3\frac{3}{8}$ **29.** $\frac{5}{14}$ **31.** $\frac{16}{21}$
33. 20 packages   **37.** $\frac{8}{29}$ **39.** 18%   **41.** C

Pages 308–311   Study Guide and Assessment

1. e   **3.** c   **5.** j   **7.** d   **9.** g   **11–17.** Sample answers are given.   **11.** $1 + 1 = 2$ **13.** $12 \div 1 = 12$ **15.** $1 - 0 = 1$ **17.** $2 \times 16 = 32$ **19.** $1\frac{1}{35}$
21. $\frac{3}{5}$ **23.** $13\frac{5}{9}$ **25.** $3\frac{1}{24}$ **27.** $\frac{1}{3}$ **29.** $34\frac{4}{7}$
31. 8   **33.** 4   **35.** 6   **37.** 16.6 m   **39.** 9.4 yd
41. 44 in.   **43.** 4 ft   **45.** $18\frac{3}{4}$ **47.** 6   **49.** 6
51. $7\frac{6}{7}$ **53.** $5\frac{11}{12}$ cups   **55.** yes

Pages 312–313   Standardized Test Practice

1. A   **3.** B   **5.** B   **7.** C   **9.** C   **11.** D   **13.** $\frac{5}{8}$
15. $Q(-6, 3)$, II; $R(-3\frac{1}{2}, -4)$, III; $S(2, 0)$, x-axis

CHAPTER 8
Using Proportional Reasoning

Page 316   Lesson 8-1A

1. Sample answer: Make equivalent fractions.
$\frac{2}{3} = \frac{4}{6} = \frac{6}{9} = \frac{10}{15}$

3.

A	R	U	X
(dots)	(dots)	2	10
B	**S**	**V**	**Y**
(dots)	24	4	(dots)
C	**T**	**W**	**Z**
(dots)	60	(dots)	50

Pages 318–320 Lesson 8-1

1. $\frac{15}{20}$, 15:20, 15 to 20 **3.** Sample answer: Write both ratios as fractions in simplest form. If the fractions are equal, then the ratios are equivalent.
5. $\frac{8}{21}$ **7.** $\frac{1}{21}$ **9.** Yes; $\frac{12}{16} = \frac{3}{4}$ and $\frac{21}{28} = \frac{3}{4}$
11. Yes; $\frac{15}{6} = \frac{5}{2}$ and $\frac{90}{36} = \frac{5}{2}$ **13.** $\frac{7}{15}$ **15.** $\frac{7}{2}$
17. $\frac{5}{1}$ **19.** $\frac{35}{36}$ **21.** $\frac{31}{44}$ **23.** $\frac{32}{9}$ **25.** $\frac{7}{40}$
27. $\frac{9}{26}$ **29.** No; 150:15 = 10:1 **31.** No; $\frac{65}{5} = \frac{13}{1}$
33. No; 14:42 = 1:3 and 58:1,218 = 1:21
35. Yes; 3 days:4 hours = 18:1 and 9 days to 12 hours = 18:1 **37.** $\frac{36}{73}$ **39.** $\frac{1}{300}$ **41a.** $\frac{1}{25}$
41b. Yes; the ratios of successive terms decrease by 1. **43.** 45; they are all 1:25. **45.** 3
47. 2, 3, 6

Pages 323–324 Lesson 8-2

1a. No; the denominator is not 1. **1b.** Yes; the denominator is 1. **1c.** Yes; the denominator is 1.
3. April is correct. The $3.99 bag has a unit price of about 24.94¢ and the $2.99 bag has a unit price of about 24.92¢. **5.** 60 miles per hour **7.** 15 people per van **9.** 6.1875¢ per ounce **11.** 40 miles per hour **13.** $70 per day **15.** 3 pounds per week
17. 5 meters per second **19.** about 29.08¢ per ounce **21.** $1.24 per pound **23.** 3 people per car
25. 9.5 feet per second **27.** 90 rotations per minute **29.** 24.5¢ per ounce **31.** 2,160,000 toys per year **33a.** adult human **33b.** adult human **35.** about 4.9 hours **37.** $a \geq 25$
39. D

Pages 327–328 Lesson 8-3

1. Compare the cross products. If they are equal, the ratios are equivalent. **3.** 6 **5.** 6 **7a.** 6 bicycles

7b. 7 feet **9.** 22.5 **11.** 2 **13.** 75 **15.** 2.5
17. 42 **19.** $\frac{1}{3}$ **21.** 210 words **23a.** tuna: $918\frac{3}{4}$ oz; pita breads: 300; dill: $37\frac{1}{2}$ t; yogurt: 50 c; celery: $18\frac{3}{4}$ c; mustard: 225 t; lettuce: 600 leaves; tomatoes: 150 **25.** Sample answers: 1 and 36; 2 and 18; 4 and 9 **27.** $17\frac{1}{2}$

Page 329 Lesson 8-3B

1. Sample answer: Yes; it uses all of the samples to find a more accurate estimate. **3.** The sample needs to be a random handful of the total number of beans.

Pages 330–331 Lesson 8-4A

1. 27 people **3.** 121 people **7.** 12 right, 2 wrong, 1 unanswered **9.** 36 handshakes
11a. $\frac{5}{8}$ feet **11b.** $28\frac{3}{4}$ feet or 345 inches

Pages 334–335 Lesson 8-4

1. the scale **5.** 9,650 km **7.** 50 inches
9. $134\frac{1}{5}$ miles **11.** $647\frac{1}{2}$ miles **13.** 4,125 km
15. 32 inches **17.** 0.6 cm **19.** 20 cm
21. 18.48 cm **23.** 1 inch:16 miles
25.

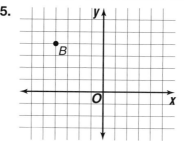

Page 335 Mid-Chapter Self Test

1. $\frac{1}{9}$ **3.** $\frac{8}{1}$ **5.** $1.19 per disk **7.** 12 **9.** 2.5

Pages 337–338 Lesson 8-5

1. Set up a proportion with a fraction with a denominator of 100.
$$\frac{3}{16} = \frac{n}{100}$$
Use cross products to solve.
$$300 = 16n$$
$$18.75 = n$$
So, $\frac{3}{16} = 18.75\%$.

3. 30% **5.** 31.25% **7.** $\frac{18}{25}$ **9.** $\frac{3}{4}$ **11.** 43.75%
13. 11.5% **15.** about 41.7% **17.** about 36.4%
19. about 3.3% **21.** $\frac{2}{5}$ **23.** $\frac{69}{200}$ **25.** $\frac{5}{8}$
27. $\frac{901}{2,000}$ **29.** 1 **31.** 32% **33.** about 8.3%
35. A **37.** −3

Pages 340–341   Lesson 8-6

1. b **3.** Sample answer: A number can be expressed as a fraction, a decimal, or a percent.
5. 6% **7.** 0.65 **9.** 0.185 **11.** 17% **13.** 85%
15. 9.9% **17.** 1% **19.** 0.9% **21.** 0.45
23. 0.16 **25.** 0.081 **27.** 0.1425 **29.** 0.9475
31. 84.8% **33.** 4%, 33% **35a.** 3.5 **35b.** 0.175
35c. 0.60 **37.** $t \div 8$ or $\frac{t}{8}$ **39.** 26-30 and 31-35

Pages 344–345   Lesson 8-7

1. 150% is greater than 100%.
3a.

3b.

5. 0.0055 **7.** 0.002 **9.** 1,550% **11.** about 0.29% **13.** Yes; This means that they can expect a price higher than the minimum guaranteed. **15.** 5
17. 1.15 **19.** 1.005 **21.** 0.00068 **23.** 0.000025
25. 0.00032 **27.** 900% **29.** 775% **31.** 3,400%
33. 0.9% **35.** 0.4% **37.** 0.01% **39.** 180%
41. No; You would be shorter than when you were born. **43.** No; The greatest possible percent of the attendance from one group would be 100%.
45. No; The population in 1990 would not be less than the population in 1790. **47.** 26.69
49. about 0.22% **51.** 73,592,376 **53.** yes; 13 to 39 = $\frac{1}{3}$ and 26 to 78 = $\frac{1}{3}$
55. $X'(2, -4)$, $Y'(1, 1)$, $Z'(3, -5)$

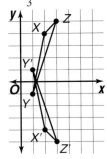

57.

Stem	Leaf
6	2 6
7	
8	1 3 5
9	0 2 5 $6|2 = 62$

Pages 347–348   Lesson 8-8

1. Sample answer: Each square represents 3. Sixty squares are shaded. So, 60% of 300 is 60 × 3 or 180. **3.** 200.0 **5.** 308 **7.** $9.10 **9.** 24
11. 9 **13.** 69 **15.** 124.8 **17.** 4 **19.** 10
21. $11.22 **23.** $0.52 billion or $520 million
25. 0.006 **27.** 32 servings

Pages 350–351   Lesson 8-9

1. P is the percentage, B is the base, and r is the number per hundred. **3.** Meredith; 30 is being compared to 20, so 20 is the base. **5.** 7.5%
7. 50% **9.** 125 **11.** 50% **13.** 29.1 **15.** 26.6
17. 174.3 **19.** 16.7% **21.** 150% **23.** 14.7
25. 10% **27.** 60 **29.** 50 grams **33.** 18.7
35. B

Pages 352–355   Study Guide and Assessment

1. e **3.** a **5.** c **7.** d **9.** b **11.** $\frac{5}{2}$ **13.** $\frac{1}{6}$
15. $\frac{30}{11}$ **17.** no **19.** 4 cups per person
21. $4.75 per pound **23.** $9.50 per hour **25.** 75
27. 1,750 **29.** 2.5 **31.** 10 inches **33.** 62.5%
35. $\frac{27}{200}$ **37.** $\frac{87}{200}$ **39.** 32.5% **41.** 0.1575
43. 0.0025 **45.** 5.63 **47.** 475% **49.** 0.95%
51. 16.2 **53.** 50.4 **55.** 0.3 **57.** 6,000 **59.** 39 students **61.** $0.00\overline{3}$

Pages 356–357   Standardized Test Practice

1. C **3.** B **5.** D **7.** D **9.** A **11.** Associative property of equality **13.** −40° **15.** 2,000 mL

CHAPTER 9
Geometry: Investigating Patterns

Pages 360–361   Lesson 9-1A

1. 27° **3.** 74° **5.** You can conveniently read the measure of an acute or obtuse angle from one scale or the other. **7.** 34°: ∠1, ∠4, ∠5, ∠8; 146°:∠2, ∠3, ∠6, ∠7 **9.** 45°:∠3, ∠6, ∠7; 135°:∠1, ∠4, ∠5, ∠8

Pages 364–365   Lesson 9-1

3. Vertical angles are congruent.   **5.** acute
7. straight   **9.** 75°   **11.** straight   **13.** acute
15. acute   **17.** obtuse   **19.** supplementary
21. complementary   **23.** 155°   **25a.** ∠EDG
25b. ∠ABC in 100,000 years or ∠DGF and ∠EDG
today   **27.** A   **29.** $3\frac{5}{12}$

Pages 366–367   Lesson 9-1B

1.

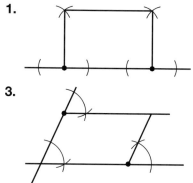

3.

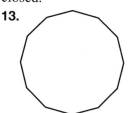

Page 369   Lesson 9-2A

1. straight   **3.** 180°   **5.** Sample answer: 360°;
Since the sum of the measures of each triangle is
180°, there should be 2 × 180° or 360°.   **7a.** 540°
7b. 720°   **7c.** 1,080°   **7d.** 1,440°

Pages 372–373   Lesson 9-2

1. No; angles are not congruent.   **5.** Not a
polygon; sides are not line segments.
7. quadrilateral, regular   **9.** Not a polygon; figure
is not closed.   **11.** Not a polygon; figure is not
closed.

13.   **15.** bleach, dry

17a. *e, f*   **17b.** *a, b* or *b, c*   **17c.** *a, c*   **17d.** *c, d*
17e. *a, d*   **17f.** *e* or *f*   **17g.** *g*   **19.** about 4.5 billion
pounds

Pages 374–375   Lesson 9-2B

1. The angles are 90° and the sides are congruent.
3. Sample answer: Follow the same steps as
inscribing a square, but fold the circle in half one
more time.   **5.** Sample answer: Follow the same
steps to inscribe a hexagon in a circle, but connect
every other intersection point with a line segment.

Pages 378–379   Lesson 9-3

1. $\frac{4}{2.4} = \frac{10}{n}$ **3.** Nikki; $\frac{2}{3} \neq \frac{3}{4}$ **5.** Yes; $\frac{2}{6} = \frac{1}{3}$ and
$\frac{3}{9} = \frac{1}{3}$ **7.** 5 cm   **9.** Yes; $\frac{2}{6} = \frac{1}{3}$ **11.** 8 m
13. 10 km   **15.** 125 feet   **19.** D

Page 379   Mid-Chapter Self Test

1. obtuse   **3.** No; figure is not closed.
5. 5.25 cm

Page 380   Lesson 9-3B

1. enlarged; doubled   **3.** one with smaller squares

Page 381   Lesson 9-4A

1. no   **3.** Triangles are rigid shapes and do not
shift.

Pages 384–385   Lesson 9-4

1.

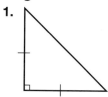

3. Sample answer: Both a rhombus and a square are
parallelograms; they are different because a square
has 4 right angles and a rhombus may not.
5. acute, equilateral   **7.** quadrilateral
9. acute, isosceles   **11.** obtuse, scalene
13. quadrilateral, parallelogram, rhombus
15.

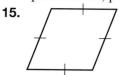

17. right

19a. 60°   **19b.** 65°   **19c.** *a* = 55°, *b* = 65°,
c = 60°, *d* = 30°   **21a.** always   **21b.** sometimes
21c. always   **21d.** sometimes   **21e.** never
23. 2 × 12 = 24

Pages 386–387   Lesson 9-4B

1. Sample answer: Vince is making a rule from a
pattern; Carlos has a rule.   **3.** Carlos   **5a.** The
diagonals are congruent.   **5b.** inductive

9.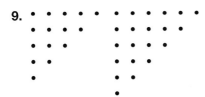

11. 101.7; All of the data are greater than 10.17.
13. C

Pages 390–391 Lesson 9-5

1. The angle measure is a factor of 360°.

3a.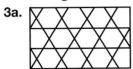

3b. $120° + 60° + 120° + 60° = 360°$ **7.** no; 140°

11. 1 square, 2 octagons;

13. Sample answer: There are no gaps.
15. Yes; the sum of the measures of angles of any triangle is 180°, which is a factor of 360°.

17.

height (thousand feet)	temperature (°F)
1	61.5
2	58.0
3	54.5

Page 394 Lesson 9-6

1. A translation is the result of sliding a figure from one place to another without turning it. **3.** Talutah; the patterns on the top and bottom will not tessellate.

5.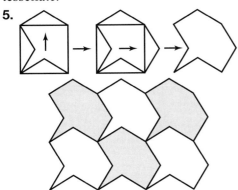

7.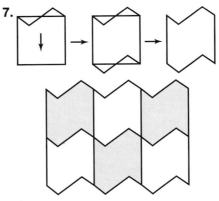

11. Sample answer: birds, fish **13.** No; There is no opposite side to translate the change. **15.** A

Pages 396–397 Lesson 9-7

1. A line of symmetry divides a figure so that one side is the reflection of the other side.

3. 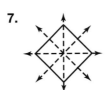 **5.** no lines of symmetry

7.

9.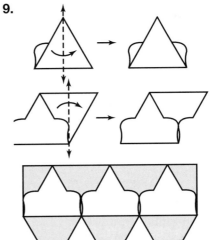

13.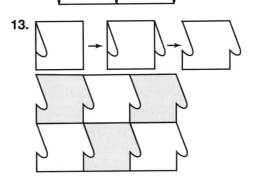

Pages 398–401 Study Guide and Assessment

1. vertex **3.** 180° **5.** regular **7.** equilateral **9.** line symmetry **11.** A translation is a slide of the same pattern over and over. Translating an image that can cover an entire surface is a tessellation. **13.** right **15.** obtuse **17.** acute **19.** Not a polygon; more than 2 sides meet at a vertex. **21.** heptagon, not regular **23.** yes; $\frac{3}{2} = \frac{3}{2}$ **25.** 6 ft **27.** obtuse, isosceles **29.** trapezoid **31.** no, 135°

33.

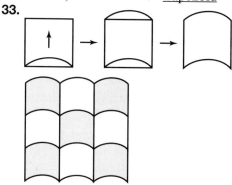

35.

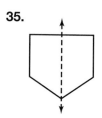

37. isosceles right **39.** 1 hexagon, 4 triangles; 2 hexagons, 2 triangles

Pages 402–403 Standardized Test Practice

1. A **3.** C **5.** C **7.** B **9.** C **11.** 18, 35 **13.** 11

CHAPTER 10
Geometry: Exploring Area

Pages 408–409 Lesson 10-1A

1. Sample answer: Charo picked 6 1-point shots and 10 2-point shots because it is easy to see that $6 + 10 = 16$. The guess results in 26 points which is less than 30. She should decrease the number of 1-point shots and increase the number of 2-point shots.
3. 2 5-card packages and 2 3-card packages
5. Sample answer: The car wash to raise money for band uniforms charged $4 for a car and $6 for a van or truck. During the first hour, they washed 16 vehicles and earned $78. How many cars did they

wash? [answer: 9 cars] **7.** 3, 4 and 5 **9.** 24 ways **11.** 441 units²

Pages 411–413 Lesson 10-1

1. The area of the square is 16 units and each side is 4 units long.

3a. 2 units 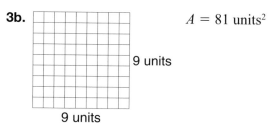 $A = 4$ units²
2 units

3b.

$A = 81$ units²
9 units
9 units

5. 225 **7.** 12 **9.** 18 **11.** 1 **13.** 1,024 **15.** 1,600 **17.** 729 **19.** 11 **21.** 17 **23.** 33 **25.** 24 **27.** 1,444 **29.** 22 in. **31.** 576 ft² **33a.** a square house that is 40 ft by 40 ft **33b.** a square house that is 35 ft by 35 ft **35a.** 50 ft by 50 ft **35b.** 3 bags **37.** C **39.** -5

Pages 416–417 Lesson 10-2

1. 34 is between 25 and 36. Since 34 is closer to 36 than 25, $\sqrt{34}$ is closer to 6 than 5.

3. 4

5. 7 **7.** 3.9 **9.** 15.2 **11.** 3 **13.** 8 **15.** 12 **17.** 11 **19.** 4.5 **21.** 8.5 **23.** 11.2 **25.** 25.4 **27.** $\sqrt{34}$ **29a.** about 23.4 ft **29b.** about 15.5 ft **31.** about 15.8 ft **33.** $\frac{7}{20}$ **35.** 2

Page 418 Lesson 10-3A

1. The sum of the area of the two smaller squares equals the area of the largest square. **3.** The sum of the squares of the lengths of the two perpendicular sides of a right triangle equals the square of the length of the side opposite the right angle.

Pages 420–422 Lesson 10-3

1. $4^2 + 3^2 = 5^2$ **3.** Ricardo; the length of the hypotenuse is 8 m, so the sum of the squares of the other two sides equals 8^2. **5.** 23.7 cm **7.** $6^2 + 3^2 = x^2$; 6.7 m **9.** yes **11.** 15.8 mm **13.** 17.6 m **15.** 7 m **17.** $12^2 + 5^2 = x^2$; 13 cm **19.** $7^2 + x^2 = 18^2$; 16.6 m **21.** $13^2 + x^2 = 18^2$; 12.4 in.

23. no   **25.** yes   **27.** yes   **29.** about 11.5 m
31. about 8.7 ft   **33.** A   **35.** $2^2 \cdot 3^2$

Pages 424–426   Lesson 10-4

1. Some figures are irregular with curved lines and corners that are not square.   **5.** about 46 units^2
7. about 51 units^2 **9.** about 44 units^2 **11.** about 53 units^2 **15.** Yes; if the figure is divided into shapes that include a 8-by-7 rectangle and a 5-by-13 rectangle, the area will be 121 in^2. However the figure is at least 2 squares less than this area, so the area of the irregular shape will be less than 120 in^2.
17a. Sample answer: **17b.** Sample answer:

17c. Sample answer: **19.** C

Page 426   Mid-Chapter Self Test

1. 324   **3.** 16   **5.** 5   **7.** 26 in.   **9.** about 394.0 ft

Page 427   Lesson 10-5A

1. 2 triangles   **5.** The area of a triangle is half the base times the height.

Pages 430–431   Lesson 10-5

1. The area of a triangle is $\frac{1}{2}$ the area of the parallelogram with the same base and height, because the two of these triangles can be formed by drawing a diagonal of the parallelogram.
3.

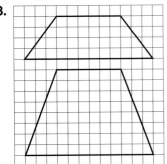

3a. 36 units^2; 72 units^2 **3b.** 1:2   **3c.** 1:2   **5.** 57 cm^2 **7.** 7.5 m^2 **9.** 96 yd^2 **11.** 30 ft^2 **13.** 36 in^2
15. 56 cm^2 **17.** 162 m^2 **19.** 220 cm^2 **21.** 30 in^2
23. 195 ft^2 **25.** $A = 2h$ **27.** hexagon

Pages 434–435   Lesson 10-6

1. Square the radius and multiply by π.   **3.** Briana; if the diameter is 20 cm, the radius is 10 cm. Since $\pi \times 10^2 \approx 314$, the area is about 314 square cm.
5. 78.5 m^2 **7.** 254.5 in^2 **9.** 3.0 m   **11a.** about 15,836.8 ft^2 **11b.** about 363,168.1 ft^2 **11c.** The area of the Superdome is about 23 times the area of the Pantheon.   **13.** 1,385.4 cm^2 **15.** 615.8 in^2
17. 18.1 cm^2 **19.** 907.9 m^2 **21.** 380.1 in^2
23. 132.7 yd^2 **25.** 4.7 in.   **27.** 7.3 cm
29a.

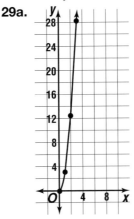

29b. 1:2; 1:4; No, $\frac{1}{2} \neq \frac{1}{4}$.   **29c.** The area is quadrupled; no.   **31.** 14-inch pizza   **33.** 154 cm^2
35.

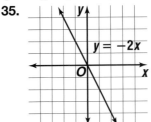

Page 437   Lesson 10-7A

1. About 20 in^2; they are about the same.   **5.** $\frac{2}{7}$

Page 439–441   Lesson 10-7

1. 29 out of 30 random landings should be dry. The sky diver has a good chance of a dry landing.
3a. Sample answer:

3b. Sample answer:

3c. Sample answer:

3d. The spinner in part b; since $\frac{1}{4} = \frac{35}{140}$, $\frac{2}{7} = \frac{40}{140}$, and $20\% = \frac{20}{100} = \frac{1}{5} = \frac{28}{140}$; $\frac{2}{7} > \frac{1}{4} > 20\%$. **5.** $\frac{6}{55}$ or about 0.109 **7.** $\frac{9}{40}$ or 0.225 **9.** $\frac{2}{7}$ or about 0.286 **11.** $\frac{6}{35}$ or about 0.171

13. Sample answer: $\frac{1}{6}$ or about 0.167

15. $\frac{1}{8}$ or 0.125 **17.** $\frac{2,842}{2,847}$ or about 0.998 **19.** $\frac{4}{25}$ or 0.16 **21.** D

Pages 442–445 Study Guide and Assessment

1. 64 **3.** parallel **5.** circle **7.** 81 **9.** radical **11.** right angle **13.** $6^2 = 36$, $7^2 = 49$, and 45 is between 36 and 45. Since 45 is closer to 49 than 36, $\sqrt{45}$ is about 7. **15.** 484 **17.** 2,500 **19.** 0 **21.** 16 **23.** 6 **25.** 7 **27.** 20 **29.** 6.6 yd **31.** 13.3 m **33.** 21 units² **35.** 75 yd² **37.** 36 ft² **39.** 153.9 mm² **41.** 1,963.5 in² **43.** $\frac{1}{4}$ **45.** 36.1 ft **47.** $\frac{1}{5}$

Pages 446–447 Standardized Test Practice

1. B **3.** C **5.** C **7.** C **9.** B **11.** C **13.** $n + 5$ **15.** 120 ft²

CHAPTER 11
Applying Percents

Pages 452–453 Lesson 11-1

1. Sample answer: You can estimate the percent of the shaded portion of the area model. **3.** 40% **5.** 0.4 **7.** $\frac{1}{4}$ **9.** Sample answer: 56(100% + 20%) = 56 + 12, or $68 **11.** $\frac{1}{100} \cdot 2,000 = 20$ lb **13-33.** Sample answers are given. **13.** 20%, 26% **15.** 0.9 **17.** 0.1 **19.** 0.01 **21.** 0.2 **23.** 3.5 **25.** $\frac{1}{4} \cdot 400 = 100$ **27.** $\frac{3}{4} \cdot 120 = 90$ **29.** $0.3 \cdot 50 = 15$ **31.** 50(100% + 50%) =

50 + 25, or 75 **33.** $\frac{1}{5} \cdot 20 = \$4$ **35.** $\frac{7}{10} \cdot 800 = 560$ **37.** $\frac{2}{5} \cdot 120 = 48$ lb **39.** $\frac{16}{40} = \frac{2}{5}$ **41.** 8% **43.** $\frac{1}{16}, \frac{1}{2}, \frac{2}{3}, \frac{5}{6}, \frac{7}{8}$

Pages 454–455 Lesson 11-1B

3. $\frac{1}{20} \cdot 740 = \$37$ billion **7.** $2,500 \cdot \frac{1}{5} = 500$ women **9.** about $4 **11.** C

Pages 457–458 Lesson 11-2

1. Sample answer: The percent equation uses the decimal form of rate; the percent proportion uses rate as a number out of 100. **3.** Sample answer: If the rate and the base are known, it is easier to use the percent equation. **5.** $27 = 0.30 \cdot B$; 90.0 **7.** $P = 0.08 \cdot 38$; 3.0 **9.** $P = 0.16 \cdot 32$; 5.1 **11.** $75 = 0.78 \cdot B$; 96.2 **13.** $45 = R \cdot 36$; 125% **15.** $17 = 0.4 \cdot B$; 42.5 **17.** $P = 0.26 \cdot 48$; 12.5 **19.** $30 = R \cdot 500$; 6% **21.** $3.97 **23.** 26.9% **27.** $0.2 \cdot 40 = 8$ **29.** 40 ft/min

Pages 462–463 Lesson 11-3

1. Change percents to decimal form. Multiply by 360° to obtain the number of degrees in the sections of the circle graph. **3.** No; the sum of the percents does not equal 100. **5a.** Park A, 0.211; Park B, 0.192; Park C, 0.176; Park D, 0.146; Park E, 0.146; Park F, 0.129 **5b.** Park A, 76.0°; Park B, 69.1°; Park C, 63.4°; Park D, 52.6°; Park E, 52.6°; Park F, 46.4°

5c. **Park Tourists**

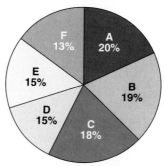

7a. **Life on Other Planets? What Women Think**

7b. Sample answer: A greater percentage of men believe in life on other planets.   **7c.** Sample answer: A graph provides a visual representation that allows you to easily compare data.   **11.** 11.84

Pages 465–467   Lesson 11-4

1. To make a prediction based on a sample, write a ratio to express the results of surveying the sample. Then multiply that ratio by the total population.
3a. 1.86 million   **3b.** 13.02 million   **5a.** Sample answer: No; Students at the University of California may not be representative of all college students because if their college specializes in a field, they will draw more people interested in that field.
5b. about 208   **7.** Sample answer: *Organic Gardening* asked households and NGA asked individuals. They are more likely to find someone in a household who gardens, than to find a particular individual who gardens.   **9a.** 310 people
9b. 16.65 million adults   **11.** 8,799,000 girls and 9,685,000 boys   **13.** B

Page 467   Mid-Chapter Self Test

1. $\frac{1}{5} \cdot 40 = 8$ **3.** $36(100\% + 10\%) = 36 + 3.6$ or 39.6   **5.** $P = 0.09 \cdot 72$; 6.5   **7.** $13.12 = 0.16 \cdot B$; 82.0   **9.** No, this sample is not random, because the students may have all been from the same social group. Their opinion cannot be used to predict what the student body as a whole might believe.

Page 468   Lesson 11-5A

1. 75%
3.

5. 20%; if the rectangle is divided into 5 sections, then 1 of the sections was removed. The decrease is $\frac{1}{5}$, or 20%.

Pages 471–472   Lesson 11-5

1. original amount
3. Sample answer:

5. 10%   **7a.** 38% increase   **7b.** 53% decrease
7c. 112% increase   **9.** 37%   **11.** 29%   **13.** 20%
15. 28%   **17.** 50   **19a.** 1992 and 1993
19b. 60%   **21.** about 810   **23.** E

Pages 475–477   Lesson 11-6

1. The $38 watch would cost $32.30. It would be cheaper than the $50 watch, which would cost $35.
3. Method 1: Multiply the price by 0.06, then add the two amounts. Method 2: Multiply the price by 1.06. Method 2 is more efficient, since it can be done in one step rather than two.   **5.** $47.85
7. $69.55   **9.** vest   **11.** $5.78   **13.** $5.22
15. $18   **17.** $132.81   **19.** $51.20   **21.** $33.71
23. 30%   **25.** $343.64   **27.** $633.61   **29.** about 10%   **31.** 480%

Pages 479–480   Lesson 11-7

1. Multiply $900 \cdot 0.045 \cdot 1$.   **3.** Sample Answer: When you borrow money, the principal is the amount that you borrow, and the interest is the additional amount you pay. When you save money, the principal is the amount you loan to the bank and the interest is what they pay you.   **5.** $43.75
7. $1,732.64   **9.** $61.20   **11.** $1,165.50
13. $22.49   **15.** $176.19   **17.** $171
19. $208.32   **21.** $1,768   **23.** $12   **25.** 10.5%
27. $153   **29.** $\frac{11}{36}$

Page 481   Lesson 11-7B

1. to change the percent to a decimal   **3.** $200
5. B2 = 7, C2 = 0.75; $1,578.75

Pages 482–485   Study Guide and Assessment Answers

1. true   **3.** false; 360°   **5.** false; original   **7.** true
9. 0.8   **11–15.** Samples answers are given.
11. $\frac{1}{10} \cdot 80 = 8$ **13.** $30(100\% + 50\%) = 30 + 15$ or 45   **15.** $\frac{1}{2} \cdot 1,000 = 500$ **17.** $32 = R \cdot 50$; 64%   **19.** $P = 0.62 \cdot 300$; 186   **21.** $108.5 = 0.155 \cdot B$; 700   **23.** 7,800 voters   **25.** 13%
27. 33%   **29.** $1.75   **31.** $440   **33.** $1,500
35. $412.50   **37.** Sample answer: $600,000
39. $5.75

Pages 486–487   Standardized Test Practice

1. A   **3.** D   **5.** C   **7.** B   **9.** E   **11.** $0.085

CHAPTER 12
Geometry: Finding Volume and Surface Area

Pages 490–491 Lesson 12-1A

1.

3. Sample answer:

7. Exercise 1: Yes; you don't need the side view. Exercises 2-4: No, all views are necessary.

9. Sample answer:

top   side   front

Pages 493–495 Lesson 12-1

1. top   side

5. top   side   front

7.

9. top   side   front

11. top   side   front

13. top   side   front

15. **17.**

23. Sample answer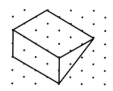

25. D

Pages 496–497 Lesson 12-1B

3.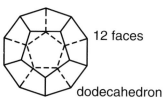

8 faces octahedron

20 faces icosahedron

12 faces dodecahedron

5. Sample answer: They often make small models before starting work on their projects. **7.** Sample answer: 20 in. × 8 in. × 8 in. **9.** 20 boxes **11.** 3.75 hours **13.** D

Pages 500–501 Lesson 12-2

1. m^3

3. Sample answer:

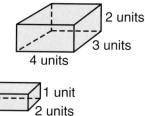

2 units / 2 units / 6 units

2 units / 3 units / 4 units

1 unit / 2 units / 12 units

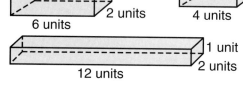

5. 135 in³ **7.** $27 **9.** 1,176 cm³ **11.** 19.8 cm³
13. 6.6 mm³ **15.** 343 in³ **17.** 1,728 in³
19. 3 in. **21.** 8 times greater; 27 times greater
23. 30%

Page 502 Lesson 12-2B

1. Sample answer: about 3 **3.** They are equal.
5. $V = \frac{1}{3}\ell wh$

Pages 505–506 Lesson 12-3

1. Sample answer: In both, you multiply the area of the base by the height. **3.** Cleveland; radius of 3: $V = 169.6$ in³; radius of 6: $V = 339.3$ in³
5. 1,583.4 in³ **7.** 76.0 cm³ **9.** 100.5 in³
11. 11,781.0 ft³ **13.** 162.0 ft³ **15.** 19.6 ft³
17. about 41 feet **19.** 152.88 m³

21.

Page 506 Mid-Chapter Self Test

1.   top   side   front

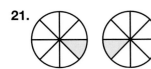

3. 84.8 in³   **5.** rectangular box

Pages 508–509   Lesson 12-4A

1. 6   **3.** 94 square units

5. 18 cm²

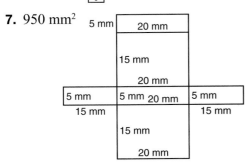

7. 950 mm²

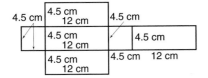

9. 256.5 cm²

11.

13. surface area = $\ell w + \ell w + wh + wh + \ell h + \ell h$

Pages 512–513   Lesson 12-4

1. Find the sum of the areas of the faces.
5. 247.9 in²   **7.** 22 ft²   **9.** 4,200 mm²

11. 167.4 m²   **13.** 235.6 ft² or $235\frac{5}{8}$ ft²
15. surface area = $6x^2$ **19.** 326.7 cm³

Pages 516–517   Lesson 12-5

1. The circumference of the base is the same as the length of the rectangle.   **5.** 942.5 cm²
7. 603.2 mm²   **9.** 149.2 in²   **11.** 92.3 in²
13. 1,105.8 in²   **17a.** 562.0 cm²   **17b.** 653.5 cm²
17c. 562.3 cm²   **17d.** cylinder from Exercise 17a
19. A
21a.

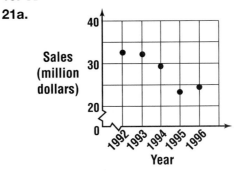

21b. Tickets sales per film have decreased.

Pages 518–521   Study Guide and Assessment

1. perspective   **3.** volume   **5.** multiplying
7. 100.5   **9.** 40
11.   top   side   front

13.   **15.** 168.8 ft³

17. 120 yd³   **19.** 91.2 cm³   **21.** 47.2 in³
23. 302.5 m²   **25.** 1,407.4 cm²   **27.** 79.5 in²
29. 231 in³   **31.** 13.75 ft²

Pages 522–523   Standardized Test Practice

1. D   **3.** C   **5.** B   **7.** C   **9.** C   **11.** $\frac{4}{15}$
13. 300 lunches   **15.** 1,200 ft³

CHAPTER 13
Exploring Discrete Math and Probability

Pages 528–529   Lesson 13-1A

1. 5.75   **3.** Sample answer: Results would

probably vary.

5. Sample answer:

9. 14 inches, 6 inches   **11.** Mauna Kea   **13.** 88, 70

Pages 532–533   Lesson 13-1

1. Experimental probability is the result of collecting data; theoretical probability is the ratio of the number of favorable outcomes to the total number of outcomes.   **3.** Sample answer: rolling a "1" on a number cube   **5.** $\frac{1}{36}$ **7.** $\frac{16}{30} = \frac{8}{15}$ **9.** $\frac{1}{6}$
11. 0   **13.** $\frac{15}{36} = \frac{5}{12}$ **15.** $\frac{2}{9}$ **17.** $\frac{1}{2}$ **19a.** $\frac{1}{10}$
19b. $\frac{1}{25}$ **21.** 44%   **23.** Sample answer: no, however, 60 would be a good estimate of the number of black cards because you would expect that in any large enough sample, about $\frac{1}{4}$ of the cards would be red.   **25.** 7

Pages 535–536   Lesson 13-2

1. Start by listing the choices for the first event. From each choice, draw branches for the choices for the next event. Continue until you have listed all the choices for the final event.   **3.** No; $P(\text{Player 1}) = \frac{3}{4}$, $P(\text{Player 2}) = \frac{1}{4}$ **5.** 12 outcomes
7. 4 outcomes   **9.** 8 outcomes   **11.** 48 outcomes
13. 40 areas
15.

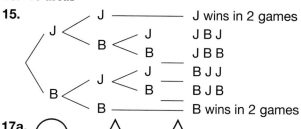

17a.

17b.

17c.

Pages 540–541   Lesson 13-3

1. 12   **5.** 12 outcomes   **7.** 48 outcomes   **9.** 260 outcomes   **11.** 27 outcomes   **13.** No; the number of selections is 312, which is less than 365.
15a. 2   **15b.** 4   **15c.** 8   **15d.** 2^n **17.** D

Pages 544–545   Lesson 13-4

1. Sample answer: choosing a card from a deck of playing cards then choosing a second without replacing the first   **3.** Jared; the result on one number cube does not affect the result on the other.
5. $\frac{1}{4}$ **7.** $\frac{1}{1,296}$ **9.** independent   **11.** $\frac{1}{12}$ **13.** $\frac{1}{24}$
15. $\frac{1}{56}$ **17b.** 3   **17c** 0.096   **19.** B

Page 545   Mid-Chapter Self Test

1. $\frac{1}{10}$
3.
orange — wheat —— orange, wheat
orange — corn —— orange, corn
orange — rice —— orange, rice
apple — wheat —— apple, wheat
apple — corn —— apple, corn
apple — rice —— apple, rice

5. 0.3

Page 546   Lesson 13-5A

1. 3   **3.** 1   **5.** $3 \times 2 \times 1 = 6$

Pages 548–549   Lesson 13-5

1. the product of all the counting numbers beginning with n and counting backward to 1
5. 60   **7.** 15,600 ways   **9.** 24   **11.** 120   **13.** 2
15. 120 ways   **17.** 336 ways   **19.** 6 ways
21. 7,140 minutes, or about 5 days   **23.** A

Page 550   Lesson 13-6A

1. 6 combinations   **3.** 2 ways   **5.** 10, 60, 6, $60 \div 6 = 10$

Pages 552–553   Lesson 13-6

1. The order of the numbers is important in a lock.
3. Miyoki; order is not important in choosing this group.   **5.** combination; 15 ways   **7.** combination; 15 ways   **9.** combination; 120 ways
11. combination; 35 ways   **13.** $\frac{1}{252}$ **15.** 28 line segments   **17.** 120 signals

Pages 554–557 Study Guide and Assessment

1. sample space **3.** experimental
5. permutation **7.** outcomes **9.** $\frac{10}{25}$ or $\frac{2}{5}$
11. $\frac{15}{25}$ or $\frac{3}{5}$ **13.** 12 outcomes **15.** 27 outcomes
17. $\frac{5}{51}$ **19.** $\frac{5}{36}$ **21.** 24 **23.** 720 **25.** 126
27. 10 jogging suits **29.** 45 ways

Pages 558–559 Standardized Test Practice

1. D **3.** B **5.** D **7.** A **9.** C **11.** $\frac{2}{5}$
13. $\frac{1}{4}$

Photo Credits

Veneklasen; **359** Matt Bradley/Tom Stack & Assoc.; **362** Mark E. Gibson; **368** (t)Michael Hirst, (bc)Dan Lecca, (br)Timothy Fuller, (bkgd)Thomas Veneklasen; **370** (b)Jeff Smith/FOTOSMITH; **376** WorldSat International, Science Source/Photo Researchers; **379** Rich Iwasaki/Tony Stone Images; **383** Peter Pearson/Tony Stone Images; **384** Tony Stone Images; **385** Scott Berner/Visuals Unlimited; **386** Timothy Fuller; **388** Thomas Veneklasen; **390** Mark E. Gibson; **392** *Symmetry Drawing E72* by M.C. Escher. ©1997 Cordon Art - Baarn - Holland. All rights reserved; **395** Steve Dunwell/The Image Bank; **397** (l)Fred Bavendam/Peter Arnold, Inc., (r)Secret Sea Visions/Peter Arnold, Inc.; **404-405** Scott Camazine/Photo Researchers; **405** L. West/Photo Researchers; **406** (t)Sean Ellis/Tony Stone Images, (b)Benelux Press B.V./Photo Researchers; **406-407** Earth Imaging/Tony Stone Images; **407** Mary Evans Picture Library/Photo Researchers; **408** Jeff Smith/FOTOSMITH; **409** Allsport USA/Al Bello; **410** Dan Ham/Tony Stone Images; **411 412** David Madison; **413** Larry Lefever/Grant Heilman Photography; **416** Brian Seed/Tony Stone Images; **417** David Madison; **419** (t)William Katz/Photo Researchers, (b)Hulton Getty/Tony Stone Images; **421** Timothy Fuller; **422** Jim Steinberg/Photo Researchers; **423** A.L. Parnes/Photo Researchers; **427** Timothy Fuller; **432** AP/Wide World Photos; **436** Dominic Oldershaw; **438** AP/Wide World Photos; **439** Dominic Oldershaw; **441** Jeff Smith/FOTOSMITH; **448** (l)Hulton Getty Images/Tony Stone Images, (r)Jay Thomas/International Stock; **448-449** Timothy Fuller; **450** James A. Sugar/National Geographic Image Collection; **451** Timothy Fuller; **453** The Stock Shop/Medichrome/Vincent Perez; **454** Timothy Fuller; **456** Guido A. Rossi/The Image Bank; **457** Michael Zito/SportsChrome; **459** Timothy Fuller; **461** Phil Degginger/Color-Pic; **464** courtesy U.S. Census Bureau;

465 Timothy Fuller; **466** D. Cavagnaro/Visuals Unlimited; **469** Phil Schofield/National Geographic Image Collection; **470** Dominic Oldershaw; **472** Dominic Oldershaw; **473** (t)Timothy Fuller, (bl)Tracy Aiguier, (br)Timothy Fuller; **474** Dominic Oldershaw; **475** Timothy Fuller; **476** (t)Dominic Oldershaw, (b)Franklin Over; **478** Dominic Oldershaw; **480** Jeff Hunter/The Image Bank; **488** (t)Art Wolfe/Tony Stone Images, (c)S.J. Krasemann/Peter Arnold, Inc., (b)Walter H. Hodge/Peter Arnold, Inc.; **488-489** Jeff Smith/FOTOSMITH; **489** (t)Catherine Ursillo/Photo Researchers, (b)E.R. Degginger; **490** E.R. Degginger; **491** Dominic Oldershaw; **492** (l)Mandolin Brothers/Photo Researchers, (r)G. Randall/FPG; **495** (t)Jose L. Pelaez/Stock Market, (b)Tribune Media Services, Inc. All Rights Reserved. Reprinted with permission; **496** Dominic Oldershaw; **498** Jeff Smith/FOTOSMITH; **499** SuperStock; **502** John Lawrence/Tony Stone Images; **503** Jeff Smith/ FOTOSMITH; **504** Dominic Oldershaw; **507** (t,b)Dominic Oldershaw, (c)Charles Harrington, (bkgd)Frans Lanting/Tony Stone Images; **510** (l)Alon Reininger/Leo de Wys, Inc., (r)Peter Bennett/The Viesti Collection; **524-525** Dominic Oldershaw; **526** (l)Dominic Oldershaw; (r)CBS/Monty Brinton; **526-527** Dominic Oldershaw; 527 (t)Aaron Haupt, (b)Dominic Oldershaw; **528 530** Dominic Oldershaw; **535** Dominic Oldershaw; **537** (bl)Rex Features, (others) Timothy Fuller; **538 540** Timothy Fuller; **542** (t)Archive Photos, (b)Nigel Cattlin/Holt Studios International/Photo Researchers; **544** Dominic Oldershaw; **547** PEANUTS reprinted by permission of United Feature Syndicate, Inc.; **548** E.R. Degginger; **549** AP/Wide World Photos; **551 552** Dominic Oldershaw; **553** Uniphoto; **561** (t)NASA, (cl)The Denver Art Museum, (c)Jeff Smith/FOTOSMITH, (cr)Jeff Hunter/The Image Bank, (bl)Mark Steinmetz/Amanita Pictures, (br)Ron Kimball.

Glossary

GLOSSARY

A

absolute value (185)   The number of units a number is from zero on the number line.

acute (362)   An angle with a measure greater than 0° and less than 90°.

addition property of equality (229)   If you add the same number to each side of an equation, the two sides remain equal. For any numbers a, b, and c, if $a = b$, then $a + c = b + c$.

additive inverse (197)   An integer and its opposite. The sum of an integer and its additive inverse is zero.

algebra (12)   A mathematical language that uses variables along with numbers. The variables stand for numbers that are unknown. $10n - 3 = 17$ is an example of an algebra problem.

algebraic expression (12)   A combination of variables, numbers, and at least one operation.

angle (362)   Two rays with a common endpoint form an angle. The rays and vertex are used to name an angle.

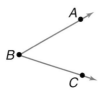

angle *ABC* or ∠*ABC*

area (30)   The number of square units needed to cover a surface enclosed by a geometric figure.

arithmetic average (102)   The mean of a set of data.

arithmetic sequence (142)   A sequence in which the difference between any two consecutive terms is the same.

associative property of addition (301)   For any numbers a, b, and c, $(a + b) + c = a + (b + c)$.

associative property of multiplication (301)   For any numbers a, b, and c, $(ab)c = a(bc)$.

B

back-to-back stem-and-leaf plot (109)   Used to compare two sets of data. The leaves for one set of data are on one side of the stem and the leaves for the other set of data are on the other side.

bar graph (94)   A graphic form using bars to make comparisons of statistics.

bar notation (70)   In repeating decimals the line or bar placed over the digits that repeat. For example, $2.\overline{63}$ indicates the digits 63 repeat.

base (17)   In a power, the number used as a factor. In 5^3, the base is 5. That is, $5^3 = 5 \times 5 \times 5$.

base (349)   In a percent proportion, the number to which the percentage is compared.

base (31)   Any side of a parallelogram.

box-and-whisker plot (114)   A diagram that summarizes data using the median, the upper and lower quartiles, and the extreme values. A box is drawn around the quartile value and whiskers extend from each quartile to the extreme data points.

C

capture-recapture technique (329)   A method used to estimate animal populations. The following proportion is used to estimate the entire population by first capturing a part of the population, tagging the sample, releasing the sample, and then recapturing another sample.

$$\frac{\text{original number captured}}{\text{total population (P)}} = \frac{\text{tagged in sample}}{\text{recaptured}}$$

cell (137)   The basic unit of a spreadsheet. A cell can contain data, labels, or formulas.

center (297)   The given point from which all points on a circle or a sphere are the same distance.

circle (297)   The set of all points in a plane that are the same distance from a given point called the center.

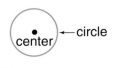

circle graph (460)   A type of statistical graph used to compare parts of a whole.

circumference (297)   The distance around a circle.

cluster (98)   Data that are grouped closely together.

clustering (51)   A method used to estimate decimal sums and differences by rounding a group of closely related numbers to the same whole number.

coefficient (234)   The numerical part of an expression.

combination (551)   An arrangement or listing of objects in which order is not important.

common denominator (172)   A common multiple of the denominators of two or more fractions. 24 is a common denominator for $\frac{1}{3}$, $\frac{5}{8}$, and $\frac{3}{4}$ because 24 is the LCM of 3, 8, and 4.

commutative property of addition (301)   For any numbers a and b, $a + b = b + a$.

commutative property of multiplication (301)   For any numbers a and b, $ab = ba$.

complementary (362)   Two angles are complementary if the sum of their measures is 90°.

composite number (138)   Any whole number greater than 1 that has more than two factors.

compound event (542)   A compound event consists of two or more simple events.

congruent (371)   Line segments that have the same length, or angles that have the same measure, or figures that have the same size and shape.

coordinate system (191)   A plane in which a horizontal number line and a vertical number line intersect at their zero points.

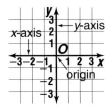

Counting Principle (538)   A method for finding the number of ways that two or more events can occur by multiplying the number of ways that each event can occur.

cross products (325)   The products of the terms on the diagonals when two ratios are compared. If the cross products are equal, then the ratios form a proportion. In the proportion $\frac{3}{6} = \frac{4}{8}$, the cross products are 3×8 and 6×4.

cubed (17)   The product in which a number is a factor three times. Two cubed is 8 because $2 \times 2 \times 2 = 8$.

cup (290)   A customary unit of capacity equal to 8 fluid ounces.

cylinder (503)   A three-dimensional figure with two parallel congruent circular bases.

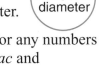

D

decagon (370)   A polygon having ten sides.

defining a variable (243)   Choosing a variable and a quantity for the variable to represent in an equation.

degree (362)   The most common unit of measure for angles.

dependent events (543)   Two or more events in which the outcome of one event does affect the outcome of the other event(s).

diameter (297)   The distance across a circle through its center.

distributive property (302)   For any numbers a, b, and c, $a(b + c) = ab + ac$ and $(b + c)a = ba + ca$.

divisible (133)   A number is divisible by another if, upon division, the remainder is zero.

division property of equality (234)   For any numbers a, b, and c, with $c \neq 0$, if $a = b$, then $\frac{a}{c} = \frac{b}{c}$.

equation (21)   A mathematical sentence that contains the equal sign, =.

equilateral (382)   All sides of a figure are congruent.

equivalent ratios (318)   Two ratios that have the same value.

evaluate (12)   To find the value of an expression by replacing variables with numerals.

event (165)   A specific outcome or type of outcome.

experimental probability (531)   An estimated probability based on the relative frequency of positive outcomes occurring during an experiment.

exponent (17)   In a power, the number of times the base is used as a factor. In 5^3, the exponent is 3. That is, $5^3 = 5 \times 5 \times 5$.

factor (17, 133)   A number that divides into a whole number with a remainder of zero.

factorial (547)   The expression $n!$ is the product of all counting numbers beginning with n and counting backward to 1.

factor tree (138)   A diagram showing the prime factorization of a number. The factors branch out from the previous factors until all the factors are prime numbers.

fair game (535)   A game in which players have an equal chance of winning.

fractal (24)   An geometric figure that is made when a rule is applied to smaller and smaller parts. The parts of a fractal are similar to the whole figure.

frequency table (88)   A table for organizing a set of data that shows the number of times each item or number appears.

function (249)   A relation in which each element of the input is paired with exactly one element of the output according to a specified rule.

gallon (290)   A customary unit of capacity equal to 4 quarts.

geometric sequence (142)   A sequence of numbers in which you can find the next term by multiplying the previous term by the same number.

gram (75)   A unit of mass in the metric system.

greatest common factor (GCF) (150)   The greatest of the common factors of two or more numbers. The GCF of 18 and 24 is 6.

height (31)   The shortest distance from the base of a parallelogram to its opposite side.

heptagon (370)   A polygon having seven sides.

hexagon (370)   A polygon having six sides.

hypotenuse (419)   The side opposite the right angle in a right triangle.

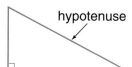

identity property of addition (301)   For any number a, $a + 0 = a$.

identity property of multiplication (301)   For any number a, $a \times 1 = a$.

independent events (542)   Two or more events in which the outcome of one event does *not* affect the outcome of the other event(s).

indirect measurement (377)   Finding a measurement by using similar triangles and writing a proportion.

inequality (246)   A mathematical sentence that contains $<$, $>$, $\neq$, $\leq$, or $\geq$.

inner measure (423)   The number of whole squares within a figure.

integer (184)   The whole numbers and their opposites.
$$\ldots, -3, -2, -1, 0, 1, 2, 3, \ldots$$

GLOSSARY

interquartile range (112, 115)   The range of the middle half of a set of numbers.
Interquartile range = $UQ - LQ$.

interval (88)   The difference between successive values on a scale.

irregular figure (423)   A figure that does not have straight sides and square corners.

isosceles (382)   An isosceles triangle has two congruent sides.

leaf (108)   The second greatest place value of data in a stem-and-leaf plot.

least common denominator (LCD) (172)   The least common multiple of the denominators of two or more fractions.

least common multiple (LCM) (169)   The least of the common multiples of two or more numbers. The LCM of 2 and 3 is 6.

leg (419)   Either of the two sides that form the right angle of a right triangle.

linear equation (255)   An equation for which the graph is a straight line.

line graph (94)   A type of statistical graph using lines to show how values change over a period of time.

line of symmetry (395)   A line that divides a figure into two halves that are reflections of each other.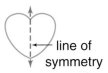

line plot (98)   A graph that uses an × above a number on a number line each time that number occurs in a set of data.

line symmetry (395)   Figures that match exactly when folded in half have line symmetry.

liter (75)   The basic unit of capacity in the metric system. A liter is a little more than a quart.

lower extreme (114)   The least number of a set of data.

lower quartile (114)   The median of the lower half of a set of numbers indicated by *LQ*.

mean (102)   The sum of the numbers in a set of data divided by the number of pieces of data.

median (102)   The middle number in a set of data when the data are arranged in numerical order. If the data has an even number, the median is the mean of the two middle numbers.

meter (74)   The basic unit of length in the metric system.

metric system (74)   A base-ten system of measurement using the basic units: meter for length, gram for mass, and liter for capacity.

mode (102)   The number(s) or item(s) that appear most often in a set of data.

modeling (22)   Writing an equation that represents a real-world problem.

multiple (169)   The product of the number and any whole number.

multiplication property of equality (302)   If each side of an equation is multiplied by the same number, then the two sides remain equal. If $a = b$, then $ac = bc$.

multiplicative inverse (301)   A number times its multiplicative inverse is equal to 1. The multiplicative inverse of $\frac{2}{3}$ is $\frac{3}{2}$.

negative integer (184)   Integer that is less than zero.

nonagon (370)   A polygon having nine sides.

obtuse (362)   Any angle that measures greater than 90° but less than 180°.

octagon (370)   A polygon having eight sides.

opposite (184) Two integers are opposite if they are represented on the number line by points that are the same distance from zero, but on opposite sides of zero. The sum of opposites is zero.

order of operations (18) The rules to follow when more than one operation is used.
1. Do all operations within grouping symbols first.
2. Do all powers before other operations.
3. Multiply and divide in order from left to right.
4. Add and subtract in order from left to right.

ordered pair (191) A pair of numbers used to locate a point in the coordinate system. The ordered pair is written in this form: (x-coordinate, y-coordinate).

origin (191) The point of intersection of the x-axis and y-axis in a coordinate system.

ounce (289) A customary unit of weight. 16 ounces equals 1 pound.

outcome (531) One possible result of a probability event. For example, 4 is an outcome when a number cube is rolled.

outer measure (423) The number of squares within and containing part of the figure.

outlier (115) Data that is more than 1.5 times the interquartile range from the quartiles.

P

parallelogram (31) A quadrilateral with two pairs of parallel sides.

pentagon (370) A polygon having five sides.

percent (158) A ratio that compares a number to 100.

percentage (349) In a percent proportion, a number that is compared to another number called the base.

percent proportion (349) $\frac{P}{B} = \frac{r}{100}$ where P represents the percentage, B represents the base, and r represents the number per hundred.

perfect square (411) A number whose square root is a whole number. 25 is a perfect square because $\sqrt{25} = 5$.

perimeter (292) The distance around a geometric figure.

permutation (547) An arrangement or listing in which order is important.

perspective (492) A perspective view of a three-dimensional figure is an angled view that shows the three dimensions of the figure.

pint (290) A customary unit of capacity equal to two cups.

polygon (370) A simple closed figure in a plane formed by three or more line segments.

population (5, 464) The entire group of items or individuals from which the samples under consideration are taken.

population density (322) The population per square mile.

positive integer (184) Integer that is greater than zero.

pound (289) A customary unit of weight equal to 16 ounces.

power (17) A number that can be written using an exponent. The power 7^3 is read *seven to the third power*, or *seven cubed*.

prime factorization (138) Expressing a composite number as the product of prime numbers. For example, the prime factorization of 63 is $3 \times 3 \times 7$.

prime number (138) A whole number greater than 1 that has exactly two factors, 1 and itself.

principal (478) The amount of an investment or a debt.

probability (165) The chance that some event will happen. It is the ratio of the number of ways a certain event can occur to the number of possible outcomes.

property of proportions (325) If $\frac{a}{b} = \frac{c}{d}$, then $ad = bc$. If $ad = bc$, then $\frac{a}{b} = \frac{c}{d}$.

proportion (325) An equation that shows that two ratios are equivalent, $\frac{a}{b} = \frac{c}{d}$, $b \neq 0$, $d \neq 0$.

Pythagorean Theorem (419) In a right triangle, the square of the length of the hypotenuse is equal to the sum of the squares of the lengths of the legs. $c^2 = a^2 + b^2$

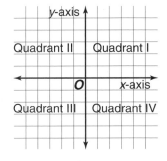

Q

quadrant (191) One of the four regions into which two perpendicular number lines separate the plane.

quadrilateral (370) A polygon having four sides.

quart (290) A customary unit of capacity equal to two pints.

quartile (112) One of four equal parts of data from a large set of numbers.

R

radical sign (411) The symbol used to indicate a nonnegative square root is $\sqrt{}$.

radius (297) The distance from the center of a circle to any point on the circle.

random (165) Outcomes occur at random if each outcome is equally likely to occur.

random (464) A sample is called random if the members of the sample are selected purely on the basis of chance.

range (88) The difference between the greatest number and the least number in a set of data.

rate (321) A ratio of two measurements having different units.

rate (349) In a percent proportion, the ratio of a number to 100.

rate (478) The percent charged or paid for the use of money.

ratio (154, 317) A comparison of two numbers by division. The ratio of 2 to 3 can be stated as 2 out of 3, 2 to 3, 2:3, or $\frac{2}{3}$.

reciprocal (301) The multiplicative inverse of a number.

rectangle (30) A quadrilateral with four equal angles.

rectangular prism (498) A prism with rectangular bases.

reflection (215, 395) A type of transformation where a figure is flipped over a line of symmetry.

regular polygon (371) A polygon having all sides congruent and all angles congruent.

repeating decimal (70) A decimal whose digits repeat in groups of one or more. Examples are 0.181818... and 0.8333... .

rhombus (383) A parallelogram with four congruent sides.

right (362) An angle that measures 90°.

S

sample (5, 329, 464) A randomly-selected group that is used to represent a whole population.

sample space (531) The set of all possible outcomes.

scale (88) The set of all possible values of a given measurement, including the least and greatest numbers in the set, separated by the intervals used.

scale (332) On a map, intervals used representing the ratio of distance on the map to the actual distance.

scale drawing (332) A drawing that is similar but either larger or smaller than the actual object.

scalene (382) A triangle with no congruent sides.

scatter plot (92, 95)   In a scatter plot, two sets of related data are plotted as ordered pairs on the same graph.

scientific notation (77)   A way of expressing a number as the product of a number that is at least 1 but less than 10 and a power of 10. For example, $687{,}000 = 6.87 \times 10^5$.

sequence (142)   A list of numbers in a certain order, such as, 0, 1, 2, 3, or 2, 4, 6, 8.

similar (376)   Figures that have the same shape but not necessarily the same size.

simple interest (478)   The amount paid for the use of money. The formula for simple interest is $I = prt$.

simplest form (154)   A fraction is in simplest form when the GCF of the numerator and the denominator is 1.

solids (493)   Three-dimensional figures. Prisms, pyramids, cones, and cylinders are examples of solids.

solution (21)   A value for the variable that makes an equation true. The solution for $12 = x + 7$ is 5.

solve (21)   To replace a variable with a number that makes an equation true.

spreadsheet (137)   A tool used for organizing and analyzing data.

square (410)   The product of a number and itself. 36 is the square of 6×6.

squared (17)   A number multiplied by itself. 7^2 is read *7 squared*.

square root (411)   One of the two equal factors of a number. If $a^2 = b$, then a is the square root of b. The square root of 144 is 12 since $12^2 = 144$.

stem (108)   The greatest place value common to all the data values is used for the stem of a stem-and-leaf plot.

stem-and-leaf plot (108)   A system used to condense a set of data where the greatest place value of the data forms the stem and the next greatest place value forms the leaves.

straight (362)   An angle is straight if it measures exactly 180°.

subtraction property of equality (228)   If you subtract the same number from each side of an equation, the two sides remain equal. For any numbers a, b, and c, if $a = b$, then $a - c = b - c$.

supplementary (362)   Two angles are supplementary if the sum of their measures is 180°.

surface area (510)   The sum of the areas of all the surfaces (faces) of a three-dimensional figure.

term (239)   A number, a variable, or a product of numbers and variables.

term (142)   Each number within a sequence is called a term.

terminating decimal (70)   A quotient in which the division ends with a remainder of zero. 0.25 and 0.125 are terminating decimals.

tessellation (388)   A repetitive pattern of polygons that fit together with no holes or gaps.

theoretical probability (531)   The ratio of the number of ways an event can occur to the number of possible outcomes.

time (478)   When used to calculate interest, time is given in years.

ton (289)   A customary unit of weight equal to 2,000 pounds.

transformation (215)   Movements of geometric figures.

translation (215, 392)   One type of transformation where a figure is slid horizontally, vertically, or both.

trapezoid (383, 429)   A quadrilateral with exactly one pair of parallel sides.

tree diagram (534)   A diagram used to show the total number of possible outcomes in a probability experiment.

GLOSSARY

triangle (370, 428) A polygon that has three angles.

unit rate (321) A rate with denominator of 1.

upper extreme (114) The greatest number of a set of data.

upper quartile (114) The median of the upper half of a set of numbers.

variable (12) A symbol, usually a letter, used to represent a number in mathematical expressions or sentences. In $3 + a = 6$, a is a variable.

vertex (362) A vertex of an angle is the common endpoint of the rays forming the angle.

vertex

volume (498) The number of cubic units needed to fill the space occupied by a solid.

x-axis (191) The horizontal number line which helps to form the coordinate system.

x-coordinate (191) The first number of an ordered pair.

y-axis (191) The vertical number line which helps to form the coordinate system.

y-coordinate (191) The second number of an ordered pair.

zero pair (196, 227) The result of pairing one positive counter with one negative counter.

GLOSSARY

Spanish Glossary

SPANISH GLOSSARY

absolute value / valor absoluto (185)
Número de unidades en la recta numérica que un número dista de cero.

acute / agudo (362)   Ángulo que mide más de 0° y menos de 90°.

addition property of equality / propiedad de adición de la igualdad (229)   Si sumas el mismo número a ambos lados de una ecuación, los lados permanecen iguales. Para números a, b y c cualesquiera, si $a = b$, entonces $a + c = b + c$.

additive inverse / inverso aditivo (197)
Opuesto de un entero. La suma de un entero y su inverso aditivo es cero.

algebra / álgebra (12)   Lenguaje matemático que usa letras y números. Las letras representan números desconocidos. $10n - 3 = 17$ es un ejemplo de un problema de álgebra.

algebraic expression / expresión algebraica (12)   Combinación de variables, números y al menos una operación.

angle / ángulo (362)   Dos rayos con un extremo común forman un ángulo. Los rayos y el vértice se usan para nombrar o identificar el ángulo.

ángulo ABC o $\angle ABC$

area / área (30)   Número de unidades cuadradas que se requieren para cubrir la superficie encerrada por una figura geométrica.

arithmetic average / promedio aritmético (102)   La media de un conjunto de datos.

arithmetic sequence / sucesión aritmética (142)   Sucesión en que la diferencia entre dos términos consecutivos es constante.

associative property of addition / propiedad asociativa de la adición (301)
Para números a, b y c cualesquiera, $(a + b) + c = a + (b + c)$.

associative property of multiplication / propiedad asociativa de la multiplicación (301)   Para números a, b y c cualesquiera, $(ab)c = a(bc)$.

back-to-back stem-and-leaf plot /diagrama de tallo y hojas consecutivo (109)   El que se usa para comparar dos conjuntos de datos. Las hojas de uno de los conjuntos de datos se escriben a un lado del tallo y las del segundo conjunto de datos al otro lado del tallo.

bar graph / gráfica de barras (94)   Tipo de gráfica que usa barras para comparar estadísticas.

bar notation / notación de barra (70)   En los decimales periódicos, la línea o barra que se escribe encima de los dígitos que se repiten. En 2.6$\overline{3}$, por ejemplo, la barra encima de 63 indica que el bloque de dos dígitos, 63, se repite indefinidamente.

base / base (17)   Número que se usa como factor en una potencia. En 5^3, la base es 5, es decir, $5^3 = 5 \times 5 \times 5$.

base / base (349)   Número con que se compara el porcentaje en una proporción porcentual.

base / base (31)   Cualquier lado de un paralelogramo.

box-and-whisker plot / diagrama de caja y patillas (114)   Diagrama que resume información usando la mediana, los cuartiles superior e inferior y los valores extremos. Se dibuja una caja alrededor de los cuartiles y se trazan patillas que los unan a los valores extremos respectivos.

capture-recapture technique / técnica de captura-recaptura (329)   Método que se usa para estimar poblaciones de animales. El procedimiento a seguir es capturar una muestra, marcarla y devolverla a su hábitat. Más tarde, se captura otra muestra. La siguiente proporción se usa para estimar el tamaño de la población.

$$\frac{\text{número capturado inicialmente}}{\text{población total } (p)} = \frac{\text{número marcado en la muestra}}{\text{recapturados}}$$

cell / celda (137)   Unidad básica de una hoja de cálculos. Las celdas pueden contener datos, rótulos o fórmulas.

center / centro (297)   Punto en el plano, del cual equidistan todos los puntos de un círculo o de una esfera.

circle / círculo (297)
Conjunto de todos los puntos en un plano que equidistan de un punto dado llamado centro.

circle graph / gráfica circular (460)   Tipo de gráfica estadística que se usa para comparar las partes de un todo.

circumference / circunferencia (297)   La distancia alrededor de un círculo.

cluster / agrupamiento (98)   Datos estrechamente agrupados.

clustering / agrupar (51)   Método que se usa para estimar sumas y restas de decimales, redondeando al mismo número entero un grupo de números estrechamente relacionados.

coefficient / coeficiente (234)   Parte numérica de un término.

combination / combinación (551)   Arreglo o lista de objetos en que el orden no es importante.

common denominator / denominador común (172)   Múltiplo común de los denominadores de dos o más fracciones. 24 es un denominador común de $\frac{1}{3}$, $\frac{5}{8}$ y $\frac{3}{4}$, porque 24 el mcm de 3, 8 y 4.

commutative property of addition / propiedad conmutativa de la adición (301)   Para números a y b cualesquiera, $a + b = b + a$.

commutative property of multiplication / propiedad conmutativa de la multiplicación (301)   Para números a y b cualesquiera, $ab = ba$.

complementary / complementarios (362)   Dos ángulos son complementarios si la suma de sus medidas es 90°.

composite number / número compuesto (138)   Cualquier número entero mayor que 1 que posee más de dos factores.

compound event / evento compuesto (542)   Un evento compuesto consiste en dos o más eventos simples.

congruent / congruentes (371)   Segmentos de recta que tienen la misma longitud; ángulos que tienen la misma medida; figuras que tienen la misma forma y tamaño.

coordinate system / sistema de coordenadas (191)   Plano en el cual se han trazado dos rectas numéricas, una horizontal y una vertical, que se intersecan en sus puntos cero.

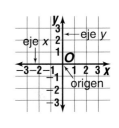

Counting Principle / Principio de Conteo (538)   Método para calcular el número de maneras en que dos o más eventos pueden ocurrir, lo cual se logra multiplicando entre sí el número de maneras en que puede ocurrir cada evento individualmente.

cross products / productos cruzados (325)   Los productos que resultan de la comparación de los términos de las diagonales de dos razones. Las razones forman una proporción si y sólo si los productos son iguales. En la proporción $\frac{3}{6} = \frac{4}{8}$, los productos cruzados son 3×8 y 6×4.

cubed / al cubo (17)   Producto de un número por sí mismo tres veces. Dos al cubo es 8 ya que $2 \times 2 \times 2 = 8$.

cup / taza (290)   Unidad de capacidad del sistema inglés de medidas que equivale a 8 onzas líquidas.

cylinder / cilindro (503)   Figura tridimensional que tiene dos bases circulares congruentes y paralelas.

decagon / decágono (370)   Polígono de diez lados.

defining a variable / definir una variable (243)   El elegir una variable y una cantidad que esté representada por la variable en una ecuación.

degree / grado (362)   La unidad de medida angular más común.

dependent events / eventos dependientes (543)   Dos o más eventos en que el resultado de uno de ellos afecta el resultado de otros eventos.

diameter / diámetro (297)   La longitud de cualquier segmento de recta cuyos extremos yacen en un círculo y que pasa por su centro.

diámetro

distributive property / propiedad distributiva (302)   Para números a, b y c cualesquiera, $a(b + c) = ab + ac$ y $(b + c)a = ba + ca$.

divisible / divisible (133)   Un número es divisible entre otro si, después de dividirlos, el residuo es cero.

division property of equality / propiedad de división de la igualdad (234)   Para números a, b y c cualesquiera, con $c \neq 0$, si $a = b$, entonces $\frac{a}{c} = \frac{b}{c}$.

equation / ecuación (21)   Enunciado matemático que contiene el signo de igualdad, $=$.

equilateral / equilátero (382)   Figura en el plano que tiene todos sus lados congruentes entre sí.

equivalent ratios / razones equivalentes (318)   Dos razones que tienen el mismo valor.

evaluate / evaluar (12)   Calcular el valor de una expresión sustituyendo las variables con números.

event / evento (165)   Resultado específico o tipo de resultado de un experimento probabilístico.

experimental probability / probabilidad experimental (531)   Probabilidad de un evento que se calcula o estima basándose en la frecuencia relativa de los resultados favorables al evento en cuestión, que ocurren durante un experimento probabilístico.

exponent / exponente (17)   Número de veces que la base de una potencia se usa como factor. En 5^3, el exponente es 3, o sea, $5^3 = 5 \times 5 \times 5$.

factor / factor (17, 133)   Número entero que divide otro número entero con un residuo de 0.

factorial / factorial (547)   La expresión $n!$ es el producto de los n primeros números de contar, contando al revés.

factor tree / árbol de factores (138)   Diagrama que sirve para encontrar la factorización prima de un número. Los factores se ramifican de los factores anteriores hasta que todos los factores son primos.

fair game / juego justo (535)   Juego en que los jugadores tienen la misma oportunidad de ganar.

fractal / fractal (24)   Figura geométrica que se construye aplicando una regla a partes más y más pequeñas de la figura. Las partes de un fractal son semejantes a la figura entera.

frequency table / tabla de frecuencia (88)   Tabla que se usa para organizar un conjunto de datos y que muestra cuántas veces aparece cada dato.

function / función (249)   Relación en que cada elemento de entrada es apareado con un único elemento de salida, según una regla específica.

gallon / galón (290)   Unidad de capacidad del sistema inglés de medidas que equivale a 4 cuartos de galón.

geometric sequence / sucesión geométrica (142)   Sucesión de números en la cual se puede calcular cualquier término, a partir del segundo, multiplicando el término anterior por el mismo número.

gram / gramo (75)   Unidad de masa del sistema métrico.

greatest common factor (GCF) / máximo común divisor (MCD) (150)   El mayor factor común de dos o más números. El MCD de 18 y 24 es 6.

height / altura (31)   La distancia más corta desde la base de un paralelogramo hasta su lado opuesto.

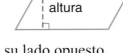

heptagon / heptágono (370)   Polígono de siete lados.

hexagon / hexágono (370)   Polígono de seis lados.

hypotenuse / hipotenusa (419)   El lado de un triángulo rectángulo opuesto a su ángulo recto.

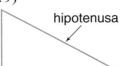

identity property of addition / propiedad de identidad de la adición (301)   Para cualquier número a, $a + 0 = a$.

identity property of multiplication / propiedad de identidad de la multiplicación (301)   Para cualquier número a, $a \times 1 = a$.

independent events / eventos independientes (542)   Dos o más eventos en que el resultado de uno de ellos *no* afecta el resultado de los otros eventos.

indirect measurement / medida indirecta (377)   Cálculo de una medida a partir de triángulos semejantes y proporciones.

inequality / desigualdad (246)   Enunciado matemático que contiene $<$, $>$, $\neq$, $\leq$ o $\geq$.

inner measure / medida interior (423)   El número de cuadrados enteros que contiene una figura.

integer / entero (184)   Los números enteros no negativos y sus opuestos.
$\ldots, -3, -2, -1, 0, 1, 2, 3, \ldots$

interquartile range / amplitud intercuartílica (112, 115)   El rango de la mitad central de un conjunto de datos o números.
Amplitud intercuartílica $= CS - CI$

interval / intervalo (88)   Diferencia entre valores sucesivos en una escala.

irregular figure / figura irregular (423)   Figura que carece de lados rectos y esquinas cuadradas.

isosceles / isósceles (382)   Triángulo que tiene dos lados congruentes.

leaf / hoja (108)   El segundo valor de posición mayor en un diagrama de tallo y hojas.

least common denominator (LCD) / mínimo común denominador (mcd) (172)   El menor múltiplo común de los denominadores de dos o más fracciones.

least common multiple (LCM) / mínimo común múltiplo (mcm) (169)   El menor múltiplo común de dos o más números. El mcm de 2 y 3 es 6.

leg / cateto (419)   Cualquiera de los lados que forman el ángulo recto de un triángulo rectángulo.

linear equation / ecuación lineal (255)   Ecuación cuya gráfica es una recta.

line graph / gráfica lineal (94)   Tipo de gráfica estadística que usa segmentos de recta para mostrar cómo cambian los valores durante un período de tiempo.

line of symmetry / eje de simetría (395)   Recta que divide una figura en dos mitades que son reflexiones una de la otra.

eje de simetría

line plot / esquema lineal (98)   Gráfica que usa una recta numérica y un $\times$ sobre un número en la recta numérica cada vez que el número aparece en un conjunto de datos.

line symmetry / simetría lineal (395)   Exhiben simetría lineal las figuras que coinciden exactamente cuando se doblan.

liter / litro (75)   Unidad fundamental de capacidad del sistema métrico. Un litro es un poco más de un cuarto de galón.

lower extreme / extremo inferior (114)   El número menor de un conjunto de datos.

lower quartile / cuartil inferior (114)   La mediana de la mitad inferior de un conjunto de datos o números, la cual se denota por *CI*.

mean / media (102)   La suma de los números de un conjunto de datos dividida entre el número total de datos.

median / mediana (102)   El número central de un conjunto de datos, una vez que los datos han sido ordenados numéricamente. Si hay un número par de datos, la mediana es el promedio de los dos datos centrales.

meter / metro (74)   Unidad fundamental de longitud del sistema métrico.

metric system / sistema métrico (74)   Sistema de medidas de base diez que usa las siguientes unidades fundamentales: metro para longitud, gramo para masa y litro para capacidad.

mode / modal (102)   Número(s) de un conjunto de datos que aparece(n) más frecuentemente.

modeling / hacer un modelo (22)   La escritura de una ecuación que represente un problema de la vida real.

multiple / múltiplo (169)   El múltiplo de un número entero es el producto del número por cualquier otro número entero.

multiplication property of equality / propiedad de multiplicación de la igualdad (302)   Si cada lado de una ecuación se multiplica por el mismo número, entonces los dos lados permanecen iguales. Si $a = b$, entonces, $ac = bc$.

multiplicative inverse / inverso multiplicativo (301)   El producto de un número por su inverso multiplicativo es igual a 1. El inverso multiplicativo de $\frac{2}{3}$ es $\frac{3}{2}$ y viceversa.

negative integer / entero negativo (184)   Entero que es menor que cero.

nonagon / eneágono (370)   Polígono de nueve lados.

obtuse / obtuso (362)   Cualquier ángulo que mide más de 90° pero menos de 180°.

octagon / octágono (370)   Polígono de ocho lados.

opposite / opuestos (184)   Dos enteros son opuestos si, en la recta numérica, están representados por puntos que equidistan de cero, pero en direcciones opuestas. La suma de opuestos es cero.

order of operations / orden de las operaciones (18)   Reglas a seguir cuando hay más de una operación involucrada.

1. Primero ejecuta todas las operaciones dentro de los símbolos de agrupamiento.

2. Ejecuta todos las potencias antes que cualquier otra opercíon.

3. Multiplica y divide, ordenadamente, de izquierda a derecha.

4. Suma y resta, ordenadamente, de izquierda a derecha.

ordered pair / par ordenado (191)   Par de números que se usa para ubicar un punto en un plano de coordenadas. Se escribe de la siguiente forma: (coordenada *x*, coordenada *y*).

origin / origen (191)   Punto de intersección axial en un plano de coordenadas.

ounce / onza (289)   Unidad de peso del sistema inglés de medidas. 16 onzas equivalen a una libra.

outcome / resultado (531)   Uno de los resultados posibles de un experimento probabilístico. Por ejemplo, 4 es un resultado posible cuando se lanza un dado.

outer measure / medida exterior (423)   Número de cuadrados dentro de una figura y que contienen parte de la figura.

outlier / valor atípico (115)   Dato o datos que dista(n) de los cuartiles respectivos más de 1.5 veces la amplitud intercuartílica.

P

parallelogram / paralelogramo (31)   Cuadrilátero con dos pares de lados paralelos.

pentagon / pentágono (370)   Polígono de cinco lados.

percent / tanto por ciento (158)   Razón que compara un número con 100.

percent proportion / proporción porcentual (349)   La proporción $\frac{P}{B} = \frac{r}{100}$ en que P representa el porcentaje, B representa la base y r representa el número por cada 100.

percentage / porcentaje (349)   Número de una proporción porcentual que se compara con otro número llamado base.

perfect square / cuadrado perfecto (411)   Número cuya raíz cuadrada es un número entero. 25 es un cuadrado perfecto porque $\sqrt{25} = 5$.

perimeter / perímetro (292)   La medida del contorno de una figura geométrica cerrada.

permutation / permutación (547)   Arreglo o lista en que el orden es importante.

perspective / perspectiva (492)   Una vista de perspectiva de una figura tridimensional es una vista de esquina que muestra las tres dimensiones de la figura.

pint / pinta (290)   Unidad de capacidad del sistema inglés de medidas que equivale a dos tazas.

polygon / polígono (370)   Figura simple cerrada en un plano, formada por tres o más segmentos de recta.

population / población (5, 464)   El grupo total de individuos del cual se toman las muestras bajo estudio.

population density / densidad demográfica (322)   Población por milla cuadrada.

positive integer / entero positivo (184)   Entero que es mayor que cero.

pound / libra (289)   Unidad de peso del sistema inglés de medidas que equivale a 16 onzas.

power / potencia (17)   Número que se puede escribir usando un exponente. La potencia 7^3 se lee *siete a la tercera potencia* o *siete al cubo*.

prime factorization / factorización prima (138)   Escritura de un número compuesto como el producto de números primos. La factorización prima de 63, por ejemplo, es $3 \times 3 \times 7$.

prime number / número primo (138)   Número entero mayor que 1 que sólo tiene dos factores, 1 y sí mismo.

principal / capital (478) Cantidad de dinero invertido o adeudado.

probability / probabilidad (165) La posibilidad de que suceda un evento. Es la razón del número de maneras en que puede ocurrir un evento al número total de resultados posibles.

property of proportions / propiedad de las proporciones (325) Si $\frac{a}{b} = \frac{c}{d}$, entonces $ad = bc$. Si $ad = bc$, entonces $\frac{a}{b} = \frac{c}{d}$.

proportion / proporción (325) Ecuación que demuestra la igualdad de dos razones, $\frac{a}{b} = \frac{c}{d}$, $b \neq 0$, $d \neq 0$.

Pythagorean Theorem / Teorema de Pitágoras (419) En un triángulo rectángulo, el cuadrado de la longitud de la hipotenusa es igual a la suma de los cuadrados de las longitudes de los catetos. $c^2 = a^2 + b^2$

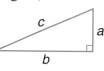

quadrant / cuadrante (191) Una de las cuatro regiones en que dos rectas perpendiculares dividen un plano.

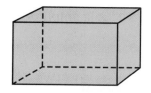

quadrilateral / cuadrilátero (370) Polígono de cuatro lados.

quart / cuarto de galón (290) Unidad de capacidad del sistema inglés de medidas que equivale a dos pintas.

quartile / cuartil (112) Una de las cuatro partes iguales en que están divididos los datos de un conjunto grande de números.

radical sign / signo radical (411) El símbolo con que se indica la raíz cuadrada no negativa es $\sqrt{}$.

radius / radio (297) Distancia desde el centro del un círculo hasta cualquier punto del mismo.

random / al azar (165) Los resultados ocurren al azar si cada resultado tiene la misma posibilidad de ocurrir.

random / aleatoria (464) Una muestra recibe el nombre de aleatoria si sus miembros han sido seleccionados basándose puramente en el azar.

range / rango (88) La diferencia entre los valores máximo y mínimo de un conjunto de datos.

rate / tasa (321) Razón de dos medidas que tienen distintas unidades de medida.

rate / tasa (349) Razón de un número a 100 en una proporción porcentual.

rate / tasa (478) El tanto por ciento que se cobra o se paga por el uso del dinero.

ratio / razón (154, 317) Comparación de dos números mediante división. La razón de 2 a 3 puede escribirse como 2 de cada 3, 2 a 3, 2:3 ó $\frac{2}{3}$.

reciprocal / recíproco (301) Inverso multiplicativo de un número.

rectangle / rectángulo (30) Cuadrilátero cuyos cuatro ángulos son congruentes entre sí.

rectangular prism / prisma rectangular (498) Prisma con bases rectangulares.

reflection / reflexión (215, 395) Transformación en que a una figura se le da vuelta de campana por encima de un eje de simetría.

regular polygon / polígono regular (371) Polígono cuyos lados, así como sus ángulos, son todos congruentes.

repeating decimal / decimal periódico (70) Decimal en el cual los dígitos, en algún momento, comienzan a repetirse en bloques de uno o más números. Por ejemplo, 0.181818... y 0.8333....

rhombus / rombo (383)   Paralelogramo cuyos lados son todos congruentes.

right / recto (362)   Ángulo que mide 90°.

S

sample / muestra (5, 329, 464)   Grupo escogido al azar o aleatoriamente que se usa para representar la población entera.

sample space / espacio muestral (531)   Conjunto de todos los resultados posibles de un experimento probabilístico.

scale / escala (88)   Conjunto de todos los valores posibles de una medida dada, el cual incluye los valores máximo y mínimo del conjunto, separados mediante los intervalos que se han usado.

scale / escala (332)   Intervalos que se usan en un mapa para representar la razón de las distancias en el mapa a las distancias verdaderas.

scale drawing / dibujo a escala (332)   Dibujo que es semejante, pero más grande o más pequeño que el objeto real.

scalene / escaleno (382)   Triángulo sin ningún par de lados congruentes.

scatter plot / diagrama de dispersión (92, 95)   Diagrama en que dos conjuntos de datos relacionados aparecen graficados como pares ordenados en la misma gráfica.

scientific notation / notación científica (77)   Escritura de un número como el producto de un número que es al menos igual a 1, pero menor que 10, multiplicado por una potencia de diez. Por ejemplo, $687{,}000 = 6.87 \times 10^5$.

sequence / sucesión (142)   Lista de números en cierto orden, como, por ejemplo, 0, 1, 2, 3 ó 2, 4, 6, 8.

similar / semejantes (376)   Figuras que tienen la misma forma, pero no necesariamente el mismo tamaño.

simple interest / interés simple (478)   Cantidad que se paga por el uso del dinero. La fórmula para calcular el interés simple es $I = prt$.

simplest form / forma reducida (154)   Una fracción está escrita en forma reducida si el MCD de su numerador y denominador es 1.

solids / sólidos (493)   Figuras tridimensionales. Los prismas, las pirámides, los conos y los cilindros son algunos ejemplos de sólidos.

solution / solución (21)   Valor de la variable de una ecuación que hace verdadera la ecuación. La solución de $12 = x + 7$ es 5.

solve / resolver (21)   Proceso de encontrar el número o números que satisfagan una ecuación.

spreadsheet / hoja de cálculos (137)   Herramienta que se usa para organizar y analizar datos.

square / cuadrado (410)   Número multiplicado por sí mismo; 36 es el cuadrado de 6 porque $6^2 = 6 \times 6 = 36$.

squared / al cuadrado (17)   Número multiplicado por sí mismo. 7^2 se lee *7 al cuadrado*.

square root / raíz cuadrada (411)   Uno de dos factores iguales de un número. Si $a^2 = b$, entonces a es una raíz cuadrada de b. La raíz cuadrada no negativa de 144 es 12 porque 12 es un número no negativo y $12^2 = 144$.

stem / tallo (108)   El mayor valor de posición común a todos los datos es el que se usa como tallo en un diagrama de tallo y hojas.

stem-and-leaf plot / diagrama de tallo y hojas (108) Sistema que se usa para condensar un conjunto de datos y en el cual el mayor valor de posición de los datos forma el tallo y el segundo mayor valor de posición de los datos forma las hojas.

straight / llano (362)   Ángulo que mide 180°.

subtraction property of equality / propiedad de sustracción de la igualdad (228)   Si sustraes el mismo número de ambos lados de una ecuación, los lados permanecen iguales. Para números *a*, *b* y *c* cualesquiera, si $a = b$, entonces $a - c = b - c$.

supplementary / suplementarios (362)
Dos ángulos son suplementarios si la suma de sus medidas es 180°.

surface area / área de superficie (510)   Suma de las áreas de todas las superficies de una figura tridimensional.

term / término (239)   Número, variable o producto de números y variables.

term / término (142)   Nombre que recibe cada número de una sucesión.

terminating decimal / decimal terminal (70)   Cociente en que la división termina, es decir, tiene un residuo de cero. 0.25 y 0.125 son ejemplos de decimales terminales.

tessellation / teselado (388)   Un patrón repetitivo de polígonos que encajan perfectamente, sin dejar huecos o espacios.

theoretical probability / probabilidad teórica (531)   Razón del número de maneras en que puede ocurrir un evento al número total de resultados posibles.

time / tiempo (478)   Cuando se usa para calcular interés, el tiempo se da en años.

ton / tonelada (289)   Unidad de peso del sistema inglés de medidas que equivale a 2,000 libras.

transformation / transformación (215)   Movimientos de figuras geométricas.

translation /traslación (215, 392)   Tipo de transformación en que una figura se desliza horizontalmente, verticalmente o de ambas maneras.

trapezoid / trapecio (383, 429)   Cuadrilátero con un único par de lados paralelos.

tree diagram / diagrama de árbol (534)   Diagrama que se usa para encontrar y mostrar el número total de resultados posibles de un experimento probabilístico.

triangle / triángulo (370, 428)   Polígono que posee tres ángulos.

unit rate / tasa unitaria (321)   Tasa cuyo denominador es 1.

upper extreme / extremo superior (114)   El número máximo de un conjunto de datos o números.

upper quartile / cuartil superior (114)   La mediana de la mitad superior de un conjunto de números o datos.

variable / variable (12)   Un símbolo, por lo general, una letra, que se usa para representar números en expresiones o enunciados matemáticos. En $3 + a = 6$, *a* es una variable.

vertex / vértice (362)
El vértice de un ángulo es el extremo común de los rayos que lo forman.

volume / volumen (498)   Número de unidades cúbicas que se requieren para llenar el espacio que ocupa un sólido.

x-axis / eje *x* (191)   La recta numérica horizontal que ayuda a formar el sistema de coordenadas.

x-coordinate / coordenada *x* (191)   Primer número de un par ordenado.

y-axis / eje *y* (191)   La recta numérica vertical que ayuda a formar el sistema de coordenadas.

y-coordinate / coordenada *y* (191)   Segundo número de un par ordenado.

zero pair / par nulo (196, 227)   Resultado de aparear una ficha positiva con una negativa.

Index

INDEX

Number and Operations

$+$	plus or positive
$-$	minus or negative
$a \cdot b$ $a \times b$ ab or $a(b)$	a times b
$\div$	divided by
$\pm$	positive or negative
$=$	is equal to
$\neq$	is not equal to
$<$	is less than
$>$	is greater than
$\leq$	is less than or equal to
$\geq$	is greater than or equal to
$\approx$	is approximately equal to
$\%$	percent
$a{:}b$	the ratio of a to b, or $\frac{a}{b}$

Geometry and Measurement

$\cong$	is congruent to
$\sim$	is similar to
$^\circ$	degree(s)
$\overleftrightarrow{AB}$	line AB
$\overline{AB}$	segment AB
$\overrightarrow{AB}$	ray AB
$\llcorner$	right angle
$\perp$	is perpendicular to
$\parallel$	is parallel to
AB	length of $\overline{AB}$, distance between A and B
$\triangle ABC$	triangle ABC
$\angle ABC$	angle ABC
$\angle B$	angle B
$m\angle ABC$	measure of angle ABC
$\odot C$	circle C
$\overset{\frown}{AB}$	arc AB
π	pi $\left(\text{approximately } 3.14159 \text{ or } \frac{22}{7}\right)$
(a, b)	ordered pair with x-coordinate a and y-coordinate b
$\sin A$	sine of angle A
$\cos A$	cosine of angle A
$\tan A$	tangent of angle A

Algebra and Functions

a'	a prime
a^n	a to the nth power
a^{-n}	$\frac{1}{a^n}$ (one over a to the nth power)
$\lvert x \rvert$	absolute value of x
$\sqrt{x}$	principal (positive) square root of x
$f(n)$	function, f of n

Probability and Statistics

$P(A)$	the probability of event A
$n!$	n factorial
$P(n, r)$	permutation of n things taken r at a time
$C(n, r)$	combination of n things taken r at a time